Student's Solutions Manual to accompany Stein/Barcellos

CALCULUS

AND ANALYTIC GEOMETRY

FIFTH EDITION

ANTHONY BARCELLOS
American River College
Sacramento, California

Volume 2

McGRAW-HILL, INC.

New York St. Louis San Francisco Auckland Bogotá
Caracas Lisbon London Madrid Mexico Milan Montreal
New Delhi Paris San Juan Singapore Sydney Tokyo Toronto

Student's Solutions Manual
to accompany Stein/Barcellos:
CALCULUS AND ANALYTIC GEOMETRY, Fifth Edition,
Volume 2

9 10 11 12 13 14 15 16 17 QSRQSR 9 8 7 6 5 4

ISBN 0-07-061207-2

The editor was Maggie Lanzillo;
the production supervisor was Annette Mayeski.
Malloy Lithographing, Inc., was printer and binder.

This book is printed on recycled paper containing a minimum of 50% total recycled fiber with 10% postconsumer de-inked fiber.

Table of Contents

A Note on Notation

This manual uses the same notation and techniques as the text by Stein and Barcellos. However, for the sake of convenience and to save space, we also take two shortcuts. The symbol $\underset{H}{=}$ represents an equality that follows from an application of l'Hôpital's rule (Sec. 6.8 and thereafter). Also, the text writes out units of measurement in full, such as "meters per second per second," whereas we usually employ the standard abbreviations, such as "m/sec^2." We may occasionally use mag **A** in place of $\|\mathbf{A}\|$ for the magnitude of the vector **A** in those cases where the expression for **A** is complicated.

Preface

Der Irrnis und der
Leiden Pfade kam ich.
Act III, Scene 1, *Parsifal*

This manual is supposed to help you learn calculus. It contains worked-out solutions to approximately half of the exercises in the fifth edition of *Calculus and Analytic Geometry* by Sherman K. Stein and Anthony Barcellos. Volume 1 covers Chapters 1 through 11 and Volume 2 deals with Chapters 12 through 17, plus appendices.

Calculus is learned only through practice. No one would expect to learn to ride a bicycle by listening to lectures and reading a book. Calculus is probably somewhat harder than learning to ride a bike, so be prepared to work over the examples in the text and to do the homework exercises. *Always attempt to solve the exercises yourself before turning to this manual.* (Looking at a problem for 30 seconds is *not* the same as trying to solve it.) Refer also to the section *To the Student* in your textbook (p. *xxvii*), which gives detailed advice on studying and learning mathematics.

We provide solutions for practically all of the odd-numbered exercises and all of the Guide Quizzes. Many solutions are accompanied by figures and graphs, although in several cases we just refer to those graphs you already have in the answer section of the text book. Some exercises, such as exploration problems (marked with a ⊕ in the text), are included in the *Instructor's Manual* instead because giving their solutions would make them pointless. Don't worry; there aren't too many of these.

The solutions are complete, providing the steps necessary to permit you to follow the line of reasoning. Quite naturally, the solutions tend to be more detailed in the earlier parts of solution sets, where you are presumably just learning the material. Near the end of a set we may take somewhat bigger steps. In all cases, however, you should be able to take pencil and paper and fill in as many intermediate steps as you may need.

Be aware that numerical answers may vary slightly, depending on the accuracy of measurements, differing assumptions, and round-off errors in calculations. In such cases, do not take the solutions in the manual as the only "right" answers. Your results are just as good. In particular, a one-unit difference in the last decimal place is nothing to get exercised about.

All of the solutions have been checked and double-checked in hopes of eliminating all errors. With over three thousand solutions, however, it is quite possible that a few errors have managed to slip through anyway. Please bring any errors that you might find to the attention of Anthony Barcellos at P.O. Box 2249, Davis, CA 95617.

About This Book. Camera-ready copy was produced on a Hewlett-Packard LaserJet III from documents created with WordPerfect 5.1. The illustrations were created by CoPlot and CoDraw from CoHort Software (P.O. Box 1149, Berkeley, CA 94701, 415/524-9878), which generated .WPG files for direct import into WordPerfect.

Acknowledgments. We wish to thank the editorial staff at McGraw-Hill for their assistance with this student manual, including former math editors Robert Weinstein and Richard Wallis. Maggie Lanzillo was a constant source of encouragement even while gently reminding us about deadlines.

Many of the people who participated in preparation of the Stein/Barcellos textbook also contributed in various ways to this manual. We acknowledge in particular the efforts of **Keith Sollers** and **Mallory Austin** who drafted solutions to thousands of exercises over the past two years and were key contributors to this manual. Other significant contributions of solutions came from Dean Hickerson and Timothy Thayer. Keith, Mallory, and Dean also provided the bulk of the proofreading, which they performed painstakingly, but are not to be held responsible for any errors left by the author. In addition to those who worked on the text, Travis Andrews and Kelly Riddle contributed to the answer-checking. Heroic efforts were made by **Richard Kinter**, **Judith Kinter**, and **Michael Kinter**, who typed the reams of solutions and in the process became among the world's greatest experts in WordPerfect 5.1's equation editor.

—Anthony Barcellos

12 Vectors

12.1 The Algebra of Vectors

1

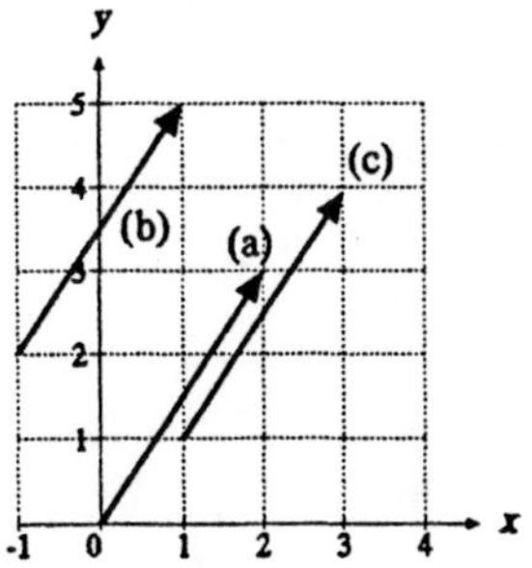

3 (a) Head located at (2, 1, 3)

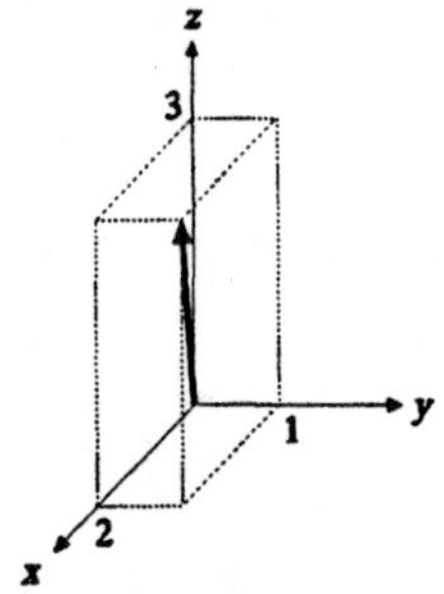

(b) Head located at (3, 2, 4)

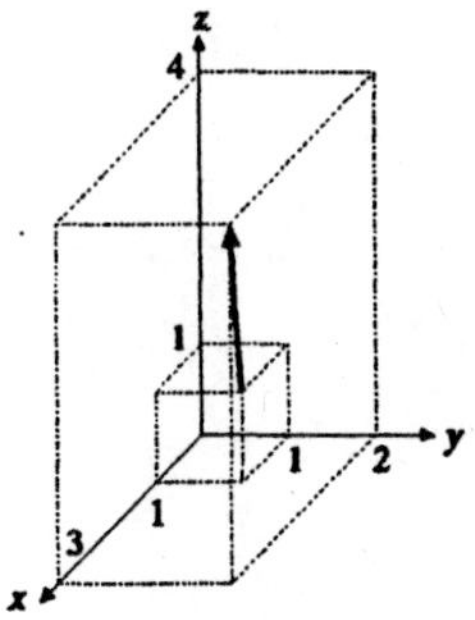

5 (a) Head located at $(-1,\ -2,\ 2)$

(b) Head located at $(0,\ -1,\ 1)$

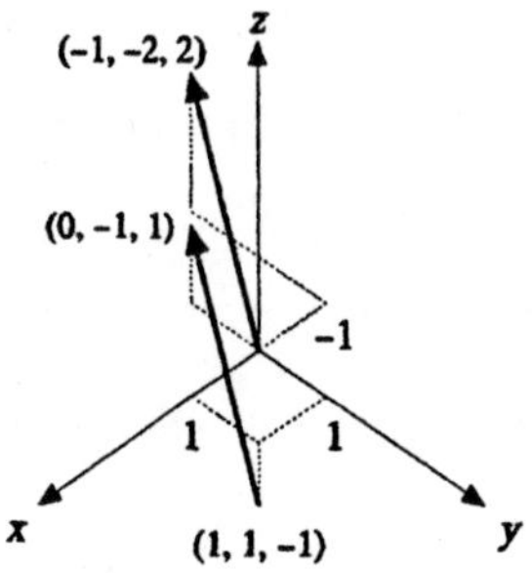

7

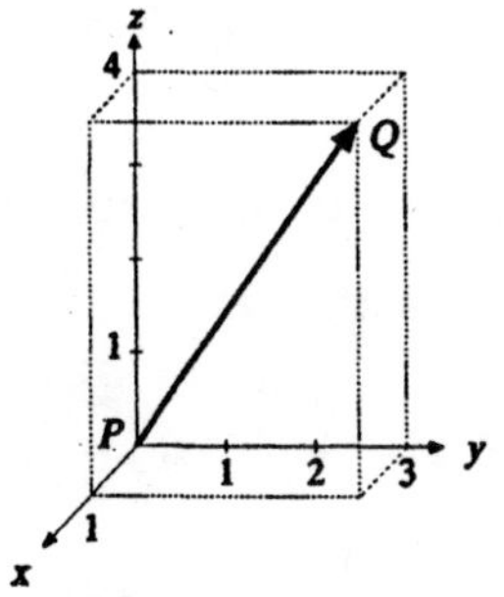

$\overrightarrow{PQ} = (1-0)\mathbf{i} + (3-0)\mathbf{j} + (4-0)\mathbf{k}$

$= \mathbf{i} + 3\mathbf{j} + 4\mathbf{k}$. The length of $\overrightarrow{PQ}$ is $|\overrightarrow{PQ}| =$

$\sqrt{x^2 + y^2 + z^2} = \sqrt{1^2 + 3^2 + 4^2} = \sqrt{26}.$

9

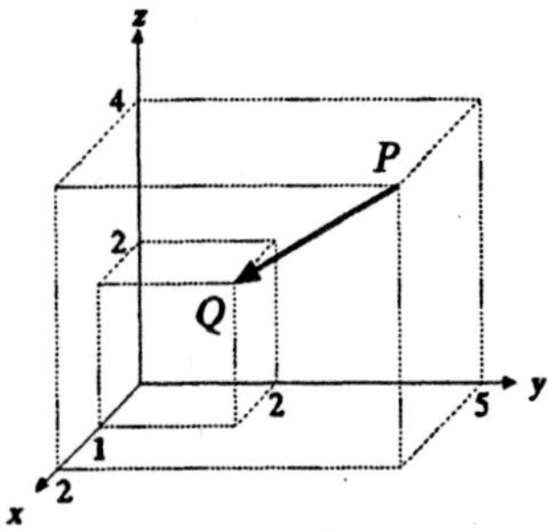

$\overrightarrow{PQ} = (1-2)\mathbf{i} + (2-5)\mathbf{j} + (2-4)\mathbf{k} = -\mathbf{i} - 3\mathbf{j} - 2\mathbf{k}$, so $|\overrightarrow{PQ}| = \sqrt{x^2+y^2+z^2} = \sqrt{(-1)^2+(-3)^2+(-2)^2} = \sqrt{14}$.

11 (a) The vector $\mathbf{A} = x\mathbf{i} + y\mathbf{j}$ forms a 45° angle with respect to the negative x axis, so $x < 0$ and $x = -y$. By the Pythagorean theorem, $x^2 + y^2 = 10^2$. Hence $2x^2 = 10^2$, so $x^2 = 50$ and therefore $x = -5\sqrt{2}$ and $y = 5\sqrt{2}$. That is, $\mathbf{A} = -5\sqrt{2}\mathbf{i} + 5\sqrt{2}\mathbf{j}$.

(b) The vector $\mathbf{A}$ has no x component and is oriented in the negative y direction. Therefore, $\mathbf{A} = -6\mathbf{j}$.

(c) $\mathbf{A}$ lies in quadrant IV so that $x > 0$ and $y = -x$. Then $x^2 + y^2 = 81$ implies that $2x^2 = 81$, so $x = \frac{9\sqrt{2}}{2}$ and $y = -\frac{9\sqrt{2}}{2}$. Therefore

$$\mathbf{A} = \frac{9\sqrt{2}}{2}\mathbf{i} - \frac{9\sqrt{2}}{2}\mathbf{j}.$$

(d) Since $\mathbf{A}$ has no y component, $\mathbf{A} = 5\mathbf{i}$.

13 The wind vector $\mathbf{W}$ has length 30 and points northeast. Hence, if $\mathbf{W} = \langle x, y\rangle$, we must have $x = y$ and $\sqrt{x^2+y^2} = 30$, so $x = y = \frac{30}{\sqrt{2}}$ and $\mathbf{W} = \frac{1}{\sqrt{2}}\langle 30, 30\rangle$. Similarly, the vector describing the plane's motion relative to the wind is $\mathbf{A} = \frac{1}{\sqrt{2}}\langle 100, -100\rangle$. Hence the vector describing the plane's motion relative to the ground is $\mathbf{W} + \mathbf{A} = \frac{1}{\sqrt{2}}\langle 130, -70\rangle$.

(a) The speed is $\|\mathbf{W} + \mathbf{A}\|$

$$= \left[\left(\frac{130}{\sqrt{2}}\right)^2 + \left(-\frac{70}{\sqrt{2}}\right)^2\right]^{1/2} = \sqrt{10{,}900}$$

≈ 104.40 mi/hr.

(b)

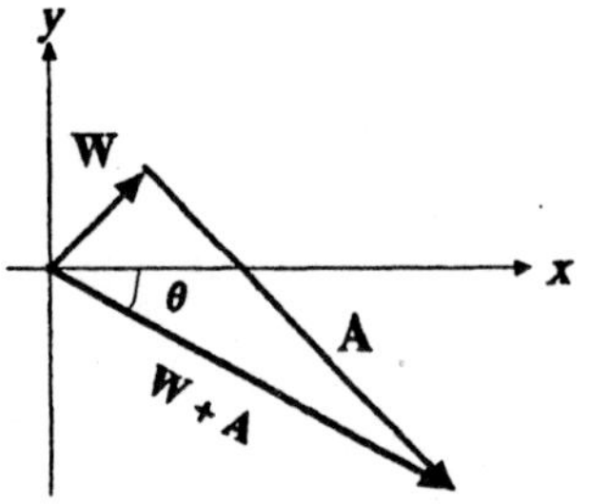

Let θ be the angle shown in the figure. Then $\tan\theta = \frac{70/\sqrt{2}}{130/\sqrt{2}} = \frac{7}{13}$, so $\theta = \tan^{-1}\frac{7}{13} \approx 28.30°$. The plane is flying about 28.30° south of east.

15 (a) $\mathbf{A} + \mathbf{B} = \langle -1+7, 2+0, 3+2\rangle = \langle 6, 2, 5\rangle$ and $\mathbf{A} - \mathbf{B} = \langle -1-7, 2-0, 3-2\rangle = \langle -8, 2, 1\rangle$.

(b) $\mathbf{A} + \mathbf{B} = (0+6)\mathbf{i} + (3+7)\mathbf{j} + (4+0)\mathbf{k} = 6\mathbf{i} + 10\mathbf{j} + 4\mathbf{k}$; $\mathbf{A} - \mathbf{B} = (0-6)\mathbf{i} + (3-7)\mathbf{j} + (4-0)\mathbf{k} = -6\mathbf{i} - 4\mathbf{j} + 4\mathbf{k}$.

17 (a) $2\mathbf{A} = 4\mathbf{i} + 6\mathbf{j} + 2\mathbf{k}$

(b) $-2\mathbf{A} = -4\mathbf{i} - 6\mathbf{j} - 2\mathbf{k}$

(c) $\frac{1}{2}\mathbf{A} = \mathbf{i} + \frac{3}{2}\mathbf{j} + \frac{1}{2}\mathbf{k}$

(d) $-\frac{1}{2}\mathbf{A} = -\mathbf{i} - \frac{3}{2}\mathbf{j} - \frac{1}{2}\mathbf{k}$

19 (a) Since $\|c\mathbf{A}\| = |c|\,\|\mathbf{A}\|$, $\|-2\mathbf{A}\| = |-2|\,\|\mathbf{A}\| = 2\cdot 6 = 12$.

(b) $\left\|\frac{1}{3}\mathbf{A}\right\| = \frac{1}{3}\cdot 6 = 2$

(c) By Example 2, this is a unit vector.

Therefore, the length of $\frac{\mathbf{A}}{\|\mathbf{A}\|}$ is 1.

(d) $\|-\mathbf{A}\| = |-1|\,\|\mathbf{A}\| = 6$.

(e) $\mathbf{A} + 2\mathbf{A} = 3\mathbf{A}$. Therefore, $\|\mathbf{A} + 2\mathbf{A}\| = \|3\mathbf{A}\| = 3\cdot 6 = 18$.

21 (a) To find **u**, divide **A** by its length: $\mathbf{u} = \frac{\mathbf{A}}{\|\mathbf{A}\|}$

$$= \frac{\mathbf{i} + 2\mathbf{j} + 3\mathbf{j}}{\sqrt{1^2 + 2^2 + 3^2}}$$

$$= \frac{1}{\sqrt{14}}\mathbf{i} + \frac{2}{\sqrt{14}}\mathbf{j} + \frac{3}{\sqrt{14}}\mathbf{k}.$$

(b)

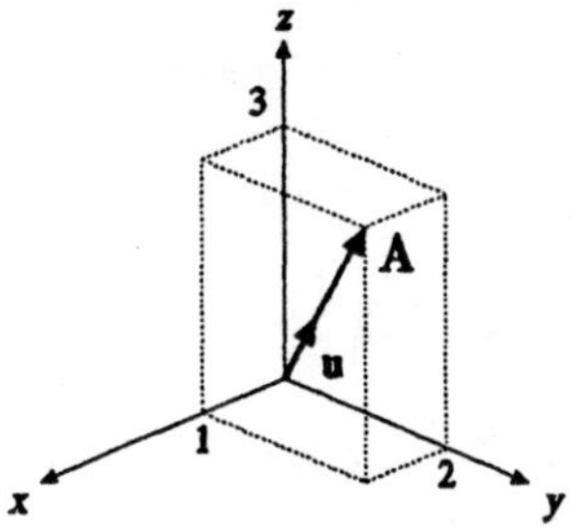

23 (a)

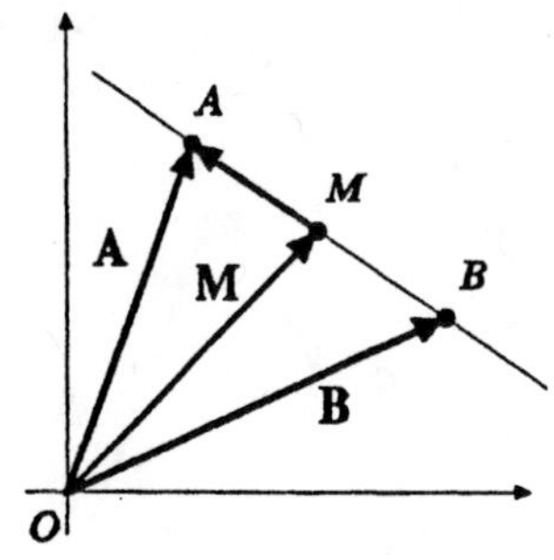

We show $\mathbf{M} = \mathbf{A} + \frac{1}{2}(\mathbf{B} - \mathbf{A})$. First, vector $\overrightarrow{MA} = \frac{1}{2}\overrightarrow{BA}$, since M is the midpoint of $\overrightarrow{BA}$. But $\overrightarrow{BA} = \mathbf{A} - \mathbf{B}$ and $\mathbf{M} + \overrightarrow{MA} = \mathbf{A}$, so $\mathbf{M} = \mathbf{A} - \overrightarrow{MA} = \mathbf{A} - \frac{1}{2}(\mathbf{A} - \mathbf{B}) = \mathbf{A} + \frac{1}{2}(\mathbf{B} - \mathbf{A})$.

(b) By (a), $\mathbf{M} = \mathbf{A} + \frac{1}{2}(\mathbf{B} - \mathbf{A})$, so $\mathbf{M} = \mathbf{A} + \frac{1}{2}\mathbf{B} - \frac{1}{2}\mathbf{A} = \frac{1}{2}\mathbf{A} + \frac{1}{2}\mathbf{B} = \frac{\mathbf{A} + \mathbf{B}}{2}$.

25 $6\mathbf{i} + 9\mathbf{j} + 12\mathbf{k} = 3(2\mathbf{i} + 3\mathbf{j} + 4\mathbf{k})$, so one vector is a multiple of the other. Hence these two vectors are parallel.

27 The proofs for this exercise will be done in two dimensions. The three-dimensional versions are the same.

(a) Let $\mathbf{A} = \langle a_1, a_2\rangle$ and $\mathbf{B} = \langle b_1, b_2\rangle$.

Therefore $c(\mathbf{A} + \mathbf{B}) = c[\langle a_1, a_2\rangle + \langle b_1, b_2\rangle]$

$= c\langle a_1 + b_1, a_2 + b_2\rangle$

$= \langle c(a_1 + b_1), c(a_2 + b_2)\rangle$

$= \langle ca_1 + cb_1, ca_2 + cb_2\rangle$

$= \langle ca_1, ca_2\rangle + \langle cb_1, cb_2\rangle$

$= c\langle a_1, a_2\rangle + c\langle b_1, b_2\rangle = c\mathbf{A} + c\mathbf{B}$.

(b)

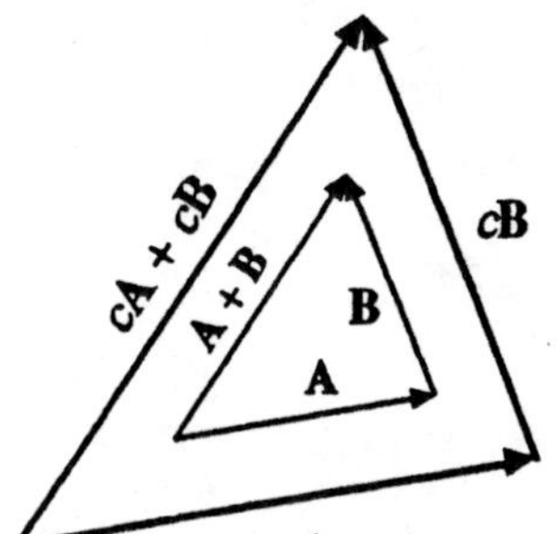

The scalar $c > 0$ serves as a scaling factor that expands (or contracts) all sides of the triangle equally, showing that $\mathbf{A} + \mathbf{B} = \mathbf{C}$ becomes $c\mathbf{A} + c\mathbf{B} = c\mathbf{C}$; that is, $c\mathbf{A} + c\mathbf{B} = c(\mathbf{A} + \mathbf{B})$. If $c < 0$, the same result holds, except that all of the vectors are reversed in direction.

29 (a) Note that $\|\mathbf{u}_1\| = \sqrt{\left(\frac{1}{2}\right)^2 + \left(\frac{\sqrt{3}}{2}\right)^2} = \sqrt{\frac{1}{4} + \frac{3}{4}} = 1$ and $\|\mathbf{u}_2\| = \sqrt{\left(\frac{\sqrt{3}}{2}\right)^2 + \left(-\frac{1}{2}\right)^2} = 1$, so $\mathbf{u}_1$ and $\mathbf{u}_2$ are unit vectors. To find the angle between the x axis and $\mathbf{u}_1$, note that $\frac{\sqrt{3}/2}{1/2} = \sqrt{3} = \tan\frac{\pi}{3}$. Hence the desired angle is $\pi/3$. Similarly, for $\mathbf{u}_2$, $\frac{-1/2}{\sqrt{3}/2} = -\frac{1}{\sqrt{3}} = \tan\left(-\frac{\pi}{6}\right)$, so $\mathbf{u}_2$ makes an angle of $-\frac{\pi}{6}$ with the x axis. The angle between $\mathbf{u}_1$ and $\mathbf{u}_2$ is the difference between these, $\frac{\pi}{3} - \left(-\frac{\pi}{6}\right) = \frac{\pi}{2}$, so the vectors are perpendicular.

(b) We have $\mathbf{i} = x\left(\frac{1}{2}\mathbf{i} + \frac{\sqrt{3}}{2}\mathbf{j}\right) + y\left(\frac{\sqrt{3}}{2}\mathbf{i} - \frac{1}{2}\mathbf{j}\right) = \left(\frac{1}{2}x + \frac{\sqrt{3}}{2}y\right)\mathbf{i} + \left(\frac{\sqrt{3}}{2}x - \frac{1}{2}y\right)\mathbf{j}$, so we must solve the equations $1 = \frac{1}{2}x + \frac{\sqrt{3}}{2}y$ and $0 = \frac{\sqrt{3}}{2}x - \frac{1}{2}y$. The second equation gives $y = \sqrt{3}x$, so $1 = \frac{1}{2}x + \frac{\sqrt{3}}{2}(\sqrt{3}x) = 2x$ and $x = 1/2$. Then $y = \sqrt{3}\cdot\frac{1}{2} = \frac{\sqrt{3}}{2}$.

31 (a)

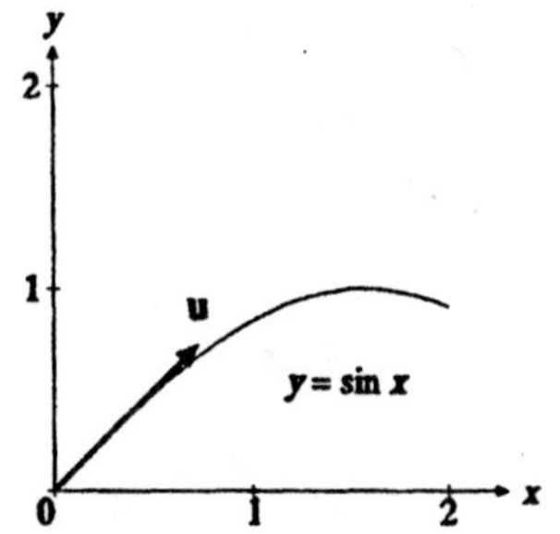

$-\mathbf{u}$ also works.

(b) Since $\frac{dy}{dx} = \cos x = 1$ at $x = 0$, the tangent at (0, 0) has slope 1. So $\mathbf{u} = a\mathbf{i} + a\mathbf{j}$ for some a. Since $\|\mathbf{u}\| = 1$, $1 = \sqrt{a^2 + a^2} = \sqrt{2}|a|$ and $a = \pm\frac{1}{\sqrt{2}}$. Thus $\mathbf{u} =$

$\pm\left(\frac{1}{\sqrt{2}}\mathbf{i} + \frac{1}{\sqrt{2}}\mathbf{j}\right).$

33 (a)

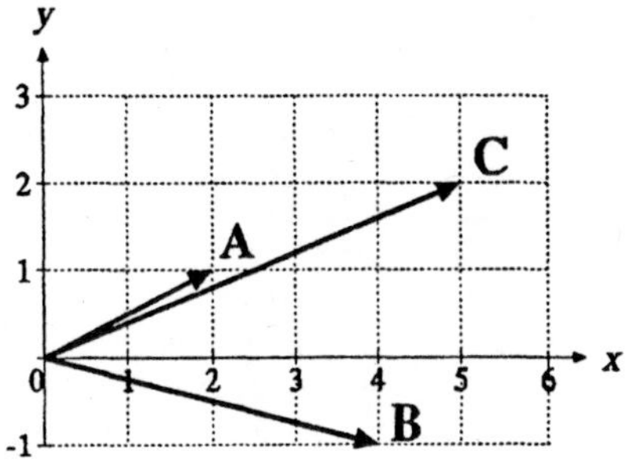

(b)

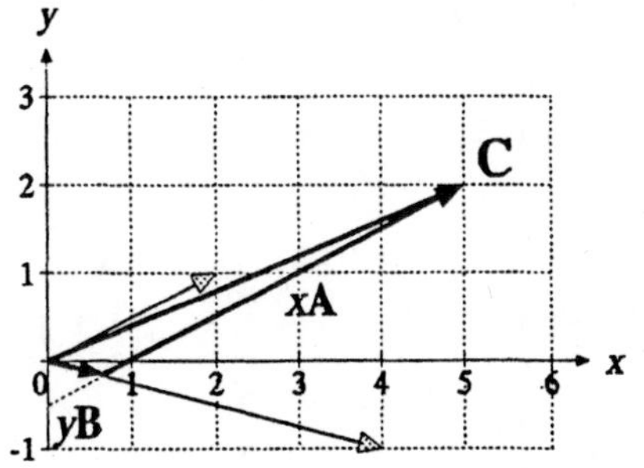

(d) $\mathbf{C} = x\mathbf{A} + y\mathbf{B}$ if and only if corresponding components are equal. Thus $\langle 5, 2\rangle = x\langle 2, 1\rangle + y\langle 4, -1\rangle = \langle 2x + 4y, x - y\rangle$ implies that $5 = 2x + 4y$ and $2 = x - y$. Solving these equations gives $x = 13/6$, $y = 1/6$.

35

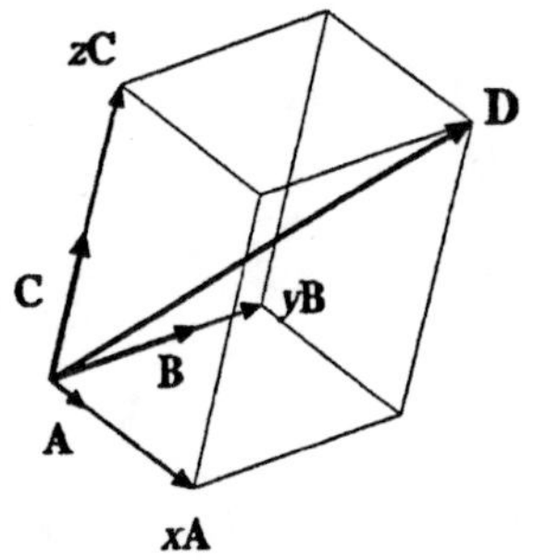

37 (a) The vectors form a closed path, so their sum is **0**.

(b)

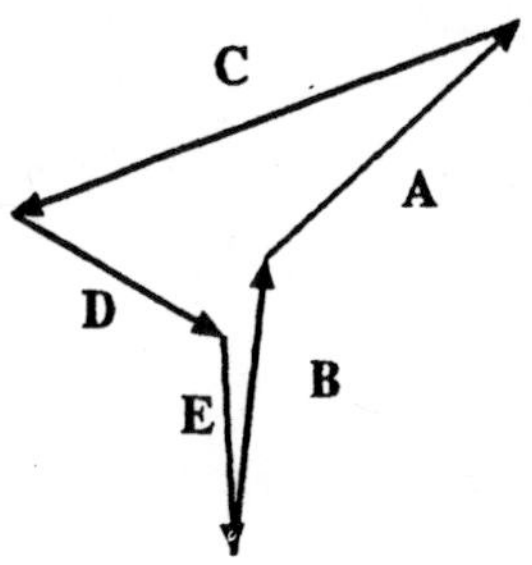

39 (a)

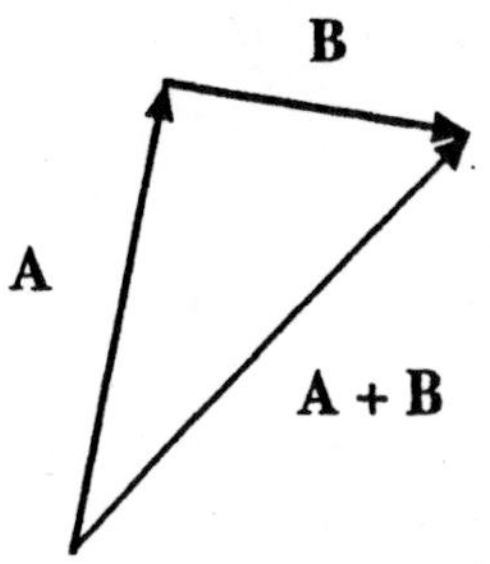

The length $\|\mathbf{A} + \mathbf{B}\|$ is the straight-line distance between the tail of **A** and the head of **B** and is therefore the shortest distance between the two points. $\|\mathbf{A}\| + \|\mathbf{B}\|$ is the distance covered by traveling along the vectors themselves—a greater distance.

(b) For equality to occur, **A** and **B** must point in the same direction.

12.2 Projections

1 (a)

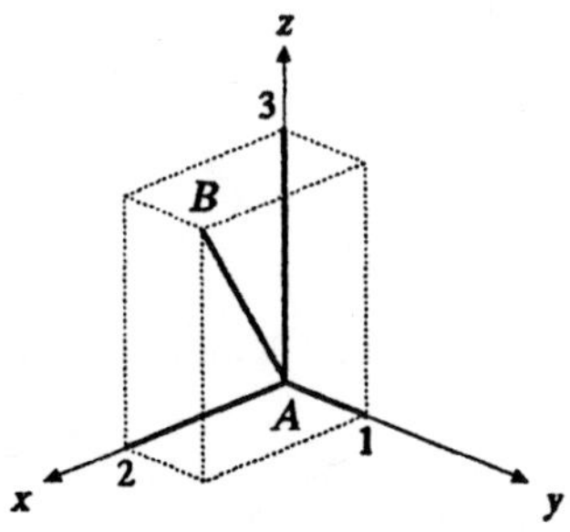

(b) By inspection of (a), $\mathbf{proj_i}(2\mathbf{i} + \mathbf{j} + 3\mathbf{k})$ has length 2, $\mathbf{proj_j}(2\mathbf{i} + \mathbf{j} + 3\mathbf{k})$ has length 1, and

$\mathrm{proj}_{\mathbf{k}}(2\mathbf{i} + \mathbf{j} + 3\mathbf{k})$ has length 3.

(c) $|\overrightarrow{AB}| = \sqrt{x^2 + y^2 + z^2} = \sqrt{2^2 + 1^2 + 3^2}$

$= \sqrt{14}$

3

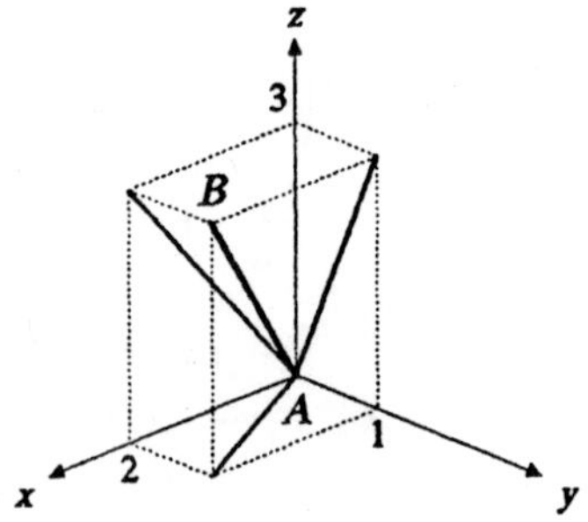

The line segment AB has projections of the same length as those of the vector $\overrightarrow{AB} = 2\mathbf{i} + \mathbf{j} + 3\mathbf{k}$:

$\|\text{proj of } 2\mathbf{i} + \mathbf{j} + 3\mathbf{k} \text{ on } xy \text{ plane}\| = \|2\mathbf{i} + \mathbf{j}\|$

$= \sqrt{2^2 + 1^2} = \sqrt{5}$;

$\|\text{proj of } 2\mathbf{i} + \mathbf{j} + 3\mathbf{k} \text{ on } xz \text{ plane}\| = \|2\mathbf{i} + 3\mathbf{k}\|$

$= \sqrt{2^2 + 3^2} = \sqrt{13}$;

$\|\text{proj of } 2\mathbf{i} + \mathbf{j} + 3\mathbf{k} \text{ on } yz \text{ plane}\| = \|\mathbf{j} + 3\mathbf{k}\|$

$= \sqrt{1^2 + 3^2} = \sqrt{10}$.

5 (a)

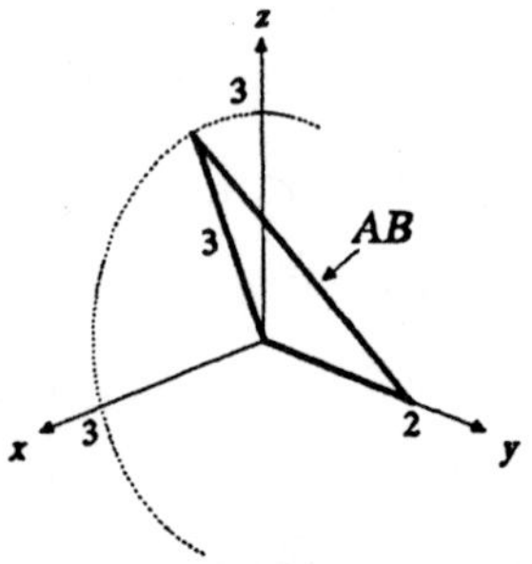

(b) Since the two projections are perpendicular to each other, the Pythagorean theorem shows that the length of AB is $\sqrt{3^2 + 2^2} = \sqrt{13}$.

7 (a)

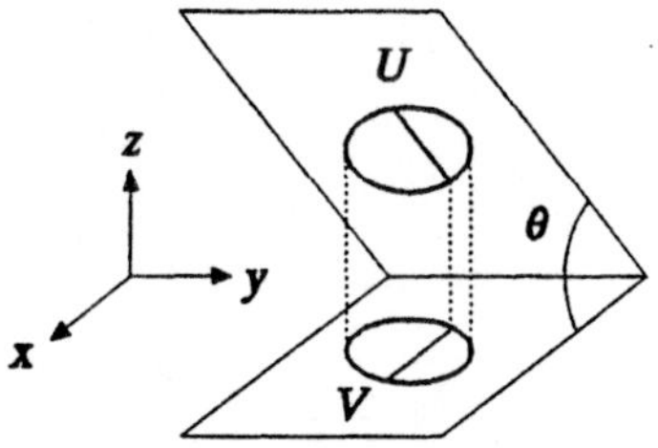

9 (a)

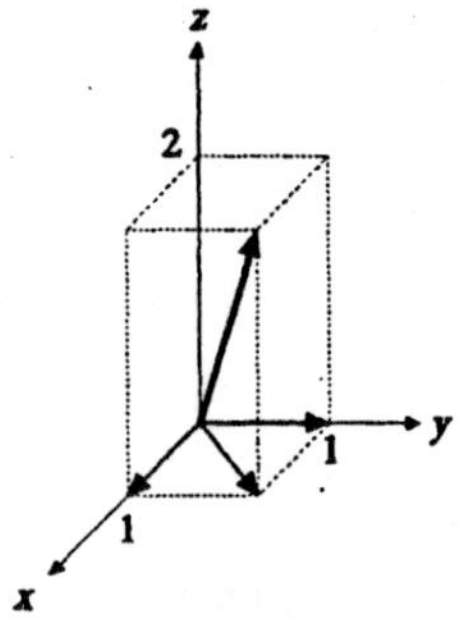

(b) $\mathrm{proj}_{\mathbf{i}}(\mathbf{i} + \mathbf{j} + 2\mathbf{k}) = \mathbf{i}$, $\mathrm{proj}_{\mathbf{j}}(\mathbf{i} + \mathbf{j} + 2\mathbf{k}) = \mathbf{j}$, and the projection of $\mathbf{i} + \mathbf{j} + 2\mathbf{k}$ on the xy plane is $\mathbf{i} + \mathbf{j}$.

11 (a)

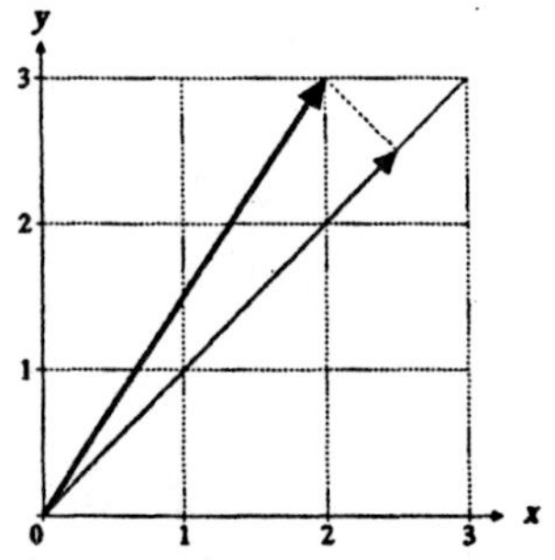

(b) See (a).

(c) The projection is $\frac{5}{2}\mathbf{i} + \frac{5}{2}\mathbf{j}$, since the line from (5/2, 5/2) to (2, 3) is perpendicular to the line $y = x$. Thus $x = y = 5/2$.

13 (a) $\mathrm{proj}_{\mathbf{i}}(\mathbf{i} - 2\mathbf{j} + 4\mathbf{k}) = \mathbf{i}$

(b) $\mathrm{proj}_{-\mathbf{i}}(\mathbf{i} - 2\mathbf{j} + 4\mathbf{k}) = \mathbf{i}$

(c) The projection of $\mathbf{i} - 2\mathbf{j} + 4\mathbf{k}$ on the z axis is $\mathrm{proj}_{\mathbf{k}}(\mathbf{i} - 2\mathbf{j} + 4\mathbf{k}) = 4\mathbf{k}$.

(d) $\mathbf{proj_j(i - 2j + 4k) = -2j}$

(e) The projection of $\mathbf{i - 2j + 4k}$ on the yz plane is $\mathbf{-2j + 4k}$

15 $\mathbf{proj_B\ A}$ will be equal to $\mathbf{A}$ if $\mathbf{B}$ is parallel to $\mathbf{A}$, that is, if $\mathbf{B} = c\mathbf{A}$.

17 (a) The relation between $\mathbf{proj_B\ A}$ and $\mathbf{proj_B(-A)}$ is $\mathbf{proj_B(-A) = -proj_B(A)}$.

(b) $\mathbf{proj_B\ A = proj_{-B}\ A}$

19 Note that $\mathbf{proj_k(2i - 3j + 2k) = 2k}$. Since $\mathbf{A} =$ $\mathbf{proj_B A + orth_B\ A}$, $\mathbf{orth_B\ A = A - proj_B\ A}$, so $\mathbf{orth_k(2i - 3j + 2k) = (2i - 3j + 2k) - 2k}$ $\mathbf{= 2i - 3j}$.

21

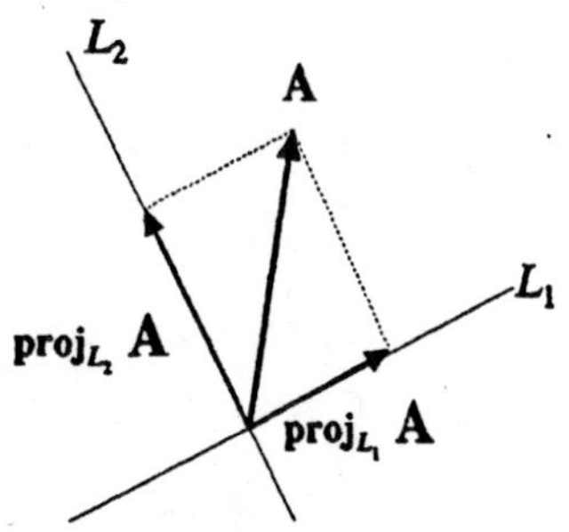

From the diagram, $\mathbf{proj}_{L_1}\mathbf{A} + \mathbf{proj}_{L_2}\mathbf{A} = \mathbf{A}$.

23 By the distance formula, its length is $\sqrt{a^2 + b^2 + c^2}$.

25

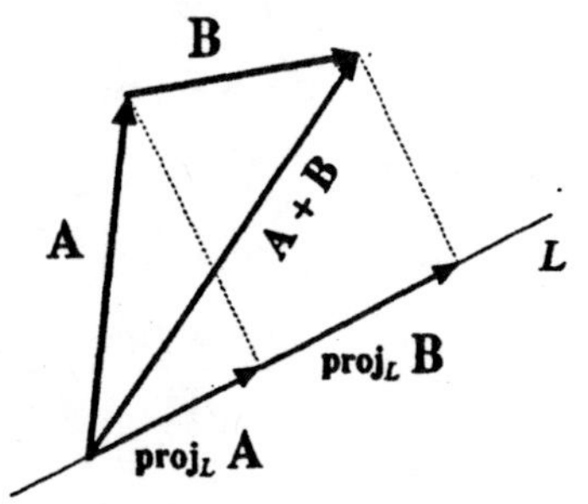

From the figure we see that $\mathbf{proj}_L\mathbf{A} + \mathbf{proj}_L\mathbf{B} =$ $\mathbf{proj}_L(\mathbf{A} + \mathbf{B})$.

12.3 The Dot Product of Two Vectors

1 By definition, $\mathbf{A \cdot B} = \|\mathbf{A}\|\,\|\mathbf{B}\| \cos\theta$

$= 3 \cdot 4 \cos \pi/4 = 12 \cdot \frac{\sqrt{2}}{2} = 6\sqrt{2}.$

3 $\mathbf{A \cdot B} = \|\mathbf{A}\|\,\|\mathbf{B}\| \cos\theta = 5 \cdot \frac{1}{2} \cos\frac{\pi}{2} = \frac{5}{2} \cdot 0$

$= 0$

5 $\mathbf{A \cdot B} = (-2)(3) + (3)(4) = -6 + 12 = 6$

7 $\mathbf{A \cdot B} = (2)(3) + (-3)(4) + (-1)(-1)$

$= 6 - 12 + 1 = -5$

9 (a)

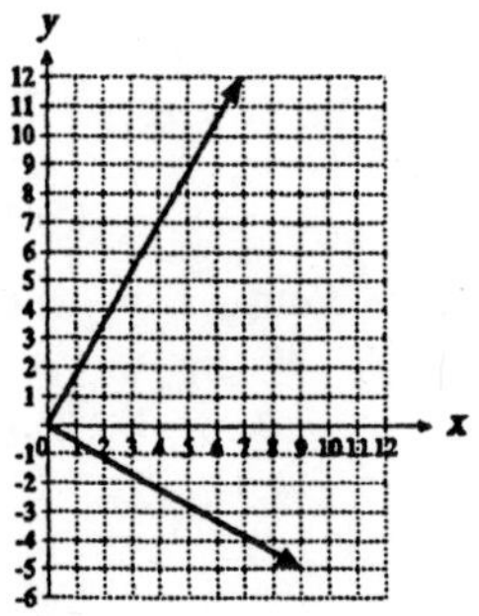

(c) $(7\mathbf{i} + 12\mathbf{j}) \cdot (9\mathbf{i} - 5\mathbf{j}) = (7)(9) + (12)(-5)$ $= 63 - 60 = 3 \neq 0$, so the vectors are not perpendicular.

11 (a)

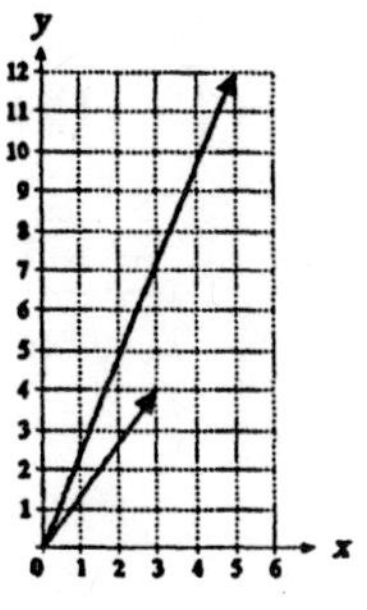

(b) $\cos\theta = \frac{\mathbf{A \cdot B}}{|\mathbf{A}|\,|\mathbf{B}|} = \frac{3 \cdot 5 + 4 \cdot 12}{\sqrt{3^2 + 4^2}\sqrt{5^2 + 12^2}}$

$= \frac{15 + 48}{\sqrt{25}\sqrt{169}} = \frac{63}{65}$. Therefore

$$\theta = \cos^{-1}\frac{63}{65} \approx 0.2487 \text{ radians} \approx 14.25°.$$

13 $\theta = \cos^{-1}\dfrac{\mathbf{A}\cdot\mathbf{B}}{|\mathbf{A}||\mathbf{B}|} = \cos^{-1}\dfrac{12}{\sqrt{56}\sqrt{14}} = \cos^{-1}\dfrac{3}{7} \approx$

1.1279 radians $\approx$ 64.6°

15 $\overrightarrow{AB} = \langle 7-1, 4-3\rangle = \langle 6, 1\rangle$ and $\overrightarrow{CD} = \langle 1-2, -5-8\rangle = \langle -1, -13\rangle$, so

$$\theta = \cos^{-1}\frac{\overrightarrow{AB}\cdot\overrightarrow{CD}}{|\overrightarrow{AB}||\overrightarrow{CD}|} = \cos^{-1}\frac{-19}{\sqrt{37}\sqrt{170}}$$

$\approx$ 1.8127 radians $\approx$ 103.9°.

17 Let $\mathbf{A} = 3\mathbf{i} - 4\mathbf{j}$ and $\mathbf{B} = \mathbf{i} + 2\mathbf{j}$. Then $\mathbf{u} = \dfrac{1}{\sqrt{5}}(\mathbf{i} + 2\mathbf{j})$ is the unit vector in the same direction as $\mathbf{B}$. We have $\text{comp}_{\mathbf{B}}\,\mathbf{A} = \text{comp}_{\mathbf{u}}\,\mathbf{A} = \mathbf{A}\cdot\mathbf{u} = \dfrac{3\cdot 1 - 4\cdot 2}{\sqrt{5}} = -\dfrac{5}{\sqrt{5}} = -\sqrt{5}$ and $\mathbf{proj}_{\mathbf{B}}\,\mathbf{A} =$

$$\mathbf{proj}_{\mathbf{u}}\,\mathbf{A} = (\mathbf{A}\cdot\mathbf{u})\mathbf{u} = -\sqrt{5}\left(\frac{\mathbf{i} + 2\mathbf{j}}{\sqrt{5}}\right) = -\mathbf{i} - 2\mathbf{j}.$$

19 Let $\mathbf{A} = 2\mathbf{i} + 3\mathbf{j}$ and $\mathbf{B} = -\mathbf{i} - \mathbf{j}$ and note that $\mathbf{C} = \mathbf{proj}_{\mathbf{B}}\,\mathbf{A}$ while $\mathbf{D} = \mathbf{orth}_{\mathbf{B}}\,\mathbf{A} = \mathbf{A} - \mathbf{C}$.

Therefore $\mathbf{C} = \left(\mathbf{A}\cdot\dfrac{\mathbf{B}}{|\mathbf{B}|}\right)\dfrac{\mathbf{B}}{|\mathbf{B}|}$

$$= \left((2\mathbf{i} + 3\mathbf{j})\cdot\frac{-\mathbf{i} - \mathbf{j}}{\sqrt{2}}\right)\frac{-\mathbf{i} - \mathbf{j}}{\sqrt{2}} = \left(-\frac{5}{\sqrt{2}}\right)\frac{-\mathbf{i} - \mathbf{j}}{\sqrt{2}}$$

$$= \frac{5}{2}\mathbf{i} + \frac{5}{2}\mathbf{j} \text{ and } \mathbf{D} = 2\mathbf{i} + 3\mathbf{j} - \left(\frac{5}{2}\mathbf{i} + \frac{5}{2}\mathbf{j}\right)$$

$$= -\frac{1}{2}\mathbf{i} + \frac{1}{2}\mathbf{j}.$$

21 (a) Let $\mathbf{A} = -3\mathbf{i} + 4\mathbf{j}$ and $\mathbf{B} = \mathbf{i} + 2\mathbf{j}$. Then $\mathbf{C} = \mathbf{proj}_{\mathbf{B}}\,\mathbf{A}$ and $\mathbf{D} = \mathbf{orth}_{\mathbf{B}}\,\mathbf{A}$, so $\mathbf{C} =$

$$\left(\mathbf{A}\cdot\frac{\mathbf{B}}{|\mathbf{B}|}\right)\frac{\mathbf{B}}{|\mathbf{B}|} = \left((-3\mathbf{i} + 4\mathbf{j})\cdot\frac{\mathbf{i} + 2\mathbf{j}}{\sqrt{5}}\right)\frac{\mathbf{i} + 2\mathbf{j}}{\sqrt{5}}$$

$$= \frac{5}{\sqrt{5}}\frac{\mathbf{i} + 2\mathbf{j}}{\sqrt{5}} = \mathbf{i} + 2\mathbf{j} \text{ and } \mathbf{D} = \mathbf{A} - \mathbf{C}$$

$$= (-3\mathbf{i} + 4\mathbf{j}) - (\mathbf{i} + 2\mathbf{j}) = -4\mathbf{i} + 2\mathbf{j}.$$

(b)

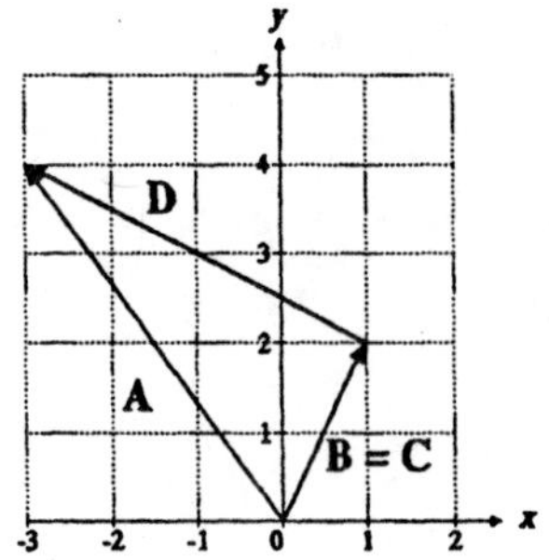

23 (a) The component of $\mathbf{A}$ on $2\mathbf{B}$ is the same as the component of $\mathbf{A}$ on $\mathbf{B}$; that is, the answer is 15.

(b) $-\mathbf{B}$ has the opposite direction from $\mathbf{B}$, so the component of $\mathbf{A}$ on $-\mathbf{B}$ will have the opposite sign from the component of $\mathbf{A}$ on $\mathbf{B}$; that is, the answer is -15.

25 If $\mathbf{proj}_{\mathbf{B}}\,\mathbf{A} = \mathbf{proj}_{\mathbf{A}}\,\mathbf{B} = \mathbf{0}$, then $\mathbf{A}$ and $\mathbf{B}$ are perpendicular. If the common projection is not $\mathbf{0}$, then $\mathbf{A}$ and $\mathbf{B}$ must be parallel, since $\mathbf{A}$ is parallel to $\mathbf{proj}_{\mathbf{A}}\,\mathbf{B}$ and $\mathbf{B}$ is parallel to $\mathbf{proj}_{\mathbf{B}}\,\mathbf{A} = \mathbf{proj}_{\mathbf{A}}\,\mathbf{B}$. But then $\mathbf{proj}_{\mathbf{A}}\,\mathbf{B} = \mathbf{B}$ and $\mathbf{proj}_{\mathbf{B}}\,\mathbf{A} = \mathbf{A}$, so $\mathbf{A} = \mathbf{B}$. Thus either $\mathbf{A}$ and $\mathbf{B}$ are perpendicular or they are equal.

27 (a) The component of $3\mathbf{i} - \mathbf{j} + 2\mathbf{k}$ on $3\mathbf{j} + 2\mathbf{k}$ is

$$(3\mathbf{i} - \mathbf{j} + 2\mathbf{k})\cdot\frac{3\mathbf{j} + 2\mathbf{k}}{\sqrt{13}} = \frac{1}{\sqrt{13}}.$$

(b) The component of $3\mathbf{i} - \mathbf{j} + 2\mathbf{k}$ on $-3\mathbf{j} - 2\mathbf{k}$ is $(3\mathbf{i} - \mathbf{j} + 2\mathbf{k})\cdot\dfrac{-3\mathbf{j} - 2\mathbf{k}}{\sqrt{13}} = -\dfrac{1}{\sqrt{13}}.$

29 $\text{proj}_{\mathbf{B}}\,\mathbf{A} = (3\mathbf{i} + 2\mathbf{j})\cdot\dfrac{4\mathbf{i} - 3\mathbf{j}}{5}\left(\dfrac{4\mathbf{i} - 3\mathbf{j}}{5}\right) =$

$\dfrac{6}{25}(4\mathbf{i} - 3\mathbf{j}) = \dfrac{24}{25}\mathbf{i} - \dfrac{18}{25}\mathbf{j}$, so $\text{orth}_{\mathbf{B}}\,\mathbf{A} =$

$3\mathbf{i} + 2\mathbf{j} - \dfrac{24}{25}\mathbf{i} + \dfrac{18}{25}\mathbf{j} = \dfrac{51}{25}\mathbf{i} + \dfrac{68}{25}\mathbf{j}.$

31

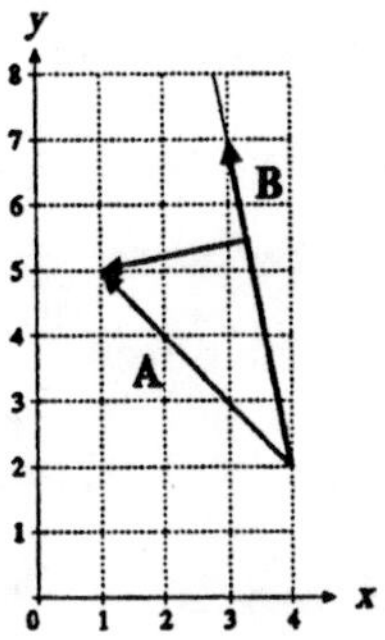

Note that $\mathbf{A} = -3\mathbf{i} + 3\mathbf{j}$ and $\mathbf{B} = -\mathbf{i} + 5\mathbf{j}$, and that the distance we want is $\|\text{orth}_{\mathbf{B}}\,\mathbf{A}\|$

$= \|\mathbf{A} - \text{proj}_{\mathbf{B}}\,\mathbf{A}\| = \left\|\mathbf{A} - \left(\mathbf{A}\cdot\dfrac{\mathbf{B}}{\|\mathbf{B}\|}\right)\dfrac{\mathbf{B}}{\|\mathbf{B}\|}\right\|$

$= \left\|-3\mathbf{i} + 3\mathbf{j} - \left[(-3\mathbf{i} + 3\mathbf{j})\cdot\dfrac{-\mathbf{i} + 5\mathbf{j}}{\sqrt{26}}\right]\dfrac{-\mathbf{i} + 5\mathbf{j}}{\sqrt{26}}\right\|$

$= \left\|-3\mathbf{i} + 3\mathbf{j} - \dfrac{18}{26}(-\mathbf{i} + 5\mathbf{j})\right\| = \left\|-\dfrac{30}{13}\mathbf{i} - \dfrac{6}{13}\mathbf{j}\right\|$

$= \dfrac{\sqrt{936}}{13} = \dfrac{6\sqrt{26}}{13} \approx 2.3534.$

33 We need to find a vector $a\mathbf{i} + b\mathbf{j}$ such that $(a\mathbf{i} + b\mathbf{j})\cdot(3\mathbf{i} - 2\mathbf{j}) = 0$; that is, $3a - 2b = 0$. Any choice where $b = \dfrac{3}{2}a$ will work, $2\mathbf{i} + 3\mathbf{j}$ being just one example.

35 $A = (0, 0, 0)$, $B = (2, 2, 0)$, $C = (2, 2, 2)$, and $D = (0, 0, 2)$, so $\overrightarrow{AC} = 2\mathbf{i} + 2\mathbf{j} + 2\mathbf{k}$ and $\overrightarrow{BD}$ $= -2\mathbf{i} - 2\mathbf{j} + 2\mathbf{k}$. Hence $\cos\theta = \dfrac{\overrightarrow{AC}\cdot\overrightarrow{BD}}{|\overrightarrow{AC}|\,|\overrightarrow{BD}|}$

$= \dfrac{2(-2) + 2(-2) + 2\cdot 2}{\sqrt{12}\sqrt{12}} = -\dfrac{1}{3}.$

37 $\overrightarrow{AC} = 2\mathbf{i} + 2\mathbf{j} + 2\mathbf{k}$ and $\overrightarrow{AM} = \mathbf{i} + 2\mathbf{j}$, so $\cos\theta$

$= \dfrac{\overrightarrow{AC}\cdot\overrightarrow{AM}}{|\overrightarrow{AC}|\,\|\overrightarrow{AM}\|} = \dfrac{2\cdot 1 + 2\cdot 2 + 2\cdot 0}{\sqrt{12}\sqrt{5}} = \dfrac{3}{\sqrt{15}}.$

39 $E = (2, 0, 0)$ and $F = (0, 2, 2)$, so $\overrightarrow{EF}$ $= -2\mathbf{i} + 2\mathbf{j} + 2\mathbf{k}$. By Exercise 35, $\overrightarrow{BD} =$ $-2\mathbf{i} - 2\mathbf{j} + 2\mathbf{k}$, so $\cos\theta = \dfrac{\overrightarrow{EF}\cdot\overrightarrow{BD}}{|\overrightarrow{EF}|\,|\overrightarrow{BD}\|} =$

$\dfrac{(-2)^2 + 2(-2) + 2^2}{\sqrt{12}\sqrt{12}} = \dfrac{1}{3}.$

41 No. Let $\mathbf{A} = \mathbf{i}$, $\mathbf{B} = 2\mathbf{i} + 3\mathbf{j}$, and $\mathbf{C} = 2\mathbf{i} + 17\mathbf{j}$. We have $\mathbf{A}\cdot\mathbf{B} = \mathbf{A}\cdot\mathbf{C} = 2$, but $\mathbf{B} \neq \mathbf{C}$, even though $\mathbf{A} \neq \mathbf{0}$.

43 $\mathbf{A} = x_1\mathbf{i} + y_1\mathbf{j} + z_1\mathbf{k}$ and $\mathbf{B} = x_2\mathbf{i} + y_2\mathbf{j} + z_2\mathbf{k}$, so the distributive rule yields $\mathbf{A}\cdot\mathbf{B} =$

$(x_1\mathbf{i} + y_1\mathbf{j} + z_1\mathbf{k})\cdot(x_2\mathbf{i} + y_2\mathbf{j} + z_2\mathbf{k}) = x_1x_2\mathbf{i}\cdot\mathbf{i} + x_1y_2\mathbf{i}\cdot\mathbf{j} + x_1z_2\mathbf{i}\cdot\mathbf{k} + y_1x_2\mathbf{j}\cdot\mathbf{i} + y_1y_2\mathbf{j}\cdot\mathbf{j} + y_1z_2\mathbf{j}\cdot\mathbf{k} + z_1x_2\mathbf{k}\cdot\mathbf{i} + z_1y_2\mathbf{k}\cdot\mathbf{j} + z_1z_2\mathbf{k}\cdot\mathbf{k} = x_1x_2\cdot 1 + x_1y_2\cdot 0 + x_1z_2\cdot 0 + y_1x_2\cdot 0 + y_1y_2\cdot 1 + y_1z_2\cdot 0 + z_1x_2\cdot 0 + z_1y_2\cdot 0 + z_1z_2\cdot 1 = x_1x_2 + y_1y_2 + z_1z_2.$

45 (a)

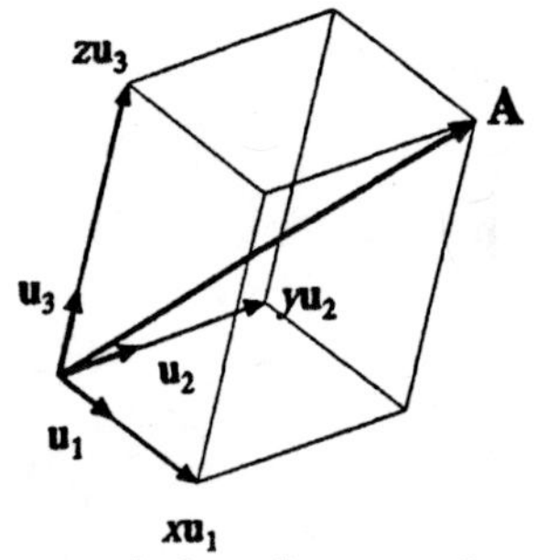

For some choice of x, y, and z, **A** will be the diagonal of the box.

(b) We have $\mathbf{A} = x\mathbf{u}_1 + y\mathbf{u}_2 + z\mathbf{u}_3$. Since the unit vectors $\mathbf{u}_1$, $\mathbf{u}_2$, and $\mathbf{u}_3$ are perpendicular, $\mathbf{A}\cdot\mathbf{u}_1 = (x\mathbf{u}_1 + y\mathbf{u}_2 + z\mathbf{u}_3)\cdot\mathbf{u}_1 = x\mathbf{u}_1\cdot\mathbf{u}_1 + y\mathbf{u}_2\cdot\mathbf{u}_1 + z\mathbf{u}_3\cdot\mathbf{u}_1 = x\cdot 1 + y\cdot 0 + z\cdot 0 = x$. (Similarly, $y = \mathbf{A}\cdot\mathbf{u}_2$ and $z = \mathbf{A}\cdot\mathbf{u}_3$.)

47 (a) Income from chair sales

(b) Income from sales of chairs and desks

(c) Cost of production of chairs and desks

(d) The firm makes a profit from the sales of chairs and desks.

49 $\langle P_1, P_2, P_3, P_4, P_5\rangle\cdot\langle x_1, x_2, x_3, x_4, x_5\rangle$ is the total profit. The vectors are perpendicular when the company is breaking even—making neither a net profit nor a net loss.

51

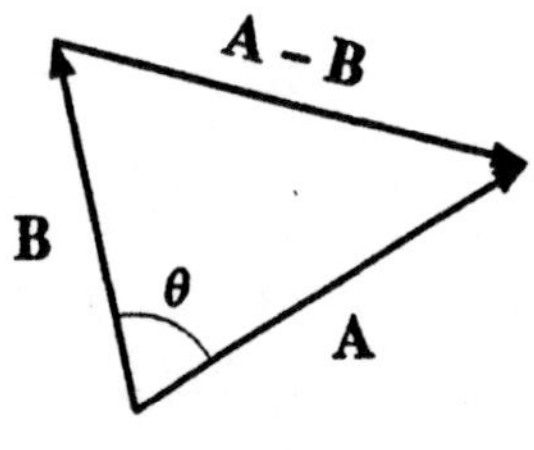

(a) By the law of cosines, $\|\mathbf{A} - \mathbf{B}\|^2 = \|\mathbf{A}\|^2 + \|\mathbf{B}\|^2 - 2\|\mathbf{A}\|\,\|\mathbf{B}\| \cos\theta$. But $\|\mathbf{A}\|\,\|\mathbf{B}\| \cos\theta = \mathbf{A}\cdot\mathbf{B}$, so we have $\|\mathbf{A} - \mathbf{B}\|^2 = \|\mathbf{A}\|^2 + \|\mathbf{B}\|^2 - 2\mathbf{A}\cdot\mathbf{B}$, as claimed.

(b) $\mathbf{A} - \mathbf{B} = (x_1 - x_2)\mathbf{i} + (y_1 - y_2)\mathbf{j}$, so $\|\mathbf{A} - \mathbf{B}\|^2 = (x_1 - x_2)^2 + (y_1 - y_2)^2 = x_1^2 - 2x_1x_2 + x_2^2 + y_1^2 - 2y_1y_2 + y_2^2$. Also, $\|\mathbf{A}\|^2 = x_1^2 + y_1^2$ and $\|\mathbf{B}\|^2 = x_2^2 + y_2^2$. By (a), $2\mathbf{A}\cdot\mathbf{B} = \|\mathbf{A}\|^2 + \|\mathbf{B}\|^2 - \|\mathbf{A} - \mathbf{B}\|^2$ $= x_1^2 + y_1^2 + x_2^2 + y_2^2 - (x_1^2 - 2x_1x_2 + x_2^2 + y_1^2 - 2y_1y_2 + y_2^2) = 2x_1x_2 + 2y_1y_2$, so $\mathbf{A}\cdot\mathbf{B} = x_1x_2 + y_1y_2$, as expected.

53 (a) If both components are negative, the vector **C** has the opposite direction in Figure 9 and we write $\text{comp}_\mathbf{C}\,\mathbf{A} = -\overline{PQ}$, $\text{comp}_\mathbf{C}\,\mathbf{B} = -\overline{QR}$, and $\text{comp}_\mathbf{C}(\mathbf{A} + \mathbf{B}) = -\overline{PR}$. Then $-\overline{PR} = -\overline{PQ} + (-\overline{QR})$, so the theorem still holds.

(b)

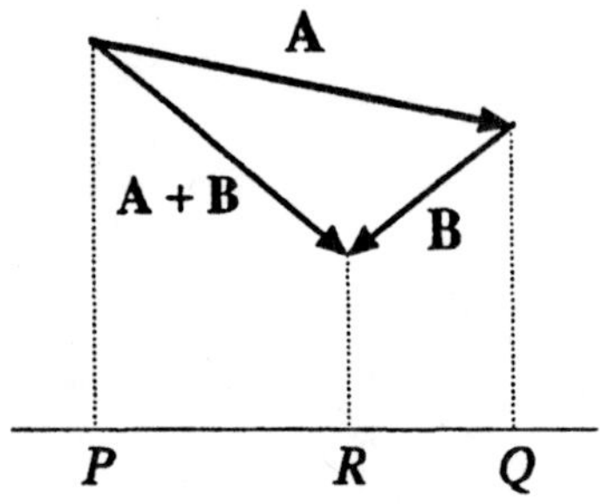

In this case, **A** and **B** have opposite directions relative to **C**, so $\text{comp}_\mathbf{C}\,\mathbf{A} = \overline{PQ}$ while $\text{comp}_\mathbf{C}\,\mathbf{B} = -\overline{QR}$ and $\text{comp}_\mathbf{C}(\mathbf{A} + \mathbf{B}) = \overline{PR}$. Then $\overline{PR} = \overline{PQ} + (-\overline{QR})$, so the theorem holds.

12.4 Lines and Planes

1 By Theorem 1, the line through (2, 3) perpendicular to 4**i** + 5**j** is $4(x - 2) + 5(y - 3) = 0$, that is, $4x + 5y = 23$.

3 By Theorem 1, an the equation of the line is $2(x - 4) + 3(y - 5) = 0$, or $2x + 3y = 23$.

5 Use Theorem 1 "backwards": The line with equation $2x - 3y + 8 = 0$ has **N** = 2**i** − 3**j** for a normal vector.

7 $y = 3x + 7$, so $3x - y + 7 = 0$. Hence, by Theorem 1, 3**i** − **j** is perpendicular to the line.

9 By Theorem 2, the distance is $\dfrac{|3 \cdot 0 + 4 \cdot 0 - 10|}{\sqrt{3^2 + 4^2}}$

$= \dfrac{10}{5} = 2.$

11 By Theorem 4, $2x - 3y + 4z + 11 = 0$ has a normal vector of 2**i** − 3**j** + 4**k**. The corresponding unit normal is $\dfrac{2\mathbf{i} - 3\mathbf{j} + 4\mathbf{k}}{\sqrt{2^2 + (-3)^2 + 4^2}} =$

$\dfrac{1}{\sqrt{29}}(2\mathbf{i} - 3\mathbf{j} + 4\mathbf{k}).$

13 By Theorem 5, the distance between (0, 0, 0) and $2x - 4y + 3z + 2 = 0$ is

$\dfrac{|2 \cdot 0 + (-4) \cdot 0 + 3 \cdot 0 + 2|}{\sqrt{2^2 + (-4)^2 + 3^2}} = \dfrac{2}{\sqrt{29}}.$

15 By Theorem 3, the plane has equation $2(x - 1) + (-7)(y - 4) + 2(z - 3) = 0$; that is, $2x - 7y + 2z + 20 = 0$. Therefore, by Theorem 5, the distance is $\dfrac{|2 \cdot 2 - 7 \cdot 2 + 2(-1) + 20|}{\sqrt{2^2 + (-7)^2 + 2^2}} = \dfrac{8}{\sqrt{57}}.$

17 Note that for any vector **A**, $\mathbf{A} \cdot \mathbf{i} = \|\mathbf{A}\| \, \|\mathbf{i}\| \cos \alpha$, where α, β, and γ are the direction angles of **A**. Since $\|\mathbf{i}\| = 1$, we have $\cos \alpha = \dfrac{\mathbf{A} \cdot \mathbf{i}}{|\mathbf{A}|}$. Similarly, $\cos \beta = \dfrac{\mathbf{A} \cdot \mathbf{j}}{|\mathbf{A}|}$ and $\cos \gamma = \dfrac{\mathbf{A} \cdot \mathbf{k}}{|\mathbf{A}\|}$. If **A** = 2**i** + 3**j** + 4**k**, then $\|\mathbf{A}\| = \sqrt{2^2 + 3^2 + 4^2} = \sqrt{29}$, so

$\cos \alpha = \dfrac{2}{\sqrt{29}}$, $\cos \beta = \dfrac{3}{\sqrt{29}}$, and $\cos \gamma = \dfrac{4}{\sqrt{29}}$.

19 (a) The direction cosines of a vector are unique. They are the components of the unit vector in the same direction as the given vector.

(b) The direction numbers of a line are not unique; they are the components of any vector with the same direction as the given line.

21 (a) The vector 2**i** − 5**j** + 8**k** is parallel to the line. The line thus has parametric equations

$x = \dfrac{1}{2} + 2t$, $y = \dfrac{1}{3} - 5t$, $z = \dfrac{1}{2} + 8t$.

(b) In vector form, the line is **P** =

$\left(\dfrac{1}{2}\mathbf{i} + \dfrac{1}{3}\mathbf{j} + \dfrac{1}{2}\mathbf{k}\right) + t(2\mathbf{i} - 5\mathbf{j} + 8\mathbf{k}).$

23 The line is parallel to $\langle 2, 1, -1\rangle - \langle 1, 0, 3\rangle = \langle 1, 1, -4\rangle$, so $\dfrac{x - 1}{1} = \dfrac{y - 0}{1} = \dfrac{z - 3}{-4}$.

25

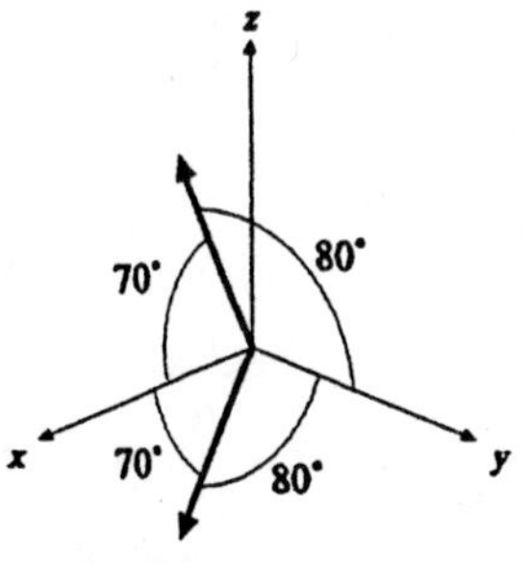

By Theorem 6, $\cos^2 \alpha + \cos^2 \beta + \cos^2 \gamma = 1$.

Hence $\cos \gamma = \pm\sqrt{1 - \cos^2 \alpha - \cos^2 \beta} =$

$\pm\sqrt{1 - \cos^2 70° - \cos^2 80°} \approx \pm 0.9235$. Since γ must be between 0° and 180°, we have either $\gamma \approx 22.6°$ or $\gamma \approx 157.4°$.

27 A vector in the direction of the first line is given by $\mathbf{A} = \langle 4 - 3, 3 - 2, 1 - 2\rangle = \langle 1, 1, -1\rangle$. A vector in the direction of the second line is given by $\mathbf{B} = \langle 5 - 3, 2 - 2, 7 - 2\rangle = \langle 2, 0, 5\rangle$. To find the angle between the lines, use the dot product: $\cos\theta = \dfrac{\mathbf{A}\cdot\mathbf{B}}{|\mathbf{A}||\mathbf{B}|} = \dfrac{2 + 0 - 5}{\sqrt{3}\sqrt{29}} = -\dfrac{3}{\sqrt{87}}$, so $\theta = \cos^{-1}\left(\dfrac{-3}{\sqrt{87}}\right) \approx 1.898$ radians $\approx 108.76°$.

29 Use the point-slope form with slope $m_1 = 3/2$ and point (1, 2) to obtain $y - 2 = \frac{3}{2}(x - 1)$, or $3x - 2y = -1$, as an equation for line L_1; use $m_2 = 4$ and point (2, 3) to obtain $4x - y = 5$ as an equation for line L_2. Multiply the second equation by -2 and add it to the first equation to obtain $-5x = -11$, so $x = 11/5$; hence $y = 19/5$. The lines therefore cross at $\left(\dfrac{11}{5}, \dfrac{19}{5}\right)$.

31 Observe that points on the line through (1, 2, −3) and (1, 6, 2) should have the form (1, y, z). However, (7, 14, 11) does not. Hence the points are not collinear.

33 If line is perpendicular to plane, then it is parallel to a normal to the plane. One such normal is $\langle 2, -3, 4\rangle$. Using this normal and the given point (1, 3, −5), we obtain parametric equations of the line: $x = 1 + 2t$, $y = 3 - 3t$, and $z = -5 + 4t$.

35 Recall from Sec. 12.2 that the area of a flat region U is related to the area of its projection V onto a plane by

$$\text{Area of } V = (\text{Area of } U)\cos\theta,$$

where θ is the dihedral angle. Let $\mathbf{N}$ be a normal to U and $\mathbf{B}$ a normal to V. Since θ is the angle between the normals to U and V, we have $\cos\theta = \dfrac{\mathbf{N}\cdot\mathbf{B}}{|\mathbf{N}||\mathbf{B}|}$. U is a square of area a^2 in the plane $2x + 3y + 2z = 8$, so $\mathbf{N} = 2\mathbf{i} + 3\mathbf{j} + 2\mathbf{k}$ and $\|\mathbf{N}\| = \sqrt{17}$.

(a) For the xy plane, let $\mathbf{B} = \mathbf{k}$. Thus Area of V

$$= a^2\frac{\mathbf{N}\cdot\mathbf{k}}{\|\mathbf{N}\|} = \frac{2a^2}{\sqrt{17}}.$$

(b) For the yz plane, let $\mathbf{B} = \mathbf{i}$. Thus Area of V

$$= a^2\frac{\mathbf{N}\cdot\mathbf{i}}{|\mathbf{N}|} = \frac{2a^2}{\sqrt{17}}.$$

(c) For the xz plane, let $\mathbf{B} = \mathbf{j}$. Thus Area of V

$$= a^2\frac{\mathbf{N}\cdot\mathbf{j}}{|\mathbf{N}|} = \frac{3a^2}{\sqrt{17}}.$$

37 $\sin^2\alpha + \sin^2\beta + \sin^2\gamma$
$= (1 - \cos^2\alpha) + (1 - \cos^2\beta) + (1 - \cos^2\gamma)$
$= 3 - (\cos^2\alpha + \cos^2\beta + \cos^2\gamma) = 3 - 1 = 2$

39 See the solution to Exercise 35. The area of U in this case is πa^2 and it lies in the plane $x + 3y + 4z = 5$, so $\mathbf{N} = \mathbf{i} + 3\mathbf{j} + 4\mathbf{k}$. The plane $2x + y - z = 6$ has normal $\mathbf{B} = 2\mathbf{i} + \mathbf{j} - \mathbf{k}$, so Area of V =

$$\pi a^2\frac{(\mathbf{i} + 3\mathbf{j} + 4\mathbf{k})\cdot(2\mathbf{i} + \mathbf{j} - \mathbf{k})}{\sqrt{1^2 + 3^2 + 4^2}\sqrt{2^2 + 1^2 + (-1)^2}} = \frac{\pi a^2}{\sqrt{26}\sqrt{6}} = \frac{\pi a^2}{2\sqrt{39}}.$$

41 The vector pointing from (3, 4, 5) to (5, 7, 10) is $2\mathbf{i} + 3\mathbf{j} + 5\mathbf{k}$, so the line through these points has

the parametric equations $x = 3 + 2t$, $y = 4 + 3t$, and $z = 5 + 5t$. The vector from (1, 4, 0) to (3, 6, 4) is $2\mathbf{i} + 2\mathbf{j} + 4\mathbf{k}$, so the line through these points has the parametric equations $x = 1 + 2s$, $y = 4 + 2s$, and $z = 4s$. If these lines intersect, then there must be numbers t and s such that $3 + 2t = 1 + 2s$, $4 + 3t = 4 + 2s$, and $5 + 5t = 4s$. The first equation gives $s = t + 1$. Combining this with the second equation gives $4 + 3t = 4 + 2(t + 1) = 6 + 2t$, so $t = 2$ and $s = 3$. But then $5 + 5t = 15 \neq 12 = 4s$. Hence the lines do not meet.

43 Let $P = (1, 2, -1)$, $Q = (1, 3, 5)$, and $R = (2, 1, -3)$. As usual, $O = (0, 0, 0)$.

(a) The line through Q and R has parametric equation $x\mathbf{i} + y\mathbf{j} + z\mathbf{k} = \overrightarrow{OQ} + t\overrightarrow{QR} = \mathbf{i} + 3\mathbf{j} + 5\mathbf{k} + t(\mathbf{i} - 2\mathbf{j} - 8\mathbf{k})$, so $x = 1 + t$, $y = 3 - 2t$, and $z = 5 - 8t$. The square of the distance between (x, y, z) and P is $(1 + t - 1)^2 + (3 - 2t - 2)^2 + (5 - 8t + 1)^2$ $= t^2 + (1 - 2t)^2 + (6 - 8t)^2$ $= 69t^2 - 100t + 37$. The first derivative is $138t - 100$, which equals 0 for $t = 50/69$. Substitute this value into $69t^2 - 100t + 37$ and take the square root to find that the minimum distance is $\dfrac{\sqrt{3657}}{69} \approx 0.8764$.

(b) The desired distance is $\|\text{orth}_{\overrightarrow{QR}}\overrightarrow{QP}\|$

$$= \|\overrightarrow{QP} - \text{proj}_{\overrightarrow{QR}}\overrightarrow{QP}\|$$

$$= \left\|\overrightarrow{QP} - \left(\frac{\overrightarrow{QP}\cdot\overrightarrow{QR}}{|\overrightarrow{QR}|}\right)\frac{\overrightarrow{QR}}{|\overrightarrow{QR}|}\right\| =$$

$$\left\|-\mathbf{j} - 6\mathbf{k} - \frac{(-\mathbf{j} - 6\mathbf{k})\cdot(\mathbf{i} - 2\mathbf{j} - 8\mathbf{k})}{69}(\mathbf{i} - 2\mathbf{j} - 8\mathbf{k})\right\|$$

$$= \left\|-\mathbf{j} - 6\mathbf{k} - \frac{50}{69}(\mathbf{i} - 2\mathbf{j} - 8\mathbf{k})\right\|$$

$$= \left\|\frac{1}{69}(-50\mathbf{i} + 31\mathbf{j} - 14\mathbf{k})\right\| = \frac{\sqrt{3657}}{69},$$ as before.

45 By Theorem 6, at least one of the direction cosines is less than or equal to $1/\sqrt{3}$. The corresponding direction angle is thus at least $\cos^{-1}\frac{1}{\sqrt{3}}$. On the other hand, all of the direction angles of $\mathbf{i} + \mathbf{j} + \mathbf{k}$ are equal to $\cos^{-1}\frac{1}{\sqrt{3}}$.

12.5 Determinants

1 $\begin{vmatrix} 3 & 4 \\ 7 & 2 \end{vmatrix} = 6 - 28 = -22$

3 $\begin{vmatrix} 4 & 2 & 0 \\ 5 & 6 & -1 \\ 1 & -1 & 2 \end{vmatrix} = 4\begin{vmatrix} 6 & -1 \\ -1 & 2 \end{vmatrix} - 2\begin{vmatrix} 5 & -1 \\ 1 & 2 \end{vmatrix} + 0\begin{vmatrix} 5 & 6 \\ 1 & -1 \end{vmatrix}$

$= 4(12 - 1) - 2(10 + 1) = 22$

5 $\begin{vmatrix} 2 & 1 & 4 \\ 2 & 1 & 4 \\ 3 & 5 & 7 \end{vmatrix} = 0$ by Theorem 1, since the first two rows are identical.

7 By Theorem 1, since rows one and three are identical, the determinant is 0.

9 $\begin{vmatrix} 1 & 1 & 1 \\ 3 & 4 & 3 \\ 5 & 2 & -1 \end{vmatrix} = 1\begin{vmatrix} 4 & 3 \\ 2 & -1 \end{vmatrix} - 1\begin{vmatrix} 3 & 3 \\ 5 & -1 \end{vmatrix} + 1\begin{vmatrix} 3 & 4 \\ 5 & 2 \end{vmatrix}$

$= (1)(-4 - 6) - (1)(-3 - 15) + (1)(6 - 20)$

$= -10 + 18 - 14 = -6$

11 Area of parallelogram spanned by **A** and **B** is

$\text{abs}\begin{vmatrix} 3 & 4 \\ 2 & -1 \end{vmatrix} = |-3 - 8| = |-11| = 11$

13 Area $= \text{abs}\begin{vmatrix} 3 & 5 \\ 5 & -3 \end{vmatrix} = |-9 - 25| = 34$

15 (a) Note that $\begin{vmatrix} x & y & 1 \\ x_1 & y_1 & 1 \\ x_2 & y_2 & 1 \end{vmatrix} =$

$x(y_1 - y_2) - y(x_1 - x_2) + 1(x_1y_2 - y_1x_2)$

$= x(y_1 - y_2) - y(x_1 - x_2) - x_1y_1 + x_1y_2 + y_1x_1 - y_1x_2$

$= x(y_1 - y_2) - y(x_1 - x_2) - x_1(y_1 - y_2) + y_1(x_1 - x_2)$

$= (x - x_1)(y_1 - y_2) - (y - y_1)(x_1 - x_2)$.

Thus the given equation is equivalent to

$(x - x_1)(y_1 - y_2) = (y - y_1)(x_1 - x_2)$, or

$\dfrac{y - y_1}{x - x_1} = \dfrac{y_1 - y_2}{x_1 - x_2}$, which is an equation of

the line through (x_1, y_1) and (x_2, y_2); that is, (x, y) is on the line determined by the given points.

(b) From (a), $0 = \begin{vmatrix} x & y & 1 \\ 2 & 3 & 1 \\ 1 & -4 & 1 \end{vmatrix}$

$= x(3 - (-4)) - y(2 - 1) + 1(-8 - 3)$

$= 7x - y - 11$; that is, the line's equation is $7x - y = 11$.

17 If D is the value of the determinant, then Theorem 2 says that $-D$ is the value of the determinant after two rows (or columns) have been reversed. But if two rows (or columns) are identical, then we can reverse them and still have the same determinant that we started with. In this case, $D = -D$, so $2D = 0$ and we find that $D = 0$, as stated by Theorem 1.

19 $\begin{vmatrix} ka_1 & ka_2 \\ b_1 & b_2 \end{vmatrix} = ka_1b_2 - ka_2b_1 = k(a_1b_2 - a_2b_1)$

$= k\begin{vmatrix} a_1 & a_2 \\ b_1 & b_2 \end{vmatrix}$

21 $\begin{vmatrix} a_1 + kb_1 & a_2 + kb_2 \\ b_1 & b_2 \end{vmatrix} = (a_1 + kb_1)b_2 - (a_2 + kb_2)b_1$

$= a_1b_2 + kb_1b_2 - a_2b_1 - kb_2b_1 = a_1b_2 - a_2b_1$

$= \begin{vmatrix} a_1 & a_2 \\ b_1 & b_2 \end{vmatrix}$

23 By Theorem 3 we know that the vectors $\mathbf{C} = \langle a_1, b_1 \rangle$ and $\mathbf{D} = \langle a_2, b_2 \rangle$ span a parallelogram whose area is $\text{abs}\begin{vmatrix} a_1 & b_1 \\ a_2 & b_2 \end{vmatrix} = |a_1b_2 - b_1a_2|$. On the other hand, $\begin{vmatrix} a_1 & a_2 \\ b_1 & b_2 \end{vmatrix} = a_1b_2 - a_2b_1$, which has the same absolute value. Hence it too gives the area of the parallelogram.

12.6 The Cross Product of Two Vectors

1

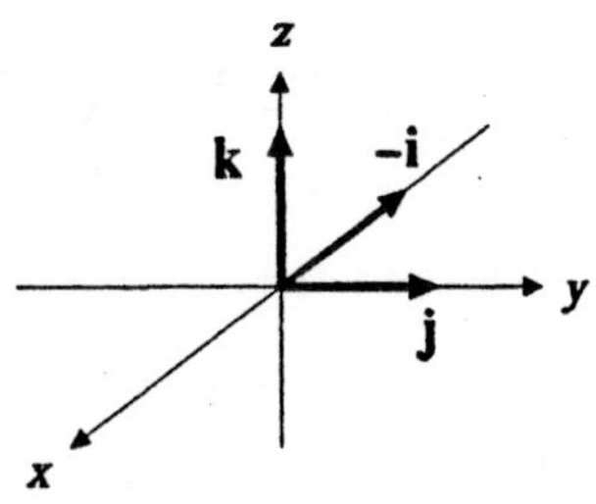

$$\mathbf{A} \times \mathbf{B} = \begin{vmatrix} \mathbf{i} & \mathbf{j} & \mathbf{k} \\ 0 & 0 & 1 \\ 0 & 1 & 0 \end{vmatrix} = \mathbf{i}\begin{vmatrix} 0 & 1 \\ 1 & 0 \end{vmatrix} - \mathbf{j}\begin{vmatrix} 0 & 1 \\ 0 & 0 \end{vmatrix} + \mathbf{k}\begin{vmatrix} 0 & 0 \\ 0 & 1 \end{vmatrix}$$

$= -\mathbf{i}$

3

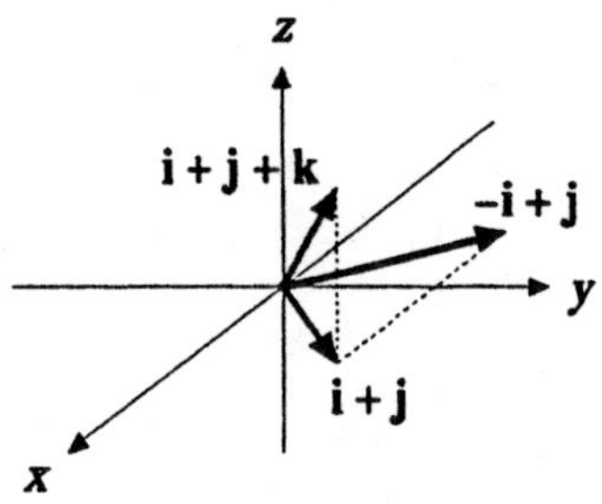

$$\mathbf{A} \times \mathbf{B} = \begin{vmatrix} \mathbf{i} & \mathbf{j} & \mathbf{k} \\ 1 & 1 & 1 \\ 1 & 1 & 0 \end{vmatrix} = \mathbf{i}\begin{vmatrix} 1 & 1 \\ 1 & 0 \end{vmatrix} - \mathbf{j}\begin{vmatrix} 1 & 1 \\ 1 & 0 \end{vmatrix} + \mathbf{k}\begin{vmatrix} 1 & 1 \\ 1 & 1 \end{vmatrix}$$

$= -\mathbf{i} + \mathbf{j}$

5 $\mathbf{A} = 2\mathbf{i} - 3\mathbf{j} + \mathbf{k}$ and $\mathbf{B} = \mathbf{i} + \mathbf{j} + 2\mathbf{k}$, so $\mathbf{A} \times \mathbf{B}$

$$= \begin{vmatrix} \mathbf{i} & \mathbf{j} & \mathbf{k} \\ 2 & -3 & 1 \\ 1 & 1 & 2 \end{vmatrix} = \mathbf{i}(-6 - 1) - \mathbf{j}(4 - 1) + \mathbf{k}(2 + 3) =$$

$-7\mathbf{i} - 3\mathbf{j} + 5\mathbf{k}$. Then $\mathbf{A}\cdot(\mathbf{A} \times \mathbf{B}) = 2(-7) - 3(-3) + 5 = 0$ and $\mathbf{B}\cdot(\mathbf{A} \times \mathbf{B}) = (-7) + (-3) + 2\cdot 5 = 0$, as claimed.

7 Let $\mathbf{A}$ be the vector from (0, 0, 0) to (1, 5, 4) and $\mathbf{B}$ the vector from (0, 0, 0) to (2, −1, 3). By Theorem 3, the area of the parallelogram is $\|\mathbf{A} \times \mathbf{B}\|$. But $\mathbf{A} \times \mathbf{B} = (\mathbf{i} + 5\mathbf{j} + 4\mathbf{k}) \times$

$$(2\mathbf{i} - \mathbf{j} + 3\mathbf{k}) = \begin{vmatrix} \mathbf{i} & \mathbf{j} & \mathbf{k} \\ 1 & 5 & 4 \\ 2 & -1 & 3 \end{vmatrix} = 19\mathbf{i} + 5\mathbf{j} - 11\mathbf{k},$$

so $\|\mathbf{A} \times \mathbf{B}\| = \sqrt{19^2 + 5^2 + (-11)^2} = \sqrt{507}$ ≈ 22.52. Note that for any three points *A*, *B*, and *C*, there are three parallelograms that have *A*, *B*, and *C* as vertices. (In the figure, these are *ABDC*, *BCEA*, and *CAFB*.) These parallelograms all have the same area, since each is twice the area of triangle *ABC*.

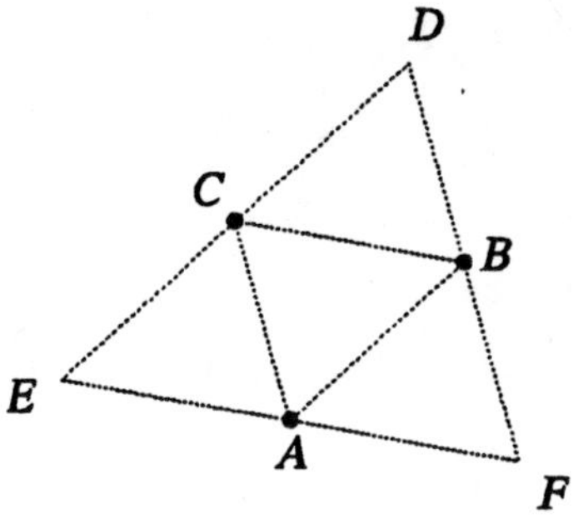

9 $$\text{Area} = \frac{1}{2}\,\text{mag}\begin{vmatrix} \mathbf{i} & \mathbf{j} & \mathbf{k} \\ 1 & 1 & 0 \\ 3 & -1 & 0 \end{vmatrix} = \frac{1}{2}\|\langle 0, 0, -4\rangle\|$$

$$= \frac{1}{2}\cdot 4 = 2$$

11 $$\text{Volume} = \text{abs}\begin{vmatrix} 2 & 1 & 3 \\ 3 & -1 & 2 \\ 4 & 0 & 3 \end{vmatrix}$$

$$= |2(-3) - 1(1) + 3(4)| = 5$$

13 $\overrightarrow{PQ} = \langle 1, 0, -3 \rangle$, $\overrightarrow{PR} = \langle 2, 4, 1 \rangle$, and $\overrightarrow{PS} = \langle 0, -2, 1 \rangle$, so the volume is abs $\begin{vmatrix} 1 & 0 & -3 \\ 2 & 4 & 1 \\ 0 & -2 & 1 \end{vmatrix}$

$= |1(6) - 3(-4)| = 18.$

15 Let $P = (1, 2, 1)$, $Q = (2, 1, -3)$, and $R = (0, 1, 5)$. Then $\overrightarrow{PQ} \times \overrightarrow{PR}$ is one such vector. Then $\overrightarrow{PQ} = \mathbf{i} - \mathbf{j} - 4\mathbf{k}$ and $\overrightarrow{PR} = -\mathbf{i} - \mathbf{j} + 4\mathbf{k}$, so

$$\overrightarrow{PQ} \times \overrightarrow{PR} = \begin{vmatrix} \mathbf{i} & \mathbf{j} & \mathbf{k} \\ 1 & -1 & -4 \\ -1 & -1 & 4 \end{vmatrix} = -8\mathbf{i} - 0\mathbf{j} + (-2)\mathbf{k}$$

$= -8\mathbf{i} - 2\mathbf{k}$. So any vector parallel to $4\mathbf{i} + \mathbf{k}$ is perpendicular to the plane.

17 A vector in the direction of the first line is $\langle 1, -1, -1 \rangle$. A vector in the direction of the second line is $\langle 1, 3, 1 \rangle$. To find a vector perpendicular to both, apply the cross product:

$\begin{vmatrix} \mathbf{i} & \mathbf{j} & \mathbf{k} \\ 1 & -1 & -1 \\ 1 & 3 & 1 \end{vmatrix} = \langle 2, -2, 4 \rangle$. So any vector parallel to $\langle 1, -1, 2 \rangle$ is perpendicular to both lines.

19 Let $\mathbf{A} = \langle a_1, a_2, a_3 \rangle$ and $\mathbf{B} = \langle b_1, b_2, b_3 \rangle$, so the cross product is given by $\mathbf{A} \times \mathbf{B} = \langle a_2b_3 - a_3b_2, a_3b_1 - a_1b_3, a_1b_2 - a_2b_1 \rangle$. Then the scalar triple product is $\mathbf{A} \cdot (\mathbf{A} \times \mathbf{B}) = \langle a_1, a_2, a_3 \rangle \cdot \langle a_2b_3 - a_3b_2, a_3b_1 - a_1b_3, a_1b_2 - a_2b_1 \rangle$

$= a_1(a_2b_3 - a_3b_2) + a_2(a_3b_1 - a_1b_3) + a_3(a_1b_2 - a_2b_1) = 0.$

21 (a) Suppose that $\mathbf{A} = x_1\mathbf{i} + y_1\mathbf{j} + z_1\mathbf{k}$ and $\mathbf{B} = x_2\mathbf{i} + y_2\mathbf{j} + z_2\mathbf{k}$. It follows that $\mathbf{B} \times \mathbf{A} =$

$$\begin{vmatrix} \mathbf{i} & \mathbf{j} & \mathbf{k} \\ x_2 & y_2 & z_2 \\ x_1 & y_1 & z_1 \end{vmatrix} = -\begin{vmatrix} \mathbf{i} & \mathbf{j} & \mathbf{k} \\ x_1 & y_1 & z_1 \\ x_2 & y_2 & z_2 \end{vmatrix} = -(\mathbf{A} \times \mathbf{B}),$$

where we used Theorem 2 of Sec. 12.5.

(b) When the right-hand rule is applied for $\mathbf{A} \times \mathbf{B}$ and for $\mathbf{B} \times \mathbf{A}$ the thumb points in opposite directions. The areas of the parallelograms are equal. Hence $\mathbf{B} \times \mathbf{A} = -(\mathbf{A} \times \mathbf{B})$.

23 Let $\mathbf{A} = a_1\mathbf{i} + a_2\mathbf{j} + a_3\mathbf{k}$ and $\mathbf{B} = b_1\mathbf{i} + b_2\mathbf{j} + b_3\mathbf{k}$. Then $\mathbf{B} \cdot (\mathbf{A} \times \mathbf{B}) = \mathbf{B} \cdot \begin{vmatrix} \mathbf{i} & \mathbf{j} & \mathbf{k} \\ a_1 & a_2 & a_3 \\ b_1 & b_2 & b_3 \end{vmatrix} = (b_1\mathbf{i} +$

$$b_2\mathbf{j} + b_3\mathbf{k}) \cdot \left(\mathbf{i} \begin{vmatrix} a_2 & a_3 \\ b_2 & b_3 \end{vmatrix} - \mathbf{j} \begin{vmatrix} a_1 & a_3 \\ b_1 & b_3 \end{vmatrix} + \mathbf{k} \begin{vmatrix} a_1 & a_2 \\ b_1 & b_2 \end{vmatrix} \right) =$$

$$b_1 \begin{vmatrix} a_2 & a_3 \\ b_2 & b_3 \end{vmatrix} - b_2 \begin{vmatrix} a_1 & a_3 \\ b_1 & b_3 \end{vmatrix} + b_3 \begin{vmatrix} a_1 & a_2 \\ b_1 & b_2 \end{vmatrix} = \begin{vmatrix} b_1 & b_2 & b_3 \\ a_1 & a_2 & a_3 \\ b_1 & b_2 & b_3 \end{vmatrix}$$

$= 0$, since the first and third rows are equal.

25 (a) $\mathbf{i} \times 2\mathbf{i} = \begin{vmatrix} \mathbf{i} & \mathbf{j} & \mathbf{k} \\ 1 & 0 & 0 \\ 2 & 0 & 0 \end{vmatrix} = \langle 0, 0, 0 \rangle = \mathbf{0}$

(b) $\mathbf{i} \times \mathbf{k} = \begin{vmatrix} \mathbf{i} & \mathbf{j} & \mathbf{k} \\ 1 & 0 & 0 \\ 0 & 0 & 1 \end{vmatrix} = \langle 0, -1, 0 \rangle = -\mathbf{j}$

(c) $\mathbf{j} \times \mathbf{i} = \begin{vmatrix} \mathbf{i} & \mathbf{j} & \mathbf{k} \\ 0 & 1 & 0 \\ 1 & 0 & 0 \end{vmatrix} = \langle 0, 0, -1 \rangle = -\mathbf{k}$

27 If $\mathbf{A} \times \mathbf{B} = \mathbf{0}$, for nonzero vectors **A** and **B**, then we must have $\sin\theta = 0$, where θ is the angle between **A** and **B**. Hence **A** and **B** either point in the same direction ($\theta = 0$) or point in opposite directions ($\theta = \pi$).

29 (a) Yes. The direction of $\mathbf{A} \times \mathbf{B}$ is the same as the direction of $\mathbf{A} \times \mathbf{C}$ because you get the same result from the right-hand rule when you apply it to parallel vectors **B** and **C**.

(b) Not necessarily. Let $\mathbf{A} = \mathbf{i} + \mathbf{j}$, $\mathbf{B} = \mathbf{i}$, and $\mathbf{C} = \mathbf{j}$. Then **B** and **C** are perpendicular, but $\mathbf{A} \times \mathbf{B} = -\mathbf{k}$ and $\mathbf{A} \times \mathbf{C} = \mathbf{k}$, which aren't.

31 Use the formula for the vector triple product, substituting **A**, **A**, and **B** in place of **A**, **B**, and **C**: $\mathbf{A} \times (\mathbf{A} \times \mathbf{B}) = (\mathbf{A}\cdot\mathbf{B})\mathbf{A} - (\mathbf{A}\cdot\mathbf{A})\mathbf{B}$.

33 Dot both sides of the result of Exercise 31 by **B** to get $\mathbf{B}\cdot[\mathbf{A} \times (\mathbf{A} \times \mathbf{B})] = \mathbf{B}\cdot[(\mathbf{A}\cdot\mathbf{B})\mathbf{A} - (\mathbf{A}\cdot\mathbf{A})\mathbf{B}]$. The left side is $(\mathbf{B} \times \mathbf{A})\cdot(\mathbf{A} \times \mathbf{B}) = -\|\mathbf{A} \times \mathbf{B}\|^2$ while the right side equals $(\mathbf{A}\cdot\mathbf{B})^2 - (\mathbf{A}\cdot\mathbf{A})(\mathbf{B}\cdot\mathbf{B}) = (\mathbf{A}\cdot\mathbf{B})^2 - \|\mathbf{A}\|^2\|\mathbf{B}\|^2$. Hence $\|\mathbf{A} \times \mathbf{B}\|^2 = \|\mathbf{A}\|^2\|\mathbf{B}\|^2 - (\mathbf{A}\cdot\mathbf{B})^2$, as claimed.

35 The vector $\mathbf{u} \times \mathbf{B}$ is perpendicular to both **u** and **B**. Furthermore, $\mathbf{u} \times (\mathbf{u} \times \mathbf{B})$ is perpendicular to both **u** and $\mathbf{u} \times \mathbf{B}$. Additional cross products by **u** will always produce vectors that are perpendicular to **u**, so they lie in the plane for which **u** is a normal vector. Since the resulting vectors are always perpendicular to each other, they are spaced by 90° in this plane, so there can be only four of them (one of them being $\mathbf{u} \times \mathbf{B}$ itself).

37 (a) $\mathbf{A} \times \mathbf{B} = (x_1\mathbf{i} + y_1\mathbf{j} + z_1\mathbf{k}) \times (x_2\mathbf{i} + y_2\mathbf{j} + z_2\mathbf{k})$

$= x_1x_2(\mathbf{i} \times \mathbf{i}) + x_1y_2(\mathbf{i} \times \mathbf{j}) + x_1z_2(\mathbf{i} \times \mathbf{k})$
$+ y_1x_2(\mathbf{j} \times \mathbf{i}) + y_1y_2(\mathbf{j} \times \mathbf{j}) + y_1z_2(\mathbf{j} \times \mathbf{k})$
$+ z_1x_2(\mathbf{k} \times \mathbf{i}) + z_1y_2(\mathbf{k} \times \mathbf{j}) + z_1z_2(\mathbf{k} \times \mathbf{k})$
$= x_1x_2(\mathbf{0}) + x_1y_2\mathbf{k} + x_1z_2(-\mathbf{j}) + y_1x_2(-\mathbf{k}) + y_1y_2(\mathbf{0}) + y_1z_2\mathbf{i} + z_1x_2\mathbf{j} + z_1y_2(-\mathbf{i}) + z_1z_2(\mathbf{0})$
$= (y_1z_2 - z_1y_2)\mathbf{i} + (z_1x_2 - x_1z_2)\mathbf{j} + (x_1y_2 - y_1x_2)\mathbf{k}$

(b) Yes.

(c) No. The geometric definition suffices to provide the various cross products of the unit vectors, but does not show that the cross product is distributive—which property is necessary from the very first step of (a).

39 (a) The right-hand rule gives the same direction for $\mathbf{A} \times \mathbf{B}$ and $\mathbf{A} \times \mathbf{B}_1$. The length of $\mathbf{A} \times \mathbf{B}$

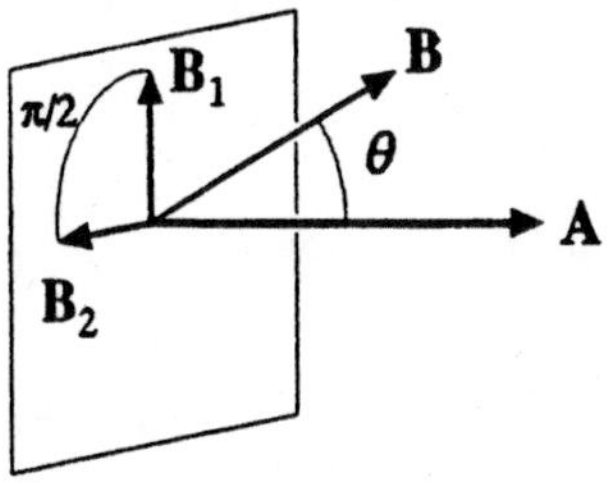

is $\|\mathbf{A}\|\,\|\mathbf{B}\|\sin\theta$. Since $\|\mathbf{B}_1\| = \|\mathbf{B}\|\sin\theta$, we have $\|\mathbf{A} \times \mathbf{B}_1\| = \|\mathbf{A}\|\,\|\mathbf{B}_1\|\sin\frac{\pi}{2} = \|\mathbf{A}\|\,\|\mathbf{B}\|\sin\theta = \|\mathbf{A} \times \mathbf{B}\|$. Thus $\mathbf{A} \times \mathbf{B}$ and $\mathbf{A} \times \mathbf{B}_1$ have the same magnitude as well as the same direction. They must be equal.

(b) By (a), $\|\mathbf{A} \times \mathbf{B}\| = \|\mathbf{A}\|\,\|\mathbf{B}_1\|$. Since $\|\mathbf{B}_1\| = \|\mathbf{B}_2\|$, $\mathbf{A} \times \mathbf{B}$ and $\|\mathbf{A}\|\mathbf{B}_2$ have the same magnitude. Since $\mathbf{B}_2$ points in the direction given by the right-hand rule for **A** and **B**, $\|\mathbf{A}\|\mathbf{B}_2$ is equal to $\mathbf{A} \times \mathbf{B}$.

41 (a) Using the left-hand distributive rule from Exercise 40, we have $(\mathbf{B} + \mathbf{C}) \times \mathbf{A}$ $= -\mathbf{A} \times (\mathbf{B} + \mathbf{C}) = -(\mathbf{A} \times \mathbf{B} + \mathbf{A} \times \mathbf{C})$ $= -(\mathbf{A} \times \mathbf{B}) - (\mathbf{A} \times \mathbf{C})$ $= \mathbf{B} \times \mathbf{A} + \mathbf{C} \times \mathbf{A}$.

(b) $\mathbf{A} \times (\mathbf{B} + \mathbf{C} + \mathbf{D}) = \mathbf{A} \times (\mathbf{B} + \mathbf{C}) + \mathbf{A} \times \mathbf{D}$ $= \mathbf{A} \times \mathbf{B} + \mathbf{A} \times \mathbf{C} + \mathbf{A} \times \mathbf{D}$

(c) Now that the distributive law has been established for cross products, the argument in Exercise 37 is valid.

43 The point P is inside the tetrahedron with vertices P_1, P_2, P_3, and P_4 if and only if four conditions hold: (1) P and P_1 are on the same side of the plane determined by P_2, P_3, and P_4, (2) P and P_2 are on the same side of the plane determined by P_1, P_3, and P_4, (3) P and P_3 are on the same side of the plane determined by P_1, P_2, and P_4, and (4) P and P_4 are on the same side of the plane determined by P_1, P_2, and P_3. To test the first condition, compute the component of $\overrightarrow{P_2P}$ on $\mathbf{N}_1$ and see if it has the same sign as the component of $\overrightarrow{P_2P_1}$ on $\mathbf{N}_1$, where $\mathbf{N}_1 = \overrightarrow{P_2P_3} \times \overrightarrow{P_2P_4}$ is a normal to the plane determined by P_2, P_3, and P_4. If the signs are the same, then condition (1) is satisfied. The other conditions are tested similarly.

12.7 More on Lines and Planes

1 To find a vector perpendicular to the plane, find two vectors in the plane. Two such vectors are $\langle 4 - 2, 5 - 1, 1 - 3\rangle = \langle 2, 4, -2\rangle$ and $\langle 4 + 2, 5 - 2, 1 - 3\rangle = \langle 6, 3, -2\rangle$. Since the cross product generates a value perpendicular to two vectors, $\langle 2, 4, -2\rangle \times \langle 6, 3, -2\rangle =$

$$\begin{vmatrix} \mathbf{i} & \mathbf{j} & \mathbf{k} \\ 2 & 4 & -2 \\ 6 & 3 & -2 \end{vmatrix} = \langle -2, -8, -18\rangle \text{ is a vector}$$

perpendicular to the two vectors and to the plane. Any scalar multiple of this vector also works.

3 A unit normal to the plane is $\frac{1}{\sqrt{14}}\langle 1, 2, 3\rangle$. To find the projection of $\mathbf{A}$ in the plane, calculate $\mathbf{A} - \text{proj}_{\mathbf{u}}\,\mathbf{A}$. First, $\text{proj}_{\mathbf{u}}\,\mathbf{A} = (\mathbf{A}\cdot\mathbf{u})\mathbf{u}$

$$= \left(\langle 3, 4, 5\rangle \cdot \frac{1}{\sqrt{14}}\langle 1, 2, 3\rangle\right)\frac{1}{\sqrt{14}}\langle 1, 2, 3\rangle$$

$$= \left[\frac{1}{\sqrt{14}}(3 + 8 + 15)\right]\frac{1}{\sqrt{14}}\langle 1, 2, 3\rangle = \frac{13}{7}\langle 1, 2, 3\rangle$$

$= \left\langle \frac{13}{7}, \frac{26}{7}, \frac{39}{7}\right\rangle$. Now $\mathbf{A} - \text{proj}_{\mathbf{u}}\,\mathbf{A} =$

$$\langle 3, 4, 5\rangle - \left\langle \frac{13}{7}, \frac{26}{7}, \frac{39}{7}\right\rangle = \left\langle \frac{8}{7}, \frac{2}{7}, -\frac{4}{7}\right\rangle.$$

5 Since the line is perpendicular to the plane, it is parallel to a normal to the plane. One such normal is $\langle 3, -1, 1\rangle$. The equations of the line are $x = 1 + 3t$, $y = 1 - t$, $z = 2 + t$.

7 The line through (1, 2, 1) and (2, −1, 3) is parallel to the vector $\mathbf{A} = \mathbf{i} - 3\mathbf{j} + 2\mathbf{k}$ and has parametric equations $x = 1 + t$, $y = 2 - 3t$, and $z = 1 + 2t$ while the line through (3, 0, 3) that is parallel to $\mathbf{B}$

$= \mathbf{i} + 2\mathbf{j} + 5\mathbf{k}$ has equations $x = 3 + s$, $y = 2s$, $z = 3 + 5s$. The vector $\mathbf{C} = [(1 + t) - (3 + s)]\mathbf{i} + [(2 - 3t) - 2s]\mathbf{j} + [(1 + 2t) - (3 + 5s)]\mathbf{k} =$ $(-2 + t - s)\mathbf{i} + (2 - 3t - 2s)\mathbf{j} + (-2 + 2t - 5s)\mathbf{k}$ connects the two lines and will be perpendicular to both **A** and **B** when s and t correspond to the points that are closest to each other (the minimum distance is the perpendicular distance). Hence we want $\mathbf{A}\cdot\mathbf{C} = -2 + t - s - 6 + 9t + 6s - 4 + 4t - 10s = -12 + 14t - 5s = 0$ and $\mathbf{B}\cdot\mathbf{C} = -2 + t - s + 4 - 6t - 4s - 10 + 10t - 25s = -8 + 5t - 30s = 0$. Solving the two equations, we obtain $t = 64/79$. Substituting this value of t into the parametric equations gives

$$\left(\frac{143}{79}, -\frac{34}{79}, \frac{207}{79}\right).$$

9 (a) The lines are parallel if and only if the vectors $\overrightarrow{P_1P_2}$ and $\overrightarrow{P_3P_4}$ are parallel. Two vectors are parallel if and only if one vector is a nonzero scalar multiple of the other.

(b) The vector from $(1, 2, -3)$ to $(5, 9, 4)$ is $4\mathbf{i} + 7\mathbf{j} + 7\mathbf{k}$ and the vector from $(-1, -1, 2)$ to $(1, 3, 5)$ is $2\mathbf{i} + 4\mathbf{j} + 3\mathbf{k}$. These vectors are not parallel, so neither are the lines.

11 (a) The line through P_1 and P_2 is parallel to the plane through Q_1, Q_2, and Q_3 if and only if $\overrightarrow{P_1P_2}$ is perpendicular to the plane's normal. A normal to the plane is given by $\overrightarrow{Q_1Q_2} \times \overrightarrow{Q_1Q_3}$, so $\overrightarrow{P_1P_2}\cdot(\overrightarrow{Q_1Q_2} \times \overrightarrow{Q_1Q_3})$ must equal 0.

(b) $P_1 = (0, 0, 0)$ and $P_2 = (1, 1, -1)$, so $\overrightarrow{P_1P_2} = \mathbf{i} + \mathbf{j} - \mathbf{k}$. The points on the plane are $Q_1 = (1, 0, 1)$, $Q_2 = (2, 1, 0)$, and $Q_3 = (1, 3, 4)$, so a normal to the plane is provided by $\overrightarrow{Q_1Q_2} \times \overrightarrow{Q_1Q_3} = (\mathbf{i} + \mathbf{j} - \mathbf{k})\times(3\mathbf{j} + 3\mathbf{k}) = 6\mathbf{i} - 3\mathbf{j} + 3\mathbf{k}$. Now $\overrightarrow{P_1P_2}\cdot(\overrightarrow{Q_1Q_2} \times \overrightarrow{Q_1Q_3}) = 6 - 3 - 3 = 0$, so the line is parallel to the plane.

13 (a) Find the angle between the normal vectors $\mathbf{N}_1 = A_1\mathbf{i} + B_1\mathbf{j} + C_1\mathbf{k}$ and $\mathbf{N}_2 = A_2\mathbf{i} + B_2\mathbf{j} + C_2\mathbf{k}$.

(b) $\mathbf{N}_1 = \mathbf{i} - \mathbf{j} - \mathbf{k}$ and $\mathbf{N}_2 = \mathbf{i} + \mathbf{j} + \mathbf{k}$ are normals to the planes. The angle between the vectors is $\theta = \cos^{-1}\dfrac{\mathbf{N}_1\cdot\mathbf{N}_2}{\|\mathbf{N}_1\|\|\mathbf{N}_2\|} = \cos^{-1}\left(-\dfrac{1}{3}\right) \approx 1.9106 \approx 109.47°$, so the angle between the planes is $180° - 109.47° = 70.53°$.

15 (a) If P_1, P_2, P_3, and P_4 lie in the same plane, then $\overrightarrow{P_1P_2}$, $\overrightarrow{P_1P_3}$, and $\overrightarrow{P_1P_4}$ also lie in that plane and span a parallelepiped of zero volume. Thus the points will lie in a plane if and only if the scalar triple product of $\overrightarrow{P_1P_2}$, $\overrightarrow{P_1P_3}$, and $\overrightarrow{P_1P_4}$ is 0.

(b) $P_1 = (1, 2, 3)$, $P_2 = (4, 1, -5)$, $P_3 = (2, 1, 6)$, and $P_4 = (3, 5, 3)$, so $\overrightarrow{P_1P_2} = 3\mathbf{i} - \mathbf{j} - 8\mathbf{k}$, $\overrightarrow{P_1P_3} = \mathbf{i} - \mathbf{j} + 3\mathbf{k}$, and $\overrightarrow{P_1P_4} = 2\mathbf{i} + 3\mathbf{j} + 0\mathbf{k}$. Hence

$$\overrightarrow{P_1P_2}\cdot(\overrightarrow{P_1P_3}\times\overrightarrow{P_1P_4}) = \begin{vmatrix} 3 & -1 & -8 \\ 1 & -1 & 3 \\ 2 & 3 & 0 \end{vmatrix} = -73,$$

so the points do not lie in the same plane.

17 (a) Use the parametric equations for L to find a vector **A** that is parallel to L. Then use a point on L and the given point P to find a vector **B** connecting the line and the point. Then **N** = **A** × **B** is normal to the plane containing L and P. Use P and the components of **N** to write an equation for the plane.

(b) The line $x = 2 + t$, $y = 3 - t$, and $z = 4 + 2t$ is parallel to the vector **A** = **i** − **j** + 2**k**. The point (2, 3, 4) lies on the line, so the vector **B** = **i** + 2**j** + 3**k** goes from the point (1, 1, 1) to the line. Then **N** = (**i** − **j** + 2**k**)×(**i** + 2**j** + 3**k**) = −7**i** − **j** + 3**k** is normal to the required plane. Therefore $-7(x - 1) - (y - 1) + 3(z - 1) = 0$, or $7x + y - 3z = 5$, is an equation for the plane.

19 (a) Any unit vector perpendicular to **A** = a_1**i** + a_2**j** + a_3**k** is perpendicular to the line. There are infinitely many such vectors.

(b) Cross **A** with any nonzero vector that is not parallel to **A** in order to get a vector which is perpendicular to **A**. Then divide the result by its magnitude to get a unit vector.

(c) The unit vector **u** must be perpendicular to both **A** = (2 − 1)**i** + (1 − 0)**j** + (−1 − (−3))**k** = **i** + **j** + 2**k** and the normal to the plane $4x + 5y + 6z = 0$, **N** = 4**i** + 5**j** + 6**k**. Note that **A** × **N** = −4**i** + 2**j** + **k**, so one possibility for **u** is $\dfrac{-4\mathbf{i} + 2\mathbf{j} + \mathbf{k}}{\sqrt{21}}$. (The other choice is −**u**.)

21 (a) We have two equations in three unknowns. Substitute an arbitrary value for one of the variables x, y, or z. Then solve the resulting two equations for the remaining variables. If the planes intersect, at least one choice of an arbitrary variable must produce equations that can be solved.

(b) The planes are $3x + z + 2 = 0$ and $x - y - z + 5 = 0$. If we choose to let $y = 0$, the equations become $3x + z + 2 = 0$ and $x - z + 5 = 0$. Add the equations to eliminate z, producing $4x + 7 = 0$, so $x = -7/4$. Solving for z yields $z = 13/4$, so one point common to both planes is $(-7/4, 0, 13/4)$. Infinitely many other choices also work.

23 The parallelogram spanned by 2**i** + 3**j** + 4**k** and **i** − **j** + **k** has an area equal to the magnitude of the cross product of the two vectors:

$$\text{mag}\begin{vmatrix} \mathbf{i} & \mathbf{j} & \mathbf{k} \\ 2 & 3 & 4 \\ 1 & -1 & 1 \end{vmatrix} = \|7\mathbf{i} + 2\mathbf{j} - 5\mathbf{k}\| = \sqrt{78}.$$ Note

also that $\mathbf{N}_1 = 7\mathbf{i} + 2\mathbf{j} - 5\mathbf{k}$ is a normal vector for the parallelogram. Similarly, $\mathbf{N}_2 = \mathbf{i} + \mathbf{j} + \mathbf{k}$ is a normal vector for the plane $x + y + z = 62$. The area of the projection on the plane of the parallelogram is $\sqrt{78}\cos\theta$, where θ is the angle between $\mathbf{N}_1$ and $\mathbf{N}_2$. Hence we have Area =

$$\sqrt{78}\,\frac{\mathbf{N}_1\cdot\mathbf{N}_2}{|\mathbf{N}_1|\,|\mathbf{N}_2|} = \sqrt{78}\,\frac{4}{\sqrt{78}\sqrt{3}} = \frac{4}{\sqrt{3}}.$$

25 The disk has area πa^2 and $N_1 = 2i + 4j + 3k$ is a normal to the plane $2x + 4y + 3z = 5$ in which it lies. $N_2 = i - j + 2k$ is a normal vector to the plane $x - y + 2z = 6$ onto which the disk is projected. The area of the projection is $\pi a^2 \cos\theta$

$$= \pi a^2 \frac{N_1 \cdot N_2}{\|N_1\|\|N_2\|} = \pi a^2 \frac{4}{\sqrt{29}\sqrt{6}} = \frac{4\pi a^2}{\sqrt{174}}.$$

27 Clearly R must lie in the plane determined by O, P, and Q; thus $R = (-x + 2y, -2x + 2y, 2x + 3y)$ for some x and y. The equal angles property implies that the angle between $\overrightarrow{OR}$ and $\overrightarrow{PR}$ equals the angle between $\overrightarrow{OR}$ and $\overrightarrow{QR}$, thus $\frac{\overrightarrow{OR}\cdot\overrightarrow{PR}}{|\overrightarrow{OR}||\overrightarrow{PR}|}$

$= \frac{\overrightarrow{OR}\cdot\overrightarrow{QR}}{|\overrightarrow{OR}||\overrightarrow{QR}|}$, so $(\overrightarrow{OR}\cdot\overrightarrow{PR})|\overrightarrow{QR}\| =$

$(\overrightarrow{OR}\cdot\overrightarrow{QR})|\overrightarrow{PR}|$. That is, $(\langle -x + 2y, -2x + 2y, 2x + 3y\rangle \cdot \langle -x + 2y + 1, -2x + 2y + 2, 2x + 3y - 2\rangle)|\overrightarrow{QR}| = (\langle -x + 2y, -2x + 2y, 2x + 3y\rangle \cdot \langle -x + 2y - 2, -2x + 2y - 2, 2x + 3y - 3\rangle)|\overrightarrow{PR}|$, where $|\overrightarrow{QR}| =$

$\sqrt{(-x+2y-2)^2 + (-2x+2y-2)^2 + (2x+3y-3)^2}$

and $|\overrightarrow{PR}| =$

$\sqrt{(-x+2y+1)^2 + (-2x+2y+2)^2 + (2x+3y-2)^2}$.

Expanding and using the fact that $(-x + 2y)^2 + (-2x + 2y)^2 + (2x + 3y)^2 = 1$ (call this Equation (A)), we obtain Equation (B):

$(1 - 9x)\sqrt{18 - 34y} = (1 - 17y)\sqrt{10 - 18x}$.

Expanding (A) yields (C): $9x^2 + 17y^2 = 1$, so $y^2 = \frac{1 - 9x^2}{17}$, which we label as (D). Squaring (B) and substitution with (D) gives Equation (E):

$(1 - 9x)^2(9 - 17y) = (18 - 34y - 153x^2)(5 - 9x)$,

so $y = \frac{9 - 166x^2 + 153x^3}{17(1 - 9x^2)}$. Substituting this in (C) yields an equation in x: $17(1 - 9x^2)^3 = (9 - 166x^2 + 153x^3)^2$. Solving this numerically gives four real roots: $x \approx -0.22933$, -0.18461, 0.24501, or 0.28004. Computing y from (E), we find that the first and third roots satisfy (B). (For the others, the left side of (B) is the negative of the right side.) Next we compute the coordinates of R. For $x \approx -0.22933$ we have $y \approx -0.17601$ and $R \approx (-0.12269, 0.10664, -0.98670)$ and for $x \approx 0.24501$ we have $y \approx 0.16444$ and $R \approx (0.083876, -0.16114, 0.98336)$. From the geometry, it is clear that the z coordinate of R must be positive, so $R \approx (0.083876, -0.16114, 0.98336)$.

29 Let $\mathbf{n}$, $\mathbf{u}$, and $\mathbf{v}$ be outward-pointing unit normals to the planes ABD, ABC, and BCD, respectively. We know that $\mathbf{u}\cdot\mathbf{n} = \cos(180° - 80°) = -\cos 80°$, since the angle between $\mathbf{u}$ and $\mathbf{n}$ is 180° minus the angle between the planes. Similarly, $\mathbf{v}\cdot\mathbf{n} = -\cos 70°$. Also, since AB is perpendicular to BD, we have $(\mathbf{n}\times\mathbf{u})\cdot(\mathbf{n}\times\mathbf{v}) = 0$. But this can be rewritten as $0 = \mathbf{n}\cdot[\mathbf{u}\times(\mathbf{n}\times\mathbf{v})]$

$= \mathbf{n}\cdot[(\mathbf{u}\cdot\mathbf{v})\mathbf{n} - (\mathbf{u}\cdot\mathbf{n})\mathbf{v}] = (\mathbf{u}\cdot\mathbf{v})(\mathbf{n}\cdot\mathbf{n}) - (\mathbf{u}\cdot\mathbf{n})(\mathbf{v}\cdot\mathbf{n})$

$= \mathbf{u}\cdot\mathbf{v} - (\mathbf{u}\cdot\mathbf{n})(\mathbf{v}\cdot\mathbf{n})$, so $\mathbf{u}\cdot\mathbf{v} = (\mathbf{u}\cdot\mathbf{n})(\mathbf{v}\cdot\mathbf{n})$

$= \cos 80° \cos 70°$. But $\mathbf{u}\cdot\mathbf{v} = \cos\theta$, where θ is the angle between $\mathbf{u}$ and $\mathbf{v}$, so $\theta = \cos^{-1}(\cos 80° \cos 70°)$. The angle between the planes ABC and BCD is $180° - \theta$

$= 180° - \cos^{-1}(\cos 80° \cos 70°) \approx 93.405°$.

31

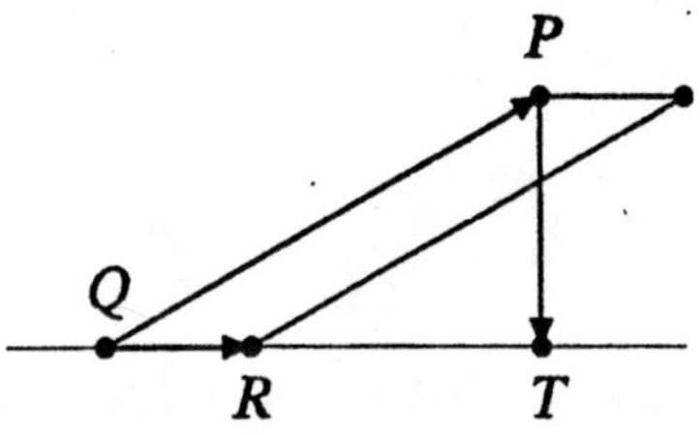

(a) The area of the parallelogram spanned by $\overrightarrow{QR}$ and $\overrightarrow{QP}$ is $|\overrightarrow{QR} \times \overrightarrow{QP}|$. But the area of a parallelogram is also equal to the product of its base, $|\overrightarrow{QR}|$, and its height, $|\overrightarrow{PT}|$, so

$$|\overrightarrow{QR} \times \overrightarrow{QP}| = |\overrightarrow{QR}|\,|\overrightarrow{PT}|.$$

(b) The desired distance is $|\overrightarrow{PT}|$ which, by (a), equals $\dfrac{|\overrightarrow{QR} \times \overrightarrow{QP}|}{|\overrightarrow{QR}|}$. Using $P = (2, 1, 3)$, $Q = (1, 5, 2)$, and $R = (2, 3, 4)$, we have $\overrightarrow{QR} = \mathbf{i} - 2\mathbf{j} + 2\mathbf{k}$ and $\overrightarrow{QP} = \mathbf{i} - 4\mathbf{j} + \mathbf{k}$.

Hence $\dfrac{|\overrightarrow{QR} \times \overrightarrow{QP}|}{|\overrightarrow{QR}|} = \dfrac{|6\mathbf{i} + \mathbf{j} - 2\mathbf{k}|}{\sqrt{9}}$

$= \dfrac{\sqrt{41}}{3}$.

12.S Guide Quiz

1 (a) $\mathbf{A}\cdot\mathbf{B} = 1\cdot 2 + 2(-1) + (-1)\cdot 3 = -3$

(b) $\|\mathbf{A}\| = \sqrt{1^2 + 2^2 + (-1)^2} = \sqrt{6}$

(c) The unit vector is $\dfrac{\mathbf{A}}{|\mathbf{A}|} = \dfrac{\mathbf{A}}{\sqrt{6}}$

$= \dfrac{1}{\sqrt{6}}\mathbf{i} + \dfrac{2}{\sqrt{6}}\mathbf{j} - \dfrac{1}{\sqrt{6}}\mathbf{k}$.

(d) $\text{comp}_{\mathbf{A}}\,\mathbf{B} = \mathbf{B}\cdot\dfrac{\mathbf{A}}{|\mathbf{A}\|} = -\dfrac{3}{\sqrt{6}}$

(e) $\text{proj}_{\mathbf{A}}\,\mathbf{B} = \left(\mathbf{B}\cdot\dfrac{\mathbf{A}}{|\mathbf{A}|}\right)\dfrac{\mathbf{A}}{|\mathbf{A}|}$

$= \left(-\dfrac{3}{\sqrt{6}}\right)\dfrac{\mathbf{i} + 2\mathbf{j} - \mathbf{k}}{\sqrt{6}} = -\dfrac{1}{2}\mathbf{i} - \mathbf{j} + \dfrac{1}{2}\mathbf{k}$

(f) $\text{comp}_{\mathbf{B}}\,\mathbf{A} = \mathbf{A}\cdot\dfrac{\mathbf{B}}{|\mathbf{B}\|} = \dfrac{-3}{\sqrt{2^2 + (-1)^2 + 3^2}}$

$= -\dfrac{3}{\sqrt{14}}$

(g) $\text{orth}_{\mathbf{A}}\,\mathbf{B} = \mathbf{B} - \text{proj}_{\mathbf{A}}\,\mathbf{B} = (2\mathbf{i} - \mathbf{j} + 3\mathbf{k}) -$

$\left(-\dfrac{1}{2}\mathbf{i} - \mathbf{j} + \dfrac{1}{2}\mathbf{k}\right) = \dfrac{5}{2}\mathbf{i} + \dfrac{5}{2}\mathbf{k}$

(h) Let θ be the angle between $\mathbf{A}$ and $\mathbf{B}$. Then

$\cos\theta = \dfrac{\mathbf{A}\cdot\mathbf{B}}{|\mathbf{A}|\,|\mathbf{B}|} = \dfrac{-3}{\sqrt{6}\sqrt{14}} = -\dfrac{3}{2\sqrt{21}}$.

(i) From (h), $\theta = \cos^{-1}\left(-\dfrac{3}{2\sqrt{21}}\right) \approx 1.9043$ radians $\approx 109.11°$.

(j) $\mathbf{A}\times\mathbf{B} = \begin{vmatrix} \mathbf{i} & \mathbf{j} & \mathbf{k} \\ 1 & 2 & -1 \\ 2 & -1 & 3 \end{vmatrix} = 5\mathbf{i} - 5\mathbf{j} - 5\mathbf{k}$

(k) $\mathbf{B} \times \mathbf{A} = -(\mathbf{A} \times \mathbf{B}) = -5\mathbf{i} + 5\mathbf{j} + 5\mathbf{k}$

(l) One such vector is $\dfrac{\mathbf{A} \times \mathbf{B}}{|\mathbf{A} \times \mathbf{B}|}$

$= \dfrac{5\mathbf{i} - 5\mathbf{j} - 5\mathbf{k}}{\sqrt{5^2 + (-5)^2 + (-5)^2}} = \dfrac{\mathbf{i} - \mathbf{j} - \mathbf{k}}{\sqrt{3}}$. The

other is $\dfrac{-\mathbf{i} + \mathbf{j} + \mathbf{k}}{\sqrt{3}}$.

(m) The area is $\|\mathbf{A} \times \mathbf{B}\| = 5\sqrt{3}$.

2 See Theorem 2 of Sec. 12.4.

3 By Theorem 4 of Sec. 12.4, the vector $\mathbf{i} - 2\mathbf{j} + 2\mathbf{k}$ is a normal to the plane. Its direction cosines are the components of the unit vector

$\dfrac{\mathbf{i} - 2\mathbf{j} + 2\mathbf{k}}{\sqrt{1^2 + (-2)^2 + 2^2}} = \dfrac{1}{3}\mathbf{i} - \dfrac{2}{3}\mathbf{j} + \dfrac{2}{3}\mathbf{k}$; that is,

$\cos\alpha = \dfrac{1}{3}$, $\cos\beta = -\dfrac{2}{3}$, and $\cos\gamma = \dfrac{2}{3}$.

(Negating all three also works.)

4 $\begin{vmatrix} a_1 & a_2 & a_3 \\ b_1 & b_2 & b_3 \\ c_1 + b_1 & c_2 + b_2 & c_3 + b_3 \end{vmatrix}$

$= a_1[b_2(c_3 + b_3) - b_3(c_2 + b_2)] - a_2[b_1(c_3 + b_3) - b_3(c_1 + b_1)] + a_3[b_1(c_2 + b_2) - b_2(c_1 + b_1)]$
$= a_1(b_2c_3 - b_3c_2) - a_2(b_1c_3 - b_3c_1) + a_3(b_1c_2 - b_2c_1) = \begin{vmatrix} a_1 & a_2 & a_3 \\ b_1 & b_2 & b_3 \\ c_1 & c_2 & c_3 \end{vmatrix}$

5 (a) The parallelepiped is spanned by the vectors $\mathbf{i} + 2\mathbf{j} + 3\mathbf{k}$, $\mathbf{i} + 4\mathbf{j} + 5\mathbf{k}$, and $2\mathbf{i} + 3\mathbf{j} + \mathbf{k}$.

Therefore its volume is abs$\begin{vmatrix} 1 & 2 & 3 \\ 1 & 4 & 5 \\ 2 & 3 & 1 \end{vmatrix} = |-8|$

$= 8$.

(b) See Theorem 4 of Sec. 12.6.

6 (a)

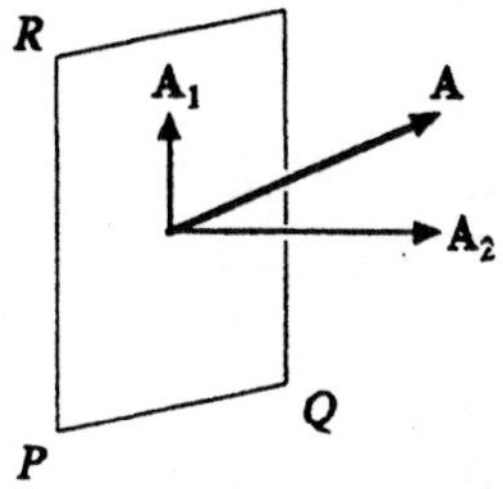

(b) $\mathbf{A}_2$ is the projection of $\mathbf{A}$ onto a normal to the plane. Such a normal is given by $\mathbf{N} = \overrightarrow{PQ} \times \overrightarrow{PR}$, so $\mathbf{A}_2 = \text{proj}_{\mathbf{N}}\,\mathbf{A}$

$= \left(\mathbf{A} \cdot \dfrac{\mathbf{N}}{|\mathbf{N}|}\right)\dfrac{\mathbf{N}}{|\mathbf{N}|} = \left(\dfrac{\mathbf{A} \cdot \mathbf{N}}{|\mathbf{N}|^2}\right)\mathbf{N} =$

$\left(\dfrac{\mathbf{A} \cdot (\overrightarrow{PQ} \times \overrightarrow{PR})}{|\overrightarrow{PQ} \times \overrightarrow{PR}|^2}\right)\overrightarrow{PQ} \times \overrightarrow{PR}$ and $\mathbf{A}_1 = \mathbf{A} - \mathbf{A}_2$

$= \mathbf{A} - \left(\dfrac{\mathbf{A} \cdot (\overrightarrow{PQ} \times \overrightarrow{PR})}{|\overrightarrow{PQ} \times \overrightarrow{PR}|^2}\right)\overrightarrow{PQ} \times \overrightarrow{PR}$.

7 The line through (1, 2, 1) and (3, 1, 1) has parametric equations $x = 1 + (3 - 1)t = 1 + 2t$, $y = 2 + (1 - 2)t = 2 - t$, and $z = 1 + (1 - 1)t = 1$. The plane through the points $P = (2, -1, 1)$, $Q = (5, 2, 3)$, and $R = (4, 1, 3)$ has a normal vector $\mathbf{N} = \overrightarrow{PQ} \times \overrightarrow{PR} = \begin{vmatrix} \mathbf{i} & \mathbf{j} & \mathbf{k} \\ 3 & 3 & 2 \\ 2 & 2 & 2 \end{vmatrix} = 2\mathbf{i} - 2\mathbf{j}$, so

an equation of the plane is $2(x - 2) - 2(y - (-1)) + 0(z - 1) = 0$; that is, $2x - 2y = 6$ or $x - y = 3$. We must have $(1 + 2t) - (2 - t) = 3$, so $3t = 4$, $t = 4/3$, and therefore $x = \frac{11}{3}$, $y = \frac{2}{3}$, and $z = 1$. The point of intersection is $\left(\frac{11}{3}, \frac{2}{3}, 1\right)$.

8 (a) See Theorem 1 of Sec. 12.3.

(b) $\mathbf{D} \cdot \mathbf{E} = \|\mathbf{D}\| \, \|\mathbf{E}\| \cos\theta$, where θ is the angle between **D** and **E**.

(c) See Theorem 2 of Sec. 12.3.

(d) See Exercise 43 of Sec. 12.3.

9 $\text{proj}_{\mathbf{n}} \mathbf{A} = (\mathbf{A}\cdot\mathbf{n})\mathbf{n}$, $\text{proj}_{\mathcal{P}} \mathbf{A} = \mathbf{A} - \text{proj}_{\mathbf{n}} \mathbf{A} = \mathbf{A} - (\mathbf{A}\cdot\mathbf{n})\mathbf{n}$

10 See Sec. 12.2.

11 We need to solve simultaneously the three equations $x + 2y + z = 4$, $2x - y + z = 1$, and $3x + 2y + z = 6$. Subtracting the first from the third gives $2x = 2$, so $x = 1$. Hence the first two equations become $2y + z = 3$ and $-y + z = -1$. Subtracting these two equations yields $3y = 4$, so $y = 4/3$. Then $z = 1/3$, so the point of intersection is $(1, 4/3, 1/3)$.

12 $\mathbf{N}_1 = \mathbf{i} + \mathbf{j} + 3\mathbf{k}$ and $\mathbf{N}_2 = 2\mathbf{i} - \mathbf{j} + \mathbf{k}$ are normals to the planes $x + y + 3z = 5$ and $2x - y + z = 2$, respectively.

(a) $\mathbf{N}_1 \times \mathbf{N}_2 = \begin{vmatrix} \mathbf{i} & \mathbf{j} & \mathbf{k} \\ 1 & 1 & 3 \\ 2 & -1 & 1 \end{vmatrix} = 4\mathbf{i} + 5\mathbf{j} - 3\mathbf{k}$ is parallel to L.

(b) Pick one variable and assign it an arbitrary value, say, $z = 1$. Then the equations for the planes become $x + y = 2$ and $2x - y = 1$. Adding these equations together yields $3x = 3$, so $x = 1$ and $y = 1$. The point is $(1, 1, 1)$. (Infinitely many other points also work.)

(c) Parametric equations for the line are $x = 1 + 4t$, $y = 1 + 5t$, and $z = 1 - 3t$.

13 $\mathbf{N}_1 = \mathbf{i} + 2\mathbf{j} + 3\mathbf{k}$ and $\mathbf{N}_2 = 2\mathbf{i} - 3\mathbf{j} + 4\mathbf{k}$ are normals to the planes $x + 2y + 3z = 6$ and $2x - 3y + 4z = 8$, respectively.

(a) $\mathbf{N}_1 \times \mathbf{N}_2 = \begin{vmatrix} \mathbf{i} & \mathbf{j} & \mathbf{k} \\ 1 & 2 & 3 \\ 2 & -3 & 4 \end{vmatrix} = 17\mathbf{i} + 2\mathbf{j} - 7\mathbf{k}$ is parallel to L.

(b) Pick one variable and assign it an arbitrary value, say, $y = 0$. Then the equations for the planes become $x + 3z = 6$ and $2x + 4z = 8$, or $x + 2z = 4$. Subtracting these equations yields $z = 2$, so $x = 0$. The point is $(0, 0, 2)$. (Infinitely many other points also work.)

(c) Parametric equations for the line are $x = 17t$, $y = 2t$, and $z = 2 - 7t$.

(d) Solve the parametric equations for t to obtain symmetric equations: $\frac{x}{17} = \frac{y}{2} = \frac{z - 2}{-7}$.

14 (a) The vector from $(1, 1, 1)$ to $(2, 3, 2)$, $\mathbf{i} + 2\mathbf{j} + \mathbf{k}$, and the vector from $(1, 1, 1)$ to $(1, 4, 5)$, $3\mathbf{j} + 4\mathbf{k}$, lie in the plane containing the three points.

(b) The cross product of the two vectors in (a), $5\mathbf{i} - 4\mathbf{j} + 3\mathbf{k}$, is normal to the plane.

(c) An equation of the plane is $5(x - 1) - 4(y - 1) + 3(z - 1) = 0$, or $5x - 4y + 3z = 4$.

15 (a) The direction numbers of the two lines provide vectors $2\mathbf{i} + 3\mathbf{j} + 4\mathbf{k}$ and $3\mathbf{i} + \mathbf{j} + 5\mathbf{k}$ that are parallel to the plane, so their cross product, $11\mathbf{i} + 2\mathbf{j} - 7\mathbf{k}$, is perpendicular to the plane.

(b) The angle β which the vector $11\mathbf{i} + 2\mathbf{j} - 7\mathbf{k}$ makes with the y axis is given by $\beta =$

$$\cos^{-1}\frac{(11\mathbf{i} + 2\mathbf{j} - 7\mathbf{k})\cdot\mathbf{j}}{|11\mathbf{i} + 2\mathbf{j} - 7\mathbf{k}|} = \cos^{-1}\frac{2}{\sqrt{174}} \approx$$

1.4186 radians $\approx 81.28°$.

(c) An equation of the plane is $11(x - 1) + 2(y - 1) - 7(z - 2) = 0$, or $11x + 2y - 7z = -1$.

16 (a) See Theorem 3 of Sec. 12.5.

(b) See Theorem 4 of Sec. 12.6.

17 See Sec. 12.6, p. 727.

18 (a)

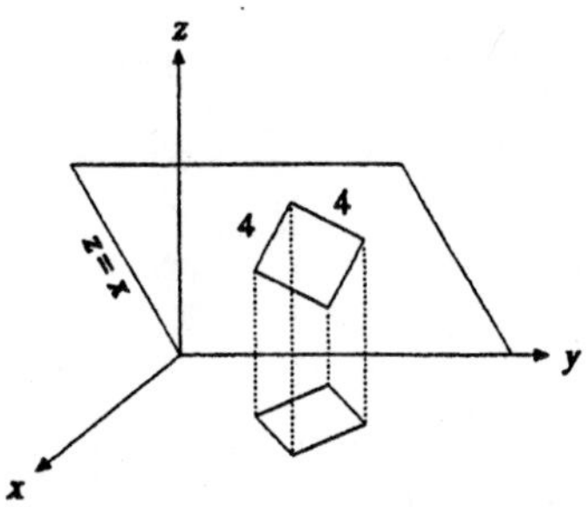

(b) See (a).

(c) The projection may be a square or a rectangle, but not necessarily. The projection must at least be a parallelogram.

(d) The area of the square is 16 and a normal to the plane $-x + z = 0$ is $-\mathbf{i} + \mathbf{k}$. The xy plane has a normal of $\mathbf{k}$, so the area of the projection is $16\cos\theta = 16\,\dfrac{(-\mathbf{i} + \mathbf{k})\cdot\mathbf{k}}{\sqrt{2}}$

$= 8\sqrt{2}$.

(e) The plane $2x - y + 3z = 8$ has normal $\mathbf{N} = 2\mathbf{i} - \mathbf{j} + 3\mathbf{k}$, so the area of the projection of the square on the plane is

$$16\,\frac{(-\mathbf{i} + \mathbf{k})\cdot(2\mathbf{i} - \mathbf{j} + 3\mathbf{k})}{\sqrt{2}\sqrt{14}} = \frac{8}{\sqrt{7}}.$$

19 (a) Observe that the projection of $\mathbf{A}$ onto the xy plane has length $4\cos 60° = 2$ and is in the same direction as the unit vector $\dfrac{1}{\sqrt{2}}(\mathbf{i} + \mathbf{j})$, so the projection is $2\cdot\dfrac{1}{\sqrt{2}}(\mathbf{i} + \mathbf{j}) =$

$\sqrt{2}\mathbf{i} + \sqrt{2}\mathbf{j}$.

(b) The projection of $\mathbf{A}$ on the z axis is $(4\cos 30°)\mathbf{k} = 2\sqrt{3}\mathbf{k}$. Combining this result with (a), we have $\mathbf{A} = \sqrt{2}\mathbf{i} + \sqrt{2}\mathbf{j} + 2\sqrt{3}\mathbf{k}$, so the projection on the x axis is $\sqrt{2}\mathbf{i}$ and on the y axis is $\sqrt{2}\mathbf{j}$.

(c) The unit vector in the same direction as $\mathbf{A}$ is

$$\frac{\mathbf{A}}{|\mathbf{A}|} = \frac{\sqrt{2}\mathbf{i} + \sqrt{2}\mathbf{j} + 2\sqrt{3}\mathbf{k}}{\sqrt{2 + 2 + 12}} =$$

$\dfrac{\sqrt{2}}{4}\mathbf{i} + \dfrac{\sqrt{2}}{4}\mathbf{j} + \dfrac{\sqrt{3}}{2}\mathbf{k}$, so the direction cosines are $\sqrt{2}/4$, $\sqrt{2}/4$, and $\sqrt{3}/2$.

(d) By (c), the direction angles are $\alpha = \beta = \cos^{-1}\dfrac{\sqrt{2}}{4} \approx 69.30°$ and $\gamma = \cos^{-1}\dfrac{\sqrt{3}}{2} =$ $30°$. (We already knew γ from the original information.)

(e) To obtain the projection on the yz plane, we subtract from $\mathbf{A}$ the projection of $\mathbf{A}$ onto the x

axis: $\sqrt{2}\mathbf{i} + \sqrt{2}\mathbf{j} + 2\sqrt{3}\mathbf{k} - \sqrt{2}\mathbf{i} =$
$\sqrt{2}\mathbf{j} + 2\sqrt{3}\mathbf{k}$.

12.S Review Exercises

1 (a) See Sec. 12.3.

(b) See Sec. 12.3.

(c) If $\mathbf{A}\cdot\mathbf{B} = 0$, then $\|\mathbf{A}\|\,\|\mathbf{B}\| \cos\theta = 0$. If $\mathbf{A} \neq \mathbf{0}$ and $\mathbf{B} \neq \mathbf{0}$, then $\|\mathbf{A}\| \neq 0$ and $\|\mathbf{B}\| \neq 0$, so $\cos\theta = 0$ and $\theta = \cos^{-1}0 = \pi/2$. That is, A is perpendicular to B.

3 In general, the projection of A on B is

$$\left(\mathbf{A}\cdot\frac{\mathbf{B}}{\|\mathbf{B}\|}\right)\frac{\mathbf{B}}{\|\mathbf{B}\|} = \frac{\mathbf{A}\cdot\mathbf{B}}{\|\mathbf{B}\|^2}\mathbf{B}.$$

(a) $$\frac{(-2\mathbf{i} + 3\mathbf{j})\cdot\mathbf{i}}{\|\mathbf{i}\|^2}\mathbf{i} = \frac{-2}{1}\mathbf{i} = -2\mathbf{i}$$

(b) $$\frac{(-2\mathbf{i} + 3\mathbf{j})\cdot\mathbf{j}}{\|\mathbf{j}\|^2}\mathbf{j} = \frac{3}{1}\mathbf{j} = 3\mathbf{j}$$

(c) $$\frac{(-2\mathbf{i} + 3\mathbf{j})\cdot(0.6\mathbf{i} + 0.8\mathbf{j})}{\|0.6\mathbf{i} + 0.8\mathbf{j}\|^2}(0.6\mathbf{i} + 0.8\mathbf{j})$$

$$= \frac{-1.2 + 2.4}{(0.6)^2 + (0.8)^2}(0.6\mathbf{i} + 0.8\mathbf{j})$$

$$= 1.2(0.6\mathbf{i} + 0.8\mathbf{j}) = 0.72\mathbf{i} + 0.96\mathbf{j}$$

(d) $$\frac{(-2\mathbf{i} + 3\mathbf{j})\cdot(4\mathbf{i} - 5\mathbf{j})}{\|4\mathbf{i} - 5\mathbf{j}\|^2}(4\mathbf{i} - 5\mathbf{j})$$

$$= \frac{-8 - 15}{4^2 + (-5)^2}(4\mathbf{i} - 5\mathbf{j}) = -\frac{23}{41}(4\mathbf{i} - 5\mathbf{j})$$

5 See the solution of Exercise 43 in Sec. 12.3.

7 See Sec. 12.6.

9 (a) Since A and B are nonparallel nonzero vectors, they determine a plane. $\mathbf{A} \times \mathbf{B}$ is normal to this plane. Also $\mathbf{A} \times (\mathbf{A} \times \mathbf{B})$ is normal to $\mathbf{A} \times \mathbf{B}$, so it lies in the plane determined by A and B. Thus there are scalars x and y such that $\mathbf{A} \times (\mathbf{A} \times \mathbf{B}) = x\mathbf{A} + y\mathbf{B}$.

(b) $\|\mathbf{A} \times (\mathbf{A} \times \mathbf{B})\| = \|\mathbf{A}\|\,\|\mathbf{A} \times \mathbf{B}\| \sin\theta$, where θ is the angle between A and $\mathbf{A} \times \mathbf{B}$. But this angle is $\pi/2$ so $\|\mathbf{A} \times (\mathbf{A} \times \mathbf{B})\| = \|\mathbf{A}\|\,\|\mathbf{A} \times \mathbf{B}\| = \|\mathbf{A}\|^2\|\mathbf{B}\| \sin\phi$, where ϕ is the angle between A and B. Since these vectors are not parallel, $\sin\phi \neq 0$. Since A and B are nonzero, $\|\mathbf{A}\| \neq 0$ and $\|\mathbf{B}\| \neq 0$; thus $\|\mathbf{A} \times (\mathbf{A} \times \mathbf{B})\| \neq 0$ and $\mathbf{A} \times (\mathbf{A} \times \mathbf{B})$ is not the zero vector.

(c) Since $\mathbf{A} \times \mathbf{A} = \mathbf{0}$, $(\mathbf{A} \times \mathbf{A}) \times \mathbf{B} = \mathbf{0} \times \mathbf{B} = \mathbf{0}$.

11 $x(3\mathbf{i} + \mathbf{j} + 2\mathbf{k}) + y(\mathbf{i} - \mathbf{j} + 2\mathbf{k}) + z(2\mathbf{i} + 2\mathbf{j} + \mathbf{k}) = (3x + y + 2z)\mathbf{i} + (x - y + 2z)\mathbf{j} + (2x + 2y + z)\mathbf{k}$, which equals $3\mathbf{i} + 3\mathbf{j} + 3\mathbf{k}$ if and only if $3x + y + 2z = 3$, $x - y + 2z = 3$, and $2x + 2y + z = 3$. Solving these equations simultaneously yields $x = -\frac{3}{2}$, $y = \frac{3}{2}$, and $z = 3$.

13 $$\begin{vmatrix} 0 & a & b \\ -a & 0 & c \\ -b & -c & 0 \end{vmatrix} = 0\begin{vmatrix} 0 & c \\ -c & 0 \end{vmatrix} - a\begin{vmatrix} -a & c \\ -b & 0 \end{vmatrix} + b\begin{vmatrix} -a & 0 \\ -b & -c \end{vmatrix}$$

$= -abc + abc = 0$ for all choices of a, b, and c.

15 (a) Note that $\overrightarrow{CA} = -\mathbf{i} - \mathbf{j} + \mathbf{k}$ and $\overrightarrow{CB} = -\mathbf{i} - \mathbf{j}$. If θ is the angle between them, then

$$\cos\theta = \frac{\overrightarrow{CA}\cdot\overrightarrow{CB}}{|\overrightarrow{CA}|\,|\overrightarrow{CB}|} =$$

$$\frac{(-1)(-1) + (-1)(-1) + 1\cdot 0}{\sqrt{(-1)^2 + (-1)^2 + 1^2}\sqrt{(-1)^2 + (-1)^2}} = \frac{2}{\sqrt{6}}.$$

(b) We have $\overrightarrow{EA} = -\mathbf{j} + \mathbf{k}$ and $\overrightarrow{EB} = -\mathbf{j}$. So

the cosine is $\dfrac{\overrightarrow{EA} \cdot \overrightarrow{EB}}{|\overrightarrow{EA}|\,|\overrightarrow{EB}|} =$

$$\frac{(-1)(-1) + 1 \cdot 0}{\sqrt{(-1)^2 + 1^2}\sqrt{(-1)^2}} = \frac{1}{\sqrt{2}}.$$

(c) $\overrightarrow{AD} = -\mathbf{i} + \mathbf{j} - \mathbf{k}$ and $\overrightarrow{AC} = \mathbf{i} + \mathbf{j} - \mathbf{k}$ so

the cosine is $\dfrac{\overrightarrow{AD} \cdot \overrightarrow{AC}}{|\overrightarrow{AD}|\,\|\overrightarrow{AC}\|} =$

$$\frac{(-1)\cdot 1 + 1\cdot 1 + (-1)(-1)}{\sqrt{(-1)^2 + 1^2 + (-1)^2}\sqrt{1^2 + 1^2 + (-1)^2}} = \frac{1}{3}.$$

17 The distance between the planes equals the distance from (x_1, y_1, z_1) to the plane through (x_2, y_2, z_2). This plane is described by the equation $A(x - x_2) + B(y - y_2) + C(z - z_2) = 0$; that is, $Ax + By + Cz - (Ax_2 + By_2 + Cz_2) = 0$. By Theorem 5 of Sec. 12.4, the distance is

$$\frac{|Ax_1 + By_1 + Cz_1 - (Ax_2 + By_2 + Cz_2)|}{\sqrt{A^2 + B^2 + C^2}}.$$

19 $\mathbf{i} + 2\mathbf{j} + 3\mathbf{k}$ is normal to $x + 2y + 3z = 0$ and $2\mathbf{i} - \mathbf{j} + 3\mathbf{k}$ is normal to $2x - y + 3z + 4 = 0$ so $(\mathbf{i} + 2\mathbf{j} + 3\mathbf{k}) \times (2\mathbf{i} - \mathbf{j} + 3\mathbf{k}) = 9\mathbf{i} + 3\mathbf{j} - 5\mathbf{k}$ is parallel to both planes. The parametric equations of the line parallel to the vector passing through $(1, 1, 2)$ are $x = 1 + 9t$, $y = 1 + 3t$, and $z = 2 - 5t$.

21 The line through $(1, 4, 7)$ and $(5, 10, 15)$ is parallel to the vector $(5 - 1)\mathbf{i} + (10 - 4)\mathbf{j} + (15 - 7)\mathbf{k} = 4\mathbf{i} + 6\mathbf{j} + 8\mathbf{k}$. If the line is perpendicular to the plane $2x + 3y + 4z = 17$, then the vector must be parallel to the plane's normal vector, $2\mathbf{i} + 3\mathbf{j} + 4\mathbf{k}$. Since the first vector is twice the second, the line is perpendicular to the plane.

23 (a) If $P = (x, y, z)$ is the point on the plane nearest the origin O, then the vector $\overrightarrow{OP}$ is perpendicular to the plane. That is, it is parallel to the normal vector $\frac{1}{2}\mathbf{i} + \frac{1}{3}\mathbf{j} + \frac{1}{4}\mathbf{k}$.

Hence, for some t, $x = \frac{1}{2}t$, $y = \frac{1}{3}t$, and $z = \frac{1}{4}t$, so $1 = \frac{x}{2} + \frac{y}{3} + \frac{z}{4} = \frac{1}{4}t + \frac{1}{9}t + \frac{1}{16}t = \frac{61}{144}t$ and $t = \frac{144}{61}$.

Therefore, $x = \frac{72}{61}$, $y = \frac{48}{61}$, $z = \frac{36}{61}$, and

$$P = \left(\frac{72}{61}, \frac{48}{61}, \frac{36}{61}\right).$$

(b) If $P = (x, y, z)$ is the desired point, then, proceeding as in (a), but using the point $(1, 2, 3)$ instead of $(0, 0, 0)$, we have $x - 1 = \frac{1}{2}t$, $y - 2 = \frac{1}{3}t$, and $z - 3 = \frac{1}{4}t$.

Hence $x = \frac{1}{2}t + 1$, $y = \frac{1}{3}t + 2$, and $z = \frac{1}{4}t + 3$, so $1 = \frac{x}{2} + \frac{y}{3} + \frac{z}{4} = \frac{1}{2}\left(\frac{1}{2}t + 1\right) + \frac{1}{3}\left(\frac{1}{3}t + 2\right) + \frac{1}{4}\left(\frac{1}{4}t + 3\right) = \frac{61}{144}t + \frac{23}{12}$

and $t = -\frac{132}{61}$. Therefore, $x = -\frac{5}{61}$, $y =$

$\frac{78}{61}$, $z = \frac{150}{61}$, and $P = \left(-\frac{5}{61}, \frac{78}{61}, \frac{150}{61}\right)$.

25 (x_1, y_1) and (x_2, y_2) are on opposite sides of the line $Ax + By + C = 0$ if and only if $Ax_1 + By_1 + C$ and $Ax_2 + By_2 + C$ have opposite signs. To see why, supposed we parameterized the line segment from (x_1, y_1) to (x_2, y_2) with some parameter t, $0 \le t \le 1$. If the linear function $Ax(t) + By(t) + C$ has one sign at $t = 0$ and the opposite sign at $t = 1$, then at some intermediate point, it will have the value 0. But this intermediate point must therefore lie on the line, showing that (x_1, y_1) and (x_2, y_2) lie on opposite sides of it. On the other hand, if $Ax(t) + By(t) + C$ has the same sign for both $t = 0$ and $t = 1$, then it must have that same sign for all $0 \le t \le 1$ (since a linear function can change sign only once), so both points must lie on the same side of the line (since no points of the line lie between).

27 The line through P_1 and P_2 is parallel to the plane if and only if it is perpendicular to a normal to the plane. But $\overrightarrow{P_3P_4} \times \overrightarrow{P_3P_5}$ is such a normal, and $\overrightarrow{P_1P_2}$ is parallel to the line, so the desired condition is $\overrightarrow{P_1P_2} \cdot (\overrightarrow{P_3P_4} \times \overrightarrow{P_3P_5}) = 0$.

29 Let **u** be parallel to the line and **v** be a normal to the plane that points to the same side of the plane as **u**. Then the desired angle is $\frac{\pi}{2} - \theta$, where θ is the angle between **u** and **v**. Choosing the vectors **u** $= \mathbf{i} + \mathbf{j} + \mathbf{k}$ and **v** $= ((4 - 1)\mathbf{i} + (1 - 2)\mathbf{j} + (5 - 3)\mathbf{k}) \times ((2 - 1)\mathbf{i} + (0 - 2)\mathbf{j} + (6 - 3)\mathbf{k}) =$

$$(3\mathbf{i} - \mathbf{j} + 2\mathbf{k}) \times (\mathbf{i} - 2\mathbf{j} + 3\mathbf{k}) = \begin{vmatrix} \mathbf{i} & \mathbf{j} & \mathbf{k} \\ 3 & -1 & 2 \\ 1 & -2 & 3 \end{vmatrix} =$$

$\mathbf{i} - 7\mathbf{j} - 5\mathbf{k}$, we find $\cos\theta = \frac{\mathbf{u} \cdot \mathbf{v}}{|\mathbf{u}||\mathbf{v}|} =$

$$\frac{1 \cdot 1 + 1(-7) + 1(-5)}{\sqrt{1^2 + 1^2 + 1^2}\sqrt{1^2 + (-7)^2 + (-5)^2}} = \frac{-11}{\sqrt{3}\sqrt{75}} =$$

$-\frac{11}{15}$. Thus $\theta = \cos^{-1}\left(-\frac{11}{15}\right) \approx 137.17°$. Since this is larger than 90°, **u** and **v** do not point to the same side of the plane. Choosing instead **u** $= -(\mathbf{i} + \mathbf{j} + \mathbf{k})$, we get $\cos\theta = \frac{11}{15}$, so $\theta = \cos^{-1}\frac{11}{15} \approx 42.83° \approx 0.7476$ radians. The angle between the line and the plane is $\pi/2 - \theta = \sin^{-1}\frac{11}{15} \approx 47.17° \approx 0.8232$ radians.

31 (a) Given (0, 0) and slope m_1, we use the slope-intercept form of the line to find the equation $y = m_1x$. If $x = 1$, $y = m_1$; therefore $(1, m_1)$ lies on L_1. Given that (0, 0) and $(1, m_1)$ lie on L_1, a vector parallel to L_1 would be $(1 - 0)\mathbf{i} + (m_1 - 0)\mathbf{j} = \mathbf{i} + m_1\mathbf{j}$.

(b) This time we have $y = m_2x$. Therefore $(1, m_2)$ lies on L_2, and $\mathbf{i} + m_2\mathbf{j}$ is a vector parallel to L_2.

(c) L_1 is perpendicular to L_2 if and only if vectors parallel to L_1 and L_2 have dot product equal to 0. Therefore $(\mathbf{i} + m_1\mathbf{j})\cdot(\mathbf{i} + m_2\mathbf{j}) = 0$, so $1 + m_1m_2 = 0$. This can happen only if $m_1m_2 = -1$, proving the usual rule for perpendicular lines from basic algebra.

33 By the identity for the vector triple product from Sec. 12.6, $\mathbf{A} \times (\mathbf{B} \times \mathbf{C}) = (\mathbf{A}\cdot\mathbf{C})\mathbf{B} - (\mathbf{A}\cdot\mathbf{B})\mathbf{C}$. Applying this three times, we find that $\mathbf{A} \times (\mathbf{B} \times \mathbf{C}) + \mathbf{B} \times (\mathbf{C} \times \mathbf{A}) + \mathbf{C} \times (\mathbf{A} \times \mathbf{B})$ $= ((\mathbf{A}\cdot\mathbf{C})\mathbf{B} - (\mathbf{A}\cdot\mathbf{B})\mathbf{C}) + ((\mathbf{B}\cdot\mathbf{A})\mathbf{C} - (\mathbf{B}\cdot\mathbf{C})\mathbf{A}) + ((\mathbf{C}\cdot\mathbf{B})\mathbf{A} - (\mathbf{C}\cdot\mathbf{A})\mathbf{B})$ $= (\mathbf{C}\cdot\mathbf{B} - \mathbf{B}\cdot\mathbf{C})\mathbf{A} + (\mathbf{A}\cdot\mathbf{C} - \mathbf{C}\cdot\mathbf{A})\mathbf{B} + (\mathbf{B}\cdot\mathbf{A} - \mathbf{A}\cdot\mathbf{B})\mathbf{C} = \mathbf{0}$, since the dot product is commutative.

35 Two planes are perpendicular if and only if their normal vectors are perpendicular. The plane through P, Q, and R has $\overrightarrow{PQ} \times \overrightarrow{PR}$ as a normal, while that through S, T, and U has $\overrightarrow{ST} \times \overrightarrow{SU}$ as a normal. Hence the desired condition is $(\overrightarrow{PQ} \times \overrightarrow{PR}) \cdot (\overrightarrow{ST} \times \overrightarrow{SU}) = 0$.

39 (a)

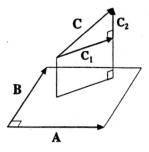

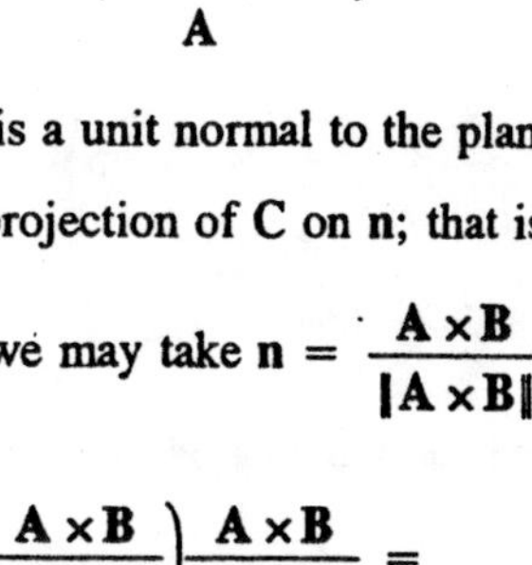

(b) If $\mathbf{n}$ is a unit normal to the plane, then $\mathbf{C}_2$ is the projection of $\mathbf{C}$ on $\mathbf{n}$; that is $\mathbf{C}_2 = (\mathbf{C}\cdot\mathbf{n})\mathbf{n}$. But we may take $\mathbf{n} = \dfrac{\mathbf{A}\times\mathbf{B}}{|\mathbf{A}\times\mathbf{B}|}$, so $\mathbf{C}_2 =$ $\left(\mathbf{C}\cdot\dfrac{\mathbf{A}\times\mathbf{B}}{|\mathbf{A}\times\mathbf{B}|}\right)\dfrac{\mathbf{A}\times\mathbf{B}}{|\mathbf{A}\times\mathbf{B}|} =$ $\dfrac{[\mathbf{C}\cdot(\mathbf{A}\times\mathbf{B})](\mathbf{A}\times\mathbf{B})}{|\mathbf{A}\times\mathbf{B}|^2}$.

41 The line L is parallel to $\overrightarrow{P_0P_1} = \langle -1, 2, 1\rangle$ and has parametric equations $x = 2 - t$, $y = 1 + 2t$, $z = 1 + t$. Given $P = (x, y, z)$ and $Q = (4, 1, 1)$, we need to determine the coordinates of P so that $\overrightarrow{PQ}$ and $\overrightarrow{P_0P_1}$ make an angle of $\pi/3$. But $\overrightarrow{PQ} = (4 - x)\mathbf{i} + (1 - y)\mathbf{j} + (1 - z)\mathbf{k} = (t + 2)\mathbf{i} - 2t\mathbf{j} - t\mathbf{k}$, with $|\overrightarrow{P_0P_1}| = \sqrt{6}$ and $|\overrightarrow{PQ}| = \sqrt{(t+2)^2 + (-2t)^2 + (-t)^2} = \sqrt{6t^2 + 4t + 4}$ so we need $\dfrac{\overrightarrow{PQ}\cdot\overrightarrow{P_0P_1}}{|\overrightarrow{PQ}|\,|\overrightarrow{P_0P_1}|} = \dfrac{-t - 2 - 4t - t}{\sqrt{6}\sqrt{6t^2+4t+4}} = \dfrac{-6t - 2}{\sqrt{6}\sqrt{6t^2+4t+4}} = \cos\dfrac{\pi}{3} = \dfrac{1}{2}$. Thus $-12t - 4 = \sqrt{6}\sqrt{6t^2+4t+4}$, so $144t^2 + 96t + 16 = 6(6t^2 + 4t + 4)$, which reduces to $27t^2 + 18t - 2 = 0$. Thus $t = \dfrac{-18 \pm \sqrt{540}}{54} = \dfrac{-3 \pm \sqrt{15}}{9}$ and $P = \left(\dfrac{21 \mp \sqrt{15}}{9}, \dfrac{3 \pm 2\sqrt{15}}{9}, \dfrac{6 \pm \sqrt{15}}{9}\right)$.

43 (a) The "flashlight" vector is $(1 - 3)\mathbf{i} + (2 - 4)\mathbf{j} + (3 - 6)\mathbf{k} = -2\mathbf{i} - 2\mathbf{j} - 3\mathbf{k}$. An equation for the line through the flashlight is $\mathbf{i} + 2\mathbf{j} + 3\mathbf{k} - t(2\mathbf{i} + 2\mathbf{j} + 3\mathbf{k})$, so $x = 1 - 2t$, $y = 2 - 2t$, and $z = 3 - 3t$. Substituting into the equation for the plane of the mirror, $x + 3y + 3z = 0$, yields $(1 - 2t) + 3(2 - 2t) + 3(3 - 3t) = 16 - 17t = 0$, so $t = 16/17$ and the point where the ray strikes the mirror is $\left(-\dfrac{15}{17}, \dfrac{2}{17}, \dfrac{3}{17}\right)$.

(b) A unit vector normal to the mirror plane is $\frac{1}{\sqrt{19}}(\mathbf{i} + 3\mathbf{j} + 3\mathbf{k})$. By Exercise 26 of Sec. 12.7, a vector parallel to the reflected beam is $\mathbf{B} = \mathbf{A} - 2(\mathbf{A}\cdot\mathbf{n})\mathbf{n}$, where $\mathbf{A} = -2\mathbf{i} - 2\mathbf{j} - 3\mathbf{k}$. Hence $\mathbf{B} = -2\mathbf{i} - 2\mathbf{j} - 3\mathbf{k} -$

$$2\left((-2\mathbf{i} - 2\mathbf{j} - 3\mathbf{k})\cdot\frac{\mathbf{i} + 3\mathbf{j} + 3\mathbf{k}}{\sqrt{19}}\right)\frac{\mathbf{i} + 3\mathbf{j} + 3\mathbf{k}}{\sqrt{19}}$$

$$= -2\mathbf{i} - 2\mathbf{j} - 3\mathbf{k} - \frac{2}{19}(-17)(\mathbf{i} + 3\mathbf{j} + 3\mathbf{k})$$

$= -\frac{4}{19}\mathbf{i} + \frac{64}{19}\mathbf{j} + \frac{45}{19}\mathbf{k}$. The line through $\left(-\frac{15}{17}, \frac{2}{17}, \frac{3}{17}\right)$ parallel to $\mathbf{B}$ has parametric equations $x = -\frac{15}{17} - \frac{4}{19}s$, $y = \frac{2}{17} + \frac{64}{19}s$, and $z = \frac{3}{17} + \frac{45}{19}s$. Substituting into the equation of the second plane, $y + z = 4$, gives $\frac{2}{17} + \frac{64}{19}s + \frac{3}{17} + \frac{45}{19}s = 4$, so $s = \frac{1197}{1853}$. Since $s > 0$, the reflected ray does in fact hit the second plane, at the point $\left(-\frac{111}{109}, \frac{250}{109}, \frac{186}{109}\right) \approx (-1.01835, 2.29358, 1.70642)$.

45 $\mathbf{D}\cdot(\mathbf{B} \times \mathbf{C}) = (x\mathbf{A} + y\mathbf{B} + z\mathbf{C})\cdot(\mathbf{B} \times \mathbf{C})$
$= x\mathbf{A}\cdot(\mathbf{B} \times \mathbf{C}) + y\mathbf{B}\cdot(\mathbf{B} \times \mathbf{C}) + z\mathbf{C}\cdot(\mathbf{B} \times \mathbf{C})$
$= x\mathbf{A}\cdot(\mathbf{B} \times \mathbf{C}) + y\cdot 0 + z\cdot 0 = x\mathbf{A}\cdot(\mathbf{B} \times \mathbf{C})$, so
$x = \frac{\mathbf{D}\cdot(\mathbf{B}\times\mathbf{C})}{\mathbf{A}\cdot(\mathbf{B}\times\mathbf{C})}$, as claimed.

47 Label the midpoints of the three sides as S, T, and U, as shown in the figure. Introduce a coordinate system so that P is at the origin and Q lies on the x axis. Then $P = (0, 0)$, $Q = (a, 0)$, $R = (b, c)$, $S = \left(\frac{a+b}{2}, \frac{c}{2}\right)$, $T = \left(\frac{b}{2}, \frac{c}{2}\right)$, and $U = \left(\frac{a}{2}, 0\right)$.

Therefore $\overrightarrow{PS} = \frac{a+b}{2}\mathbf{i} + \frac{c}{2}\mathbf{j}$, $\overrightarrow{QT} = \left(\frac{b}{2} - a\right)\mathbf{i} + \frac{c}{2}\mathbf{j}$, and $\overrightarrow{RU} = \left(\frac{a}{2} - b\right)\mathbf{i} - c\mathbf{j}$.

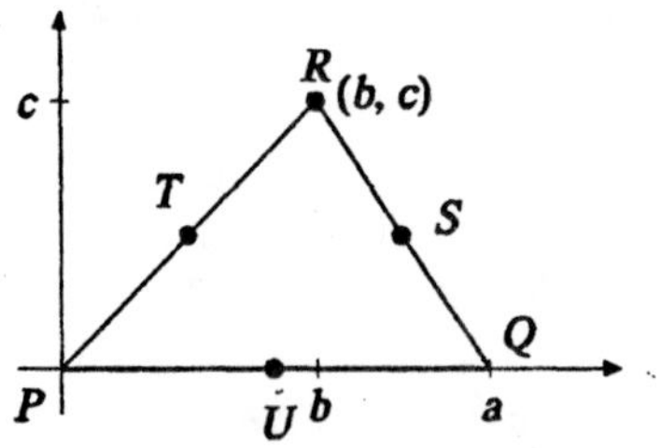

(a) Three vectors form the sides of a triangle if their sum is $\mathbf{0}$. Since $\overrightarrow{PS} + \overrightarrow{QT} + \overrightarrow{RU}$

$= \frac{a+b}{2}\mathbf{i} + \frac{c}{2}\mathbf{j} + \left(\frac{b}{2} - a\right)\mathbf{i} + \frac{c}{2}\mathbf{j} + \left(\frac{a}{2} - b\right)\mathbf{i} - c\mathbf{j} =$

$\left(\frac{a+b}{2} + \frac{b}{2} - a + \frac{a}{2} - b\right)\mathbf{i} + \left(\frac{c}{2} + \frac{c}{2} - c\right)\mathbf{j} = \mathbf{0}$, they do form a triangle.

(b) Note that the area of $PQR = \frac{1}{2}ac$. The area of the triangle formed by $\overrightarrow{PS}$, $\overrightarrow{QT}$, and $\overrightarrow{RU}$

equals $\frac{1}{2}\|\overrightarrow{PS} \times \overrightarrow{RU}\| = \frac{1}{2}\,\text{mag}\begin{vmatrix} \mathbf{i} & \mathbf{j} & \mathbf{k} \\ \frac{a+b}{2} & \frac{c}{2} & 0 \\ \frac{a}{2} - b & -c & 0 \end{vmatrix}$

$= \frac{1}{2}\left|\left(-\frac{1}{2}ac - \frac{1}{2}bc - \frac{1}{4}ac + \frac{1}{2}bc\right)\mathbf{k}\right|$

$= \frac{1}{2}\left|-\frac{3}{4}ac\mathbf{k}\right| = \frac{3}{8}ac$. The area is three-

fourths that of PQR.

49 (a) Suppose that the region of area A has a normal whose direction cosines are $\cos\alpha$, $\cos\beta$, and $\cos\gamma$. By Sec. 12.2, we have $A_1 = A\cos\alpha$, $A_2 = A\cos\beta$, and $A_3 = A\cos\gamma$. Since $1 = \cos^2\alpha + \cos^2\beta + \cos^2\gamma =$

$\left(\frac{A_1}{A}\right)^2 + \left(\frac{A_2}{A}\right)^2 + \left(\frac{A_3}{A}\right)^2 = \frac{A_1^2 + A_2^2 + A_3^2}{A^2}$,

we have $A = \sqrt{A_1^2 + A_2^2 + A_3^2}$.

(b) By the triangle inequality, $A =$

$\sqrt{A_1^2 + A_2^2 + A_3^2} \le A_1 + A_2 + A_3$.

13 The Derivative of a Vector Function

13.1 The Derivative of a Vector Function

1 (a) $\mathbf{G}(t) = t\mathbf{i} + t^2\mathbf{j}$

$\mathbf{G}(1) = 1\cdot\mathbf{i} + 1^2\mathbf{j} = \mathbf{i} + \mathbf{j}$

$\mathbf{G}(2) = 2\mathbf{i} + 2^2\mathbf{j} = 2\mathbf{i} + 4\mathbf{j}$

$\mathbf{G}(3) = 3\mathbf{i} + 3^2\mathbf{j} = 3\mathbf{i} + 9\mathbf{j}$

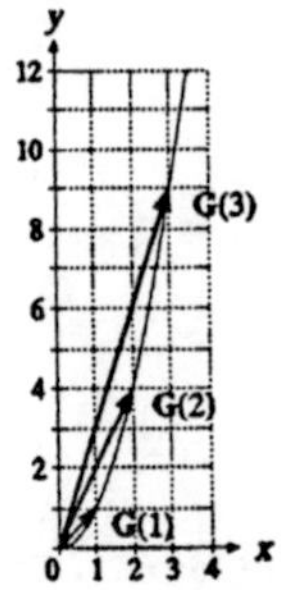

(b) At time t, we have $x = t$ and $y = t^2$. Therefore $y = x^2$ and the path is a parabola.

3 (a) $\mathbf{G}(t) = 2t\mathbf{i} + t^2\mathbf{j}$, so $\mathbf{G}(1.1) = 2.2\mathbf{i} + 1.21\mathbf{j}$, $\mathbf{G}(1) = 2\mathbf{i} + \mathbf{j}$, and $\Delta\mathbf{G} = \mathbf{G}(1.1) - \mathbf{G}(1) = (2.2 - 2)\mathbf{i} + (1.21 - 1)\mathbf{j} = 0.2\mathbf{i} + 0.21\mathbf{j}$.

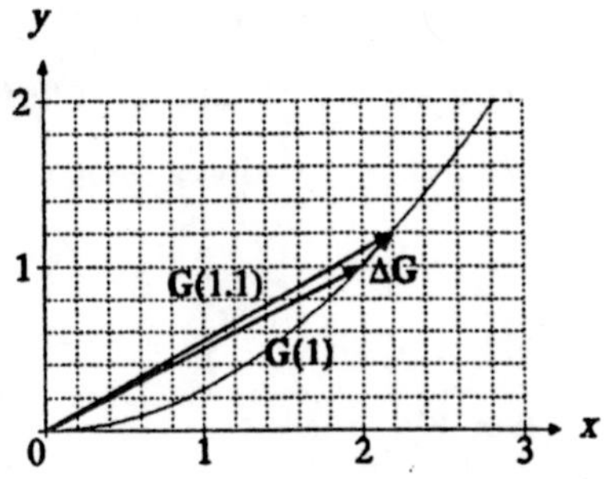

(b) By the result of (a), $\dfrac{\Delta\mathbf{G}}{0.1} = \dfrac{0.2\mathbf{i} + 0.21\mathbf{j}}{0.1}$ $= 2\mathbf{i} + 2.1\mathbf{j}$.

(c) By Theorem 1, $\mathbf{G}'(t) = (2t)'\mathbf{i} + (t^2)'\mathbf{j} = 2\mathbf{i} + 2t\mathbf{j}$, so $\mathbf{G}'(1) = 2\mathbf{i} + 2\mathbf{j}$.

5 (a) $\mathbf{r}(t) = 32t\mathbf{i} - 16t^2\mathbf{j}$, so $\mathbf{r}(1) = 32\mathbf{i} - 16\mathbf{j}$ and $\mathbf{r}(2) = 64\mathbf{i} - 64\mathbf{j}$.

(b) The ball's path is the curved line in the graph. Since $x = 32t$ and $y = -16t^2 = -16\left(\dfrac{x}{32}\right)^2 = -\dfrac{x^2}{64}$, the path is a parabola.

(c) $\mathbf{v} = \mathbf{r}' = 32\mathbf{i} - 32t\mathbf{j}$, so $\mathbf{v}(0) = 32\mathbf{i}$, $\mathbf{v}(1) = 32\mathbf{i} - 32\mathbf{j}$, and $\mathbf{v}(2) = 32\mathbf{i} - 64\mathbf{j}$.

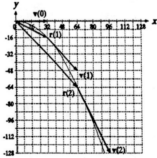

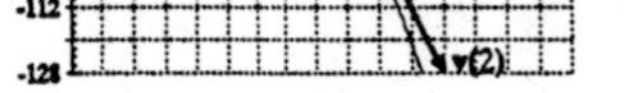

7 (a) The particle moves approximately $\|\mathbf{G}(1.2) - \mathbf{G}(1)\| = \|(2.31\mathbf{i} + 4.05\mathbf{j}) - (2.3\mathbf{i} + 4.1\mathbf{j})\| = \|0.01\mathbf{i} - 0.05\mathbf{j}\|$ $= \sqrt{(0.01)^2 + (-0.05)^2} = \dfrac{1}{100}\sqrt{26}$.

(b) The slope of the tangent vector at $\mathbf{G}(1)$ is approximately the slope of $\Delta\mathbf{G} = \mathbf{G}(1.2) - \mathbf{G}(1) = 0.01\mathbf{i} - 0.05\mathbf{j}$ (from part (a)). Its slope is $\dfrac{-0.05}{0.01} = -5$.

(c) $\mathbf{G}'(1) \approx \dfrac{\Delta \mathbf{G}}{0.2} = \dfrac{1}{0.2}(0.01\mathbf{i} - 0.05\mathbf{j})$

$= 0.05\mathbf{i} - 0.25\mathbf{j}$

(d) Speed $= \|\mathbf{G}'(1)\| \approx \|0.05\mathbf{i} - 0.25\mathbf{j}\|$

$= \sqrt{(0.05)^2 + (-0.25)^2} = \dfrac{1}{20}\sqrt{26}$

9 $\mathbf{v}(t) = \mathbf{r}'(t) = (\cos 3t)'\mathbf{i} + (\sin 3t)'\mathbf{j} + (6t)'\mathbf{k}$ $= (-3 \sin 3t)\mathbf{i} + (3 \cos 3t)\mathbf{j} + 6\mathbf{k}$; speed $=$ $\|\mathbf{v}(t)\| = \sqrt{(-3 \sin 3t)^2 + (3 \cos 3t)^2 + 6^2} = \sqrt{45}$ $= 3\sqrt{5}$.

11 $\mathbf{v}(t) = \mathbf{r}'(t) = [\ln(1 + t^2)]'\mathbf{i} + (e^{3t})'\mathbf{j} + \left(\dfrac{\tan t}{1 + 2t}\right)'\mathbf{k}$

$= \left(\dfrac{2t}{1 + t^2}\right)\mathbf{i} + 3e^{3t}\mathbf{j} + \left[\dfrac{(1 + 2t)(\sec^2 t) - (\tan t)(2)}{(1 + 2t)^2}\right]\mathbf{k} = \left(\dfrac{2t}{1 + t^2}\right)\mathbf{i} + 3e^{3t}\mathbf{j} + \left[\dfrac{(1 + 2t) \sec^2 t - 2 \tan t}{(1 + 2t)^2}\right]\mathbf{k}$; speed $=$

$$\sqrt{\left(\frac{2t}{1 + t^2}\right)^2 + (3e^{3t})^2 + \left(\frac{(1 + 2t) \sec^2 t - 2 \tan t}{(1 + 2t)^2}\right)^2}$$

$$= \sqrt{\frac{4t^2}{(1 + t^2)^2} + 9e^{6t} + \frac{[(1 + 2t) \sec^2 t - 2 \tan t]^2}{(1 + 2t)^4}}$$

13 (a) $\mathbf{G}(t) = t^2\mathbf{i} + t^3\mathbf{j}$, so $\mathbf{G}(1.1) = (1.1)^2\mathbf{i} + (1.1)^3\mathbf{j} = 1.21\mathbf{i} + 1.331\mathbf{j}$, $\mathbf{G}(1) = \mathbf{i} + \mathbf{j}$, and $\Delta\mathbf{G} = \mathbf{G}(1.1) - \mathbf{G}(1) = 0.21\mathbf{i} + 0.331\mathbf{j}$.

(b) $\dfrac{\Delta \mathbf{G}}{\Delta t} = \dfrac{0.21\mathbf{i} + 0.331\mathbf{j}}{0.1} = 2.1\mathbf{i} + 3.31\mathbf{j}$

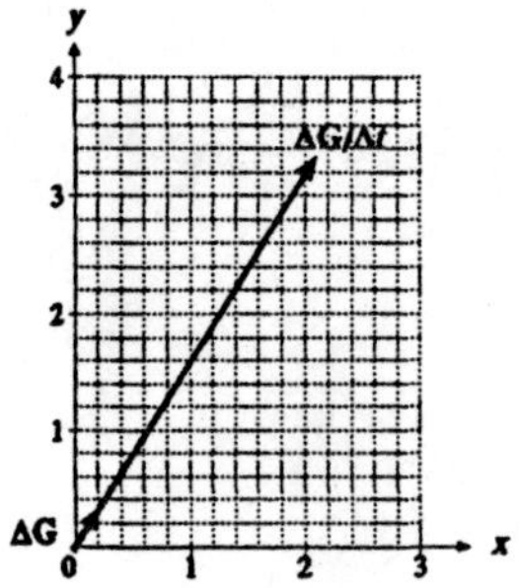

(c) $\mathbf{G}'(t) = (t^2)'\mathbf{i} + (t^3)'\mathbf{j} = 2t\mathbf{i} + 3t^2\mathbf{j}$, so $\mathbf{G}'(1) = 2\mathbf{i} + 3\mathbf{j}$.

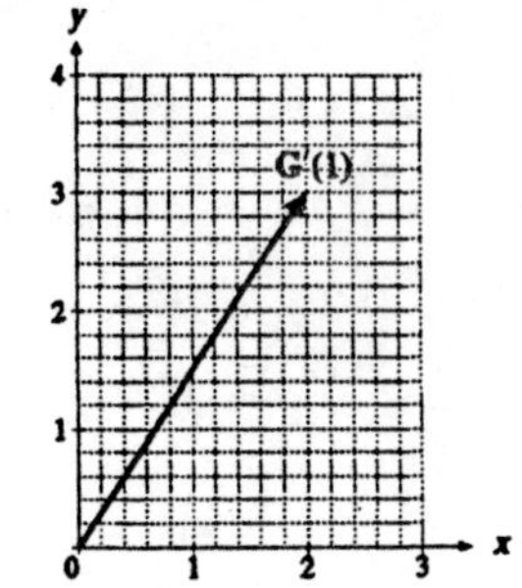

(d) $\left\|\dfrac{\Delta \mathbf{G}}{\Delta t} - \mathbf{G}'(1)\right\| = \|(2.1\mathbf{i} + 3.31\mathbf{j}) - (2\mathbf{i} + 3\mathbf{j})\|$

$= \|0.1\mathbf{i} + 0.31\mathbf{j}\| = \sqrt{(0.1)^2 + (0.31)^2}$

$= \sqrt{0.1061} \approx 0.3257$

15 We have $x = t \cos 2\pi t$, $y = t \sin 2\pi t$, and $z = t$. If, as in polar coordinates, we let r be the distance to the z axis, we have $r^2 = x^2 + y^2 = t^2 \cos^2 2\pi t + t^2 \sin^2 2\pi t = t^2 = z^2$, so $r = z$. In other words, at height z the surface has a radius equal to z. Hence the particle is tracing out a curve which lies on a conical surface of revolution. It travels in a counterclockwise direction as seen from above.

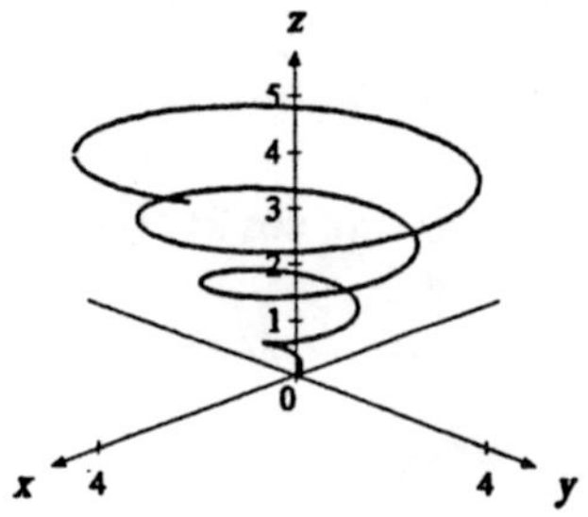

17 (a) $x = t$ and $y = t^{-1} = \frac{1}{x}$, so the path is part of the hyperbola $y = \frac{1}{x}$ for $x \geq 1$.

(b) $\mathbf{r}(t) = t\mathbf{i} + (1/t)\mathbf{j}$, so $\mathbf{r}(1) = \mathbf{i} + \mathbf{j}$, $\mathbf{r}(2) = 2\mathbf{i} + \frac{1}{2}\mathbf{j}$, and $\mathbf{r}(3) = 3\mathbf{i} + \frac{1}{3}\mathbf{j}$. These are plotted in the figure.

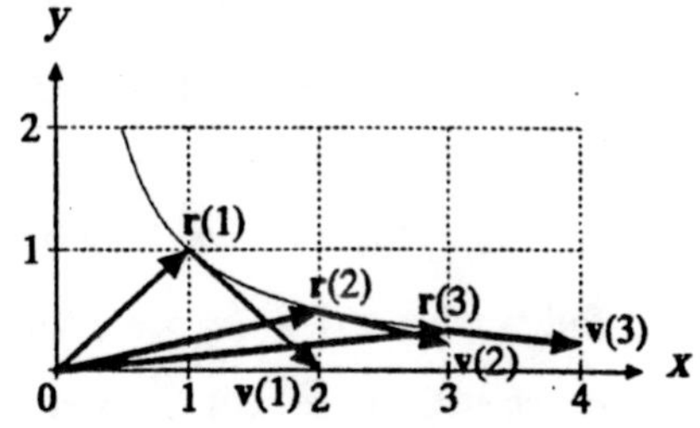

(c) $\mathbf{v}(t) = \mathbf{r}'(t) = (t)'\mathbf{i} + (t^{-1})'\mathbf{j} = \mathbf{i} - t^{-2}\mathbf{j}$, so $\mathbf{v}(1) = \mathbf{i} - \mathbf{j}$, $\mathbf{v}(2) = \mathbf{i} - \frac{1}{4}\mathbf{j}$, and $\mathbf{v}(3) = \mathbf{i} - \frac{1}{9}\mathbf{j}$. These are displayed in the figure.

(d) $\frac{dx}{dt} = 1$, $\frac{dy}{dt} = -t^{-2}$, $\mathbf{v}(t) = \mathbf{i} - t^{-2}\mathbf{j}$, and $\|\mathbf{v}(t)\| = \sqrt{1 + t^{-4}}$. As $t \to \infty$, $\frac{dx}{dt}$ stays at 1, $\frac{dy}{dt} \to 0$, $\mathbf{v}(t) \to \mathbf{i}$, and $\|\mathbf{v}(t)\| \to 1$.

19 (a) Let the polar coordinates of the particle at time t be $(r(t), \theta(t))$. Note that $r(t) = 100$ for all values of t. Also, since the particle travels at constant speed, $\theta(t)$ is a linear function of time, $\theta(t) = at + b$. Since $\theta(0) = 0$ and $\theta\left(\frac{1}{200}\right) = -2\pi$, we have $\theta(t) = -400\pi t$.

Therefore $\mathbf{r}(t) = x(t)\mathbf{i} + y(t)\mathbf{j}$, where $x(t) = r(t)\cos\theta(t) = 100\cos(-400\pi t) = 100\cos 400\pi t$ and $y(t) = r(t)\sin\theta(t) = 100\sin(-400\pi t) = -100\sin 400\pi t$. Thus $\mathbf{r}(t) = 100\cos 400\pi t\,\mathbf{i} - 100\sin 400\pi t\,\mathbf{j}$ and $\mathbf{v}(t) = \mathbf{r}'(t) = -40000\pi\sin 400\pi t\,\mathbf{i} - 40000\pi\cos 400\pi t\,\mathbf{j}$.

(b) $\mathbf{r}(0) = 100\cos 0\,\mathbf{i} - 100\sin 0\,\mathbf{j} = 100\mathbf{i}$, $\mathbf{r}\left(\frac{1}{800}\right) = 100\cos\frac{\pi}{2}\,\mathbf{i} - 100\sin\frac{\pi}{2}\,\mathbf{j} = -100\mathbf{j}$, $\mathbf{v}(0) = -40000\pi\sin 0\,\mathbf{i} - 40000\pi\cos 0\,\mathbf{j} = -40000\pi\mathbf{j}$, $\mathbf{v}\left(\frac{1}{800}\right) = -40000\pi\sin\frac{\pi}{2}\,\mathbf{i} - 40000\pi\cos\frac{\pi}{2}\,\mathbf{j} = -40000\pi\mathbf{i}$

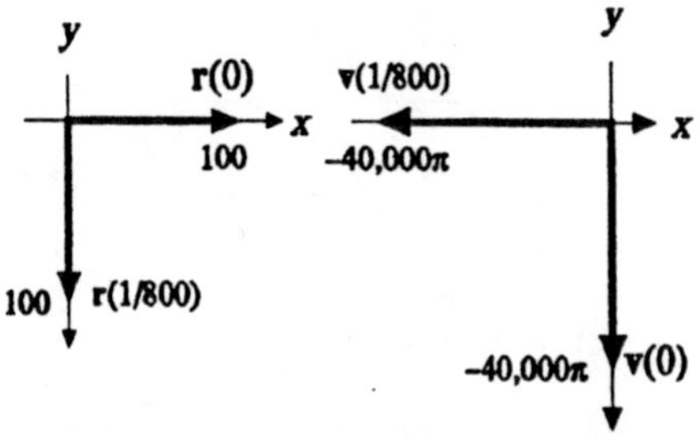

(c) $\|\mathbf{r}(t)\| = 100$ and $\|\mathbf{v}(t)\| = 40000\pi$; they are constant and do not change with time.

21 (a) $\mathbf{r} = x\mathbf{i} + y\mathbf{j}$, so $\dfrac{d\mathbf{r}}{ds} = \dfrac{dx}{ds}\mathbf{i} + \dfrac{dy}{ds}\mathbf{j}$ and $\left\|\dfrac{d\mathbf{r}}{ds}\right\|$

$= \sqrt{\left(\dfrac{dx}{ds}\right)^2 + \left(\dfrac{dy}{ds}\right)^2} = \dfrac{ds}{ds} = 1$, by Eq. (2) of

Sec. 9.4.

(b) A small section of a curve can be well approximated by a straight line. From the figure, we see that $\|\Delta\mathbf{r}/\Delta s\|$ is close to 1 when Δs is small.

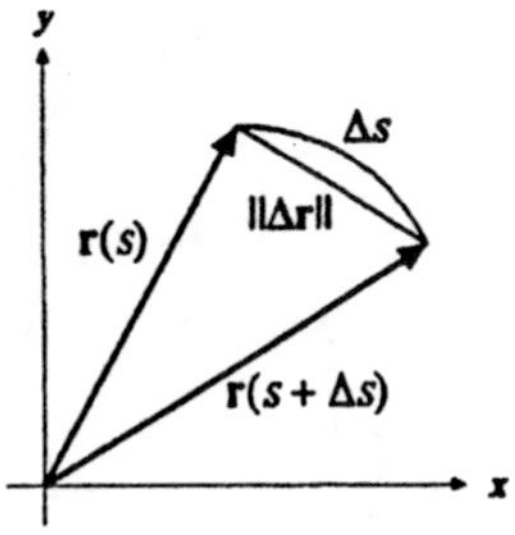

23 (a) $\mathbf{r}(t) = a\cos 2\pi t\,\mathbf{i} + a\sin 2\pi t\,\mathbf{j}$, so $\mathbf{r}(0) =$

$a\cos 0\,\mathbf{i} + a\sin 0\,\mathbf{j} = a\mathbf{i}$ and $\mathbf{r}\left(\dfrac{1}{4}\right) =$

$a\cos\dfrac{\pi}{2}\,\mathbf{i} + a\sin\dfrac{\pi}{2}\,\mathbf{j} = a\mathbf{j}.$

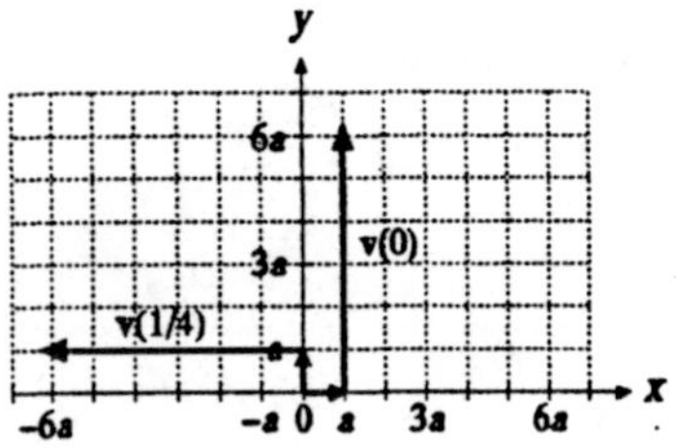

(b) $\mathbf{v}(t) = \mathbf{r}'(t) =$

$-2\pi a\sin 2\pi t\,\mathbf{i} + 2\pi a\cos 2\pi t\,\mathbf{j}$, so $\mathbf{v}(0) =$

$-2\pi a\sin 0\,\mathbf{i} + 2\pi a\cos 0\,\mathbf{j} = 2\pi a\mathbf{j}$ and

$\mathbf{v}\left(\dfrac{1}{4}\right) = -2\pi a\sin\dfrac{\pi}{2}\,\mathbf{i} + 2\pi a\cos\dfrac{\pi}{2}\,\mathbf{j}$

$= -2\pi a\mathbf{i}.$

(c) $\mathbf{r}(t)\cdot\mathbf{v}(t) = (a\cos 2\pi t)(-2\pi a\sin 2\pi t) +$

$(a\sin 2\pi t)(2\pi a\cos 2\pi t) =$

$-2\pi a^2\sin 2\pi t\cos 2\pi t +$

$2\pi a^2\sin 2\pi t\cos 2\pi t = 0$, so $\mathbf{r}(t)$ and $\mathbf{v}(t)$ are perpendicular.

25 (a) Recall from elementary physics that we can consider the vertical and horizontal components of motion separately. The magnitude of the initial velocity is v_0 and its direction is at an angle θ from the horizontal. The horizontal component of the initial velocity therefore has magnitude $v_0\cos\theta$, and the magnitude of the vertical component is $v_0\sin\theta$. No influence acts on $x(t)$, the horizontal component of $\mathbf{r}(t)$, so $x'(t)$ is constant. But initial horizontal speed is $v_0\cos\theta$, so $x'(0) = v_0\cos\theta$; since $x'(t)$ is constant, $x'(t) = v_0\cos\theta$ for all values of t. It follows that $x(t) = (v_0\cos\theta)t + C$ for some constant C; since $x(0)$ is given to be 0, we have $C = 0$ and $x(t) = (v_0\cos\theta)t$. The vertical component of $\mathbf{r}(t)$, $y(t)$, is influenced by a gravitational acceleration of -32 ft/sec^2. By Sec. 4.3, $y(t) = \dfrac{a}{2}t^2 + bt + c$, where a is gravitational acceleration, b is the initial vertical velocity, and c is initial position. We have $a = -32$, $b = v_0\sin\theta$, and $c = 0$, so $y(t) = -16t^2 + (v_0\sin\theta)t$. Hence $\mathbf{r}(t) = x(t)\mathbf{i} + y(t)\mathbf{j} = (v_0\cos\theta)t\mathbf{i} + ((v_0\sin\theta)t - 16t^2)\mathbf{j}$, as claimed.

(b) The rock returns to its initial height when $y(t) = 0$; that is, when $(v_0\sin\theta)t - 16t^2 = 0$, so $(v_0\sin\theta - 16t)t = 0$, which occurs when $t =$

0 or $t = \frac{1}{16}v_0 \sin\theta$. The solution $t = 0$ is simply the initial point, so it is the second solution that interests us. The horizontal distance traveled is $x\left(\frac{1}{16}v_0 \sin\theta\right) =$

$(v_0 \cos\theta)\left(\frac{1}{16}v_0 \sin\theta\right) = \frac{1}{16}v_0^2 \cos\theta \sin\theta.$

If instead the initial angle had been $\frac{\pi}{2} - \theta$, our result would be of the form

$\frac{1}{16}v_0^2 \cos\left(\frac{\pi}{2} - \theta\right) \sin\left(\frac{\pi}{2} - \theta\right)$; but $\cos\left(\frac{\pi}{2} - \theta\right)$

$= \sin\theta$ and $\sin\left(\frac{\pi}{2} - \theta\right) = \cos\theta$, so the new result is equal to the old.

(c) $\frac{1}{16}v_0^2 \cos\theta \sin\theta = \frac{1}{32}v_0^2 \cdot 2\cos\theta\sin\theta =$

$\frac{1}{32}v_0^2 \sin 2\theta$, which is maximized when $\sin 2\theta = 1$. Since θ is between 0 and $\frac{\pi}{2}$, the desired value is $2\theta = \frac{\pi}{2}$; that is, $\theta = \frac{\pi}{4}$.

27 Note that $\mathbf{v}(t) = \mathbf{r}'(t) = \frac{1}{t}\mathbf{i} - 3\sin 3t\,\mathbf{j}$ for $1 \le t \le 2$, so $\mathbf{v}(2) = \frac{1}{2}\mathbf{i} - 3\sin 6\,\mathbf{j}$. Also, $\mathbf{r}(2) = \ln 2\,\mathbf{i} + \cos 6\,\mathbf{j}$, so if $\mathbf{v}(t)$ is constant for $t \ge 2$, then $\mathbf{r}(t) = \mathbf{r}(2) + (t - 2)\mathbf{v}(2) =$

$\left[\ln 2 + \frac{1}{2}(t - 2)\right]\mathbf{i} + [\cos 6 - 3(t - 2)\sin 6]\mathbf{j}$ for $t \ge 2$. In particular, $\mathbf{r}(3) =$

$\left(\ln 2 + \frac{1}{2}\right)\mathbf{i} + (\cos 6 - 3\sin 6)\mathbf{j}.$

29 (a) Let the origin of the arrow's path be located at (0, 0). Let the initial position of the ball be (l, h). By the arguments in Exercise 25, it is clear that the position of the ball at time t is $(l, h - 16t^2)$. The arrow's position is $((v_0\cos\theta)t, (v_0\sin\theta)t - 16t^2)$ where $\theta = \tan^{-1} h/l$ is the angle at which the arrow was shot. The arrow and the ball have the same x coordinate when $t = \frac{l}{v_0\cos\theta}$ and the same y coordinate when $t = \frac{h}{v_0\sin\theta}$. Since $\frac{h}{\sin\theta}$

$= \frac{h/\tan\theta}{\cos\theta} = \frac{l}{\cos\theta}$, we see that these events occur at the same time, so the arrow does hit the ball.

(b) If there were no acceleration due to gravity, the arrow would strike the ball. However, since the arrow and ball both undergo the *same* acceleration, they will collide regardless of the influence of gravity.

Introduction to Exercises 31—35. Let $\mathbf{r}(t) = x(t)\mathbf{i} + y(t)\mathbf{j}$. By Theorem 1, $\mathbf{v}(t) = x'(t)\mathbf{i} + y'(t)\mathbf{j}$. Hence the components of $\mathbf{r}(t)$ are antiderivatives of those of $\mathbf{v}(t)$. The value of the "arbitrary" constant of integration is found by setting $t = 0$.

31 $x(t) = \int \frac{t}{t^2 + t + 1}\,dt$

$= \frac{1}{2}\int \frac{2t + 1}{t^2 + t + 1}\,dt - \frac{1}{2}\int \frac{dt}{t^2 + t + 1}$

$= \frac{1}{2}\ln(t^2+t+1) - \frac{1}{\sqrt{3}}\tan^{-1}\frac{2t+1}{\sqrt{3}} + C$. Since $x(0) = 1$, $C = 1 + \frac{1}{\sqrt{3}}\tan^{-1}\frac{1}{\sqrt{3}} = 1 + \frac{1}{\sqrt{3}}\cdot\frac{\pi}{6}$ $= 1 + \frac{\pi}{6\sqrt{3}}$. Hence $x(t) = \frac{1}{2}\ln(t^2+t+1) - \frac{1}{\sqrt{3}}\tan^{-1}\frac{2t+1}{\sqrt{3}} + 1 + \frac{\pi}{6\sqrt{3}}$. Next, $y(t) =$ $\int \tan^{-1}3t\,dt = \frac{1}{3}\int 3\tan^{-1}3t\,dt$

$= \frac{1}{3}\left[3t\tan^{-1}3t - \frac{1}{2}\ln(1+(3t)^2)\right] + C'$

$= t\tan^{-1}3t - \frac{1}{6}\ln(1+9t^2) + C'$. Since $y(0) =$ 1, $C' = 1$, so $y(t) = t\tan^{-1}3t - \frac{1}{6}\ln(1+9t^2) +$ 1. Hence $\mathbf{r}(t) =$

$$\left[\frac{1}{2}\ln(t^2+t+1) - \frac{1}{\sqrt{3}}\tan^{-1}\frac{2t+1}{\sqrt{3}} + 1 + \frac{\pi}{6\sqrt{3}}\right]\mathbf{i}$$

$$+ \left[t\tan^{-1}3t - \frac{1}{6}\ln(1+9t^2) + 1\right]\mathbf{j}.$$

33 $\int e^{2t}\sin 3t\,dt =$ $\frac{1}{13}e^{2t}(2\sin 3t - 3\cos 3t) + C_1$, $\int \frac{t^3\,dt}{3t+2} =$ $\frac{1}{9}t^3 - \frac{1}{9}t^2 + \frac{4}{27}t - \frac{8}{81}\ln|3t+2| + C_2$. Hence $\mathbf{r}(t)$

$$= \left[\frac{1}{13}e^{2t}(2\sin 3t - 3\cos 3t) + \frac{16}{13}\right]\mathbf{i} +$$

$$\left[\frac{1}{9}t^3 - \frac{1}{9}t^2 + \frac{4}{27}t - \frac{8}{81}\ln|3t+2| + \frac{8}{81}\ln 2 + 3\right]\mathbf{j}.$$

35 $x(t) = \int \frac{[\ln(t+1)]^3}{t+1}\,dt = \frac{1}{4}(\ln(t+1))^4 + C$.

Since $x(0) = 1$, $C = 1$. Then $x(t) =$ $\frac{1}{4}(\ln(t+1))^4 + 1$. $y(t) = \int \frac{dt}{\sqrt{1-4t^2}} =$

$\frac{1}{2}\int \frac{2\,dt}{\sqrt{1-(2t)^2}} = \frac{1}{2}\sin^{-1}2t + C'$. Since $y(0) =$ 1, $C' = 1$. Then $y(t) = \frac{1}{2}\sin^{-1}2t + 1$. Finally,

$z(t) = \int \sec^2 3t\,dt = \frac{1}{3}\int 3\sec^2 3t\,dt =$ $\frac{1}{3}\tan 3t + C''$. Since $z(0) = 1$, $C'' = 1$. Then $z(t)$ $= \frac{1}{3}\tan 3t + 1$. Hence $\mathbf{r}(t) = \left[\frac{1}{4}(\ln(t+1))^4 + 1\right]\mathbf{i}$

$$+ \left[\frac{1}{2}\sin^{-1}2t + 1\right]\mathbf{j} + \left[\frac{1}{3}\tan 3t + 1\right]\mathbf{k}.$$

37

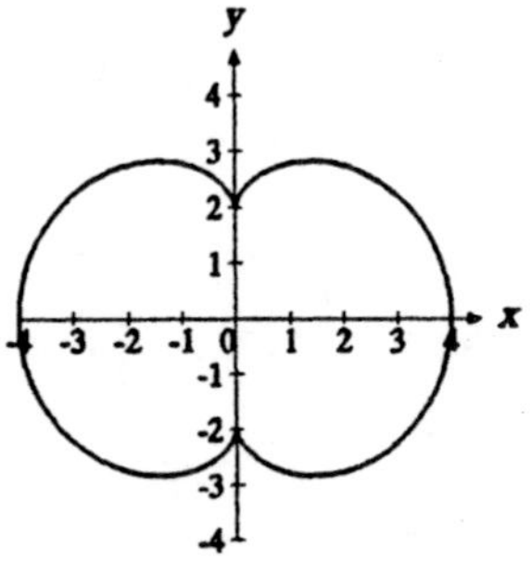

13.2 Properties of the Derivative of a Vector Function

1 Here $\mathbf{G}'(t) = (t^2)'\mathbf{i} + (t^3)'\mathbf{j} = 2t\mathbf{i} + 3t^2\mathbf{j}$ and $\mathbf{H}'(t) = (3t)'\mathbf{i} + (4t^2)'\mathbf{j} = 3\mathbf{i} + 8t\mathbf{j}$. So $(\mathbf{G}\cdot\mathbf{H})'$ $= \mathbf{G}\cdot\mathbf{H}' + \mathbf{H}\cdot\mathbf{G}' = [t^2\mathbf{i} + t^3\mathbf{j}]\cdot[3\mathbf{i} + 8t\mathbf{j}] +$ $[3t\mathbf{i} + 4t^2\mathbf{j}]\cdot[2t\mathbf{i} + 3t^2\mathbf{j}] = 3t^2 + 8t^4 + 6t^2 + 12t^4$ $= 9t^2 + 20t^4$. Also, $\mathbf{G}\cdot\mathbf{H} = [t^2\mathbf{i} + t^3\mathbf{j}]\cdot[3t\mathbf{i} + 4t^2\mathbf{j}]$

$= 3t^3 + 4t^5$. So $(\mathbf{G}\cdot\mathbf{H})' = (3t^3 + 4t^5)'$
$= 9t^2 + 20t^4$, as expected.

3 Here $\mathbf{G}'(t) = (e^{2t})'\mathbf{i} + (t^2)'\mathbf{j} = 2e^{2t}\mathbf{i} + 2t\mathbf{j}$ and $\mathbf{H}'(t) = (t)'\mathbf{i} + (\sin 3t)'\mathbf{j} = \mathbf{i} + 3\cos 3t\,\mathbf{j}$. Therefore we have $(\mathbf{G}\cdot\mathbf{H})' = \mathbf{G}\cdot\mathbf{H}' + \mathbf{H}\cdot\mathbf{G}' =$ $[e^{2t}\mathbf{i} + t^2\mathbf{j}]\cdot[\mathbf{i} + 3\cos 3t\,\mathbf{j}] + [t\mathbf{i} + \sin 3t\,\mathbf{j}]\cdot[2e^{2t}\mathbf{i} + 2t$ $= e^{2t} + 3t^2\cos 3t + 2te^{2t} + 2t\sin 3t$. Also, $\mathbf{G}\cdot\mathbf{H}$ $= [e^{2t}\mathbf{i} + t^2\mathbf{j}]\cdot[t\mathbf{i} + \sin 3t] = te^{2t} + t^2\sin 3t$. Hence $(\mathbf{G}\cdot\mathbf{H})' = (te^{2t})' + (t^2\sin 3t)' =$ $2te^{2t} + e^{2t} + 3t^2\cos 3t + 2t\sin 3t$, as expected.

5 By Theorem 1, $[t^2\mathbf{G}(t)]' = t^2\mathbf{G}'(t) + 2t\mathbf{G}(t)$. At $t = 1$, therefore, we have $1^2\mathbf{G}'(1) + 2\mathbf{G}(1)$ $= (2\mathbf{i} - 3\mathbf{j}) + 2(\mathbf{i} + \mathbf{j}) = 4\mathbf{i} - \mathbf{j}$.

7 By Theorem 1, $(f(t)\mathbf{G}(t))' = (f(t)g(t))'\mathbf{i} + (f(t)h(t))'\mathbf{j} = (f(t)g'(t) + f'(t)g(t))\mathbf{i} + (f(t)h'(t) + f'(t)h(t))\mathbf{j} = f(t)(g'(t)\mathbf{i} + h'(t)\mathbf{j}) + f'(t)(g(t)\mathbf{i} + h(t)\mathbf{j}) = f(t)\mathbf{G}'(t) + f'(t)\mathbf{G}(t)$.

9 (a) $\Delta(\mathbf{G}\times\mathbf{H}) = \mathbf{G}(t+\Delta t)\times\mathbf{H}(t+\Delta t) - \mathbf{G}(t)\times\mathbf{H}(t) = (\mathbf{G}(t) + \Delta\mathbf{G})\times(\mathbf{H}(t) + \Delta\mathbf{H}) - \mathbf{G}(t)\times\mathbf{H}(t) = \mathbf{G}(t)\times\mathbf{H}(t) + \mathbf{G}(t)\times\Delta\mathbf{H} + \Delta\mathbf{G}\times\mathbf{H}(t) + \Delta\mathbf{G}\times\Delta\mathbf{H} - \mathbf{G}(t)\times\mathbf{H}(t)$ $= \mathbf{G}(t)\times\Delta\mathbf{H} + \Delta\mathbf{G}\times\mathbf{H}(t) + \Delta\mathbf{G}\times\Delta\mathbf{H}$,

so $(\mathbf{G}\times\mathbf{H})' = \lim_{\Delta t\to 0}\dfrac{\Delta(\mathbf{G}\times\mathbf{H})}{\Delta t} =$

$$\lim_{\Delta t\to 0}\left(\mathbf{G}(t)\times\frac{\Delta\mathbf{H}}{\Delta t} + \frac{\Delta\mathbf{G}}{\Delta t}\times\mathbf{H}(t) + \Delta\mathbf{G}\times\frac{\Delta\mathbf{H}}{\Delta t}\right)$$

$= \mathbf{G}(t)\times\mathbf{H}'(t) + \mathbf{G}'(t)\times\mathbf{H}(t) + \mathbf{0}\times\mathbf{H}'(t)$

$= \mathbf{G}\times\mathbf{H}' + \mathbf{G}'\times\mathbf{H}$.

(b) The cross product is not commutative, so we cannot replace $\mathbf{G}'\times\mathbf{H}$ in the above formula by $\mathbf{H}\times\mathbf{G}'$.

11 Let $\mathbf{G}(s) = g(s)\mathbf{i} + h(s)\mathbf{j}$, where $s = f(t)$. Then

$$\frac{d}{dt}\mathbf{G}(s) = \frac{d}{dt}g(s)\mathbf{i} + \frac{d}{dt}h(s)\mathbf{j}$$

$$= \frac{dg}{ds}\cdot\frac{ds}{dt}\mathbf{i} + \frac{dh}{ds}\cdot\frac{ds}{dt}\mathbf{j} = \left(\frac{dg}{ds}\mathbf{i} + \frac{dh}{ds}\mathbf{j}\right)\frac{ds}{dt}$$

$$= \frac{d}{ds}\mathbf{G}(s)\cdot\frac{ds}{dt} = \mathbf{G}'(f(t))(f'(t)).$$

13 (a) $\mathbf{G}(t) = \cos 2t\,\mathbf{i} + \sin 2t\,\mathbf{j}$ and $\|\mathbf{G}(t)\| = \sqrt{\cos^2 2t + \sin^2 2t} = 1$, so $\mathbf{G}(t)$ is a unit vector.

(b) $\mathbf{G}'(t) = -2\sin 2t\,\mathbf{i} + 2\cos 2t\,\mathbf{j}$. $\|\mathbf{G}'(t)\| = \sqrt{4\sin^2 2t + 4\cos^2 2t} = 2$, so $\mathbf{G}'(t)$ is not a unit vector.

15 $\mathbf{r}(t) = e^t\cos t\,\mathbf{i} + e^t\sin t\,\mathbf{j}$, so $\mathbf{v}(t) =$ $(-e^t\sin t + e^t\cos t)\mathbf{i} + (e^t\cos t + e^t\sin t)\mathbf{j}$ $= e^t[(\cos t - \sin t)\mathbf{i} + (\cos t + \sin t)\mathbf{j}]$. Thus $\|\mathbf{r}(t)\| = \sqrt{(e^t\cos t)^2 + (e^t\sin t)^2} = e^t$ and $\|\mathbf{v}(t)\| = e^t\sqrt{(\cos t - \sin t)^2 + (\cos t + \sin t)^2} =$ $\sqrt{2}e^t$. Now $\cos\theta = \dfrac{\mathbf{r}\cdot\mathbf{v}}{\|\mathbf{r}\|\,\|\mathbf{v}\|} =$

$$\frac{e^{2t}(\cos t\,\mathbf{i} + \sin t\,\mathbf{j})\cdot[(\cos t - \sin t)\mathbf{i} + (\cos t + \sin t)\mathbf{j}]}{e^t\cdot\sqrt{2}e^t}$$

$$= \frac{1}{\sqrt{2}}(\cos^2 t - \sin t\cos t + \sin t\cos t + \sin^2 t) =$$

$\dfrac{1}{\sqrt{2}}$. Therefore $\theta = \cos^{-1}\dfrac{1}{\sqrt{2}} = \dfrac{\pi}{4}$ is the angle between the two vectors for all time t.

17 Let $f(t) = \|\mathbf{G}(t)\|$ and $\mathbf{H}(t) = \dfrac{\mathbf{G}(t)}{\|\mathbf{G}(t)\|}$. Then $f(t) > 0$, $\|\mathbf{H}(t)\| = 1$, and $\mathbf{G}(t) = f(t)\mathbf{H}(t)$.

19 Since $\|\mathbf{G}(t)\|$ is minimal at $t = t_0$, so is $\|\mathbf{G}(t)\|^2$ $= \mathbf{G}(t)\cdot\mathbf{G}(t)$. Hence $[\mathbf{G}(t)\cdot\mathbf{G}(t)]'$ is 0 at $t = t_0$; that is, $2\mathbf{G}(t_0)\cdot\mathbf{G}'(t_0) = 0$, so $\mathbf{G}(t_0)$ and $\mathbf{G}'(t_0)$ are perpendicular.

13.3 The Acceleration Vector

1 $\mathbf{r}(t) = t^2\mathbf{i} + t^3\mathbf{j}$, so $\mathbf{v}(t) = \mathbf{r}'(t) = 2t\mathbf{i} + 3t^2\mathbf{j}$ and $\mathbf{a}(t) = \mathbf{v}'(t) = 2\mathbf{i} + 6t\mathbf{j}$. Therefore $\mathbf{r}(1) = \mathbf{i} + \mathbf{j}$, $\mathbf{v}(1) = 2\mathbf{i} + 3\mathbf{j}$, and $\mathbf{a}(1) = 2\mathbf{i} + 6\mathbf{j}$.

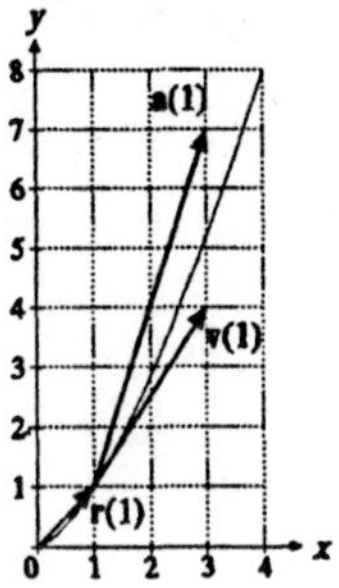

3 $\mathbf{r}(t) = 4t\mathbf{i} - 16t^2\mathbf{j}$, so $\mathbf{v}(t) = \mathbf{r}'(t) = 4\mathbf{i} - 32t\mathbf{j}$ and $\mathbf{a}(t) = \mathbf{v}'(t) = -32\mathbf{j}$. Therefore $\mathbf{r}(0) = \mathbf{0}$, $\mathbf{v}(0) = 4\mathbf{i}$, $\mathbf{r}(1) = 4\mathbf{i} - 16\mathbf{j}$, $\mathbf{v}(1) = 4\mathbf{i} - 32\mathbf{j}$, $\mathbf{r}(2) = 8\mathbf{i} - 64\mathbf{j}$, $\mathbf{v}(2) = 4\mathbf{i} - 64\mathbf{j}$, and $\mathbf{a}(0) = \mathbf{a}(1) = \mathbf{a}(2) = -32\mathbf{j}$.

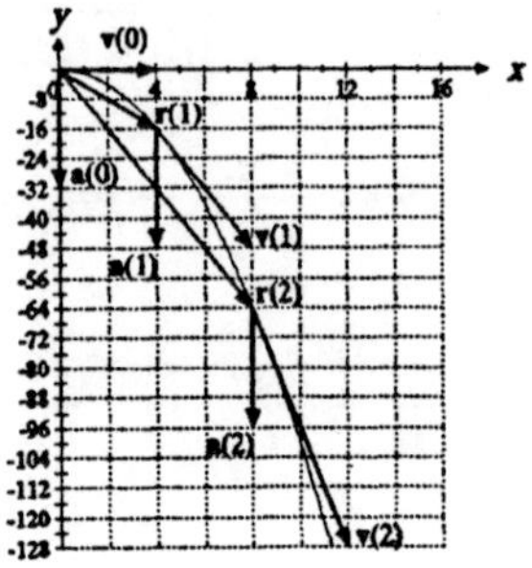

5 (a) $\mathbf{r}(t) = x\mathbf{i} + y\mathbf{j} = t\mathbf{i} + t^{-1}\mathbf{j}$, so $x = t$ and $y = \frac{1}{t} = \frac{1}{x}$. For $t > 0$ the particle travels on the

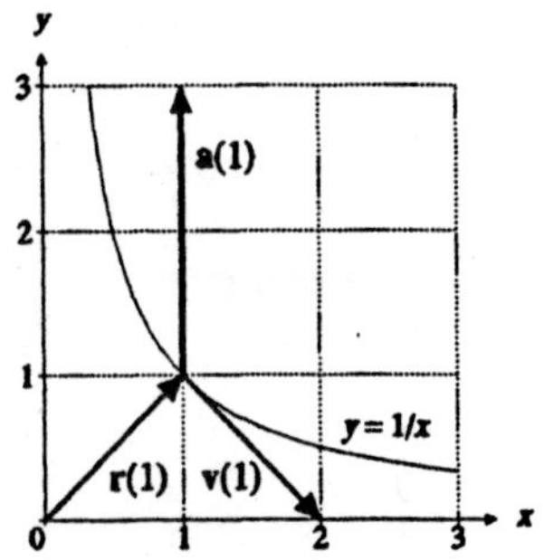

first-quadrant branch of the hyperbola $y = \frac{1}{x}$.

(b) $\mathbf{r}(1) = \mathbf{i} + \mathbf{j}$; $\mathbf{v}(t) = \mathbf{r}'(t) = \mathbf{i} - t^{-2}\mathbf{j}$ so $\mathbf{v}(1) = \mathbf{i} - \mathbf{j}$; $\mathbf{a}(t) = \mathbf{v}'(t) = 2t^{-3}\mathbf{j}$ so $\mathbf{a}(1) = 2\mathbf{j}$. $\mathbf{r}(1)$, $\mathbf{v}(1)$, and $\mathbf{a}(1)$ are plotted on the graph in (a).

(c) As $t \to \infty$, $t^{-2} \to 0$, so $\mathbf{v}(t) \to \mathbf{i} - 0\mathbf{j} = \mathbf{i}$. As $t \to \infty$, $2t^{-3} \to 0$, so $\mathbf{a}(t) \to \mathbf{0}$.

(d) Speed $= \|\mathbf{v}(t)\| \to 1$ as $t \to \infty$.

7 (a) $\mathbf{r}(t) = 5\cos 3t\,\mathbf{i} + 5\sin 3t\,\mathbf{j} + 6t\mathbf{k}$, so $\mathbf{v}(t) = \mathbf{r}'(t) = -15\sin 3t\,\mathbf{i} + 15\cos 3t\,\mathbf{j} + 6\mathbf{k}$ and $\mathbf{a}(t) = \mathbf{v}'(t) = -45\cos 3t\,\mathbf{i} - 45\sin 3t\,\mathbf{j}$.

(b) Since the acceleration vector has no component in the z direction, velocity in the z direction must be constant. (In fact, it is; from (a) we see that the component in the z direction of the velocity vector is $6\mathbf{k}$.)

9 (a) If the direction of $\mathbf{v}$ is constant, then the direction of $\mathbf{a}$ is either the same as $\mathbf{v}$ or its opposite. Consider a particle traveling a straight line, first accelerating then decelerating. Since the particle moves in one direction on a straight line, the direction of $\mathbf{v}$ is constant. However, the direction of $\mathbf{a}$ is the same as $\mathbf{v}$ during the acceleration, but is in the opposite direction during the deceleration.

(b) $\mathbf{a}$ need not have constant magnitude even if $\mathbf{v}$ has constant magnitude. Consider $\mathbf{v}(t) = \cos t^2\,\mathbf{i} + \sin t^2\,\mathbf{j}$, where $\|\mathbf{v}(t)\| = 1$ but $\mathbf{a}(t) = -2t\sin t^2\,\mathbf{i} + 2t\cos t^2\,\mathbf{j}$, so $\|\mathbf{a}(t)\| = 2t$, which is not constant.

11 We have $\frac{d}{dt}(|\mathbf{v}|^2) = (\mathbf{v}\cdot\mathbf{v})' = \mathbf{v}\cdot\mathbf{v}' + \mathbf{v}'\cdot\mathbf{v} = 2\mathbf{v}\cdot\mathbf{v}' = 2\mathbf{v}\cdot\mathbf{a}$. But $\mathbf{v}$ and $\mathbf{a}$ are perpendicular, so $\mathbf{v}\cdot\mathbf{a} = 0$. Hence $\frac{d}{dt}(|\mathbf{v}|^2) = 0$, so $\|\mathbf{v}\|^2$ is constant.

Therefore $\|\mathbf{v}\|$ is constant, as claimed.

13 (a) If the particle is confined to a circular orbit, then $\|\mathbf{r}\|$ is constant, so $0 = (\mathbf{r}\cdot\mathbf{r})' = \mathbf{r}\cdot\mathbf{r}' + \mathbf{r}'\cdot\mathbf{r} = \mathbf{r}\cdot\mathbf{v} + \mathbf{v}\cdot\mathbf{r} = 2\mathbf{r}\cdot\mathbf{v}$. Thus $\mathbf{r}\cdot\mathbf{v} = 0$, so $0 = (\mathbf{r}\cdot\mathbf{v})' = \mathbf{r}\cdot\mathbf{v}' + \mathbf{r}'\cdot\mathbf{v} = \mathbf{r}\cdot\mathbf{a} + \mathbf{v}\cdot\mathbf{v}$.

(b) Now $\mathbf{v}\cdot\mathbf{v} = \|\mathbf{v}\|^2 \geq 0$, so it follows from (a) that $\mathbf{r}\cdot\mathbf{a} = -\mathbf{v}\cdot\mathbf{v} \leq 0$.

(c) $\mathbf{F}$ points into the circle.

15 The force required to hold a car on the turn is $ma = \dfrac{mv^2}{r}$, where m is the mass of the car, v its velocity, a its acceleration, and r the radius of the curve. Therefore the force is proportional to the square of the velocity, so four times as great a force is required for a car traveling twice as fast.

17 (a) Gravitational acceleration is inversely proportional to the square of the distance r from the center of the earth, so it is equal to k/r^2 for some constant k. At the surface of the earth, where r is approximately 4000 miles, the acceleration is 32 ft/sec^2 $= \dfrac{32}{5280}$ mi/sec^2 ≈ 0.0061 mi/sec^2. Therefore $\dfrac{k}{4000^2} \approx 0.0061$, so $k \approx (0.0061)(4000^2)$ mi^3/sec^2. The desired expression for acceleration is thus $\dfrac{k}{r^2} \approx (0.0061)(4000/r)^2$, where r is in terms of miles and the result is in terms of miles per second-squared.

(b) By the theorem of this section, the magnitude of the acceleration of a satellite traveling at a constant speed v in an orbit of radius r is given by $a = \dfrac{v^2}{r}$. By (a), we also have $a = \dfrac{k}{r^2}$, so $v^2 = ar = \dfrac{kr}{r^2} = \dfrac{k}{r}$ and $v = \sqrt{k/r}$. A satellite traveling 1000 miles above the surface of the earth has an orbital radius of 5000 miles, so $v \approx \sqrt{(0.0061)(4000^2)/5000} = \sqrt{19.52} \approx 4.42$ mi/sec.

19 If the orbital radius is r miles, then the satellite travels a distance of $2\pi r$ miles in 24 hours, so its speed is $v = \dfrac{2\pi r}{24}$ mi/hr $= \dfrac{\pi r}{43{,}200}$ mi/sec. By Exercise 17(b), $v = \sqrt{k/r} = \sqrt{(0.0061)(4000^2)/r}$ mi/sec. Hence $\dfrac{\pi r}{43{,}200} = \sqrt{\dfrac{(0.0061)(4000^2)}{r}}$, so $r^{3/2} = \dfrac{43{,}200}{\pi}\sqrt{(0.0061)(4000^2)}$ and $r = \sqrt[3]{\left(\dfrac{43{,}200}{\pi}\right)^2(0.0061)(4000^2)} \approx 26{,}400$ mi. The altitude above the earth is $r - 4000 \approx 22{,}400$ mi.

21 As shown in Sec. 8.8, escape velocity from the surface of the earth is approximately $\sqrt{48.8}$ mi/sec. In this section, it was shown that the orbital velocity at the surface of the earth is approximately $\sqrt{24.4}$ mi/sec. Observe that $\sqrt{24.4} = \dfrac{1}{\sqrt{2}}\sqrt{48.8}$, as claimed. (This is a perfectly general result and is not a special characteristic of the earth.)

23 (a) Since the force $\mathbf{F}$ is always directed toward the origin, $\mathbf{F} = g(t)\mathbf{r}$, where $g(t)$ is a negative scalar function. But $\mathbf{F} = m\mathbf{a}$, where m is

mass, so $\mathbf{a} = \frac{1}{m}\mathbf{F} = \frac{1}{m}g(t)\mathbf{r} = f(t)\mathbf{r}$, where

$f(t) = \frac{1}{m}g(t)$ is a scalar function.

(b) $(\mathbf{r} \times \mathbf{v})' = \mathbf{r} \times \mathbf{v}' + \mathbf{r}' \times \mathbf{v} = \mathbf{r} \times \mathbf{a} + \mathbf{v} \times \mathbf{v} = \mathbf{r} \times f(t)\mathbf{r} + \mathbf{0} = f(t)(\mathbf{r} \times \mathbf{r}) = \mathbf{0}$

(c) Since $\mathbf{r} \times \mathbf{v} = \mathbf{C}$ is constant and nonzero, $\mathbf{r}$ and $\mathbf{v}$ must always be perpendicular to the fixed vector $\mathbf{C}$. Hence $\mathbf{r}$ and $\mathbf{v}$ both lie in the plane perpendicular to $\mathbf{C}$.

(d) If $\mathbf{r} \times \mathbf{v} = \mathbf{0}$, then $\mathbf{r}$ and $\mathbf{v}$ must be parallel. Hence the particle travels along a line through the origin.

25 Let $\mathbf{r}(t) = x(t)\mathbf{i} + y(t)\mathbf{j}$. Then $\mathbf{v}(t) = x'(t)\mathbf{i} + y'(t)\mathbf{j}$ and $\mathbf{a}(t) = x''(t)\mathbf{i} + y''(t)$; thus $x''(t) = 2\sec^2 t \tan t$, so $x'(t)$

$= x'(0) + \int_0^t 2\sec^2 u \tan u \, du$

$= 3 + \int_0^t 2\sec u (\sec u \tan u \, du)$

$= 3 + \sec^2 u \big|_0^t = 3 + \sec^2 t - 1 = 2 + \sec^2 t.$

Therefore $x(t) = x(0) + \int_0^t (2 + \sec^2 u)\, du$

$= 0 + (2u + \tan u)\big|_0^t = 2t + \tan t$. Next, $y''(t) = \sec 2t \tan 2t$, so $y'(t) =$

$y'(0) + \int_0^t \sec 2u \tan 2u \, du = 0 + \frac{1}{2}\sec 2u\Big|_0^t = \frac{1}{2}\sec 2t - \frac{1}{2}$. So $y(t)$

$= y(0) + \int_0^t \left(\frac{1}{2}\sec 2u - \frac{1}{2}\right) du$

$= 5 + \left(\frac{1}{4}\ln|\sec 2u + \tan 2u| - \frac{u}{2}\right)\Big|_0^t$

$= 5 + \frac{1}{4}\ln(\sec 2t + \tan 2t) - \frac{t}{2}$. (Since $0 \le t \le \frac{\pi}{4}$, $\sec 2t + \tan 2t > 0$, so we can drop the absolute value bars.) Hence $\mathbf{r}(t) = (2t + \tan t)\mathbf{i} + \left[5 + \frac{1}{4}\ln(\sec 2t + \tan 2t) - \frac{t}{2}\right]\mathbf{j}$.

27 (a) $\mathbf{v}(2) \approx \frac{\mathbf{r}(2.01) - \mathbf{r}(2)}{0.01} = \frac{(3.02\mathbf{i} + 3.99\mathbf{j} + 5.02\mathbf{k}) - (3\mathbf{i} + 4\mathbf{j} + 5\mathbf{k})}{0.01}$

$= \frac{0.02\mathbf{i} - 0.01\mathbf{j} + 0.02\mathbf{k}}{0.01} = 2\mathbf{i} - \mathbf{j} + 2\mathbf{k}$,

and $\mathbf{v}(2.01) \approx \frac{\mathbf{r}(2.02) - \mathbf{r}(2.01)}{0.01}$

$= 100[(3.0403\mathbf{i} + 3.9698\mathbf{j} + 5.0404\mathbf{k}) - (3.02\mathbf{i} + 3.99\mathbf{j} + 5.02\mathbf{k})]$

$= 100(0.0203\mathbf{i} - 0.0202\mathbf{j} + 0.0204\mathbf{k})$

$= 2.03\mathbf{i} - 2.02\mathbf{j} + 2.04\mathbf{k}.$

(b) $\mathbf{a}(2) \approx \frac{\mathbf{v}(2.01) - \mathbf{v}(2)}{0.01} = \frac{(2.03\mathbf{i} - 2.02\mathbf{j} + 2.04\mathbf{k}) - (2\mathbf{i} - \mathbf{j} + 2\mathbf{k})}{0.01} =$

$\frac{0.03\mathbf{i} - 1.02\mathbf{j} + 0.04\mathbf{k}}{0.01} = 3\mathbf{i} - 102\mathbf{j} + 4\mathbf{k}$

29 (a) $\mathbf{v}(1.05) \approx \mathbf{v}(1) + 0.05\mathbf{a}(1) = 3\mathbf{j} + 0.05(3\mathbf{i} + 4\mathbf{j}) = 0.15\mathbf{i} + 3.20\mathbf{j}$;

$\mathbf{v}(1.10) \approx \mathbf{v}(1.05) + 0.05\mathbf{a}(1.05) = 0.15\mathbf{i} + 3.20\mathbf{j} + 0.05(4\mathbf{i} + 5\mathbf{j}) = 0.35\mathbf{i} + 3.45\mathbf{j}$;

$\mathbf{v}(1.20) \approx \mathbf{v}(1.10) + 0.10\mathbf{a}(1.10) = 0.35\mathbf{i} + 3.45\mathbf{j} + 0.10(5\mathbf{i} + 7\mathbf{j}) = 0.85\mathbf{i} + 4.15\mathbf{j}$

(b) $\mathbf{r}(1.05) \approx \mathbf{r}(1) + 0.05\mathbf{v}(1) = \mathbf{i} + \mathbf{j} + 0.05(3\mathbf{j}) = \mathbf{i} + 1.15\mathbf{j}$;

$\mathbf{r}(1.10) \approx \mathbf{r}(1.05) + 0.05\mathbf{v}(1.05)$

$= \mathbf{i} + 1.15\mathbf{j} + 0.05(0.15\mathbf{i} + 3.20\mathbf{j}$

$= 1.0075\mathbf{i} + 1.31\mathbf{j};$

$\mathbf{r}(1.20) \approx \mathbf{r}(1.10) + 0.10\mathbf{v}(1.10)$

$= 1.0075\mathbf{i} + 1.31\mathbf{j} + 0.10(0.35\mathbf{i} + 3.45\mathbf{j})$

$= 1.0425\mathbf{i} + 1.6550\mathbf{j}$

(As with any approximation process, not all of the decimal places can be regarded as significant.)

13.4 The Components of Acceleration

1 $\mathbf{r}(t) = t\mathbf{i} - t^2\mathbf{j}$, so $\mathbf{v}(t) = \mathbf{i} - 2t\mathbf{j}$. Hence $v(t) = \|\mathbf{v}(t)\| = \sqrt{1 + 4t^2}$. Now $\mathbf{v}(1) = \mathbf{i} - 2\mathbf{j}$, $v(1) = \sqrt{1+4} = \sqrt{5}$, $\mathbf{T}(1) = \dfrac{\mathbf{v}(1)}{v(1)} = \dfrac{\mathbf{i} - 2\mathbf{j}}{\sqrt{5}}$, and $\mathbf{T}(t) = \dfrac{\mathbf{i} - 2t\mathbf{j}}{\sqrt{1+4t^2}}$, so $\mathbf{T}'(t) = (-4t(1 + 4t^2)^{-3/2})\mathbf{i} - 2(1 + 4t^2)^{-3/2}\mathbf{j}$, and $\mathbf{T}'(1) = \dfrac{-4\mathbf{i} - 2\mathbf{j}}{5\sqrt{5}}$, so

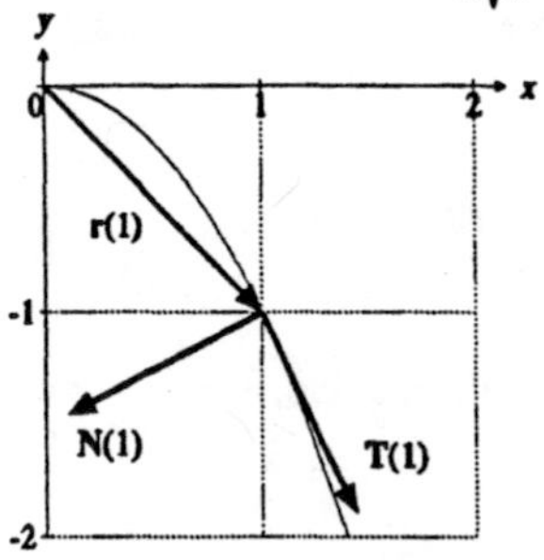

$$\|\mathbf{T}'(1)\| = \frac{\sqrt{(-4)^2 + (-2)^2}}{5\sqrt{5}} = \frac{\sqrt{20}}{5\sqrt{5}} = \frac{2}{5}.$$

Therefore $\mathbf{N}(1) = \dfrac{\mathbf{T}'(1)}{\|\mathbf{T}'(1)\|} = \dfrac{-2\mathbf{i} - \mathbf{j}}{\sqrt{5}}$. Finally, $\mathbf{r} = t\mathbf{i} - t^2\mathbf{j}$, so $x = t$ and $y = -t^2 = -x^2$; hence the path is part of a parabola, as shown in the figure.

3 $\mathbf{r}(t) = \cos t\ \mathbf{i} + \sin t\ \mathbf{j}$, so $\mathbf{v}(t) = -\sin t\ \mathbf{i} + \cos t\ \mathbf{j}$ and $\|\mathbf{v}(t)\| = [(-\sin t)^2 + (\cos t)^2]^{1/2} = 1$, so $\mathbf{T}(t) = \mathbf{v}(t)$. Then $\mathbf{T}'(t) = -\cos t\ \mathbf{i} - \sin t\ \mathbf{j}$, so $\|\mathbf{T}'(t)\| = 1$ and thus $\mathbf{N}(t) = \mathbf{T}'(t)$. We have $\mathbf{r}(1) = \cos 1\ \mathbf{i} + \sin 1\ \mathbf{j}$, $\mathbf{T}(1) = -\sin 1\ \mathbf{i} + \cos 1\ \mathbf{j}$, and $\mathbf{N}(1) = -\cos 1\ \mathbf{i} - \sin 1\ \mathbf{j}$. Observe that $\|\mathbf{r}(t)\| = 1$, so the path of the point $\mathbf{r}(t)$ is a unit circle. In the figure, $\mathbf{r}(1)$ coincides with $\mathbf{N}(1)$, but points in the opposite direction.

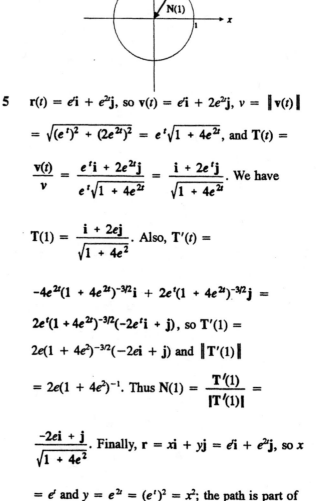

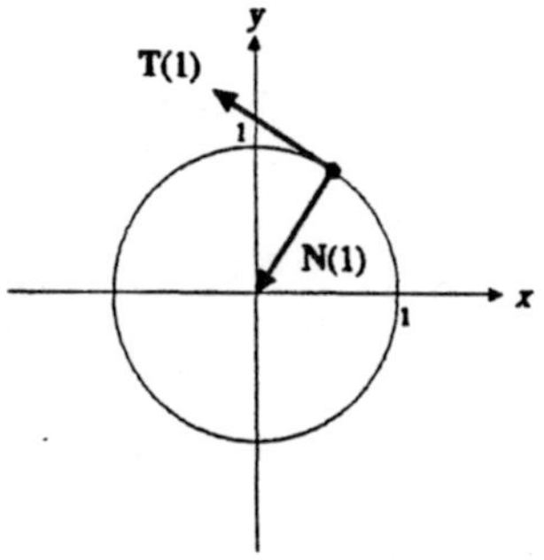

5 $\mathbf{r}(t) = e^t\mathbf{i} + e^{2t}\mathbf{j}$, so $\mathbf{v}(t) = e^t\mathbf{i} + 2e^{2t}\mathbf{j}$, $v = \|\mathbf{v}(t)\| = \sqrt{(e^t)^2 + (2e^{2t})^2} = e^t\sqrt{1 + 4e^{2t}}$, and $\mathbf{T}(t) = \dfrac{\mathbf{v}(t)}{v} = \dfrac{e^t\mathbf{i} + 2e^{2t}\mathbf{j}}{e^t\sqrt{1+4e^{2t}}} = \dfrac{\mathbf{i} + 2e^t\mathbf{j}}{\sqrt{1+4e^{2t}}}$. We have $\mathbf{T}(1) = \dfrac{\mathbf{i} + 2e\mathbf{j}}{\sqrt{1+4e^2}}$. Also, $\mathbf{T}'(t) = -4e^{2t}(1 + 4e^{2t})^{-3/2}\mathbf{i} + 2e^t(1 + 4e^{2t})^{-3/2}\mathbf{j} = 2e^t(1 + 4e^{2t})^{-3/2}(-2e^t\mathbf{i} + \mathbf{j})$, so $\mathbf{T}'(1) = 2e(1 + 4e^2)^{-3/2}(-2e\mathbf{i} + \mathbf{j})$ and $\|\mathbf{T}'(1)\| = 2e(1 + 4e^2)^{-1}$. Thus $\mathbf{N}(1) = \dfrac{\mathbf{T}'(1)}{|\mathbf{T}'(1)|} = \dfrac{-2e\mathbf{i} + \mathbf{j}}{\sqrt{1+4e^2}}$. Finally, $\mathbf{r} = x\mathbf{i} + y\mathbf{j} = e^t\mathbf{i} + e^{2t}\mathbf{j}$, so $x = e^t$ and $y = e^{2t} = (e^t)^2 = x^2$; the path is part of

the parabola $y = x^2$.

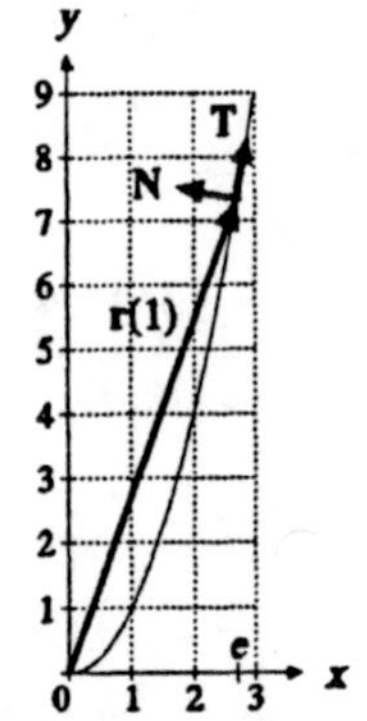

7 The statement is true. By definition, $\mathbf{T}(t) = \dfrac{\mathbf{v}(t)}{\|\mathbf{v}(t)\|}$, so **T** and **v** point in the same direction.

Since **v** points in the direction of travel, so does **T**.

9 (a) $\mathbf{r}(t) = x\mathbf{i} + y\mathbf{j} = t\mathbf{i} + t^2\mathbf{j}$, so $x = t$ and $y = t^2 = x^2$. The path of the particle is a parabola, namely, $y = x^2$, the right half of which is graphed in the accompanying figure.

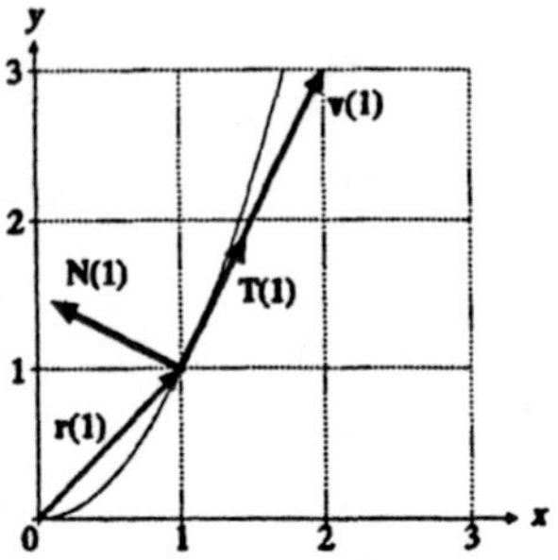

(b) $\mathbf{v}(t) = \mathbf{i} + 2t\mathbf{j}$, so $v = \|\mathbf{v}(t)\| = \sqrt{1 + 4t^2}$. We have $\mathbf{r}(1) = \mathbf{i} + \mathbf{j}$ and $\mathbf{v}(1) = \mathbf{i} + 2\mathbf{j}$, which are plotted in the graph. Now $\mathbf{T}(t) = \dfrac{\mathbf{v}(t)}{v} = \dfrac{\mathbf{i} + 2t\mathbf{j}}{\sqrt{1 + 4t^2}}$, so $\mathbf{T}'(t) =$

$[-4t(1 + 4t^2)^{-3/2}]\mathbf{i} + [2(1 + 4t^2)^{-3/2}]\mathbf{j}$,

$\|\mathbf{T}'(t)\| = (1 + 4t^2)^{-3/2}((-4t)^2 + 2^2)^{1/2}$

$= 2(1 + 4t^2)^{-1}$, and $\mathbf{N}(t) = \dfrac{\mathbf{T}'(t)}{\|\mathbf{T}'(t)\|} =$

$\dfrac{-2t\mathbf{i} + \mathbf{j}}{\sqrt{1 + 4t^2}}$. Therefore $\mathbf{T}(1) = \dfrac{\mathbf{i} + 2\mathbf{j}}{\sqrt{5}}$ and

$$\mathbf{N}(1) = \frac{-2\mathbf{i} + \mathbf{j}}{\sqrt{5}}.$$

(c) See the graph above.

11 By definition, $\mathbf{v}(1)\cdot\mathbf{T}(1) = \|\mathbf{v}(1)\|\,\|\mathbf{T}(1)\|\cos\theta$, where θ is the angle between $\mathbf{v}(1)$ and $\mathbf{T}(1)$. But $\|\mathbf{v}(1)\| = 3$, $\|\mathbf{T}(1)\| = 1$, and $\theta = 0$, so $\mathbf{v}(1)\cdot\mathbf{T}(1) = 3$.

13 By Theorem 3, $a_N = \dfrac{v^2}{r} = \dfrac{4^2}{5} = \dfrac{16}{5}$ and $a_T = \dfrac{d^2s}{dt^2} = \dfrac{dv}{dt} = 3$.

15 $a_N = \dfrac{v^2}{r} = \dfrac{1^2}{3} = \dfrac{1}{3}$, and $a_T = \dfrac{d^2s}{dt^2} = 0$.

17 The path is circular so the radius of curvature equals the radius: $r = 5$. Now $\mathbf{r}(t) = 5\cos 3\pi t\,\mathbf{i} + 5\sin 3\pi t\,\mathbf{j}$, so $\mathbf{v}(t) = -15\pi\sin 3\pi t\,\mathbf{i} + 15\pi\cos 3\pi t\,\mathbf{j}$ and $v = 15\pi$. We have $a_T = \dfrac{d^2s}{dt^2} = \dfrac{dv}{dt} = 0$ and $a_N = \dfrac{v^2}{r} =$

$\dfrac{225\pi^2}{5} = 45\pi^2$. Also, $\kappa = \dfrac{1}{r} = \dfrac{1}{5}$.

19 (a) $\mathbf{a}\cdot\mathbf{T} = \left(\dfrac{d^2s}{dt^2}\mathbf{T} + \dfrac{v^2}{r}\mathbf{N}\right)\cdot\mathbf{T}$

$= \dfrac{d^2s}{dt^2}(\mathbf{T}\cdot\mathbf{T}) + \dfrac{v^2}{r}(\mathbf{N}\cdot\mathbf{T}) = \dfrac{d^2s}{dt^2}$

(b) $\mathbf{a}\cdot\mathbf{N} = \left(\dfrac{d^2s}{dt^2}\mathbf{T} + \dfrac{v^2}{r}\mathbf{N}\right)\cdot\mathbf{N}$

$$= \frac{d^2s}{dt^2}(\mathbf{T}\cdot\mathbf{N}) + \frac{v^2}{r}(\mathbf{N}\cdot\mathbf{N}) = \frac{v^2}{r}$$

21 $\mathbf{r}(t) = t^2\mathbf{i} + t^3\mathbf{j}$, so $\mathbf{v}(t) = 2t\mathbf{i} + 3t^2\mathbf{j}$ and $v(t) = \sqrt{4t^2 + 9t^4}$. Therefore $a_T = \frac{dv}{dt} =$

$$\frac{1}{2}(4t^2 + 9t^4)^{-1/2}(8t + 36t^3) = \frac{2t(2 + 9t^2)}{(4t^2 + 9t^4)^{1/2}}.$$

Furthermore, $\mathbf{a} = 2\mathbf{i} + 6t\mathbf{j}$, so $\|\mathbf{a}\|^2 = 4 + 36t^2$; thus $a_N^2 = |\mathbf{a}|^2 - a_T^2 = (4 + 36t^2) -$

$$\frac{4t^2(2 + 9t^2)^2}{4t^2 + 9t^4} =$$

$$\frac{16t^2 + 36t^4 + 144t^4 + 324t^6 - (16t^2 + 144t^4 + 324t^6)}{4t^2 + 9t^4}$$

$$= \frac{36t^4}{4t^2 + 9t^4} = \frac{36t^2}{4 + 9t^2}, \text{ so } a_N = \frac{6|t|}{\sqrt{4 + 9t^2}}.$$

Since $a_N = \frac{v^2}{r} = v^2\kappa$, $\kappa = \frac{a_N}{v^2} = \frac{6}{|t|(4 + 9t^2)^{3/2}}$.

23 Since the particle is traveling northeast, that is the direction of **T**. It is veering to the left, so **N**, which is perpendicular to **T**, must point to the northwest. the particle is speeding up, so a_T is positive; a_N is always positive, so **a** lies somewhere between **T** and **N**. One possible example is shown.

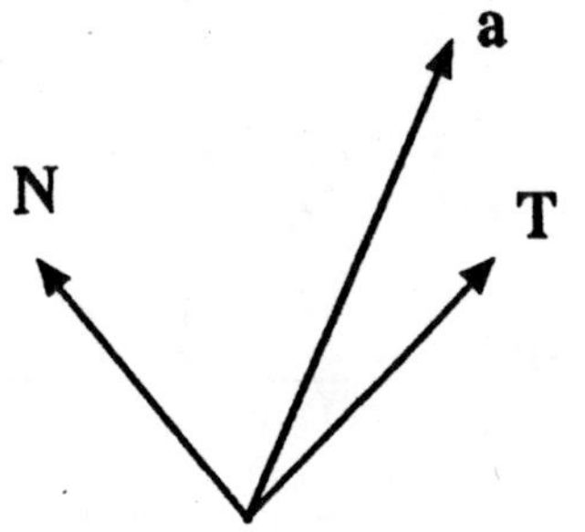

25 (a)

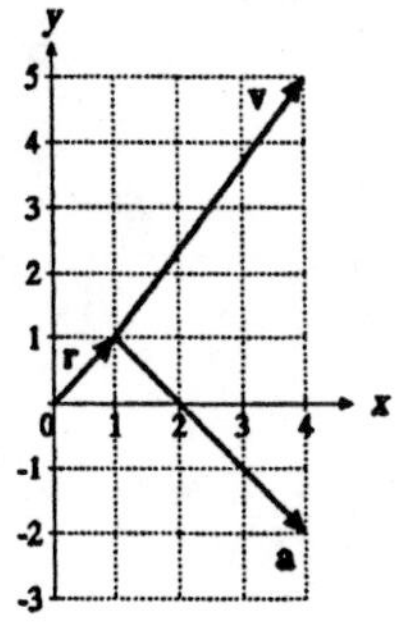

(b) $\mathbf{a}\cdot\mathbf{v} = 3\cdot3 + 4(-3) = -3 < 0$, so the particle is slowing down.

(d) $v = \|\mathbf{v}\| = \sqrt{3^2 + 4^2} = 5$, so $\mathbf{T} = \frac{\mathbf{v}}{|\mathbf{v}|}$

$= \frac{3}{5}\mathbf{i} + \frac{4}{5}\mathbf{j}$ and $a_T = \mathbf{a}\cdot\mathbf{T} =$

$(3\mathbf{i} - 3\mathbf{j})\cdot\left(\frac{3}{5}\mathbf{i} + \frac{4}{5}\mathbf{j}\right) = -\frac{3}{5}$. Hence $a_N =$

$\sqrt{|\mathbf{a}|^2 - a_T^2} = \sqrt{18 - \frac{9}{25}} = \frac{21}{5}$. Now, a_N

$= \frac{v^2}{r} = v^2\kappa$, so $\kappa = \frac{a_N}{v^2} = \frac{21/5}{5^2} = \frac{21}{125}$.

27 (a) $\mathbf{a}(t) = 2\mathbf{i} + 6t^2\mathbf{j}$, so $\mathbf{a}(1/\sqrt{6}) = 2\mathbf{i} + 6\cdot\frac{1}{6}\mathbf{j}$

$= 2\mathbf{i} + \mathbf{j}$.

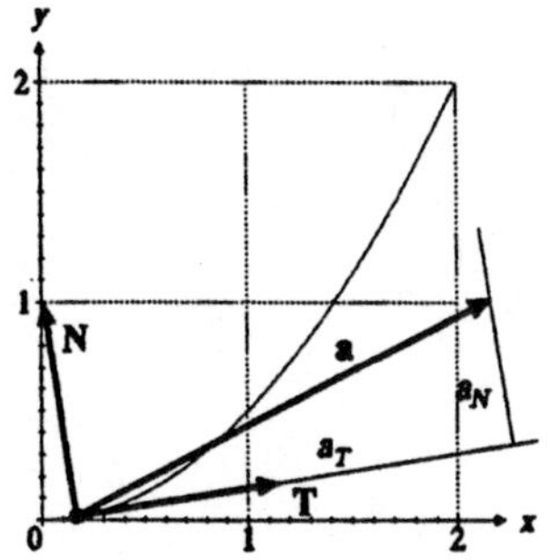

(c) At $t = \frac{1}{\sqrt{6}}$, $\mathbf{v} = 2t\mathbf{i} + 2t^3\mathbf{j} = \frac{2}{\sqrt{6}}\mathbf{i} + \frac{2}{6\sqrt{6}}\mathbf{j}$;

the magnitude of **v** is $\frac{2}{\sqrt{6}}\sqrt{1^2 + \left(\frac{1}{6}\right)^2} =$

$\frac{2\sqrt{37}}{6\sqrt{6}}$, so $\mathbf{T} = \frac{6\sqrt{6}}{2\sqrt{37}}\left(\frac{2}{\sqrt{6}}\mathbf{i} + \frac{2}{6\sqrt{6}}\mathbf{j}\right) =$

$\frac{6}{\sqrt{37}}\mathbf{i} + \frac{1}{\sqrt{37}}\mathbf{j}$. Hence $a_T = \mathbf{a}\cdot\mathbf{T} =$

$2\cdot\frac{6}{\sqrt{37}} + 1\cdot\frac{1}{\sqrt{37}} = \frac{13}{\sqrt{37}}$. Finally, $a_N =$

$\sqrt{|\mathbf{a}|^2 - a_T^2} = \sqrt{(2^2 + 1^2) - \frac{169}{37}} = \sqrt{\frac{16}{37}}$

$= \frac{4}{\sqrt{37}}$.

(d) $a_T \approx 2.13719$, $a_N \approx 0.65760$

29 (a) $\mathbf{r}(t) = e^t\mathbf{i} + e^{2t}\mathbf{j}$; $\mathbf{v}(t) = e^t\mathbf{i} + 2e^{2t}\mathbf{j}$ and $\mathbf{a}(t) =$ $e^t\mathbf{i} + 4e^{2t}\mathbf{j}$.

(b) From (a), $v(t) = e^t\sqrt{1 + 4e^{2t}}$, so $\frac{dv}{dt} =$

$e^t\left[\frac{1}{2}(1 + 4e^{2t})^{-1/2}(8e^{2t})\right] + e^t(1 + 4e^{2t})^{1/2}$

$= \frac{e^t(4e^{2t}) + e^t(1 + 4e^{2t})}{(1 + 4e^{2t})^{1/2}} = \frac{e^t(1 + 8e^{2t})}{(1 + 4e^{2t})^{1/2}}$.

(c) $a_T = \frac{dv}{dt} = \frac{e^t(1 + 8e^{2t})}{(1 + 4e^{2t})^{1/2}}$; $a_N^2 =$

$\|\mathbf{a}\|^2 - a_T^2 = (e^{2t} + 16e^{4t}) - \frac{e^{2t}(1 + 8e^{2t})^2}{1 + 4e^{2t}}$

$=$

$\frac{e^{2t} + 16e^{4t} + 4e^{4t} + 64e^{6t} - (e^{2t} + 16e^{4t} + 64e^{6t})}{1 + 4e^{2t}}$

$= \frac{4e^{4t}}{1 + 4e^{2t}}$, so $a_N = \frac{2e^{2t}}{(1 + 4e^{2t})^{1/2}}$. $\kappa = \frac{a_N}{v^2}$

$= \frac{2e^{2t}/\sqrt{1 + 4e^{2t}}}{e^{2t}(1 + 4e^{2t})} = \frac{2}{(1 + 4e^{2t})^{3/2}}$.

31 By Sec. 13.2, if $\|\mathbf{v}\|$ is constant then **v** and **v**′ are perpendicular; therefore, $\mathbf{a}\cdot\mathbf{v} = 0$.

33 We know from Sec. 13.2 that **N**′ is perpendicular to **N** because **N** has constant length. Since **N**′ and **T** are both perpendicular to **N**, they must be parallel to each other. That is, $\mathbf{N}' = c\mathbf{T}$ for some scalar c.

35 Let $\mathbf{a} = \frac{d^2x}{dt^2}\mathbf{i} + \frac{d^2y}{dt^2}\mathbf{j}$. Then from $\|\mathbf{a}\|^2 =$

$a_T^2 + a_N^2$, we have $a_N^2 = |\mathbf{a}|^2 - a_T^2$ or $\left(\frac{v^2}{r}\right)^2 =$

$\left(\frac{d^2x}{dt^2}\right)^2 + \left(\frac{d^2y}{dt^2}\right)^2 - \left(\frac{d^2s}{dt^2}\right)^2$; so $\frac{v^4}{r^2} =$

$\left(\frac{d^2x}{dt^2}\right)^2 + \left(\frac{d^2y}{dt^2}\right)^2 - \left(\frac{d^2s}{dt^2}\right)^2$.

37 $\mathbf{T} = \frac{d\mathbf{r}}{ds} = \frac{d\mathbf{r}/dt}{ds/dt} = \frac{d\mathbf{r}/dt}{|d\mathbf{r}/dt|}$, which must be a unit vector since it is a vector divided by its own magnitude. Similarly, $\mathbf{N} = \frac{d\mathbf{T}/ds}{|d\mathbf{T}/ds|}$ must also be a unit vector.

39 (a) $\mathbf{T}(t) = \frac{-6 \sin 2t\, \mathbf{i} + 6 \cos 2t\, \mathbf{j} + 5\mathbf{k}}{\sqrt{(-6 \sin 2t)^2 + (6 \cos 2t)^2 + 5^2}}$

$= \frac{1}{\sqrt{61}}(-6 \sin 2t\, \mathbf{i} + 6 \cos 2t\, \mathbf{j} + 5\mathbf{k})$,

$\mathbf{N}(t) = \frac{-12 \cos 2t\, \mathbf{i} - 12 \sin 2t\, \mathbf{j}}{\sqrt{(-12 \cos 2t)^2 + (-12 \sin 2t)^2}} =$

$-\cos 2t\, \mathbf{i} - \sin 2t\, \mathbf{j}$, $\mathbf{B}(t) = \mathbf{T}(t) \times \mathbf{N}(t)$

$= -\frac{1}{\sqrt{61}}(-6 \sin 2t\ \mathbf{i} + 6 \cos 2t\ \mathbf{j} + 5\mathbf{k}) \times$

$(\cos 2t\ \mathbf{i} + \sin 2t\ \mathbf{j})$

$= -\frac{1}{\sqrt{61}}(-5 \sin 2t\ \mathbf{i} + 5 \cos 2t\ \mathbf{j} - 6\mathbf{k})$

$= \frac{1}{\sqrt{61}}(5 \sin 2t\ \mathbf{i} - 5 \cos 2t\ \mathbf{j} + 6\mathbf{k}).$

(b) $\mathbf{T}(\pi/4) = \frac{1}{\sqrt{61}}(-6\mathbf{i} + 5\mathbf{k})$

$\mathbf{N}(\pi/4) = -\mathbf{j}$

$\mathbf{B}(\pi/4) = \frac{1}{\sqrt{61}}(5\mathbf{i} + 6\mathbf{k})$

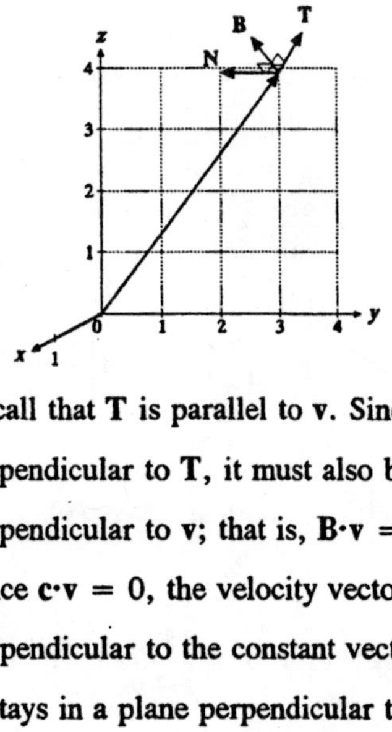

41 (a) Recall that **T** is parallel to **v**. Since **B** is perpendicular to **T**, it must also be perpendicular to **v**; that is, $\mathbf{B}\cdot\mathbf{v} = 0$.

(b) Since $\mathbf{c}\cdot\mathbf{v} = 0$, the velocity vector is always perpendicular to the constant vector **c**. Hence **v** stays in a plane perpendicular to **c** and therefore the curve must also lie in some plane perpendicular to **c**.

43 (a)

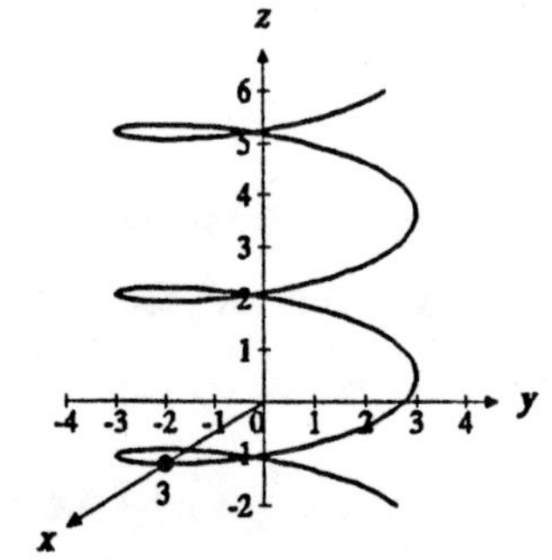

(b) $\mathbf{v} = (3 \cos 2\pi t)'\mathbf{i} + (3 \sin 2\pi t)'\mathbf{j} + (3t)'\mathbf{k}$

$= -6\pi \sin 2\pi t\ \mathbf{i} + 6\pi \cos 2\pi t\ \mathbf{j} + 3\mathbf{k};$

$\mathbf{a} = (-6\pi \sin 2\pi t)'\mathbf{i} + (6\pi \cos 2\pi t)'\mathbf{j} + (3)'\mathbf{k}$

$= -12\pi^2 \cos 2\pi t\ \mathbf{i} - 12\pi^2 \sin 2\pi t\ \mathbf{j}$

(c) First, $v = \|\mathbf{v}\|$

$= \sqrt{(-6\pi \sin 2\pi t)^2 + (6\pi \cos 2\pi t)^2 + 3^2} =$

$\sqrt{36\pi^2 + 9} = 3\sqrt{4\pi^2 + 1}$. Since v is

constant, $\kappa = \frac{1}{v}\left\|\frac{d\mathbf{T}}{dt}\right\| = \frac{1}{v}\left\|\frac{d}{dt}\left(\frac{\mathbf{v}}{v}\right)\right\|$

$= \frac{1}{v^2}\left\|\frac{d\mathbf{v}}{dt}\right\| = \frac{1}{v^2}\|\mathbf{a}\|$. Since $\|\mathbf{a}\| =$

$\sqrt{(-12\pi^2 \cos 2\pi t)^2 + (-12\pi^2 \sin 2\pi t)^2} =$

$12\pi^2$, $\kappa = \frac{12\pi^2}{9(4\pi^2 + 1)} = \frac{4\pi^2}{3(4\pi^2 + 1)}.$

(d) $a_T = \frac{d^2s}{dt^2} = \frac{d}{dt}\left(\frac{ds}{dt}\right) = \frac{dv}{dt} = 0$, since v is

constant. $a_N = \sqrt{\|\mathbf{a}\|^2 - a_T^2} = \sqrt{\|\mathbf{a}\|^2 - 0} =$

$\|\mathbf{a}\| = 12\pi^2.$

13.5 Newton's Law Implies Kepler's Three Laws

1 For small Δt, the area swept out by the position vector is approximately half the area of the parallelogram defined by $\mathbf{r}(t)$ and $\Delta\mathbf{r}$; hence the change in area, ΔA, is given by $\Delta A \approx \frac{1}{2}\|\mathbf{r} \times \Delta\mathbf{r}\|$.

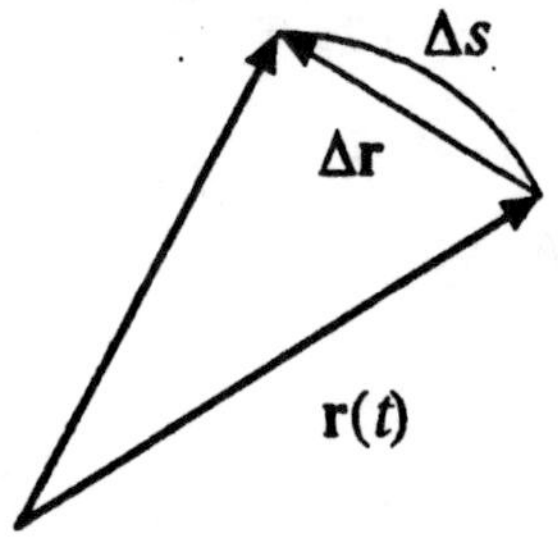

2 By Exercise 1, $\frac{\Delta A}{\Delta t} \approx \frac{1}{2}\frac{\|\mathbf{r} \times \Delta\mathbf{r}\|}{\Delta t} = \frac{1}{2}\left\|\mathbf{r} \times \frac{\Delta\mathbf{r}}{\Delta t}\right\|$, so $\frac{dA}{dt} = \lim_{\Delta t \to 0} \frac{\Delta A}{\Delta t} = \frac{1}{2}\left\|\mathbf{r} \times \frac{d\mathbf{r}}{dt}\right\|$.

3 According to (2), $\mathbf{a} = -q\mathbf{u}/r^2$. Now $\frac{d}{dt}(\mathbf{r} \times \mathbf{v})$ $= \mathbf{r}' \times \mathbf{v} + \mathbf{r} \times \mathbf{v}' = \mathbf{v} \times \mathbf{v} + \mathbf{r} \times \mathbf{a} =$ $0 + \mathbf{r} \times \left(\frac{-q\mathbf{u}}{r^2}\right) = \mathbf{r} \times \left(\frac{-q\mathbf{r}}{r^3}\right) = -\frac{q}{r^3}(\mathbf{r} \times \mathbf{r}) =$ $\mathbf{0}$. Therefore $\mathbf{r} \times \mathbf{v}$ does not vary with respect to time; it is constant.

4 $h = \|h\mathbf{k}\| = \|\mathbf{r} \times \mathbf{v}\| = \left\|\mathbf{r} \times \frac{d\mathbf{r}}{dt}\right\| = 2\frac{dA}{dt}$, by the result of Exercise 2.

5 Since $\mathbf{r}$ and $\mathbf{v}$ are both perpendicular to $\mathbf{k}$, the motion of the planet lies in a plane perpendicular to $\mathbf{k}$. The vector $\mathbf{r}$ lies along the line connecting the sun and the planet, so the sun lies in the plane.

6 (a) From polar coordinates we know that Area $= \int_{\theta_0}^{\theta} \frac{1}{2}r^2\, d\theta$. Now θ is a function of t, so we may write $d\theta = \theta'(t)\, dt$. Let $\theta_0 = \theta(t_0)$, so

Area $= \int_{t_0}^{t} \frac{1}{2}r^2 \cdot \frac{d\theta}{dt}\, dt$.

(b) Denote the second integral in (a) by $A(t)$. By the first fundamental theorem of calculus, $\frac{dA}{dt}$ $= \frac{1}{2}r^2\frac{d\theta}{dt}$.

7 By Exercises 4 and 6, $h = 2\frac{dA}{dt} = r^2\frac{d\theta}{dt}$.

Furthermore, $\mathbf{r} \times \mathbf{v} = h\mathbf{k} = r^2\frac{d\theta}{dt}\mathbf{k}$, as claimed.

8 $\mathbf{u}$ is a unit vector, so its magnitude is constant and $\dot{\mathbf{u}}$ is therefore perpendicular to $\mathbf{u}$. By the chain rule, $\dot{\mathbf{u}} = \frac{d\mathbf{u}}{d\theta} \cdot \frac{d\theta}{dt}$.

9 $\mathbf{u} \times \dot{\mathbf{u}} = \frac{\mathbf{r}}{r} \times \frac{d}{dt}\left(\frac{\mathbf{r}}{r}\right) = \frac{\mathbf{r}}{r} \times \left(\frac{r\dot{\mathbf{r}} - \mathbf{r}\dot{r}}{r^2}\right)$

$= \frac{\mathbf{r}}{r} \times \left(\frac{\mathbf{v}}{r} - \frac{\dot{r}}{r^2}\mathbf{r}\right) = \frac{1}{r^2}(\mathbf{r} \times \mathbf{v}) - \frac{\dot{r}}{r^3}(\mathbf{r} \times \mathbf{r})$

$= \frac{1}{r^2}(\mathbf{r} \times \mathbf{v}) + \mathbf{0} = \frac{1}{r^2}(h\mathbf{k})$, so $h\mathbf{k} = r^2(\mathbf{u} \times \dot{\mathbf{u}})$.

10 $\mathbf{a} \times h\mathbf{k} = \mathbf{a} \times r^2(\mathbf{u} \times \dot{\mathbf{u}}) = r^2(\mathbf{a} \times (\mathbf{u} \times \dot{\mathbf{u}}))$ $= r^2[(\mathbf{a} \cdot \dot{\mathbf{u}})\mathbf{u} - (\mathbf{a} \cdot \mathbf{u})\dot{\mathbf{u}}]$; since $\mathbf{a} = -q\mathbf{u}/r^2$, $\mathbf{a} \cdot \mathbf{u} = -q/r^2$ and $\mathbf{a} \cdot \dot{\mathbf{u}} = 0$ (because $\dot{\mathbf{u}}$ is perpendicular to $\mathbf{u}$). Thus $\mathbf{a} \times h\mathbf{k} = r^2\left[0 - \left(-\frac{q}{r^2}\right)\dot{\mathbf{u}}\right] = q\dot{\mathbf{u}}$.

11 Using the result of Exercise 10, $\frac{d}{dt}(\mathbf{v} \times h\mathbf{k}) =$

$\frac{d\mathbf{v}}{dt} \times h\mathbf{k} = \mathbf{a} \times h\mathbf{k} = q\dot{\mathbf{u}} = q\frac{d\mathbf{u}}{dt} = \frac{d}{dt}(q\mathbf{u})$.

Therefore $\frac{d}{dt}(\mathbf{v} \times h\mathbf{k} - q\mathbf{u}) = 0$. It follows that

$\mathbf{v} \times h\mathbf{k} - q\mathbf{u} = \mathbf{C}$ for some constant vector $\mathbf{C}$, so $\mathbf{v} \times h\mathbf{k} = q\mathbf{u} + \mathbf{C}$.

12 (a) $(\mathbf{r} \times \mathbf{v})\cdot h\mathbf{k} = h\mathbf{k}\cdot h\mathbf{k} = h^2(\mathbf{k}\cdot\mathbf{k}) = h^2$

(b) $\mathbf{r}\cdot(\mathbf{v} \times h\mathbf{k}) = \mathbf{r}\cdot(q\mathbf{u} + \mathbf{C}) = q\mathbf{r}\cdot\mathbf{u} + \mathbf{r}\cdot\mathbf{C}$ $= qr + \mathbf{r}\cdot\mathbf{C}$

(c) Let θ be the angle between $\mathbf{r}$ and $\mathbf{C}$; then by the equality of the scalar triple products in (a) and (b), $h^2 = qr + \mathbf{r}\cdot\mathbf{C} = qr + rc\cos\theta$, where $c = \|\mathbf{C}\|$.

13 Working directly from the expression in part (c) of Exercise 12, we have $h^2 = qr + rc\cos\theta = qr + c(r\cos\theta) = q\sqrt{x^2 + y^2} + cx$, so $q\sqrt{x^2 + y^2} = h^2 - cx$, $q^2(x^2 + y^2) = h^4 - 2h^2cx + c^2x^2$, $q^2x^2 - c^2x^2 + q^2y^2 + 2h^2cx = h^4$, $(q^2 - c^2)x^2 + 2h^2cx + q^2y^2 = h^4$, which is the equation of a conic section. (It follows from Sec. G.4 that the conic expressed in Equation (5) has one focus at the origin.)

14 The area enclosed by the orbit is πab, while the orbital period is T. Therefore the average rate at which area is swept out by the position vector is $\frac{\pi ab}{T}$. But it was established in Exercise 4 that $\frac{dA}{dt}$ is constant; in fact, $\frac{dA}{dt} = \frac{h}{2}$. The instantaneous rate of change is therefore equal to the average rate of change, so $\frac{h}{2} = \frac{\pi ab}{T}$ or $T = \frac{2\pi ab}{h}$, as claimed.

15 Equation (5) says that $r(\theta) = \frac{h^2}{q + c\cos\theta}$. In the figure we see that the planet is at B when $\theta = \frac{\pi}{2}$; that is, $f = r\left(\frac{\pi}{2}\right) = \frac{h^2}{q + c\cdot 0} = \frac{h^2}{q}$.

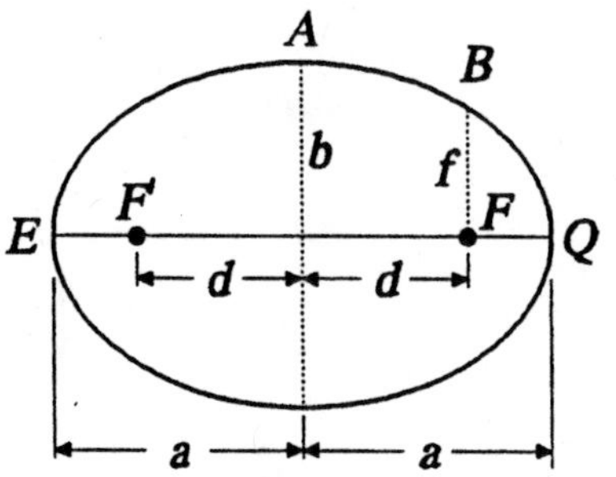

16 (a) (See the figure in Exercise 15.) $\overline{F'A} + \overline{FA} = 2a$, and $\overline{F'A} = \overline{FA}$, so $\overline{FA} = a$. By the Pythagorean theorem then, $a^2 = b^2 + d^2$.

(b) B is a point on the ellipse, so $\overline{F'B} + \overline{FB} = 2a$. Since $\overline{FB} = f$, we have $\overline{F'B} = 2a - f$. Applying the Pythagorean theorem to the triangle $F'BF$, we have $(2a - f)^2 = f^2 + (2d)^2$, or $4a^2 - 4af + f^2 = f^2 + 4d^2$, so $4a^2 - 4af = 4d^2$, or $a^2 - af = d^2$.

(c) From (a) we have $a^2 = b^2 + d^2$, while from (b) we have $a^2 = d^2 + af$. Equating the two expressions for a^2 and subtracting d^2 from both sides results in $b^2 = af$.

17 $b^2 = af = a(h^2/q) = ah^2/q$

18 From (6) we have $T^2 = \frac{4\pi^2a^2b^2}{h^2} = \frac{4\pi^2a^2}{h^2}\cdot\frac{ah^2}{q} = \frac{4\pi^2a^3}{q}$, so $\frac{T^2}{a^3} = \frac{4\pi^2}{q}$; that is, the square of the period is proportional to the cube of the semimajor axis. This is Kepler's third law.

13.S Guide Quiz

1 (a) $G'(t) = \lim_{\Delta t \to 0} \dfrac{G(t + \Delta t) - G(t)}{\Delta t}$

(b) $G'(1) \approx \dfrac{G(1.01) - G(1)}{0.01}$

$= \dfrac{3.02\mathbf{i} + 4.03\mathbf{j} - (3\mathbf{i} + 4\mathbf{j})}{0.01}$

$= \dfrac{0.02\mathbf{i} + 0.03\mathbf{j}}{0.01} = 2\mathbf{i} + 3\mathbf{j}$

2 See Sec. 13.1.

3 (a) $G'(s) = \dfrac{dx}{dx}\mathbf{i} + \dfrac{dy}{ds}\mathbf{j}$, so $\|G'(s)\|$

$= \sqrt{\left(\dfrac{dx}{ds}\right)^2 + \left(\dfrac{dy}{ds}\right)^2} = \dfrac{\sqrt{dx^2 + dy^2}}{ds} = \dfrac{ds}{ds}$

$= 1$. (See Sec. 9.4 for the planar case.)

(b) See Exercise 21 of Sec. 13.1 and also Sec. 13.4.

4 See Sec. 13.4.

5 See Sec. 13.4.

6 (a) $\kappa = \dfrac{|\mathbf{v} \times \mathbf{a}|}{\|\mathbf{v}|^3} = \dfrac{|(3\mathbf{i} + 4\mathbf{j}) \times (2\mathbf{i} + \mathbf{j})|}{|3\mathbf{i} + 4\mathbf{j}|^3}$

$= \dfrac{|-5\mathbf{k}|}{\left(\sqrt{3^2 + 4^2}\right)^3} = \dfrac{5}{5^3} = \dfrac{1}{25}$

(b) $r = \dfrac{1}{\kappa} = \dfrac{1}{1/25} = 25$

(c) $a_T = \mathbf{a} \cdot \mathbf{T} = \mathbf{a} \cdot \dfrac{\mathbf{v}}{|\mathbf{v}|} = \dfrac{(2\mathbf{i} + \mathbf{j}) \cdot (3\mathbf{i} + 4\mathbf{j})}{|3\mathbf{i} + 4\mathbf{j}|}$

$= \dfrac{2 \cdot 3 + 4 \cdot 1}{5} = 2$

(d) $a_T > 0$, so the particle is speeding up.

(e) $a_N = \dfrac{v^2}{r} = \dfrac{5^2}{25} = 1$

7 (a) $\mathbf{v} = \mathbf{r}' = 2t\mathbf{i} + \mathbf{j}$ so $\mathbf{T} = \dfrac{\mathbf{v}}{|\mathbf{v}|} = \dfrac{2t\mathbf{i} + \mathbf{j}}{\sqrt{4t^2 + 1}}$.

When $t = 1$, $\mathbf{T} = \dfrac{2\mathbf{i} + \mathbf{j}}{\sqrt{5}}$.

(b) $\mathbf{T}' = \dfrac{2\mathbf{i} - 4t\mathbf{j}}{(4t^2 + 1)^{3/2}}$, so $\mathbf{N} = \dfrac{\mathbf{T}'}{|\mathbf{T}'|} =$

$\dfrac{\mathbf{i} - 2t\mathbf{j}}{\sqrt{4t^2 + 1}}$. When $t = 1$, $\mathbf{N} = \dfrac{\mathbf{i} - 2\mathbf{j}}{\sqrt{5}}$.

(c) From (a), $\mathbf{v} = 2t\mathbf{i} + \mathbf{j}$, which equals $2\mathbf{i} + \mathbf{j}$ when $t = 1$.

(d) $v = \|\mathbf{v}\| = \sqrt{4t^2 + 1}$, so $\dfrac{d^2s}{dt^2} = \dfrac{dv}{dt} =$

$\dfrac{4t}{\sqrt{4t^2 + 1}} = \dfrac{4}{\sqrt{5}}$ when $t = 1$.

(e) $\mathbf{a} = \mathbf{v}' = 2\mathbf{i}$, so $\kappa = \dfrac{|\mathbf{v} \times \mathbf{a}|}{\|\mathbf{v}|^3} =$

$\dfrac{|(2t\mathbf{i} + \mathbf{j}) \times 2\mathbf{i}|}{(4t^2 + 1)^{3/2}} = \dfrac{|-2\mathbf{k}|}{(4t^2 + 1)^{3/2}} =$

$\dfrac{2}{(4t^2 + 1)^{3/2}} = \dfrac{2}{5\sqrt{5}}$ when $t = 1$.

(f) $r = \dfrac{1}{\kappa} = \dfrac{5\sqrt{5}}{2}$

8 See Sec. 13.4.

13.S Review Exercises

1 (a) Observe that $\mathbf{G}(t) = \cosh t\,\mathbf{i} + \sinh t\,\mathbf{j}$. Thus $x = \cosh t$ and $y = \sinh t$. We have $x^2 - y^2 = \cosh^2 t - \sinh^2 t = 1$, so the particle moves on the hyperbola $x^2 - y^2 = 1$.

(b) $\mathbf{v}(t) = \sinh t\,\mathbf{i} + \cosh t\,\mathbf{j}$ and $\mathbf{a}(t) = \cosh t\,\mathbf{i} + \sinh t\,\mathbf{j}$; observe that $v = \|\mathbf{v}(t)\| = \|\mathbf{a}(t)\| = \sqrt{\sinh^2 t + \cosh^2 t} = \sqrt{\cosh 2t}$, by Exercise 26 of Sec. 6.9. Then $a_T = \mathbf{a}\cdot\mathbf{T} = \dfrac{\mathbf{a}\cdot\mathbf{v}}{v}$

$= \dfrac{1}{\sqrt{\cosh 2t}}(2\cosh t \sinh t) = \dfrac{\sinh 2t}{\sqrt{\cosh 2t}}$, and a_N^2

$= \|\mathbf{a}\|^2 - a_T^2 = \cosh 2t - \dfrac{\sinh^2 2t}{\cosh 2t} =$

$\dfrac{\cosh^2 2t - \sinh^2 2t}{\cosh 2t} = \dfrac{1}{\cosh 2t} = \operatorname{sech} 2t$. At

$t = 1$, $a_T = \dfrac{\sinh 2}{\sqrt{\cosh 2}} = \dfrac{e^2 - e^{-2}}{\sqrt{2(e^2 + e^{-2})}}$ and a_N

$= \sqrt{\operatorname{sech} 2} = \left(\dfrac{2}{e^2 + e^{-2}}\right)^{1/2}$.

3 (a) $x = t$, $y = 3t$, and $z = 4t$, so $3x = 3t = y$; that is, the path lies in the plane $3x = y$. Also, $4y = 4(3t) = 12t = 3(4t) = 3z$; that is, the path lies in the plane $4y = 3z$.

(b)

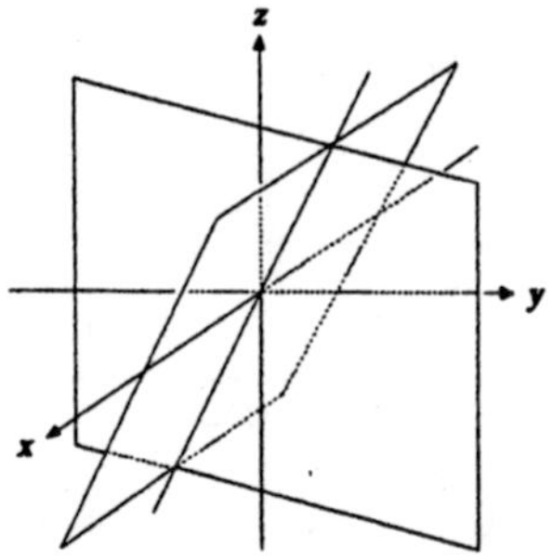

5 (a) We know that $\mathbf{v} = x'\mathbf{i} + y'\mathbf{j}$ and $\mathbf{a} = x''\mathbf{i} + y''\mathbf{j}$; we also have $\mathbf{a}(t) = 6t\mathbf{i} + e^t\mathbf{j}$, so $x''(t) = 6t$ and $y''(t) = e^t$. Therefore $x'(t) = 3t^2 + C_1$ and $y'(t) = e^t + C_2$; since $\mathbf{v}(0) = \mathbf{j}$, we know $x'(0) = 0$ and $y'(0) = 1$, so $0 = 3\cdot 0^2 + C_1 = C_1$ and $1 = e^0 + C_2 = 1 + C_2$, $C_2 = 0$. Now $x'(t) = 3t^2$ and $y'(t) = e^t$, so $\mathbf{v}(t) = 3t^2\mathbf{i} + e^t\mathbf{j}$ and $\mathbf{v}(1) = 3\mathbf{i} + e\mathbf{j}$. Finally, $\mathbf{a}(1) = 6\mathbf{i} + e\mathbf{j}$.

(b) At $t = 1$, $v = \|\mathbf{v}(1)\| = |3\mathbf{i} + e\mathbf{j}| = \sqrt{9 + e^2}$. Furthermore, $a_T = \mathbf{a}\cdot\mathbf{T} = \dfrac{\mathbf{a}\cdot\mathbf{v}}{|\mathbf{v}|} =$

$\dfrac{(6\mathbf{i} + e\mathbf{j})\cdot(3\mathbf{i} + e\mathbf{j})}{\sqrt{9 + e^2}} = \dfrac{18 + e^2}{\sqrt{9 + e^2}}$. Also,

$\|\mathbf{a}(1)\| = \sqrt{36 + e^2}$, so, at $t = 1$, $a_N^2 =$

$|\mathbf{a}|^2 - a_T^2 = (36 + e^2) - \dfrac{324 + 36e^2 + e^4}{9 + e^2}$

$= \dfrac{9e^2}{9 + e^2}$, so $a_N = \dfrac{3e}{\sqrt{9 + e^2}}$. Finally, the radius of curvature is given by $r = \dfrac{1}{a_N}v^2$, so

$r = \dfrac{\sqrt{9 + e^2}}{3e}(9 + e^2) = \dfrac{(9 + e^2)^{3/2}}{3e}$.

7 (a) $G(t) = e^t \cos t\ \mathbf{i} + e^t \sin t\ \mathbf{j} + e^t\mathbf{k}$, so $G'(t)$ $= e^t(\cos t - \sin t)\mathbf{i} + e^t(\sin t + \cos t)\mathbf{j} + e^t\mathbf{k}$; the speed $v(t)$ is given by $v(t) = \|G'(t)\|$ $= e^t\sqrt{(\cos t - \sin t)^2 + (\sin t + \cos t)^2 + 1^2}$ $= e^t\sqrt{2(\sin^2 t + \cos^2 t) + 1} = e^t\sqrt{3}$.

(b) Distance traveled is the integral of speed, so we have $\int_0^1 v(t)\,dt = \int_0^1 e^t\sqrt{3}\,dt = \sqrt{3}e^t\Big|_0^1$ $= \sqrt{3}(e - 1)$.

9 (a) In 2 units of time, the particle travels from (1, 2, 1) to (4, 1, 5). The distance traveled is $\sqrt{(4-1)^2 + (1-2)^2 + (5-1)^2} = \sqrt{9+1+16}$ $= \sqrt{26}$. The speed is therefore $\dfrac{\sqrt{26}}{2}$.

(b) The particle travels in the direction of $(4-1)\mathbf{i} + (1-2)\mathbf{j} + (5-1)\mathbf{k} = 3\mathbf{i} - \mathbf{j} + 4\mathbf{k}$. The direction cosines are the scalar components of the unit vector and the magnitude of the above vector is $\sqrt{26}$, so the direction cosines are $\dfrac{3}{\sqrt{26}}$, $-\dfrac{1}{\sqrt{26}}$, and $\dfrac{4}{\sqrt{26}}$.

(c) Since the velocity is constant, $\mathbf{r}(t) =$ $\mathbf{r}(0) + t\mathbf{v}(0) = \mathbf{r}(0) + \dfrac{t}{2}[\mathbf{r}(2) - \mathbf{r}(0)]$

$= \mathbf{i} + 2\mathbf{j} + \mathbf{k} + \dfrac{t}{2}(3\mathbf{i} - \mathbf{j} + 4\mathbf{k})$

$= \left(1 + \dfrac{3}{2}t\right)\mathbf{i} + \left(2 - \dfrac{1}{2}t\right)\mathbf{j} + (1 + 2t)\mathbf{k}$.

11 The astronaut is traveling along the path given by $\mathbf{r}(t) = \sin 3t\ \mathbf{i} + \tan t\ \mathbf{j} + e^{2t}\mathbf{k}$ so his velocity is $\mathbf{v}(t) = 3\cos 3t\ \mathbf{i} + \sec^2 t\ \mathbf{j} + 2e^{2t}\mathbf{k}$. At time $t =$ $\pi/4$ his location is $\mathbf{r}\left(\dfrac{\pi}{4}\right) = \sin\dfrac{3\pi}{4}\ \mathbf{i} + \tan\dfrac{\pi}{4}\ \mathbf{j}$ $+ e^{\pi/2}\mathbf{k} = \dfrac{1}{\sqrt{2}}\mathbf{i} + \mathbf{j} + e^{\pi/2}\mathbf{k}$ and his velocity is

$\mathbf{v}\left(\dfrac{\pi}{4}\right) = 3\cos\dfrac{3\pi}{4}\ \mathbf{i} + \sec^2\dfrac{\pi}{4}\ \mathbf{j} + 2e^{\pi/2}\mathbf{k} =$ $-\dfrac{3}{\sqrt{2}}\mathbf{i} + 2\mathbf{j} + 2e^{\pi/2}\mathbf{k}$. Hence the astronaut's path will be $\dfrac{1}{\sqrt{2}}\mathbf{i} + \mathbf{j} + e^{\pi/2}\mathbf{k} + s\left(-\dfrac{3}{\sqrt{2}}\mathbf{i} + 2\mathbf{j} + 2e^{\pi/2}\mathbf{k}\right)$

at time $\dfrac{\pi}{4} + s$ after the rockets are turned off.

Thus $x = \dfrac{1}{\sqrt{2}} - \dfrac{3}{\sqrt{2}}s$, $y = 1 + 2s$, and $z = e^{\pi/2}$ $+ 2se^{\pi/2}$. Substituting these parametric equations into $x + 2y + 3z = 100$ yields $s =$

$\dfrac{98 - 1/\sqrt{2} - 3e^{\pi/2}}{6e^{\pi/2} + 4 - 3/\sqrt{2}} \approx 2.6954$, so $x \approx -5.0107$,

$y \approx 6.3908$, and $z \approx 30.7430$.

13 If $G'(t) = x'(t)\mathbf{i} + y'(t)\mathbf{j} + z'(t)\mathbf{k} = \mathbf{0}$, then $x'(t)$ $= 0$, $y'(t) = 0$, and $z'(t) = 0$. Scalar functions whose derivatives are zero must be constant, so $x(t)$ $= a$, $y(t) = b$, and $z(t) = c$ for some constants a, b, and c. Hence $G(t) = a\mathbf{i} + b\mathbf{j} + c\mathbf{k}$; that is, $G(t)$ is a constant vector.

15 Since $a_N = \kappa v^2 = \kappa\|\mathbf{v}\|^2$ and $\kappa = \dfrac{|\mathbf{v}\times\mathbf{a}|}{|\mathbf{v}|^3}$, we

have $a_N = \dfrac{|\mathbf{v}\times\mathbf{a}|}{|\mathbf{v}|^3}|\mathbf{v}|^2 = \dfrac{|\mathbf{v}\times\mathbf{a}|}{|\mathbf{v}|}$.

14 Partial Derivatives

14.1 Graphs

1 See the graph in the answer section of the text.

3 See the graph in the answer section of the text.

5 See the graph in the answer section of the text.

7 Every vertical line through the line $y = 2x$ (in the xy plane) is contained in the plane. See the graph in the answer section of the text.

9 Every horizontal line through the line $z = x$ (in the xz plane) is contained in the plane. See the graph in the answer section of the text.

11 We will first find the x, y, and z intercepts of the plane $x + y - z = 1$ before graphing. When $y = z = 0$, we have $x = 1$; when $x = z = 0$ we see that $y = 1$; when $x = y = 0$, we find that $-z = 1$, or $z = -1$. See the graph in the answer section of the text.

13 Here $(x_0, y_0, z_0) = (0, 0, 0)$ and $r = 3$. The equation for the sphere is $(x - 0)^2 + (y - 0)^2 + (z - 0)^2 = 3^2$, or $x^2 + y^2 + z^2 = 9$. See the graph in the answer section of the text.

15 We are given that $(x_0, y_0, z_0) = (1, 2, 3)$ and $r = 1$. The equation for the sphere is then $(x - 1)^2 + (y - 2)^2 + (z - 3)^2 = 1^2$, or $x^2 + y^2 + z^2 - 2x - 4y - 6z + 13 = 0$. See the graph in the answer section of the text.

17 The cylinder is represented by all vertical lines through the circle $x^2 + y^2 = 5$ (in the xy plane). Note that the graph is that of a right circular cylinder. See the graph in the answer section of the text.

19 The cylinder is represented by all vertical lines through the line $x + y = 1$ (in the xy plane). Note that this cylinder is simply a plane. See the graph in the answer section of the text.

21 The cylinder consists of all lines perpendicular to the yz plane and through the parabola $z = y^2$ (in the yz plane). This is a parabolic cylinder. See the graph in the answer section of the text.

23 The cylinder consists of all lines perpendicular to the xy plane and through the parabola $x = y^2$ (in the xy plane). This is a parabolic cylinder. See the graph in the answer section of the text.

25 The cylinder consists of all vertical lines through the curve $y = 2^{-x}$ (in the xy plane). See the graph in the answer section of the text.

27 Let Q be a point on the line $z = 2y$; then $Q = (0, y, z') = (0, y, 2y)$ for some $y > 0$. Let $P =$

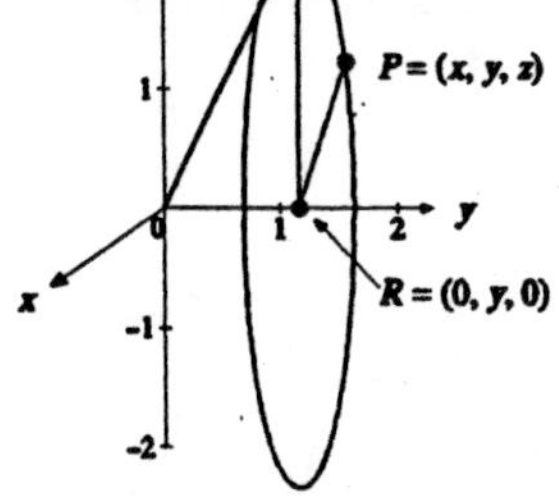

(x, y, z) be a point on the surface of revolution. Note that $\overline{PR} = \sqrt{x^2 + z^2} = \overline{QR}$ from the circle in the figure. However, $\overline{QR}$ is also the z coordinate of $Q = (0, y, 2y)$. Hence $\sqrt{x^2 + z^2} = 2y$ is the desired equation for the surface.

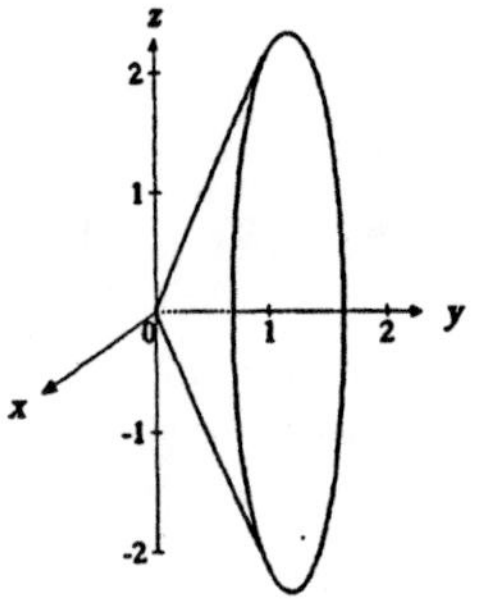

29 Let Q be a point on the semicircle $x^2 + y^2 = 1$, $y \geq 0$; then $Q = (x, y', 0) = \left(x, \sqrt{1 - x^2}, 0\right)$. Let P

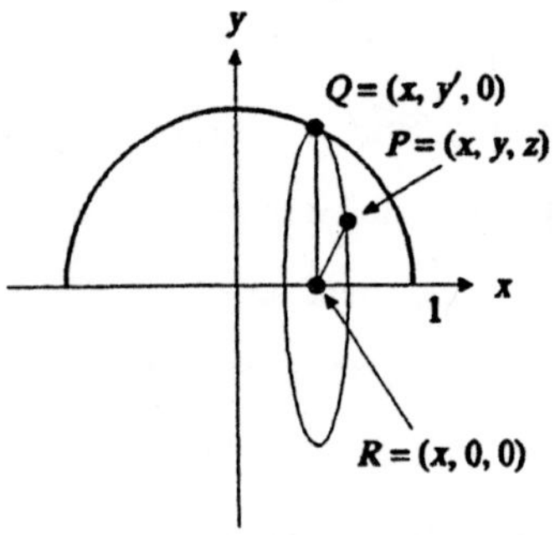

$= (x, y, z)$ be a point on the surface of revolution. Again, note that $\overline{PR} = \sqrt{z^2 + y^2} = \overline{QR}$ from the circle. Since $\overline{QR}$ is the y coordinate of $Q = \left(x, \sqrt{1 - x^2}, 0\right)$, the equation of the surface is $\sqrt{1 - x^2} = \sqrt{z^2 + y^2}$ or $x^2 + y^2 + z^2 = 1$. This is a sphere with center (0, 0, 0) and radius 1.

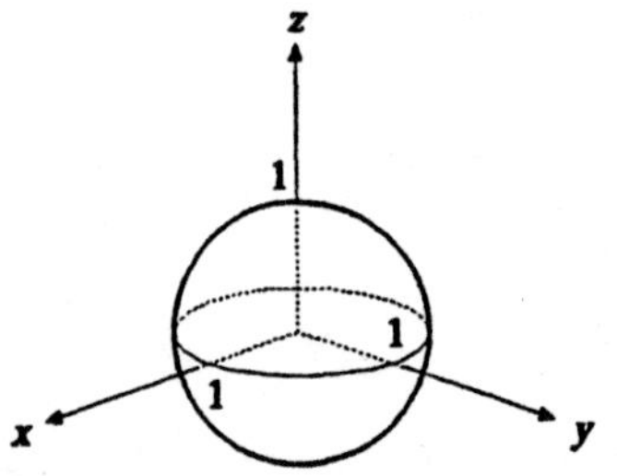

31 (a) Let $P = (x, y, z)$, $Q = (1, 2, 3)$, and $R = (4, -1, 2)$. We want $\overline{RP} = \overline{QP}$; hence, through simplification,

$\sqrt{(x-4)^2 + (y-(-1))^2 + (z-2)^2} =$

$\sqrt{(x-1)^2 + (y-2)^2 + (z-3)^2}$, so

$(x-4)^2 + (y+1)^2 + (z-2)^2 =$

$(x-1)^2 + (y-2)^2 + (z-3)^2$, $x^2 - 8x + 16 + y^2 + 2y + 1 + z^2 - 4z + 4 = x^2 - 2x + 1 + y^2 - 4y + 4 + z^2 - 6z + 9$. So the equation that tells when P is in S is

$-6x + 6y + 2z = -7$.

(b) The equation $-6x + 6y + 2z = -7$ is that of a plane.

(c) First, we find the intercepts of the plane. When $x = y = 0$, $z = -7/2$; when $y = z = 0$, $x = 7/6$; when $x = z = 0$, $y = -7/6$. See the graph in the answer section of the text.

33 See the graph in the answer section of the text.

35 (a) Looking at the cross-section, the point P must satisfy $\sqrt{x^2 + y^2} = \sqrt{-\ln z}$ or $z = e^{-(x^2+y^2)}$. See the graph in the answer section of the text.

(b) The cylindrical shell has radius r, height e^{-r^2}, and thickness dr, so its volume is about $2\pi r e^{-r^2}\, dr$.

(c) Volume $= \int_0^\infty 2\pi r e^{-r^2}\, dr$

(d) For $b > 0$, $\int_0^b 2\pi r e^{-r^2}\, dr = \left(-\pi e^{-r^2}\right)\Big|_0^b =$

$\pi\left(1 - e^{-b^2}\right)$, so $\int_0^\infty 2\pi r e^{-r^2}\, dr =$

$\lim_{b \to \infty} \pi\left(1 - e^{-b^2}\right) = \pi$.

14.2 Quadric Surfaces

1 The x intercepts (with $y = z = 0$) occur when $x^2 + 0/4 + 0/9 = 1$, or when $x^2 = 1$. So the x intercepts are ± 1. Similarly, the y intercepts are given by $\frac{y^2}{4} = 1$, or when $y = \pm 2$. Finally, the z intercepts occur when $\frac{z^2}{9} = 1$, or when $z = \pm 3$. The trace in the $z = 0$ plane is $x^2 + \frac{y^2}{4} = 1$ (an ellipse), and the trace in the $z = 2$ plane is $x^2 + \frac{y^2}{4} + \frac{4}{9} = 1$, or $\frac{x^2}{5/9} + \frac{y^2}{20/9} = 1$ (an ellipse). See the graph in the answer section of the text.

3 The x intercepts occur when $x^2 = -1$; hence, there are no x intercepts. The z intercepts do not exist for the same reason. The y intercepts are given by $y^2 = 1$, or $y = \pm 1$. The trace in the $x = 0$ plane is $y^2 - z^2 = 1$ (a hyperbola). Since there are two minuses in $-x^2 + y^2 - z^2 = 1$, this is a hyperboloid of two sheets. See the graph in the answer section of the text.

5 The x and z intercepts do not exist. The y intercepts are given by $y^2 = 1$, or $y = \pm 1$. The trace in the $z = 3$ plane is $\frac{-x^2}{4} + y^2 - \frac{3^2}{9} = 1$, or $y^2 - \frac{x^2}{4} = 2$, or $\frac{y^2}{(\sqrt{2})^2} - \frac{x^2}{(2\sqrt{2})^2} = 1$ (a hyperbola). Since there are two minuses in $-\frac{x^2}{4} + y^2 - \frac{z^2}{9} = 1$, this is a hyperboloid of two sheets. See the graph in the answer section of the text.

7 There are no y intercepts; the x intercepts are ± 2 and the z intercepts are ± 1. The trace in the $y = 0$ plane is $\frac{x^2}{4} + z^2 = 1$ (an ellipse); the trace in the $y = \pm 3$ plane is $\frac{x^2}{(2\sqrt{2})^2} + \frac{z^2}{(\sqrt{2})^2} = 1$ (an ellipse). Since there is one minus in $\frac{x^2}{4} - \frac{y^2}{9} + z^2 = 1$, this is a hyperboloid of one sheet. See the graph in the answer section of the text.

9 The general formula (from (14) in the text) is $(\tan^2 \alpha)z^2 = x^2 + y^2$; since $\alpha = \pi/6$, we have $\tan^2 \alpha = \tan^2 \pi/6 = 1/3$. Hence the equation of the double cone is $\frac{1}{3}z^2 = x^2 + y^2$.

11 Here the axis contained in the double cone is the x axis; so the general formula is $(\tan^2 \alpha)x^2 = y^2 + z^2$. Since $\alpha = \pi/4$, we have $\tan^2 \alpha = \tan^2 \pi/4 = 1$. Hence the equation of the double cone is $x^2 = y^2 + z^2$.

13 Referring to equation (14), we see that $k = \tan^2 \alpha = 3$; so the half-angle is $\alpha = \tan^{-1}\sqrt{3} = \pi/3$. The trace in the $z = 1$ plane is $3 = x^2 + y^2$ (a circle). The trace in the $x = 1$ plane is $3z^2 - y^2 = 1$ (a hyperbola). See the graph in the answer section of the text.

15 In this case, the general formula for the double cone is $(\tan^2 \alpha)y^2 = x^2 + z^2$. Here $\tan^2 \alpha = 1$, so $\alpha = \pi/4$ radians $= 45°$. The trace in the $z = 1$ plane is $y^2 - x^2 = 1$ (a hyperbola); the trace in the $x = 3$ plane is $y^2 - z^2 = 9$ (a hyperbola). See the graph in the answer section of the text.

17 Let $P = (x, y, z)$ be a point on the surface of revolution. Let $Q = (0, y', z)$ be the point on the

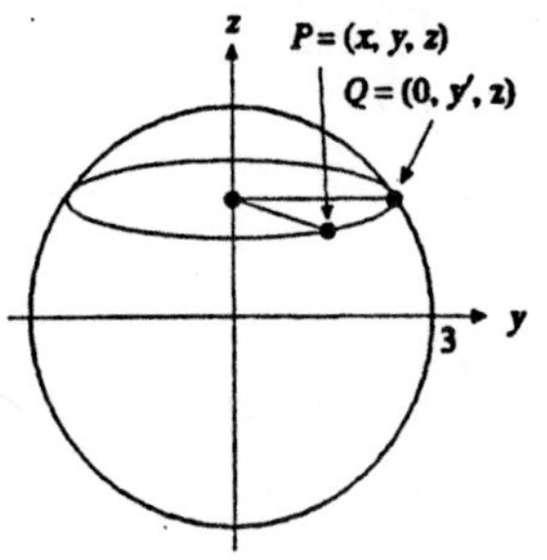

curve $y^2 + z^2 = 9$ in the yz plane from which P is derived. (See the figure.) Now $y' = \sqrt{x^2 + y^2}$; so $Q = (0, \sqrt{x^2 + y^2}, z)$. Since Q lies on $y^2 + z^2 = 9$, the equation for the surface is $\left(\sqrt{x^2 + y^2}\right)^2 + z^2 = 9$ or $x^2 + y^2 + z^2 = 9$. The surface is a sphere of radius 3.

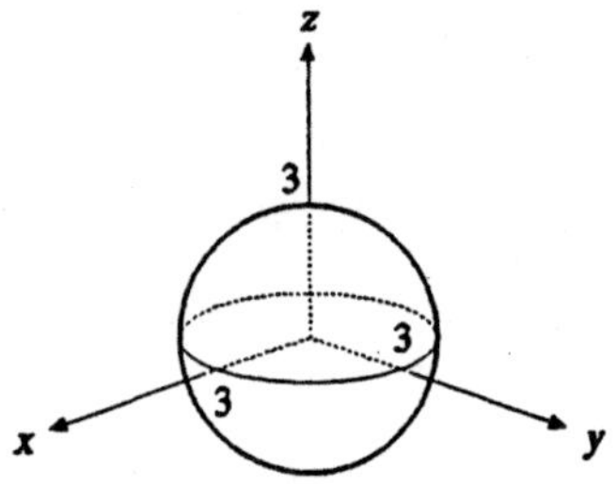

19 Let $P = (x, y, z)$ be a point on the surface of revolution. Let $Q = (x', y, 0)$ be the point on the

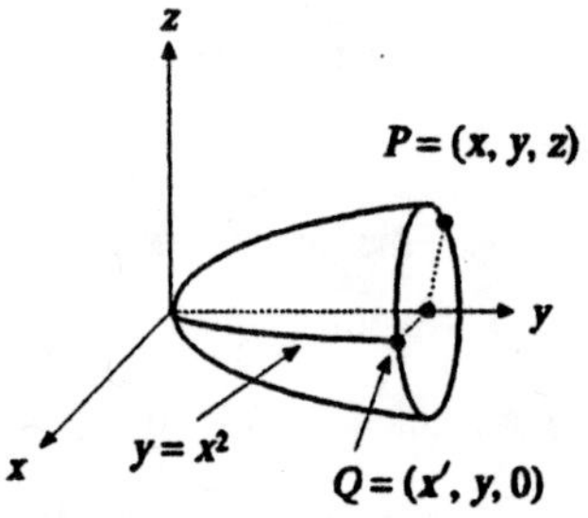

curve $y = x^2$ ($x \geq 0$ in the xy plane) from which P is derived. Now $x' = \sqrt{x^2 + z^2}$; so $Q = (\sqrt{x^2 + z^2}, y, 0)$. Since Q lies on $y = x^2$, the equation for the surface is $y = \left(\sqrt{x^2 + z^2}\right)^2$ or $y = x^2 + z^2$. The surface is a paraboloid.

21 Traces of $z = x^2 + 3y^2$ in planes parallel to the xy plane are ellipses of the form $x^2 + 3y^2 = k$, where $z = k$ is the intersecting plane. Traces parallel to the xz and yz planes are parabolas. See the graph in the answer section of the text.

23 Traces of $z = x^2/4 + y^2/9$ in planes parallel to the xy plane are ellipses of the form $x^2/4 + y^2/9 = k$, where $z = k$ is the intersecting plane. Traces parallel to the xz and yz planes are parabolas. See the graph in the answer section of the text.

25 The graph of $y = x^2$ is a parabolic cylinder. See the graph in the answer section of the text.

27 $x^2 + 2xy + y^2 = 0$ implies that $(x + y)^2 = 0$, so $x + y = 0$. The resulting graph is a cylinder (a plane, actually) consisting of all of the vertical lines through the line in the xy plane $y = -x$. See the graph in the answer section of the text.

29 The plane $z = k$ is always parallel to the xy plane. The formula for the trace of $x^2 + 2y^2 + 3z^2 = 1$ in the $z = k$ plane is given by $x + 2y^2 + 3k^2 = 1$, or $x^2 + 2y^2 = 1 - 3k^2$. Since the ellipsoid extends only from $z = -\frac{1}{\sqrt{3}}$ to $z = \frac{1}{\sqrt{3}}$, we shall concern ourselves with $|k| \leq \frac{1}{\sqrt{3}}$ only. If $|k| = \frac{1}{\sqrt{3}}$, then $x^2 + 2y^2 = 1 - 3\left(\pm\frac{1}{\sqrt{3}}\right)^2 = 0$, which has $(0, 0, \pm 1/\sqrt{3})$ as a solution. If $|k| < \frac{1}{\sqrt{3}}$, then

$$x^2 + 2y^2 = 1 - 3(|k|)^2 > 1 - 3\left(\pm\frac{1}{\sqrt{3}}\right)^2 = 0,$$

which has an ellipse as a solution.

31 The half-angle α is 45° and the double cone contains the z axis; thus the formula for the double cone is $(\tan^2 \alpha)z^2 = x^2 + y^2$, or $z^2 = x^2 + y^2$.

33 If $\frac{x^2}{a^2} + \frac{y^2}{b^2} + \frac{z^2}{c^2} = 1$ is a surface of revolution about the z axis, its trace in the plane $z = k$ must be either a circle or a point or empty. The surface passes only through values of k where $|k| \le c$; so only $|k| \le c$ will be considered. The formula for the trace is given by $\frac{x^2}{a^2} + \frac{y^2}{b^2} + \frac{k^2}{c^2} = 1$, so $\frac{x^2}{a^2} + \frac{y^2}{b^2} = 1 - \frac{k^2}{c^2}$. Since $|k| \le c$, $1 - \frac{k^2}{c^2} \ge 0$; thus the trace above is a circle (or a point) if and only if $|a| = |b|$. There is no restriction on c.

35 (a) See the graph in the answer section of the text.

(b) The curve of intersection is a parabola.

37 See the graph in the answer section of the text.

39 (a) See the graph in the answer section of the text.

(b) Using polar coordinates, a differential slab of the surface has width $r\,d\theta = 1\,d\theta = d\theta$ and height $z = x = \cos\theta$. The differential area is thus given by $\cos\theta\,d\theta$, and θ varies from $-\pi/2$ to $\pi/2$. Hence the total area of the surface is $\int_{-\pi/2}^{\pi/2} \cos\theta\,d\theta = \sin\theta\Big|_{-\pi/2}^{\pi/2} = 1 - (-1) = 2$.

41 See the graph in the answer section of the text.

43 See the graph in the answer section of the text.

45 Since $(x + y)(z - y) = 0$, we have $y = -x$ or $z = y$. Thus the graph consists of two planes. See the graph in the answer section of the text.

47 (a) The surface $4x^2 + 4y^2 - z^2 = 1$ is a hyperboloid of one sheet. See the graph in the answer section of the text.

(b) Note that the trace of $4x^2 + 4y^2 - z^2 = 1$ in the $z = k$ plane is $4x^2 + 4y^2 - k^2 = 1$, or $x^2 + y^2 = \frac{1 + k^2}{4}$, which is a circle. Hence the hyperboloid is a surface of revolution about the z axis.

(c) The trace of $4x^2 + 4y^2 - z^2 = 1$ in the plane $z = 2y$ is $4x^2 + 4y^2 - (2y)^2 = 1$, or $x^2 = 1/4$, or $x = \pm 1/2$. The graph of $x = \pm 1/2$ (in the $z = 2y$ plane) consists of two lines.

(e) The surface is made up of tilted lines, all of which pass through the circle $4x^2 + 4y^2 = 1$ in the xy plane.

14.3 Functions and Their Level Curves

1 Let $z = f(x, y) = y$. Then the graph of f consists of all points (x, y, z) such that $z = y$. Note that, for any given k, the trace of f in the plane $x = k$ is the line $z = y$. By rewriting $z = y$ as $-y + z = 0$, it is clear that the graph of f is a plane. See the graph in the answer section of the text.

3 Let $z = f(x, y) = 3$. The graph is that of a constant function; $f(x, y) = 3$ for all x and y. Hence the graph of f is a plane parallel to the xy plane, passing through the point $(0, 0, 3)$. See the figure in the answer section of the text.

5 Let $z = f(x, y) = x^2$ The graph is the set of all points (x, y, z) with the condition that $z = x^2$. For any given k, the trace of f in the plane $y = k$ is the parabola $z = x^2$. The graph of f is a parabolic cylinder. See the figure in the answer section of the text.

7 Let $z = f(x, y) = x + y + 1$. Rewriting $z = x + y + 1$ as $-x - y + z - 1 = 0$, we see that the graph of f is a plane. We now find the intercepts of this plane. When $x = y = 0$, $z = 1$; when $x = z = 0$, $y = -1$; when $y = z = 0$, $x = -1$. Thus the plane passes through the points $(-1, 0, 0)$, $(0, -1, 0)$, and $(0, 0, 1)$. Moreover, these three points determine the plane. See the figure in the answer section of the text.

9 Let $z = f(x, y) = x^2 + 2y^2$. The trace of f in the plane $x = 0$ is the parabola $z = 2y^2$, and the trace in the plane $y = 0$ is the parabola $z = x^2$. For any given k, the trace of f in the plane $z = k$ is $x^2 + 2y^2 = k$, which describes the empty set for $k < 0$, the origin for $k = 0$, and an ellipse for $k > 0$. The graph of f is an elliptic paraboloid. See the figure in the answer section of the text.

11 Since $f(x, y) = x + y$, the level curves of $f(x, y) = -1, 0, 1$ and 2 are, respectively, the lines $x + y = -1$, $x + y = 0$, $x + y = 1$, and $x + y = 2$.

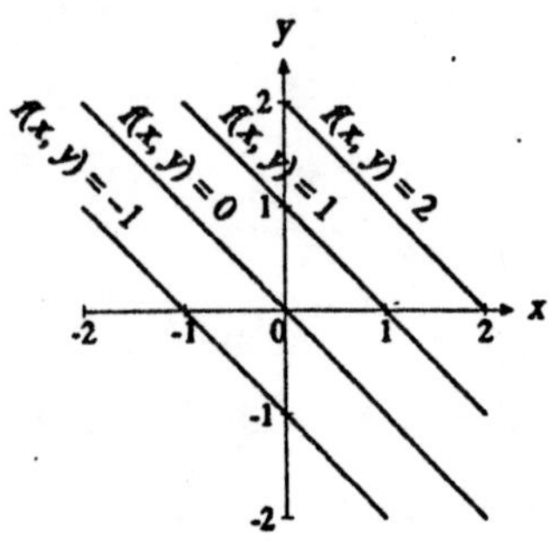

13 Since $f(x, y) = x^2 + 2y^2$, the level curves of $f(x, y) = 1$ and $f(x, y) = 2$ are, respectively, the ellipses $x^2 + 2y^2 = 1$ and $x^2 + 2y^2 = 2$. The level curve of $f(x, y) = 0$ is the origin since $x^2 + 2y^2 = 0$ if and only if $x = y = 0$. The level curve of $f(x, y) = -1$ is empty since $x^2 + 2y^2 \geq 0$ for all x and y.

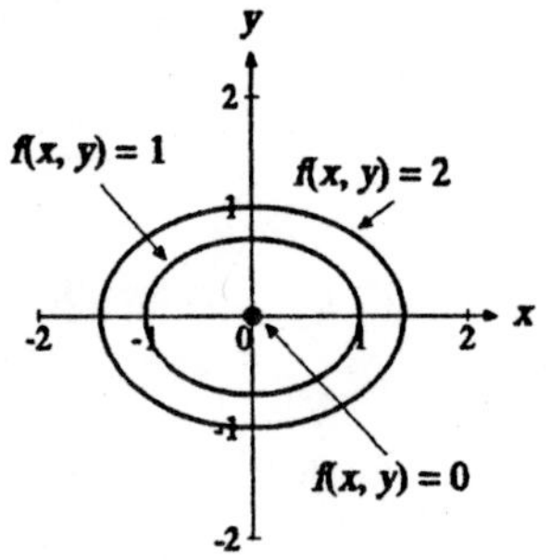

15 Observe that $f(1, 1) = 1^2 + 1^2 = 2$, so the corresponding level curve is $f(x, y) = 2$. The level curve is the circle $x^2 + y^2 = 2$.

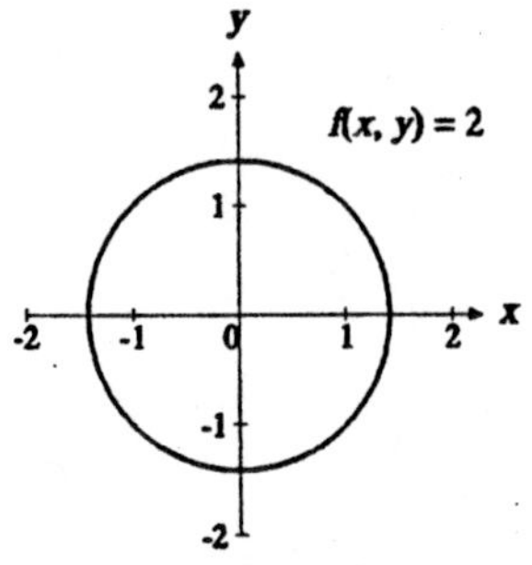

17 Note that $f(3, 2) = 3^2 - 2^2 = 5$, so the corresponding level curve is $f(x, y) = 5$. The level curve is the hyperbola $x^2 - y^2 = 5$.

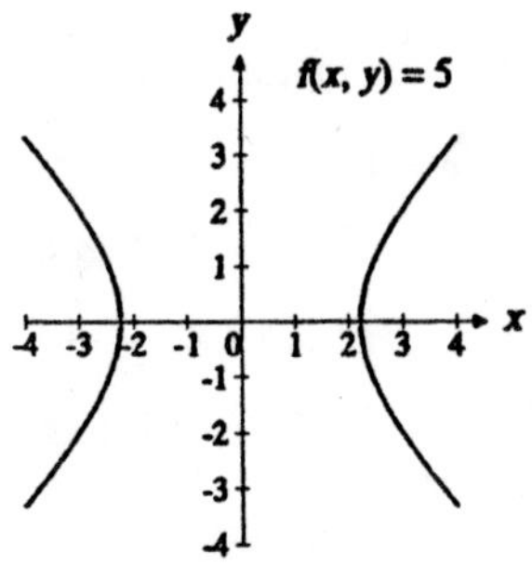

19 (a) The level curves are the circles $x^2 + y^2 = k$, where $k = 0, 1, \cdots, 9$.

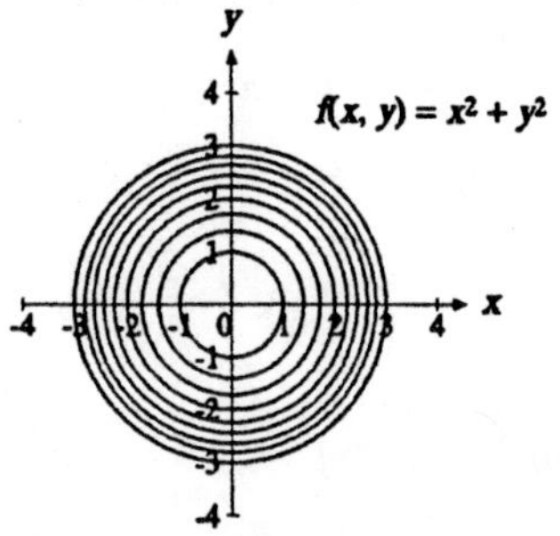

(b) The two closest level curves are the circles corresponding to $k = 8$ and $k = 9$. Hence, within the region graphed in (a), the function changes most rapidly in the plane region between the circles $x^2 + y^2 = 8$ and $x^2 + y^2 = 9$. (The function changes more rapidly the farther you get from the origin.)

21 The wind is strongest where the pressure changes most rapidly, that is, where the level curves are closest together. This occurs near the point E in the figure.

14.4 Limits and Continuity

1 $\lim\limits_{(x,y)\to(2,3)} \dfrac{x+y}{x^2+y^2} = \dfrac{2+3}{2^2+3^2} = \dfrac{5}{13}$

3 Along the line $y = x$, we have $\lim\limits_{(x,y)\to(0,0)} \dfrac{x^2}{x^2+y^2}$

$= \lim\limits_{x\to 0} \dfrac{x^2}{x^2+x^2} = \lim\limits_{x\to 0} \dfrac{x^2}{2x^2} = \lim\limits_{x\to 0} \dfrac{1}{2} = \dfrac{1}{2}$.

Along the line $y = 0$, we have $\lim\limits_{(x,y)\to(0,0)} \dfrac{x^2}{x^2+y^2}$

$= \lim\limits_{x\to 0} \dfrac{x^2}{x^2+0^2} = \lim\limits_{x\to 0} \dfrac{x^2}{x^2} = \lim\limits_{x\to 0} 1 = 1$. Since $\dfrac{1}{2} \neq 1$, $\lim\limits_{(x,y)\to(0,0)} \dfrac{x^2}{x^2+y^2}$ does not exist.

5 $\lim\limits_{(x,y)\to(2,3)} x^y = 2^3 = 8$

7 When $x = 0$ or $y = 0$, the function is undefined. Since there are such points in every disk centered at (0, 0), the limit does not exist.

9 (a) The domain of f is the set of all points $P = (x, y)$ excluding the line $x + y = 0$.

(b) For every point $P_0 = (x_0, y_0)$ in the domain of f, there exists a disk with center P_0 and positive radius that is contained in the domain of f. Also, $\lim\limits_{P\to P_0} f(P) = \lim\limits_{P\to P_0} \dfrac{1}{x+y} =$

$\dfrac{1}{x_0+y_0} = f(P_0)$ since $x_0 + y_0 \neq 0$. Thus f is continuous by definition.

11 (a) The domain of f is the set of all points $P = (x, y)$ that do not satisfy $x^2 + y^2 = 9$.

(b) For every point $P_0 = (x_0, y_0)$ in the domain of f, there exists a disk with center P_0 and positive radius that is contained in the domain of f. Also, $\lim\limits_{P\to P_0} f(P) = \lim\limits_{P\to P_0} \dfrac{1}{9-x^2-y^2} =$

$\dfrac{1}{9-x_0^2-y_0^2} = f(P_0)$ since $x_0^2 + y_0^2 \neq 9$.

Hence f is continuous throughout its domain.

13 (a) The domain of f consists of all points $P = (x, y)$ such that $x^2 + y^2 \leq 16$.

(b) Yes, f is continuous throughout its domain, since $\lim\limits_{(x,y)\to(x_0,y_0)} \sqrt{16-x^2-y^2} =$

$\sqrt{16 - x_0^2 - y_0^2} = f(x_0, y_0)$.

15 R is bordered by the circle $x^2 + y^2 = 1$. Hence the circle $x^2 + y^2 = 1$ is the boundary of R.

17 The function $\dfrac{1}{x^2 + y^2}$ is defined everywhere except at the origin. Therefore, the point (0, 0) is the boundary of R.

19 R is bordered by the parabola $y = x^2$. Thus the parabola $y = x^2$ is the boundary of R.

21 (a) If $P = (x, y)$ lies within a distance 0.01 of (1, 2), then x and y must satisfy $\sqrt{(x-1)^2 + (y-2)^2} < 0.01$ (by the distance formula). Observe that $\sqrt{(x-1)^2} \le \sqrt{(x-1)^2 + (y-2)^2} < 0.01$ since $(y-2)^2 \ge 0$ for all y. From $\sqrt{(x-1)^2} = |x-1|$, it follows that $|x - 1| < 0.01$. Similarly, $|y - 2| = \sqrt{(y-2)^2} \le \sqrt{(x-1)^2 + (y-2)^2} < 0.01$.

(b) Note that $|f(x, y) - 3| = |x + y - 3| = |x - 1 + y - 2|$. Since $|x - 1| < 0.01$ and $|y - 2| < 0.01$, we have $|f(x, y) - 3| = |x - 1 + y - 2| \le |x - 1| + |y - 2| < 0.01 + 0.01 = 0.02$, where the triangle inequality was used.

(c) Let $\delta = 0.0005 > 0$. Then $|x - 1| < \delta$ and $|y - 2| < \delta$ by the constraint on P. From (b), it is apparent that $|f(x, y) - 3| \le |x - 1| + |y - 2| < \delta + \delta = 2\delta = 2(0.0005) = 0.001$. Any $\delta \le 0.0005$ will suffice to show that $|f(x, y) - 3| < 0.001$.

(d) Let $\epsilon > 0$ be given. Choose $\delta = \epsilon/2 > 0$. Then for all P such that P is in the disk of radius δ and center (1, 2), $|x - 1| < \delta$ and $|y - 2| < \delta$. It then follows that $|f(x, y) - 3| = |x + y - 3| = |x - 1 + y - 2| \le |x - 1| + |y - 2| < \delta + \delta = 2\delta = 2 \cdot \frac{\epsilon}{2} = \epsilon$, as claimed.

(e) $\lim_{(x,y)\to(1,2)} f(x, y) = 3$

23 (a) The domain of f is all points $P = (x, y)$ excluding the origin.

(b)

(x, y)	(0.01, 0.01)	(0.01, 0.02)	(0.001, 0.003)
$f(x, y)$	0.00499975	0.00249997	0.00016667

(c) Yes. It would seem that $\lim_{P\to(0,0)} f(P) = 0$.

(d)

(x, y)	(0.5, 0.25)	(0.1, 0.01)	(0.001, 0.000001)
$f(x, y)$	1/3	1/3	1/3

(e) Yes. It would seem that $\lim_{P\to(0,0)} f(P) = 1/3$.

(f) No. Along the line $y = 0$, $f(P) \to 0$ as $P \to (0, 0)$. Along the line $y = x^2$, $f(P) \to 1/3$ as $P \to (0, 0)$. Since $0 \ne 1/3$, the limit cannot exist.

14.5 Partial Derivatives

1 $f(x, y) = 3x + 2y$, so $f_x = 3$ and $f_y = 2$.

3 $f(x, y) = x^3y^4$, so $f_x = 3x^2y^4$ and $f_y = 4x^3y^3$.

5 $f(x, y) = x \cos xy$; $f_x = x(-\sin xy)(y) + \cos xy = -xy \sin xy + \cos xy$; $f_y = x(-\sin xy)(x) = -x^2 \sin xy$

7 $f(x, y) = \tan^{-1} xy;\ f_x = \dfrac{1}{1 + (xy)^2}(y) = \dfrac{y}{1 + x^2y^2};\ f_y = \dfrac{1}{1 + (xy)^2}(x) = \dfrac{x}{1 + x^2y^2}$

9 $f(x, y) = x \sin^2(x + y);\ f_x$
$= x(2 \sin(x + y))(\cos(x + y)) + \sin^2(x + y)$
$= x \sin[2(x + y)] + \sin^2(x + y);$
$f_y = x(2 \sin(x + y))(\cos(x + y))$
$= x \sin[2(x + y)]$

11 $f(x, y) = \sqrt{x} \sec x^2y = x^{1/2} \sec x^2y;$
$f_x = x^{1/2}(\sec x^2y \tan x^2y)(2xy) + \dfrac{1}{2}x^{-1/2} \sec x^2y$
$= \dfrac{1}{2}x^{-1/2} \sec x^2y + 2x^{3/2}y \sec x^2y \tan x^2y;\ f_y =$
$x^{1/2}(\sec x^2y \tan x^2y)(x^2) = x^{5/2} \sec x^2y \tan x^2y$

13 $f(x, y) = \dfrac{e^x}{y^3};\ f_x = \dfrac{e^x}{y^3};\ f_y = e^x\left(\dfrac{-3}{y^4}\right) = -\dfrac{3e^x}{y^4}$

15 $f(x, y) = 5x^2 - 3xy + 6y^2;\ f_x = 10x - 3y;\ f_{xx} = 10;\ f_{xy} = -3;\ f_y = -3x + 12y;\ f_{yy} = 12;\ f_{yx} = -3 = f_{xy}$

17 $f(x, y) = \dfrac{1}{\sqrt{x^2 + y^2}} = (x^2 + y^2)^{-1/2};$
$f_x = -\dfrac{1}{2}(x^2 + y^2)^{-3/2}(2x) = -x(x^2 + y^2)^{-3/2};$
$f_{xx} = -x\left[-\dfrac{3}{2}(x^2 + y^2)^{-5/2}(2x)\right] - (x^2 + y^2)^{-3/2}$
$= 3x^2(x^2 + y^2)^{-5/2} - (x^2 + y^2)^{-3/2};$
$f_{xy} = -x\left[-\dfrac{3}{2}(x^2 + y^2)^{-5/2}(2y)\right] = 3xy(x^2 + y^2)^{-5/2};$
$f_y = -\dfrac{1}{2}(x^2 + y^2)^{-3/2}(2y) = -y(x^2 + y^2)^{-3/2};$
$f_{yy} = -y\left[-\dfrac{3}{2}(x^2 + y^2)^{-5/2}(2y)\right] - (x^2 + y^2)^{-3/2}$
$= 3y^2(x^2 + y^2)^{-5/2} - (x^2 + y^2)^{-3/2};\ f_{yx}$
$= -y\left[-\dfrac{3}{2}(x^2 + y^2)^{-5/2}(2x)\right] = 3xy(x^2 + y^2)^{-5/2} = f_{xy}$

19 $f(x, y) = \tan(x + 3y);\ f_x = \sec^2(x + 3y);$
$f_{xx} = 2 \sec(x + 3y) \sec(x + 3y) \tan(x + 3y)$
$= 2 \sec^2(x + 3y) \tan(x + 3y);\ f_{xy} =$
$2 \sec(x + 3y) \sec(x + 3y) \tan(x + 3y)(3)$
$= 6 \sec^2(x + 3y) \tan(x + 3y);\ f_y = 3 \sec^2(x + 3y);$
$f_{yy} = 3[2 \sec(x + 3y) \sec(x + 3y) \tan(x + 3y)(3)]$
$= 18 \sec^2(x + 3y) \tan(x + 3y);\ f_{yx} =$
$3(2 \sec(x + 3y) \sec(x + 3y) \tan(x + 3y))$
$= 6 \sec^2(x + 3y) \tan(x + 3y) = f_{xy}$

21 The desired slope is $\dfrac{\partial z}{\partial x}$ evaluated at (1, 2); that is,

$$\left.\frac{\partial z}{\partial x}\right|_{(1,2)} = \left.y^2\right|_{(1,2)} = 4.$$

23 The requested slope is $\left.\dfrac{\partial z}{\partial x}\right|_{(1,1)} = \left.\dfrac{1}{y}\right|_{(1,1)} = 1.$

25 The slope is $\left.\dfrac{\partial z}{\partial y}\right|_{(1,1)} = \left.(e^{xy} + yxe^{xy})\right|_{(1,1)} = 2e.$

27 (a)

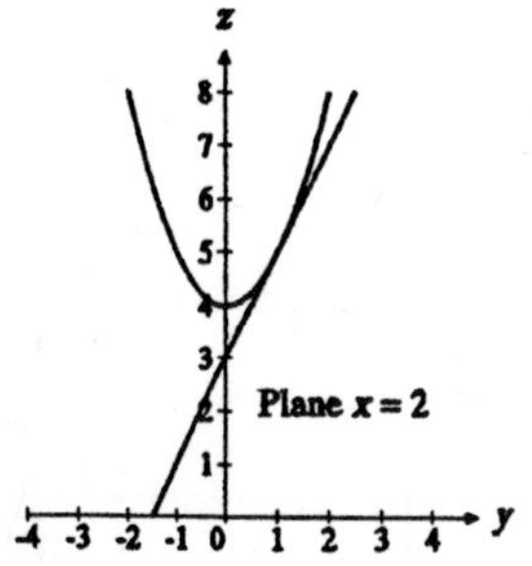

(b) The equation of the trace is $z = 2^2 + y^2 = y^2 + 4$. The slope equals $\left.\dfrac{\partial z}{\partial y}\right|_{(2,1,5)} =$

$2y\,|_{(2,1,5)} = 2.$

(c) See the graph in (a).

29 When Δx and Δy are small, it is known that

$$\left.\frac{\partial f}{\partial x}\right|_{(x_0,y_0)} \approx \frac{f(x_0+\Delta x, y_0) - f(x_0, y_0)}{\Delta x}$$ and

$$\left.\frac{\partial f}{\partial y}\right|_{(x_0,y_0)} \approx \frac{f(x_0, y_0+\Delta y) - f(x_0, y_0)}{\Delta y}$$ from the

definition of the partial derivative. We then have

$$\left.\frac{\partial f}{\partial x}\right|_{(1,1)} \approx \frac{f(1.02, 1) - f(1, 1)}{1.02 - 1} = \frac{3.05 - 3}{0.02} =$$

$$\frac{0.05}{0.02} = \frac{5}{2} \text{ and } \left.\frac{\partial f}{\partial y}\right|_{(1,1)} \approx \frac{f(1, 0.97) - f(1, 1)}{0.97 - 1}$$

$$= \frac{2.4 - 3}{-0.03} = \frac{-0.6}{-0.03} = 20.$$

31 From the definition of the partial derivative,

$$\left.\frac{\partial T}{\partial x}\right|_{(x_0,y_0)} \approx \frac{T(x_0+\Delta x, y_0) - T(x_0, y_0)}{\Delta x}$$ (when Δx

is small) and $\left.\frac{\partial T}{\partial y}\right|_{(x_0,y_0)}$

$$\approx \frac{T(x_0, y_0+\Delta y) - T(x_0, y_0)}{\Delta y}$$ (when Δy is small).

We thus have (from the table) $\left.\frac{\partial T}{\partial x}\right|_{(1,2)} \approx$

$$\frac{T(1.01, 2) - T(1, 2)}{0.01} = \frac{5.025 - 5}{0.01} = \frac{0.025}{0.01}$$

$$= \frac{5}{2} \text{ and } \left.\frac{\partial T}{\partial y}\right|_{(1,2)} \approx \frac{T(1, 2.02) - T(1, 2)}{0.02}$$

$$= \frac{5.06 - 5}{0.02} = \frac{0.06}{0.02} = 3.$$

33 (a) One such function is the constant function $f(x, y) = 3$.

(b) Integrating $\frac{\partial f}{\partial x}$ with respect to x gives $f(x, y) = C + g(y)$, where $g(y)$ is a function of y alone. Since $\frac{\partial f}{\partial y} = 0$, we have $\frac{\partial f}{\partial y}(x, y) = \frac{\partial}{\partial y}(C + g(y)) = \frac{d}{dy}g(y) = 0$. Thus $g(y)$ must be a constant, and it follows that $f(x, y) = C + g(y)$ is also a constant. To find this constant, note that $f(1, 1) = 3$. Then $f(x, y) = 3$ for all x and y. So the function given in (a) is the only such function.

35 No. If it did, then $\frac{\partial}{\partial y}(e^x \cos y) = \frac{\partial^2 f}{\partial y\,\partial x} = \frac{\partial^2 f}{\partial x\,\partial y} = \frac{\partial}{\partial x}(e^x \sin y)$. But $\frac{\partial}{\partial y}(e^x \cos y) = -e^x \sin y$, while $\frac{\partial}{\partial x}(e^x \sin y) = e^x \sin y$. Hence f does not exist.

37 (a) When computing $\frac{\partial f}{\partial x}$, y is fixed. So by the first fundamental theorem of calculus, $\frac{\partial f}{\partial x} =$

$$\frac{\partial}{\partial x}\left(\int_0^x \sqrt{y+t}\,dt\right) = \sqrt{y+x}.$$

(b) When computing $\frac{\partial f}{\partial y}$, x is fixed. Since $\sqrt{y+x}$ is a continuous function, we employ part (a) of Exercise 36 (justified in Appendix K). Thus $\frac{\partial f}{\partial y} = \int_0^x \frac{\partial}{\partial y}(\sqrt{y+t})\,dt =$

$\int_0^x \frac{dt}{2\sqrt{y+t}} = \sqrt{y+t}\Big|_0^x = \sqrt{y+x} - \sqrt{y}$ by the second fundamental theorem of calculus.

39 $f(x, t) = e^{-\pi^2 a^2 t}\sin \pi x$; therefore, $f_x = e^{-\pi^2 a^2 t}(\cos \pi x)\pi = \pi e^{-\pi^2 a^2 t}\cos \pi x$, $f_{xx} = \pi e^{-\pi^2 a^2 t}(-\sin \pi x)\pi = -\pi^2 e^{-\pi^2 a^2 t}\sin \pi x$, and $f_t = -\pi^2 a^2 e^{-\pi^2 a^2 t}\sin \pi x$. We thus have $a^2 f_{xx} = -\pi^2 a^2 e^{-\pi^2 a^2 t}\sin \pi x = f_t$, so $f(x, t)$ satisfies the heat equation.

14.6 The Chain Rule

Note: For convenience, we often use subscript notation for partial derivatives in these solutions. As a reminder, f_x means the same thing as $\frac{\partial f}{\partial x}$ and f_{xy} means the same thing as $\frac{\partial^2 f}{\partial y\, \partial x}$.

1 (a) By Theorem 2, $\frac{dz}{dt} = z_x\frac{dx}{dt} + z_y\frac{dy}{dt}$

$= (2xy^3)(2t) + (3x^2y^2)(3t^2)$
$= 2t^2(t^3)^3\cdot 2t + 3(t^2)^2(t^3)^2\cdot 3t^2 = 4t^{12} + 9t^{12}$
$= 13t^{12}$.

(b) Writing z as a function of t gives $z = x^2y^3 = (t^2)^2(t^3)^3 = t^{13}$. Therefore, $\frac{dz}{dt} = 13t^{12}$, which verifies the chain rule in Theorem 2.

3 (a) Applying Theorem 2, we have $\frac{dz}{dt} = z_x\frac{dx}{dt} + z_y\frac{dy}{dt} = [-\sin(xy^2)y^2][2e^{2t}] + [-\sin(xy^2)2xy][3\sec 3t \tan 3t]$

$= -\sin(e^{2t}\sec^2 3t)\cdot\sec^2 3t\cdot 2e^{2t} - \sin(e^{2t}\sec^2 3t)\cdot 2e^{2t}\cdot\sec 3t\cdot 3\sec 3t\tan 3t$

$= -2e^{2t}\sec^2 3t\sin(e^{2t}\sec^2 3t)[1 + 3\tan 3t]$.

(b) Writing z as a function of t yields $z = \cos(xy^2) = \cos(e^{2t}\sec^2 3t)$. Then $\frac{dz}{dt} = -\sin(e^{2t}\sec^2 3t)[e^{2t}\cdot 2\sec 3t\,(3\sec 3t\tan 3t) + 2e^{2t}\cdot\sec^2 3t]$

$= -2e^{2t}\sec^2 3t\sin(e^{2t}\sec^2 3t)[1 + 3\tan 3t]$,

and the chain rule in Theorem 2 is verified.

5 By Theorem 3 we have $\frac{\partial z}{\partial t} = \frac{\partial z}{\partial x}\frac{\partial x}{\partial t} + \frac{\partial z}{\partial y}\frac{\partial y}{\partial t} = x(5x + 6y) = (3t + 4u)[5(3t + 4u) + 6(5t - u)]$
$= (3t + 4u)(45t + 14u)$. Writing z as a function of t and u gives $z = x^2y = (3t + 4u)^2(5t - u)$. Then

$\frac{\partial z}{\partial t} = (3t + 4u)^2\cdot 5 + 2(3t + 4u)\cdot 3\cdot(5t - u)$

$= (3t + 4u)[(3t + 4u)\cdot 5 + 6(5t - u)] = (3t + 4u)(45t + 14u)$, confirming the validity of the chain rule in Theorem 3.

7 (a),(b)

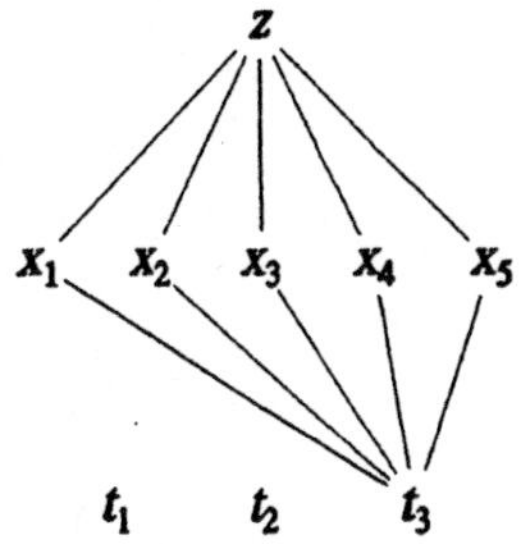

(c) $\frac{\partial z}{\partial t_3} = \sum_{i=1}^{5}\frac{\partial f}{\partial x_i}\cdot\frac{\partial x_i}{\partial t_3}$

(d) x_1, x_3, x_4, x_5

(e) t_1, t_2

9 By Theorem 2, $\frac{dz}{dt} = z_x\frac{dx}{dt} + z_y\frac{dy}{dt} = 4\cdot 4 + 3\cdot 1$

$= 19$.

11 (a) Since $z = f(x, y)$ and x and y are functions of u and v, Theorem 3 applies. Thus $\frac{\partial z}{\partial u} = z_x \cdot x_u + z_y \cdot y_u = z_x + z_y$ and $\frac{\partial z}{\partial v} = z_x \cdot x_v + z_y \cdot y_v = z_x - z_y$ (see the diagram). Therefore

$$\frac{\partial z}{\partial u} \cdot \frac{\partial z}{\partial v} = (z_x + z_y)(z_x - z_y) = (z_x)^2 - (z_y)^2,$$

as claimed.

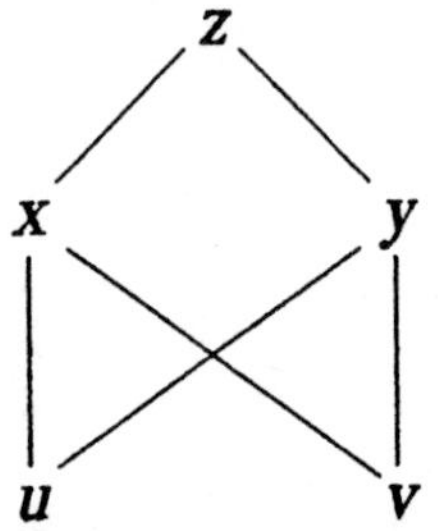

(b) $(z_x)^2 - (z_y)^2 = (2x)^2 - (6y^2)^2 = 4x^2 - 36y^4$, $z = x^2 + 2y^3 = (u + v)^2 + 2(u - v)^3$, so we have $z_u = 2(u + v) + 6(u - v)^2$ and $z_v = 2(u + v) - 6(u - v)^2$. Hence $z_u z_v = [2(u + v)]^2 - [6(u - v)^2]^2 = (2x)^2 - (6y^2)^2 = 4x^2 - 36y^4$, as expected.

13 (a) Let $x = t - u$ and $y = -t + u$. So $z = f(x, y)$ and Theorem 3 applies. Hence $\frac{\partial z}{\partial t} = z_x \cdot x_t + z_y \cdot y_t = z_x - z_y$ and $\frac{\partial z}{\partial u} = z_x \cdot x_u + z_y \cdot y_u = -z_x + z_y$ (see the diagram). Thus

$$\frac{\partial z}{\partial t} + \frac{\partial z}{\partial u} = (z_x - z_y) + (-z_x + z_y) = 0.$$

(b) $z = x^2 y = (t - u)^2(-t + u) = -(t - u)^3$, so $\frac{\partial z}{\partial t} = -3(t - u)^2$, $\frac{\partial z}{\partial u} = 3(t - u)^2$, and

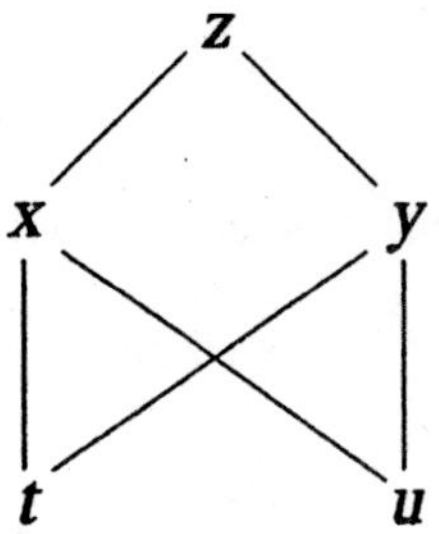

$\frac{\partial z}{\partial t} + \frac{\partial z}{\partial u} = 0.$

15 (a) By Theorem 3, $\frac{\partial z}{\partial x} = f_u \cdot u_x + f_v \cdot v_x = af_u + cf_v$, and $\frac{\partial^2 z}{\partial x^2} = \frac{\partial}{\partial x}\left(\frac{\partial z}{\partial x}\right) = a\frac{\partial f_u}{\partial x} + c\frac{\partial f_v}{\partial x}$. Since f_u and f_v are functions of u and v, Theorem 3 applies as it did for $z = f(u, v)$. Hence $\frac{\partial f_u}{\partial x} = af_{uu} + cf_{uv}$ and $\frac{\partial f_v}{\partial x} = af_{vu} + cf_{vv}$. Finally, we have $\frac{\partial^2 z}{\partial x^2} = a(af_{uu} + cf_{uv}) + c(af_{vu} + cf_{vv})$ $= a^2 f_{uu} + 2acf_{vu} + c^2 f_{vv}$ since $f_{vu} = f_{uv}$.

(b) By Theorem 3, $\frac{\partial z}{\partial y} = f_u \cdot u_y + f_v \cdot v_y = bf_u + df_v$, and $\frac{\partial z^2}{\partial y^2}. = \frac{\partial}{\partial y}\left(\frac{\partial z}{\partial y}\right) = b\frac{\partial f_u}{\partial y} + d\frac{\partial f_v}{\partial y}$. Again, by Theorem 3, we have $\frac{\partial f_u}{\partial y} = bf_{uu} + df_{uv}$ and $\frac{\partial f_v}{\partial y} = bf_{vu} + df_{vv}$.

Then $\frac{\partial^2 z}{\partial y^2} = b(bf_{uu} + df_{uv}) + d(bf_{vu} + df_{vv})$

$= b^2 f_{uu} + 2bdf_{vu} + d^2 f_{vv}$ since $f_{vu} = f_{uv}$.

(c) By the derivation for $\frac{\partial z}{\partial y}$ in (b) and $\frac{\partial f_u}{\partial x}$ and $\frac{\partial f_v}{\partial x}$ in (a), we have $\frac{\partial^2 z}{\partial x\,\partial y} = \frac{\partial}{\partial x}\left(\frac{\partial z}{\partial y}\right) = \frac{\partial}{\partial x}(bf_u + df_v) = b\frac{\partial f_u}{\partial x} + d\frac{\partial f_v}{\partial x} = b(af_{uu} + cf_{uv}) + d(af_{vu} + cf_{vv}) = abf_{uu} + (ad + bc)f_{vu} + cdf_{vv}$, since $f_{vu} = f_{uv}$.

17 (a) Let $t = x + y$ so that $z = f(t)$. Then $\frac{\partial z}{\partial x} = \frac{dz}{dt}\frac{\partial t}{\partial x} = f'(t)\cdot 1 = f'(x + y)$. Next, $\frac{\partial^2 z}{\partial x^2} = \frac{\partial}{\partial x}(f'(x+y)) = f''(x + y)$. Similarly, $\frac{\partial^2 z}{\partial x\,\partial y} = \frac{\partial^2 z}{\partial y^2} = f''(x + y)$, so $\frac{\partial^2 z}{\partial x^2} - 2\frac{\partial^2 z}{\partial x\,\partial y} + \frac{\partial^2 z}{\partial y^2} = 0$.

(b) For $z = (x + y)^3$, $\frac{\partial z}{\partial x} = 3(x + y)^2 = \frac{\partial z}{\partial y}$, $\frac{\partial^2 z}{\partial x\,\partial y} = 6(x + y)$, and $\frac{\partial^2 z}{\partial x^2} = 6(x + y) = \frac{\partial^2 z}{\partial y^2}$. So $\frac{\partial^2 z}{\partial x^2} - 2\frac{\partial^2 z}{\partial x\,\partial y} + \frac{\partial^2 z}{\partial y^2} = 6(x + y) - 2[6(x + y)] + 6(x + y) = 0$, and (a) is verified.

19 (a) Let $u = x + y$ and $v = x - y$ so that $z = f(u) + e^y g(v)$. Then by the chain rule, $\frac{\partial z}{\partial x} = \frac{df}{du}\cdot\frac{\partial u}{\partial x} + e^y\left(\frac{dg}{dv}\cdot\frac{\partial v}{\partial x}\right) = f'(u) + e^y g'(v)$, $\frac{\partial^2 z}{\partial x^2} = \frac{\partial}{\partial x}\left(\frac{\partial z}{\partial x}\right) = \frac{\partial}{\partial x}(f'(u)) + e^y\frac{\partial}{\partial x}(g'(v)) = \frac{d}{du}(f'(u))\cdot\frac{\partial u}{\partial x} + e^y\frac{d}{dv}(g'(v))\cdot\frac{\partial v}{\partial x} = f''(u) + e^y g''(v)$, $\frac{\partial z}{\partial y} = \frac{df}{du}\cdot\frac{\partial u}{\partial y} + \frac{\partial}{\partial y}(e^y g) = f'(u) + e^y\cdot\frac{dg}{dv}\cdot\frac{\partial v}{\partial y} + e^y g(v) = f'(u) + e^y(g(v) - g'(v))$, and $\frac{\partial^2 z}{\partial y^2} = \frac{\partial}{\partial y}\left(\frac{\partial z}{\partial y}\right) = \frac{\partial}{\partial y}(f'(u)) + \frac{\partial}{\partial y}[e^y(g(v) - g'(v))] = \frac{d}{du}(f'(u))\cdot\frac{\partial u}{\partial y} + e^y\left(g'(v)\cdot\frac{\partial v}{\partial y} - \frac{d}{dv}(g'(v))\cdot\frac{\partial v}{\partial y}\right) + e^y(g(v) - g'(v)) = f''(u) + e^y(g(v) - 2g'(v) + g''(v))$. Thus $\frac{\partial^2 z}{\partial x^2} - \frac{\partial^2 z}{\partial y^2} - \frac{\partial z}{\partial x} + \frac{\partial z}{\partial y} = (f''(u) + e^y g''(v)) - [f''(u) + e^y(g(v) - 2g'(v) + g''(v)] - (f'(u) + e^y g'(v)) + [f'(u) + e^y(g(v) - g'(v))] = f''(u) + e^y g''(v) - f''(u) - e^y g(v) + 2e^y g'(v) - e^y g''(v) - f'(u) - e^y g'(v) + f'(u) + e^y(g(v) - g'(v)) = 0$, as claimed.

(b) For $z = (x + y)^2 + e^y\sin(x - y)$, $\frac{\partial z}{\partial x} = 2(x + y) + e^y\cos(x - y)$, $\frac{\partial^2 z}{\partial x^2} =$

$2 - e^y \sin(x - y)$, $\frac{\partial z}{\partial y} = 2(x + y) +$

$e^y \sin(x - y) - e^y \cos(x - y)$ and $\frac{\partial^2 z}{\partial y^2} =$

$2 + e^y \sin(x - y) - e^y \cos(x - y) -$

$e^y \cos(x - y) - e^y \sin(x - y)$

$= 2(1 - e^y \cos(x - y))$. Thus

$\frac{\partial^2 z}{\partial x^2} - \frac{\partial^2 z}{\partial y^2} - \frac{\partial z}{\partial x} + \frac{\partial z}{\partial y} = [2 - e^y \sin(x - y)]$

$- 2(1 - e^y \cos(x - y)) - [2(x + y) +$

$e^y \cos(x - y)] + [2(x + y) + e^y \sin(x - y)$

$- e^y \cos(x - y)] = 2 - e^y \sin(x - y) - 2 +$

$2e^y \cos(x - y) - 2(x + y) - e^y \cos(x - y)$

$+ 2(x + y) + e^y \sin(x - y) - e^y \cos(x - y)$

$= 0$, and (a) is verified.

21 $u = f(r)$ and $r = (x^2 + y^2 + z^2)^{1/2}$, so $\frac{\partial r}{\partial x} =$

$\frac{1}{2}(x^2 + y^2 + z^2)^{-1/2}(2x) = x(x^2 + y^2 + z^2)^{-1/2} =$

$\frac{x}{r}$. Similarly, $\frac{\partial r}{\partial y} = \frac{y}{r}$ and $\frac{\partial r}{\partial z} = \frac{z}{r}$. Thus $u_x =$

$f'(r)\frac{\partial r}{\partial x} = u_r \cdot \frac{x}{r}$ and $u_{xx} = \frac{\partial}{\partial x}(u_x) = \frac{\partial}{\partial x}(xu_r/r)$

$= x \cdot \frac{\partial}{\partial x}(u_r/r) + u_r/r = x\frac{ru_{rr} - u_r}{r^2}\left(\frac{x}{r}\right) + \frac{u_r}{r} =$

$\frac{x^2}{r^3}(ru_{rr} - u_r) + \frac{u_r}{r}$. Similarly, we have $u_{yy} =$

$\frac{y^2}{r^3}(ru_{rr} - u_r) + \frac{u_r}{r}$, and $u_{zz} =$

$\frac{z^2}{r^3}(ru_{rr} - u_r) + \frac{u_r}{r}$. Putting these together, we

have $u_{xx} + u_{yy} + u_{zz} =$

$\frac{x^2 + y^2 + z^2}{r^3}(ru_{rr} - u_r) + 3 \cdot \frac{u_r}{r} =$

$\frac{r^2}{r^3}(ru_{rr} - u_r) + 3 \cdot \frac{u_r}{r} = u_{rr} + 2 \cdot \frac{u_r}{r}$.

23 $\frac{dT}{dt} = T_x \cdot \frac{dx}{dt} + T_y \cdot \frac{dy}{dt} + T_z \cdot \frac{dz}{dt}$

$= 4 \cdot 4 + 7 \cdot 4 + 9(-3) = 17°$/sec.

25 (a) Recalling the algebraic identity $a - b =$ $(c - b) + (d - c) + (a - d)$, the expression for Δf follows immediately by setting $a =$ $f(x + \Delta x, y + \Delta y, z + \Delta z)$, $b = f(x, y, z)$, $c = f(x + \Delta x, y, z)$, and $d =$ $f(x + \Delta x, y + \Delta y, z)$.

(b) By the mean-value theorem, there are numbers c_1, c_2, and c_3 between x and $x + \Delta x$, y and $y + \Delta y$, and z and $z + \Delta z$, respectively, such that

$f(x + \Delta x, y, z) - f(x, y, z) = \frac{\partial f}{\partial x}(c_1, y, z)\Delta x$,

$f(x + \Delta x, y + \Delta y, z) - f(x + \Delta x, y, z) =$

$\frac{\partial f}{\partial y}(x + \Delta x, c_2, z)\Delta y$, and $f(x + \Delta x, y + \Delta y,$

$z + \Delta z) - f(x + \Delta x, y + \Delta y, z) =$

$\frac{\partial f}{\partial z}(x + \Delta x, y + \Delta y, c_3)\Delta z$. By (a), $\Delta f =$

$\frac{\partial f}{\partial x}(c_1, y, z)\Delta x + \frac{\partial f}{\partial y}(x + \Delta x, c_2, z)\Delta y +$

$\frac{\partial f}{\partial z}(x + \Delta x, y + \Delta y, c_3)\Delta z$. Assuming $\frac{\partial f}{\partial x}$,

$\frac{\partial f}{\partial y}$, and $\frac{\partial f}{\partial z}$ are continuous at (x, y, z), we conclude that $\frac{\partial f}{\partial x}(c_1, y, z) = \frac{\partial f}{\partial x}(x, y, z) + \epsilon_1$,

$\frac{\partial f}{\partial y}(x + \Delta x, c_2, z) = \frac{\partial f}{\partial y}(x, y, z) + \epsilon_2$, and

$\frac{\partial f}{\partial z}(x + \Delta x, y + \Delta y, c_3) = \frac{\partial f}{\partial z}(x, y, z) + \epsilon_3$,

where $\epsilon_1, \epsilon_2, \epsilon_3 \to 0$ as $\Delta x, \Delta y, \Delta z \to 0$. The given expression for Δf follows.

(c) Let $u = f(x, y, z)$ have continuous partial derivatives $\frac{\partial f}{\partial x}$, $\frac{\partial f}{\partial y}$, and $\frac{\partial f}{\partial z}$. Let $x = x(t_1, t_2, t_3, \cdots, t_n)$, $y = y(t_1, t_2, t_3, \cdots, t_n)$ and $z = z(t_1, t_2, t_3, \cdots, t_n)$ have continuous partial derivatives of the first order. Then

$\frac{\partial u}{\partial t_i} = \frac{\partial u}{\partial x}\cdot\frac{\partial x}{\partial t_i} + \frac{\partial u}{\partial y}\cdot\frac{\partial y}{\partial t_i} + \frac{\partial u}{\partial z}\cdot\frac{\partial z}{\partial t_i}$ for all i between 1 and n.

27 By the chain rule, $\frac{\partial z}{\partial r} = \frac{\partial f}{\partial x}\cdot\frac{\partial x}{\partial r} + \frac{\partial f}{\partial y}\cdot\frac{\partial y}{\partial r} =$

$\frac{\partial f}{\partial x}\cdot\frac{\partial}{\partial r}(r\cos\theta) + \frac{\partial f}{\partial y}\cdot\frac{\partial}{\partial r}(r\sin\theta) =$

$\cos\theta\,\frac{\partial f}{\partial x} + \sin\theta\,\frac{\partial f}{\partial y}$ and $\frac{\partial^2 z}{\partial r^2} = \frac{\partial}{\partial r}\left(\frac{\partial z}{\partial r}\right)$

$= \cos\theta\,\frac{\partial}{\partial r}\left(\frac{\partial f}{\partial x}\right) + \sin\theta\,\frac{\partial}{\partial r}\left(\frac{\partial f}{\partial y}\right)$

$= \cos\theta\left(\frac{\partial f_x}{\partial x}\cdot\frac{\partial x}{\partial r} + \frac{\partial f_x}{\partial y}\cdot\frac{\partial y}{\partial r}\right) +$

$\sin\theta\left(\frac{\partial f_y}{\partial x}\cdot\frac{\partial x}{\partial r} + \frac{\partial f_y}{\partial y}\cdot\frac{\partial y}{\partial r}\right)$

$= \cos\theta\left(\cos\theta\,\frac{\partial^2 f}{\partial x^2} + \sin\theta\,\frac{\partial^2 f}{\partial y\,\partial x}\right) +$

$\sin\theta\left(\cos\theta\,\frac{\partial^2 f}{\partial x\,\partial y} + \sin\theta\,\frac{\partial^2 f}{\partial y^2}\right) =$

$\cos^2\theta\,\frac{\partial^2 f}{\partial x^2} + 2\sin\theta\cos\theta\,\frac{\partial^2 f}{\partial x\,\partial y} + \sin^2\theta\,\frac{\partial^2 f}{\partial y^2}$

since $\frac{\partial^2 f}{\partial x\,\partial y} = \frac{\partial^2 f}{\partial y\,\partial x}$.

29 By the chain rule $u_s = u_x \cdot x_s + u_y\, y_s$. Also, $u_{ss} = (u_x \cdot x_s)_s + (u_y \cdot y_s)_s = u_{xs}\cdot x_s + u_x \cdot x_{ss} + u_{ys}\cdot y_s + u_y \cdot y_{ss} = (u_{xx}\cdot x_s + u_{xy}\cdot y_s)x_s + u_x\cdot x_{ss} + (u_{yx}\cdot x_s + u_{yy}\cdot y_s)y_s + u_y\cdot y_{ss} = u_{xx}\cdot(x_s)^2 + 2u_{yx}x_s y_s + u_{yy}\cdot(y_s)^2 + u_x\cdot x_{ss} + u_y\cdot y_{ss}$, as claimed.

31 (a) $f(x, y) = 3x + 4y$, so $f(kx, ky) = 3kx + 4ky = k(3x + 4y) = kf(x, y)$, so f is a homogeneous function. Furthermore, $f_x = 3$ and $f_y = 4$, so $xf_x + yf_y = 3x + 4y = f(x, y)$. Euler's theorem holds.

(b) $g(x, y) = x^3y^{-2}$, so $g(kx, ky) = (kx)^3(ky)^{-2} = k^3x^3k^{-2}y^{-2} = kx^3y^{-2} = kg(x, y)$, so g is homogeneous. Furthermore, $g_x = 3x^2y^{-2}$ and $g_y = -2x^3y^{-3}$, so $xg_x + yg_y = 3x^3y^{-2} + (-2x^3y^{-2}) = x^3y^{-2} = g(x, y)$.

(c) $h(x, y) = xe^{x/y}$, so $h(kx, ky) = kxe^{kx/(ky)} = kxe^{x/y} = kh(x, y)$; h is homogeneous.

Furthermore, $h_x = x\left(e^{x/y}\cdot\frac{1}{y}\right) + e^{x/y} =$

$e^{x/y}\left(1 + \frac{x}{y}\right)$ and $h_y = xe^{x/y}\left(-\frac{x}{y^2}\right) =$

$-\frac{x^2}{y^2}e^{x/y}$, so $xh_x + yh_y =$

$e^{x/y}\left(x + \frac{x^2}{y}\right) - \frac{x^2}{y}e^{x/y} = e^{x/y}\left(x + \frac{x^2}{y} - \frac{x^2}{y}\right)$

$= xe^{x/y} = h(x, y)$.

33 Let $g(x, y, k) = f(kx, ky) = k^r f(x, y)$ and compute g_k. First, $g(x, y, k) = k^r f(x, y)$, so $g_k = rk^{r-1}f(x, y)$. Second $g(x, y, k) = f(u, v)$ for $u = kx$ and $v = ky$, so $g_k = f_1 \cdot \frac{\partial u}{\partial k} + f_2 \cdot \frac{\partial v}{\partial k} = f_1 \cdot x + f_2 \cdot y = xf_1 + yf_2$.

Combining these results, we have $rk^{r-1}f(x, y) = xf_1(kx, ky) + yf_2(kx, ky)$. Letting $k = 1$, we have $rf = xf_1 + yf_2$.

35 $k^r f(x, y) = f(kx, ky) = f(u, v)$ for $u = kx$ and $v = ky$. Differentiating with respect to x, we obtain

$$k^r f_x(x, y) = \frac{\partial f}{\partial u} \cdot \frac{\partial u}{\partial x} + \frac{\partial f}{\partial v} \cdot \frac{\partial v}{\partial x} = \frac{\partial f}{\partial u} \cdot k + 0 =$$

$kf_u(u, v) = kf_x(kx, ky)$. Hence $k^{r-1}f_x(x, y) = f_x(kx, ky)$, so f_x is homogeneous of degree $r - 1$.

14.7 Directional Derivatives and the Gradient

1 Let $f(x, y) = x^4y^5$. Then $\frac{\partial f}{\partial x} = 4x^3y^5$ and $\frac{\partial f}{\partial y} = 5x^4y^4$.

(a) By Theorem 1, the directional derivative in the direction of $\mathbf{u} = 1\mathbf{i} + 0\mathbf{j}$ at (1, 1) is

$$\frac{\partial f}{\partial x}(1, 1) \cos\theta + \frac{\partial f}{\partial y}(1, 1) \sin\theta$$

$$= \frac{\partial f}{\partial x}(1, 1) \cdot 1 + \frac{\partial f}{\partial y}(1, 1) \cdot 0 = \frac{\partial f}{\partial x}(1, 1)$$

$$= 4 \cdot 1^3 \cdot 1^5 = 4.$$

(b) The directional derivative in the direction of $\mathbf{u} = (-1)\mathbf{i} + 0\mathbf{j}$ at (1, 1) is

$$\frac{\partial f}{\partial x}(1, 1) \cos\theta + \frac{\partial f}{\partial y}(1, 1) \sin\theta$$

$$= \frac{\partial f}{\partial x}(1, 1) \cdot (-1) + \frac{\partial f}{\partial y}(1, 1) \cdot 0 = -\frac{\partial f}{\partial x}(1, 1)$$

$$= -4 \cdot 1^3 \cdot 1^5 = -4.$$

(c) The directional derivative in the direction of $\mathbf{u} = \cos\frac{\pi}{4}\,\mathbf{i} + \sin\frac{\pi}{4}\,\mathbf{j}$ is

$$\frac{\partial f}{\partial x}(1, 1) \cos\frac{\pi}{4} + \frac{\partial f}{\partial y}(1, 1) \sin\frac{\pi}{4}$$

$$= \frac{\sqrt{2}}{2}\frac{\partial f}{\partial x}(1, 1) + \frac{\sqrt{2}}{2}\frac{\partial f}{\partial y}(1, 1)$$

$$= \frac{\sqrt{2}}{2} \cdot 4 + \frac{\sqrt{2}}{2} \cdot 5 = \frac{9\sqrt{2}}{2}.$$

3 Let $f(x, y, z) = x^2yz^3$. Then $\frac{\partial f}{\partial x} = 2xyz^3$, $\frac{\partial f}{\partial y} = x^2z^3$, and $\frac{\partial f}{\partial z} = 3x^2yz^2$.

(a) By Theorem 3, the directional derivative in the direction of $\mathbf{u} = 0\mathbf{i} + 1\mathbf{j} + 0\mathbf{k}$ is

$$\frac{\partial f}{\partial x}\cos\alpha + \frac{\partial f}{\partial y}\cos\beta + \frac{\partial f}{\partial z}\cos\gamma =$$

$$\frac{\partial f}{\partial x} \cdot 0 + \frac{\partial f}{\partial y} \cdot 1 + \frac{\partial f}{\partial z} \cdot 0 = \frac{\partial f}{\partial y} = x^2z^3.$$

(b) The directional derivative in the direction of $\mathbf{u} = 0\mathbf{i} + 0\mathbf{j} + 1\mathbf{k}$ is $\frac{\partial f}{\partial x} \cdot 0 + \frac{\partial f}{\partial y} \cdot 0 + \frac{\partial f}{\partial z} \cdot 1$

$$= \frac{\partial f}{\partial z} = 3x^2yz^2.$$

(c) The directional derivative in the direction of $(-1)\mathbf{i} + 0\mathbf{j} + 0\mathbf{k}$ is $-\frac{\partial f}{\partial x} = -2xyz^3$.

5 (a) $\nabla f(2, 3) = \frac{\partial f}{\partial x}(2, 3)\,\mathbf{i} + \frac{\partial f}{\partial y}(2, 3)\,\mathbf{j} = 4\mathbf{i} + 5\mathbf{j}$

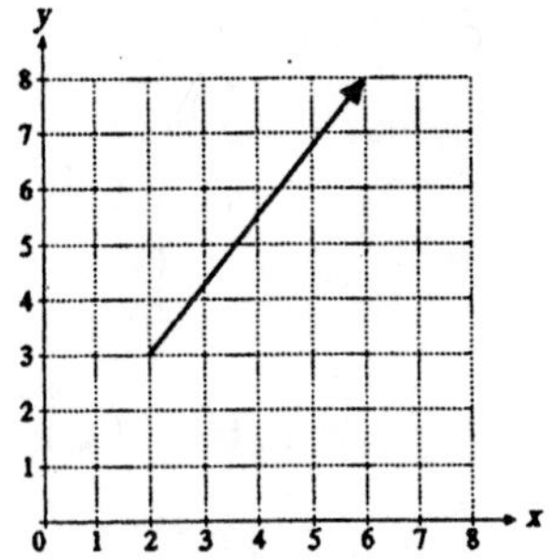

(b) By Theorem 2, the maximal directional derivative of f at (2, 3) is $|\nabla f(2, 3)| =$

$\|4\mathbf{i} + 5\mathbf{j}\| = \sqrt{4^2 + 5^2} = \sqrt{41}$.

(c) By Theorem 2, $D_{\mathbf{u}}f$ at (2, 3) is maximal in the direction of ∇f at (2, 3); that is, $\mathbf{u} =$

$\frac{\nabla f(2, 3)}{|\nabla f(2,3)|} = \frac{4\mathbf{i} + 5\mathbf{j}}{\sqrt{41}} = \frac{4}{\sqrt{41}}\mathbf{i} + \frac{5}{\sqrt{41}}\mathbf{j}$.

7 $f(x, y) = x^2y$, $f_x(x, y) = 2xy$, and $f_y(x, y) = x^2$, so $\nabla f(x, y) = 2xy\mathbf{i} + x^2\mathbf{j}$.

(a) $\nabla f(2, 5) = 20\mathbf{i} + 4\mathbf{j}$

(b) $\nabla f(3, 1) = 6\mathbf{i} + 9\mathbf{j}$

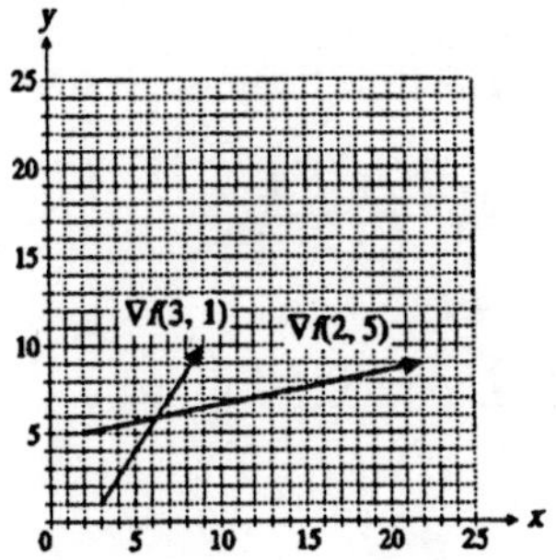

9 Let $\mathbf{u} = \frac{\nabla f(a, b)}{|\nabla f(a, b)|}$. Then $D_{\mathbf{u}}f = \nabla f \cdot \mathbf{u} =$

$|\nabla f(a, b)|\,|\mathbf{u}| \cos 0° = 5$. To achieve a minimal directional derivative of f at (a, b), we must minimize the cosine term in the dot product above. This occurs when the angle between the direction chosen and the direction of $\mathbf{u}$ is 180°; hence, choose the directional derivative of f at (a, b) in the direction of $-\mathbf{u}$. Thus $D_{-\mathbf{u}}f = \nabla f \cdot (-\mathbf{u}) =$

$|\nabla f(a, b)|\,|-\mathbf{u}| \cos 180° = -5$ is the minimal directional derivative of f at (a, b).

11 Note that $\nabla f(a, b) = 2\mathbf{i} + 3\mathbf{j}$. If $\mathbf{u}$ is a unit vector, then Theorem 1 says that $D_{\mathbf{u}}f = (2\mathbf{i} + 3\mathbf{j})\cdot\mathbf{u}$.

(a) We want $(2\mathbf{i} + 3\mathbf{j})\cdot\mathbf{u} = 0$; that is, $\mathbf{u}$ must be perpendicular to $2\mathbf{i} + 3\mathbf{j}$. If $\mathbf{u} = x\mathbf{i} + y\mathbf{j}$, then our condition is equivalent to $2x + 3y = 0$,

so $y = -\frac{2}{3}x$. Thus $\mathbf{u}$ is parallel to $3\mathbf{i} - 2\mathbf{j}$, so $\mathbf{u}$

$= \pm\frac{3\mathbf{i} - 2\mathbf{j}}{\|3\mathbf{i} - 2\mathbf{j}\|} = \pm\frac{3\mathbf{i} - 2\mathbf{j}}{\sqrt{3^2 + 2^2}} = \pm\frac{3\mathbf{i} - 2\mathbf{j}}{\sqrt{13}}$.

(b) By Theorem 2, $\mathbf{u}$ points in the same direction as ∇f; thus $\mathbf{u} = \frac{2\mathbf{i} + 3\mathbf{j}}{\sqrt{2^2 + 3^2}} = \frac{2\mathbf{i} + 3\mathbf{j}}{\sqrt{13}}$.

(c) $\mathbf{u}$ should point in the direction opposite ∇f, so

$\mathbf{u} = -\frac{2\mathbf{i} + 3\mathbf{j}}{\sqrt{13}}$.

13 (a) $\mathbf{u}$ is a unit vector pointing from (1, 2) toward (0.99, 2.01) or the reverse; that is, $\mathbf{u} =$

$\pm\frac{-0.01\mathbf{i} + 0.01\mathbf{j}}{\sqrt{(-0.01)^2 + (0.01)^2}} = \pm\frac{1}{\sqrt{2}}(-\mathbf{i} + \mathbf{j})$.

(b) The distance from (1, 2) to (0.99, 2.01) is

$\frac{\sqrt{2}}{100}$, so $D_{\mathbf{u}}f \approx \pm\frac{1.98 - 2}{\sqrt{2}/100} = \mp\sqrt{2}$.

15 Let east be in the direction of $\mathbf{i}$ and north be in the direction of $\mathbf{j}$. In addition, let $T(x, y)$ be the temperature at (x, y). Then we are given $T_x = 0.02°/\text{cm}$ and $T_y = -0.03°/\text{cm}$. Hence $\nabla T =$

$0.02°/\text{cm}\,\mathbf{i} - 0.03°/\text{cm}\,\mathbf{j}$.

(a) $D_{-\mathbf{j}}T$ is asked for. Now $D_{-\mathbf{j}}T = \nabla T \cdot (-\mathbf{j}) = 0.03°/\text{cm}$.

(b) The direction required is $\mathbf{u} = \cos 30°\,\mathbf{i} + \sin 30°\,\mathbf{j}$. Then $D_{\mathbf{u}}T = \nabla T \cdot \mathbf{u} = (0.02)(\cos 30°) + (-0.03)(\sin 30°)$

$$= \frac{0.02\sqrt{3} - 0.03}{2} \approx 0.0023°/\text{cm}.$$

(c) We wish to find a $\mathbf{v}$ such that $D_{\mathbf{v}}f = 0°/\text{cm}$.

Let $\mathbf{v} = \dfrac{0.03\mathbf{i} + 0.02\mathbf{j}}{\sqrt{0.03^2 + 0.02^2}}$. Then $D_{\mathbf{v}}f = \nabla T \cdot \mathbf{v}$

$$= \frac{0.02 \cdot 0.03 - 0.03 \cdot 0.02}{\sqrt{0.03^2 + 0.02^2}} = 0°/\text{cm}.$$ The bug should crawl in the direction of $\mathbf{v}$.

17 Let $\mathbf{u} = \dfrac{\mathbf{i} + \mathbf{j}}{\sqrt{1^2 + 1^2}} = \dfrac{1}{\sqrt{2}}(\mathbf{i} + \mathbf{j})$ and observe that $\mathbf{u}$ is a unit vector pointing from (1, 2) toward (1.1, 2.1). The distance from (1, 2) to (1.1, 2.1) is $\dfrac{\sqrt{2}}{10}$. Since $D_{\mathbf{u}}f \approx \dfrac{f(1.1, 2.1) - f(1, 2)}{\sqrt{2}/10}$, $f(1.1, 2.1)$

$$\approx (D_{\mathbf{u}}f)\frac{\sqrt{2}}{10} + f(1, 2) = \frac{0.7\sqrt{2}}{10} + 3$$

$$= \frac{300 + 7\sqrt{2}}{100} \approx 3.0990.$$

19 Letting $f(x, y, z) = xyz^2$, we have $f_x = yz^2$, $f_y = xz^2$, and $f_z = 2xyz$, so $\nabla f = yz^2\mathbf{i} + xz^2\mathbf{j} + 2xyz\mathbf{k} = \mathbf{j}$ at (1, 0, 1). The directional derivative in the direction $\mathbf{i} + \mathbf{j} + \mathbf{k}$ is $\nabla f \cdot \dfrac{\mathbf{i} + \mathbf{j} + \mathbf{k}}{|\mathbf{i} + \mathbf{j} + \mathbf{k}|} = \dfrac{1}{\sqrt{3}}$. The maximum directional derivative at (1, 0, 1) is $\|\nabla f\| = 1$.

21 Letting $f(x, y, z) = e^{xy \sin z}$, we have $f_x = y(\sin z)e^{xy \sin z}$, $f_y = x(\sin z)e^{xy \sin z}$, $f_z = xy(\cos z)e^{xy \sin z}$, and

$$\nabla f = e^{xy \sin z}(y \sin z\,\mathbf{i} + x \sin z\,\mathbf{j} + xy \cos z\,\mathbf{k})$$

$$= e^{1/\sqrt{2}} \cdot \frac{1}{\sqrt{2}}(\mathbf{i} + \mathbf{j} + \mathbf{k})$$ at $(1, 1, \pi/4)$. The directional derivative in the direction of $\mathbf{i} + \mathbf{j} + 3\mathbf{k}$ is $\nabla f \cdot \dfrac{\mathbf{i} + \mathbf{j} + 3\mathbf{k}}{|\mathbf{i} + \mathbf{j} + 3\mathbf{k}|} = \dfrac{e^{1/\sqrt{2}}}{\sqrt{2}} \cdot \dfrac{(\mathbf{i} + \mathbf{j} + \mathbf{k}) \cdot (\mathbf{i} + \mathbf{j} + 3\mathbf{k})}{\sqrt{11}}$

$$= \frac{5}{\sqrt{22}}e^{1/\sqrt{2}}.$$ The maximum directional derivative at $(1, 1, \pi/4)$ is $\|\nabla f\| = \sqrt{\dfrac{3}{2}}e^{1/\sqrt{2}}$.

23 $f(x, y, z) = \ln(1 + xyz)$, so $f_x = \dfrac{yz}{1 + xyz}$, $f_y = \dfrac{xz}{1 + xyz}$, $f_z = \dfrac{xy}{1 + xyz}$, and $\nabla f =$

$$\frac{1}{1 + xyz}(yz\mathbf{i} + xz\mathbf{j} + xy\mathbf{k}) = \frac{1}{7}(3\mathbf{i} + 2\mathbf{j} + 6\mathbf{k})$$ at (2, 3, 1). The directional derivative in the direction $-\mathbf{i} + \mathbf{j}$ is $\nabla f \cdot \dfrac{-\mathbf{i} + \mathbf{j}}{|-\mathbf{i} + \mathbf{j}|} =$

$$\frac{1}{7} \cdot \frac{(3\mathbf{i} + 2\mathbf{j} + 6\mathbf{k}) \cdot (-\mathbf{i} + \mathbf{j})}{\sqrt{2}} = -\frac{1}{7\sqrt{2}}.$$ The maximum directional derivative at (2, 3, 1) is

$$\|\nabla f\| = \frac{1}{7}\sqrt{3^2 + 2^2 + 6^2} = 1.$$

25 (a) $\nabla f(x, y, z) = 2\mathbf{i} + 3\mathbf{j} + \mathbf{k}$, so $\nabla f(0, 0, 0) = \nabla f(1, 1, 1) = 2\mathbf{i} + 3\mathbf{j} + \mathbf{k}$.

(b)

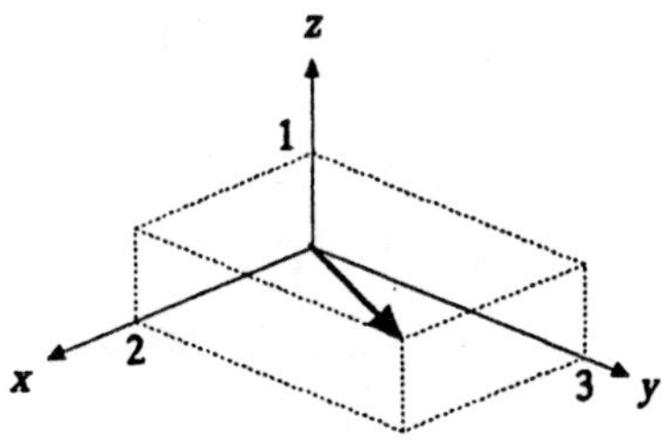

27 Suppose that $\nabla f(a, b) = A\mathbf{i} + B\mathbf{j}$. Let $\mathbf{u}_1 = \dfrac{-B\mathbf{i} + A\mathbf{j}}{\sqrt{A^2 + B^2}}$. Then $\mathbf{u}_1$ is a unit vector perpendicular to ∇f, so the directional derivative in its direction is 0. Let $\mathbf{u}_2 = -\mathbf{u}_1$ for the other vector.

29 (a) $D_{\mathbf{u}}T = \nabla T \cdot \mathbf{u} = (2\mathbf{i} + 3\mathbf{j} + 4\mathbf{k}) \cdot \dfrac{\mathbf{i} - \mathbf{j} + 2\mathbf{k}}{|\mathbf{i} - \mathbf{j} + 2\mathbf{k}|}$

$= \dfrac{2 - 3 + 8}{\sqrt{6}} = \dfrac{7}{\sqrt{6}}$

(b) The change is approximately the product of the distance traveled and the directional derivative in that direction; that is, approximately $0.2 \cdot \dfrac{7}{\sqrt{6}} = \dfrac{7}{5\sqrt{6}}$.

(c) $D_{\mathbf{u}}T = 0$ if and only if $\nabla T \cdot \mathbf{u} = 0$; that is, if and only if $\mathbf{u}$ is perpendicular to $2\mathbf{i} + 3\mathbf{j} + 4\mathbf{k}$. Using the cross product, we can construct infinitely many such vectors. For any a, b, c, $2\mathbf{i} + 3\mathbf{j} + 4\mathbf{k}$ is perpendicular to $(2\mathbf{i} + 3\mathbf{j} + 4\mathbf{k}) \times (a\mathbf{i} + b\mathbf{j} + c\mathbf{k})$

$= \begin{vmatrix} \mathbf{i} & \mathbf{j} & \mathbf{k} \\ 2 & 3 & 4 \\ a & b & c \end{vmatrix} = (3c - 4b)\mathbf{i} + (4a - 2c)\mathbf{j} + (2b - 3a)\mathbf{k}$. Dividing this vector by its length gives a unit vector. For example, letting (a, b, c) in turn be $(1, 0, 0)$, $(0, 1, 0)$, and $(0, 0, 1)$ yields, respectively, $\mathbf{u} = \dfrac{4\mathbf{j} - 3\mathbf{k}}{5}$, $\dfrac{-2\mathbf{i} + \mathbf{k}}{\sqrt{5}}$, and $\dfrac{3\mathbf{i} - 2\mathbf{j}}{\sqrt{13}}$.

31 (a) Note that $f(x, y) = \dfrac{1}{\sqrt{x^2 + y^2}} = \dfrac{1}{|x\mathbf{i} + y\mathbf{j}|} = \dfrac{1}{|\mathbf{r}|}$. We have $f_x = -\dfrac{1}{2}(x^2 + y^2)^{-3/2}(2x) = -x(x^2 + y^2)^{-3/2} = \dfrac{-x}{(x^2 + y^2)^{3/2}} = -\dfrac{x}{|\mathbf{r}|^3}$.

Similarly, $f_y = -\dfrac{y}{|\mathbf{r}|^3}$, so $\nabla f = -\dfrac{x\mathbf{i}}{\|\mathbf{r}\|^3} - \dfrac{y\mathbf{j}}{|\mathbf{r}|^3} = \dfrac{-\mathbf{r}}{|\mathbf{r}|^3}$, as claimed.

(b) $\|\nabla f\| = \left| -\dfrac{\mathbf{r}}{\|\mathbf{r}\|^3} \right| = \dfrac{|\mathbf{r}|}{|\mathbf{r}|^3} = \dfrac{1}{|\mathbf{r}|^2}$

33 We know that the slope of the curve at (a, b) is $\dfrac{dy}{dx} = -\dfrac{a}{b}$. But $\nabla f = 2x\mathbf{i} + 2y\mathbf{j} = 2a\mathbf{i} + 2b\mathbf{j}$, a vector of slope $\dfrac{b}{a}$. The gradient and the tangent line are perpendicular.

35 Suppose that $D_{\mathbf{u}}f = 0$ for $\mathbf{u} = x\mathbf{i} + y\mathbf{j} + z\mathbf{k}$. Then $0 = D_{\mathbf{u}}f = \nabla f \cdot \mathbf{u} = (2\mathbf{i} + 3\mathbf{j} + \mathbf{k}) \cdot (x\mathbf{i} + y\mathbf{j} + z\mathbf{k}) = 2x + 3y + z$.

(a) Suppose $z = 0$. Then $2x + 3y = 0$ and, as in Exercise 11(a), $\mathbf{u} = \pm\dfrac{3\mathbf{i} - 2\mathbf{j}}{\sqrt{13}}$. Suppose instead that $x = 0$. Then $3y + z = 0$ and $\mathbf{u}$

must be parallel to $\mathbf{j} - 3\mathbf{k}$, so $\mathbf{u} = \pm\dfrac{\mathbf{j} - 3\mathbf{k}}{\sqrt{10}}$.

Thus we may choose, for example, the vectors

$\dfrac{-3\mathbf{i} + 2\mathbf{j}}{\sqrt{13}}$, $\dfrac{3\mathbf{i} - 2\mathbf{j}}{\sqrt{13}}$, and $\dfrac{\mathbf{j} - 3\mathbf{k}}{\sqrt{10}}$.

(b) There are infinitely many choices for **u**. The heads of the acceptable vectors lie on the circle formed by the intersection of the plane $2x + 3y + z = 0$ and the sphere $x^2 + y^2 + z^2 = 1$.

37 Observe that $g(t) = f(x, y)$. Letting $x = x(t)$ and $y = y(t)$, we have from the chain rule that $\dfrac{\partial g}{\partial t} =$

$\dfrac{\partial f}{\partial x}\cdot\dfrac{dx}{dt} + \dfrac{\partial f}{\partial y}\cdot\dfrac{dy}{dt} = \nabla f\cdot\mathbf{v}$, where $\mathbf{v}$ is the velocity vector of the bug at (1, 1). Since the direction of **i** is east and the direction of **j** is north, $\mathbf{v} = \dfrac{1}{\sqrt{2}}(-3\mathbf{i} + 3\mathbf{j})$. It follows that $\dfrac{dg}{dt}$ at (1, 1) is $\nabla f\cdot\mathbf{v}$

$$= (2\mathbf{i} + 3\mathbf{j})\cdot\left(-\frac{3}{\sqrt{2}}\mathbf{i} + \frac{3}{\sqrt{2}}\mathbf{j}\right) = -\frac{6}{\sqrt{2}} + \frac{9}{\sqrt{2}}$$

$= \dfrac{3}{\sqrt{2}}$ degrees/sec.

39 Let $\mathbf{u}_1 = \cos\theta_1\,\mathbf{i} + \sin\theta_1\,\mathbf{j}$ and $\mathbf{u}_2 = \cos\theta_2\,\mathbf{i} + \sin\theta_1\,\mathbf{j}$. Then $D_{\mathbf{u}_1}f = \nabla f\cdot\mathbf{u}_1 = \cos\theta_1 f_x + \sin\theta_1 f_y$, and $D_{\mathbf{u}_2}(D_{\mathbf{u}_1}f) = \nabla(\cos\theta_1 f_x + \sin\theta_1 f_y)\cdot\mathbf{u}_2 = \cos\theta_2(\cos\theta_1 f_{xx} + \sin\theta_1 f_{yx}) + \sin\theta_2(\cos\theta_1 f_{xy} + \sin\theta_1 f_{yy}) = \cos\theta_1\cos\theta_2 f_{xx} + \sin\theta_1\cos\theta_2 f_{yx} + \cos\theta_1\sin\theta_2 f_{xy} + \sin\theta_1\sin\theta_2 f_{yy}$ while $D_{\mathbf{u}_2}f = \nabla f\cdot\mathbf{u}_2 = \cos\theta_2 f_x + \sin\theta_2 f_y$ and $D_{\mathbf{u}_1}(D_{\mathbf{u}_2}f) = \nabla(\cos\theta_2 f_x + \sin\theta_2 f_y)\cdot\mathbf{u}_1 = \cos\theta_1(\cos\theta_2 f_{xx} + \sin\theta_2 f_{yx}) + \sin\theta_1(\cos\theta_2 f_{xy} + \sin\theta_2 f_{yy}) = \cos\theta_1\cos\theta_2 f_{xx} + \cos\theta_1\sin\theta_2 f_{yx} + \sin\theta_1\cos\theta_2 f_{xy} + \sin\theta_1\sin\theta_2 f_{yy}$. Since all partial derivatives of f of all orders are continuous, $f_{xy} = f_{yx}$ and it follows that $D_{\mathbf{u}_2}(D_{\mathbf{u}_1}f) = D_{\mathbf{u}_1}(D_{\mathbf{u}_2}f)$.

41 (a) Since (1.02, 2) appears to lie on the curve

$f(x, y) = 3.02$, $D_{\mathbf{i}}f \approx \dfrac{f(1.02, 2) - f(1, 2)}{1.02 - 1}$

$\approx \dfrac{3.02 - 3}{0.02} = 1$ at (1, 2).

(b) Since $f(1, 2.01)$ appears to be 3.02, $D_{\mathbf{j}}f \approx$

$\dfrac{f(1, 2.01) - f(1, 2)}{2.01 - 2} \approx \dfrac{3.02 - 3}{0.01} = 2$ at

(1, 2).

(c) $\nabla f = (D_{\mathbf{i}}f)\mathbf{i} + (D_{\mathbf{j}}f)\mathbf{j} \approx \mathbf{i} + 2\mathbf{j}$ at (1, 2).

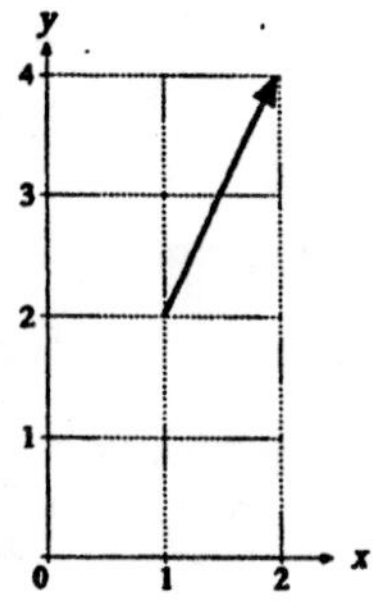

14.8 Normals and the Tangent Plane

1 Let $f(x, y) = xy$. Then $xy = 8$ is the level curve of f passing through (2, 4). By Theorem 1, a normal is given by $\nabla f = y\mathbf{i} + x\mathbf{j} = 4\mathbf{i} + 2\mathbf{j}$ at (2, 4). See the figure in the answer section of the text.

3 Let $f(x, y) = x^2/4 + y^2/9$. Then $\nabla f = \frac{x}{2}\mathbf{i} + \frac{2y}{9}\mathbf{j}$ $= \mathbf{i} + \frac{2}{3}\mathbf{j}$ at (2, 3) is such a vector. See the figure in the answer section of the text.

5 Proceed as in Example 3: The given surface is a level surface of the function $f(x, y, z) = x^2 + 2y^2 + 3z^2$. The gradient of f is $\nabla f = 2x\mathbf{i} + 4y\mathbf{j} + 6z\mathbf{k}$. At (1, 1, 1), $\nabla f = 2\mathbf{i} + 4\mathbf{j} + 6\mathbf{k}$, which is the desired normal vector.

7 The surface is a level surface of $f(x, y, z) = x^2 + y^2 - z$, so $\nabla f = 2x\mathbf{i} + 2y\mathbf{j} - \mathbf{k} = 2\mathbf{i} + 2\mathbf{j} - \mathbf{k}$ is a normal vector at (1, 1, 2).

9 The surface is a level surface of $f(x, y, z) = x^2 + y^2 + z^2$, so $\nabla f = 2x\mathbf{i} + 2y\mathbf{j} + 2z\mathbf{k} = 2\mathbf{i} + 2\mathbf{j} + 2\mathbf{k}$ is a normal vector at (1, 1, 1). The plane tangent to the surface at (1, 1, 1) is

$$\frac{\partial f}{\partial x}(1,1,1)(x-1) + \frac{\partial f}{\partial y}(1,1,1)(y-1) + \frac{\partial f}{\partial z}(1,1,1)(z-1) = 2(x-1) + 2(y-1) + 2(z-1) = 0,$$

or $x + y + z = 3$. See the figure in the answer section of the text.

11 The surface is a level surface of $f(x, y, z) = x^2 + y^2 - z$, so $\nabla f = 2x\mathbf{i} + 2y\mathbf{j} - \mathbf{k} = 2\mathbf{i} + 4\mathbf{j} - \mathbf{k}$ is a normal vector at (1, 2, 5). The plane tangent to the surface at (1, 2, 5) is

$$\frac{\partial f}{\partial x}(1,2,5)(x-1) + \frac{\partial f}{\partial y}(1,2,5)(y-2) + \frac{\partial f}{\partial z}(1,2,5)(z-5) = 2(x-1) + 4(y-2) - (z-5) = 0,$$

or $2x + 4y - z = 5$. See the figure in the answer section of the text.

13 This is like Example 4. Let $f(x, y, z) = xyz$, so $\nabla f = yz\mathbf{i} + xz\mathbf{j} + xy\mathbf{k}$; $\nabla f(1, 2, 3) = 6\mathbf{i} + 3\mathbf{j} + 2\mathbf{k}$ is a normal to the surface and is thus perpendicular to the tangent plane. An equation of the tangent plane is $6(x-1) + 3(y-2) + 2(z-3) = 0$, or $6x + 3y + 2z = 18$.

15 Let $f(x, y, z) = \frac{e^{xy}}{1+z^2}$. Then a normal to the surface, hence to the plane, is given by $\nabla f =$

$$\frac{ye^{xy}}{1+z^2}\mathbf{i} + \frac{xe^{xy}}{1+z^2}\mathbf{j} - \frac{2ze^{xy}}{(1+z^2)^2}\mathbf{k} = \frac{e^{xy}}{1+z^2}\left(y\mathbf{i} + x\mathbf{j} - \frac{2z}{1+z^2}\mathbf{k}\right).$$

Hence $y\mathbf{i} + x\mathbf{j} - \frac{2z}{1+z^2}\mathbf{k}$ is also a normal vector, which equals $2\mathbf{i} - \mathbf{k}$ at (0, 2, 1), so an equation of the plane is $2(x-0) + 0(y-2) + (-1)(z-1) = 0$; that is, $z = 2x + 1$.

17 Here $\Delta z = \cos(1.01 + (2.03)^2) - \cos(1 + 2^2) \approx 0.1227$. On the other hand, dz

$$= \frac{\partial z}{\partial x}(1,2)(x-1) + \frac{\partial z}{\partial y}(1,2)(y-2)$$

$$= \left(-\sin(x+y^2)\big|_{(1,2)}\right)(x-1) + \left(-\sin(x+y^2)\cdot 2y\big|_{(1,2)}\right)(y-2)$$

$$= (-\sin 5)(x-1) - 4(\sin 5)(y-2)$$

$= (-\sin 5)(0.01) - 4(\sin 5)(0.03) \approx 0.1247$ when

$(x, y) = (1.01, 2.03)$.

19 The volume of a right circular cylinder is $V = \pi r^2 h$, where r is the radius of its base and h is its height. Since the diameter of the base is exactly $2r$, a 5 percent error in estimating the diameter is equivalent to a 5 percent error in estimating the radius. Now apply the procedure in Example 7. We are given $\left|\frac{\Delta r}{r}\right| \le 0.05$ and $\left|\frac{\Delta h}{h}\right| \le 0.03$. The differential of V is $dV = V_r \Delta r + V_h \Delta h$; that is, $dV = 2\pi rh\ \Delta r + \pi r^2 \Delta h$. Then $\frac{dV}{V} =$

$$\frac{2\pi rh\,\Delta r + \pi r^2\,\Delta h}{\pi r^2 h} = 2\frac{\Delta r}{r} + \frac{\Delta h}{h} \text{ and } \left|\frac{dV}{V}\right| =$$

$$\left|2\frac{\Delta r}{r} + \frac{\Delta h}{h}\right| \le 2\left|\frac{\Delta r}{r}\right| + \left|\frac{\Delta h}{h}\right| \le 2(0.05) + 0.03$$

$= 0.13$. Therefore, he may make about a 13 percent error in measuring the volume.

21 $\Delta A = (x + \Delta x)(y + \Delta y) - xy = y\Delta x + x\Delta y + \Delta x\Delta y$; $A_x = y$ and $A_y = x$, so $dA = y\Delta x + x\Delta y$.

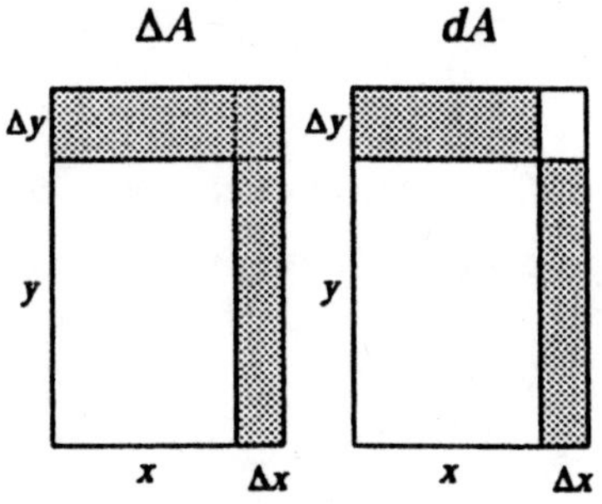

23 We have $u = f(x, y, z)$ where x, y, and z are functions of t (given by $\mathbf{r} = \mathbf{G}(t) = x(t)\mathbf{i} + y(t)\mathbf{j} + z(t)\mathbf{k}$). By the chain rule, $\frac{du}{dt} =$

$$\frac{\partial f}{\partial x}\cdot\frac{dx}{dt} + \frac{\partial f}{\partial y}\cdot\frac{dy}{dt} + \frac{\partial f}{\partial z}\cdot\frac{dz}{dt}.$$ Since $\mathbf{G}'(t) =$

$$\frac{dx}{dt}\mathbf{i} + \frac{dy}{dt}\mathbf{j} + \frac{dz}{dt}\mathbf{k} \text{ and } \nabla f = \frac{\partial f}{\partial x}\mathbf{i} + \frac{\partial f}{\partial y}\mathbf{j} + \frac{\partial f}{\partial z}\mathbf{k},$$

it follows that $\frac{du}{dt} = \nabla f\cdot\mathbf{G}'(t)$, where ∇f is evaluated at $\mathbf{G}(t)$.

25 (a) If $\mathbf{u}$ is tangent to the level surface at (a, b, c), it lies in the tangent plane to the level surface at (a, b, c). Hence ∇f and $\mathbf{u}$ are perpendicular. Therefore $D_{\mathbf{u}}f = \nabla f\cdot\mathbf{u} = 0$.

(b) If $\mathbf{u}$ is normal to the level surface at (a, b, c), it is parallel to ∇f. Hence $D_{\mathbf{u}}f = \pm|\nabla f|\,|\mathbf{u}|$

$= \pm|\nabla f| = \pm\sqrt{2^2 + 3^2 + (-4)^2} = \pm\sqrt{29}$.

27 Let $f(x, y) = 2x^2 + y^2$ and $g(x, y) = y^2/x$.

(a),(b)

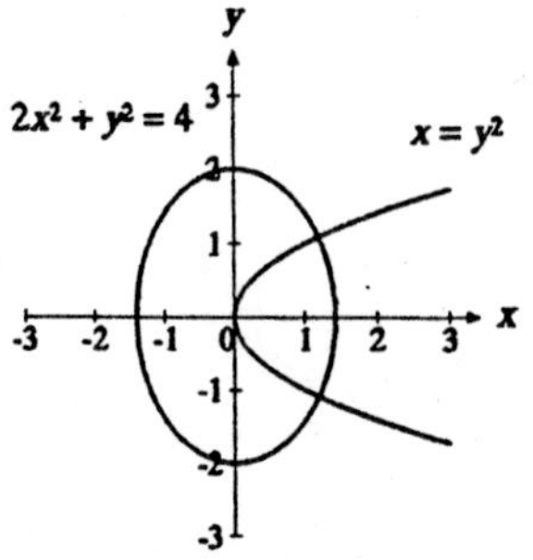

(c) Two curves are perpendicular to each other at a point of intersection if their respective tangent lines at that point are perpendicular. Since the gradient of a function at a point is normal to the graph of that function at that point, showing that ∇f and ∇g are perpendicular at points of intersection will also prove that graphs of any level curves of f and g cross at right angles. Since $\nabla f\cdot\nabla g =$

$$(4x\mathbf{i} + 2y\mathbf{j})\cdot\left(-\frac{y^2}{x^2}\mathbf{i} + \frac{2y}{x}\mathbf{j}\right) = -\frac{4y^2}{x} + \frac{4y^2}{x}$$

$= 0$, ∇f and ∇g are always perpendicular.

Hence the graphs of any level curves of f and g cross at right angles.

29 According to the three-dimensional analog of Theorem 2 of Sec. 14.7, the temperature changes most rapidly in the direction of ∇T. By Theorem 2 of this section, ∇T is perpendicular to the level surface of T passing through (a, b, c); that is, ∇T is perpendicular to the isotherm passing through (a, b, c).

31 Let $f(x, y, z) = xy - z$; then $f_x = y, f_y = x$, and $f_z = -1$. An equation of the tangent plane is

$$\frac{\partial f}{\partial x}(3,4,12)(x-3) + \frac{\partial f}{\partial y}(3,4,12)(y-4) + \frac{\partial f}{\partial z}(3,4,12)(z-12) = 0,\ 4(x-3) + 3(y-4) + (-1)(z-12) = 0,$$

or $4x + 3y - z - 12 = 0$. The distance from the point $(2, 1, 3)$ to the plane $4x + 3y - z - 12 = 0$ is

$$\frac{|4\cdot 2 + 3\cdot 1 + (-1)\cdot 3 + (-12)|}{\sqrt{4^2 + 3^2 + (-1)^2}} = \frac{4}{\sqrt{26}}.$$

33 (a) Since $1\cdot 1\cdot 2 = 2$ and $1^3\cdot 1\cdot 2^2 = 4$, the point $(1, 1, 2)$ lies on the surfaces $f(x, y, z) = xyz = 2$ and $g(x, y, z) = x^3yz^2 = 4$.

(b) Let $f(x, y, z) = xyz$ and $g(x, y, z) = x^3yz^2$. Then $\nabla f = yz\mathbf{i} + xz\mathbf{j} + xy\mathbf{k}$ and $\nabla g = 3x^2yz^2\mathbf{i} + x^3z^2\mathbf{j} + 2x^3yz\mathbf{k}$, so $\nabla f(1, 1, 2) = 2\mathbf{i} + 2\mathbf{j} + \mathbf{k}$ and $\nabla g(1, 1, 2) = 12\mathbf{i} + 4\mathbf{j} + 4\mathbf{k}$. Then $\cos\theta = \dfrac{\nabla f\cdot\nabla g}{|\nabla f||\nabla g|}$

$$= \frac{3}{\sqrt{11}}, \text{ so } \theta = \cos^{-1}\frac{3}{\sqrt{11}} \approx 25.2°.$$

35 (a) $(2, 1, 4)$ lies on $z = x^2y^3$ because $4 = 2^2 1^3$ and on $z = 2xy$ because $4 = 2\cdot 2\cdot 1$.

(b) Let $f(x, y, z) = x^2y^3 - z$ and $g(x, y, z) = 2xy - z$. Then $\nabla f = 2xy^3\mathbf{i} + 3x^2y^2\mathbf{j} - \mathbf{k}$, $\nabla f(2, 1, 4) = 4\mathbf{i} + 12\mathbf{j} - \mathbf{k}$, $\nabla g = 2y\mathbf{i} + 2x\mathbf{j} - \mathbf{k}$, and $\nabla g(2, 1, 4) = 2\mathbf{i} + 4\mathbf{j} - \mathbf{k}$. Hence

$$\cos\theta = \frac{\nabla f\cdot\nabla g}{|\nabla f||\nabla g|} = \frac{57}{7\sqrt{69}}, \text{ so } \theta = \cos^{-1}\frac{57}{7\sqrt{69}} \approx 11.4°.$$

14.9 Critical Points and Extrema

1 First determine the critical points of $f(x, y) = x^2 + 3xy + y^2$: $f_x = 2x + 3y$ and $f_y = 3x + 2y$, so the only critical point is $(0, 0)$. We have $D = f_{xx}f_{yy} - (f_{xy})^2 = 2\cdot 2 - 3^2 = -5 < 0$, so, by Theorem 2, $(0, 0)$ is a saddle point.

3 $f(x, y) = x^2 - 2xy + 2y^2 + 4x$, so $f_x = 2x - 2y + 4$ and $f_y = -2x + 4y$. Setting these equal to 0, we find that $x = 2y$ and $0 = 2(2y) - 2y + 4 = 2y + 4$, so $y = -2$ and $x = -4$. The only critical point is $(-4, -2)$. Therefore $D = f_{xx}f_{yy} - (f_{xy})^2 = 2\cdot 4 - (-2)^2 = 4 > 0$ and $f_{xx} = 2 > 0$, so Theorem 2 shows that f has a relative minimum at $(-4, -2)$. The value of this minimum is $f(-4, -2) = -8$.

5 $f(x, y) = x^2 - xy + y^2$, so $f_x = 2x - y$ and $f_y = -x + 2y$; hence $(0, 0)$ is the only critical point. $D = f_{xx}f_{yy} - (f_{xy})^2 = 2\cdot 2 - (-1)^2 = 3 > 0$ and $f_{xx} = 2 > 0$, so $(0, 0)$ is a relative minimum. The value is $f(0, 0) = 0$.

7 $f(x, y) = 2x^2 + 2xy + 5y^2 + 4x$, so $f_x = 4x + 2y + 4$ and $f_y = 2x + 10y$. Setting these equal to 0 gives $x = -5y$ and $0 = 4(-5y) + 2y + 4 = -18y + 4$; thus $y = \dfrac{2}{9}$ and $x = -\dfrac{10}{9}$. The only

critical point is $\left(-\frac{10}{9}, \frac{2}{9}\right)$. Now $D = f_{xx}f_{yy} - (f_{xy})^2$ $= 4\cdot 10 - 2^2 = 36 > 0$ and $f_{xx} = 4 > 0$, so f has a relative minimum at $\left(-\frac{10}{9}, \frac{2}{9}\right)$. Its value is

$$f\left(-\frac{10}{9}, \frac{2}{9}\right) = -\frac{20}{9}.$$

9 $f(x, y) = \frac{4}{x} + \frac{2}{y} + xy$, so $f_x = -\frac{4}{x^2} + y$ and $f_y = -\frac{2}{y^2} + x$. Setting these equal to 0 gives $y = 4/x^2$ and $x = 2/y^2$. Hence $x = \frac{2}{(4/x^2)^2} = \frac{x^4}{8}$, so $x(x^3 - 8) = 0$ and thus either $x = 0$ or $x = 2$. f is not defined for $x = 0$, so $x = 2$ and $y = 1$; $(2, 1)$ is the only critical point. At $(2, 1)$ we have $f_{xx} = 8/x^3 = 8/2^3 = 1$, $f_{yy} = 4/y^3 = 4/1^3 = 4$, and $f_{xy} = 1$, so $D = 1\cdot 4 - 1^2 = 3$. Since $D > 0$ and $f_{xx} > 0$, f has a relative minimum at $(2, 1)$. The value of the minimum is $f(2, 1) = 6$.

11 $D = 2\cdot 8 - 4^2 = 0$; there is insufficient information.

13 $D = 2\cdot 4 - 3^2 = -1 < 0$; f has a saddle point at (a, b), neither a maximum nor minimum.

15 $D = (-3)(-4) - (-2)^2 = 8 > 0$ and $f_{xx} = -3$; f has a relative maximum at (a, b).

17 $f(x, y) = x + y - x^{-1}y^{-1}$, $f_x = 1 + x^{-2}y^{-1}$, $f_y = 1 + x^{-1}y^{-2}$, $f_{xx} = -2x^{-3}y^{-1}$, $f_{yy} = -2x^{-1}y^{-3}$, $f_{xy} = -x^{-2}y^{-2}$. Setting $f_x = f_y = 0$ yields $x = y = -1$, so $(-1, -1)$ is the only critical point. At $(-1, -1)$, $f_{xx} = f_{yy} = -2$ and $f_{xy} = -1$, so $D = 3$. Since $D > 0$ and $f_{xx} < 0$, there is a relative maximum of -3 at $(-1, -1)$.

19 $f_x = 3(4y - x^2)$, $f_y = 3(4x - y^2)$, $f_{xx} = -6x$, $f_{yy} = -6y$, $f_{xy} = 12$. Setting $f_x = f_y = 0$ gives $x^2 = 4y$ and $y^2 = 4x$, so $x^4 = (4y)^2 = 64x$ and $x(x^3 - 64) = 0$. Thus either $x = 0$ or $x = 4$; the critical points are $(0, 0)$ and $(4, 4)$. At $(0, 0)$, $f_{xx}f_{yy} = 0$ and $f_{xy} = 12$, so $D = -144 < 0$. At $(4, 4)$, $f_{xx} = f_{yy} = -24$ and $f_{xy} = 12$, so $D = (-24)^2 - 12^2 = 432 > 0$. $(0, 0)$ is a saddle point and there is a relative maximum of 64 at $(4, 4)$.

21 $f_x = 3x^2e^{x^3+y^3}$, $f_y = 3y^2e^{x^3+y^3}$, $f_{xx} = 3x(3x^3 + 2)e^{x^3+y^3}$, $f_y = 3y^2e^{x^3+y^3}$, $f_{yy} = 3y(3y^3 + 2)e^{x^3+y^3}$, $f_{xy} = 9x^2y^2e^{x^3+y^3}$. Letting $f_x = f_y = 0$, we find that $(0, 0)$ is the only critical point. Since $f_{xx} = f_{yy} = f_{xy} = 0$, $D = 0$ and the second-partial-derivative test yields no information. But note that $f(x, 0) = e^{x^3}$ is an increasing function of x. Hence f has neither a maximum nor a minimum at $(0, 0)$; the origin is a saddle point.

23 $f_x = 3 + y + 2xy$, $f_y = x + x^2 - 2$, $f_{xx} = 2y$, $f_{yy} = 0$, $f_{xy} = 1 + 2x$. Letting $f_y = 0$ gives $(x - 1)(x + 2) = 0$, so $x = 1$ or $x = -2$. Letting $f_x = 0$ yields $y = -\frac{3}{2x + 1}$, so $y = -1$ or 1, respectively; that is, the critical points are $(1, -1)$ and $(-2, 1)$. Since $D = 2y\cdot 0 - (1 + 2x)^2 = -(1 + 2x)^2$ is negative for $x \neq -\frac{1}{2}$, both $(1, -1)$ and $(-2, 1)$ are saddle points.

25 Let the base have dimensions x and y and let the height be z; then $xyz = 1$, so $z = \frac{1}{xy}$. The surface area is $xy + 2xz + 2yz = xy + \frac{2}{y} + \frac{2}{x} =$

$S(x, y)$, so $S_x = y - \frac{2}{x^2}$ and $S_y = x - \frac{2}{y^2}$. At a critical point, $y = 2/x^2$ and $x = 2/y^2 = \frac{2}{(2/x^2)^2} = \frac{x^4}{2}$. Since $x \neq 0$, we get $1 = \frac{x^3}{2}$ or $x = 2^{1/3}$ and $y = 2^{1/3}$. That is, $(\sqrt[3]{2}, \sqrt[3]{2})$ is the only critical point. At this point, $S_{xx} = \frac{4}{x^3} = \frac{4}{2} = 2$, $S_{yy} = \frac{4}{y^3} = 2$, $S_{xy} = 1$, and $D = 2 \cdot 2 - 1^2 = 3$. Since $D > 0$ and $S_{xx} > 0$, S has a relative minimum. The box of smallest area has a square base with sides of length $\sqrt[3]{2}$ and height $z = \frac{1}{xy} = 2^{-2/3} = \frac{\sqrt[3]{2}}{2}$.

27 Let $P = (x, y)$ and minimize $f(x, y) = (x - x_1)^2 + (y - y_1)^2 + (x - x_2)^2 + (y - y_2)^2 + (x - x_3)^2 + (y - y_3)^2 + (x - x_4)^2 + (y - y_4)^2$. Now $f_x = 2(x - x_1) + 2(x - x_2) + 2(x - x_3) + 2(x - x_4)$, $f_y = 2(y - y_1) + 2(y - y_2) + 2(y - y_3) + 2(y - y_4)$, $f_{xx} = 8$, $f_{yy} = 8$, and $f_{xy} = 0$. $f_x = 0$ when $x = \frac{1}{4}(x_1 + x_2 + x_3 + x_4)$ and $f_y = 0$ when $y = \frac{1}{4}(y_1 + y_2 + y_3 + y_4)$. $D = 8 \cdot 8 - 0^2 > 0$ and $f_{xx} = 8 > 0$, so f has a minimum at $\left(\frac{1}{4}(x_1 + x_2 + x_3 + x_4), \frac{1}{4}(y_1 + y_2 + y_3 + y_4)\right)$.

29 (a) x, y, and z satisfy $x + y + z = 1$. Let $f = x^2 + y^2 + z^2 = x^2 + y^2 + (1 - x - y)^2$. Then $f_x = 2x - 2(1 - x - y)$, $f_y = 2y - 2(1 - x - y)$, $f_{xx} = 2 + 2 = 4$, $f_{yy} = 4$, and $f_{xy} = 2$. f has a critical point at $\left(\frac{1}{3}, \frac{1}{3}, \frac{1}{3}\right)$. $D = 4 \cdot 4 - 2^2 > 0$ and $f_{xx} = 4 > 0$, so f has a minimum of $\frac{1}{3}$ at $\left(\frac{1}{3}, \frac{1}{3}, \frac{1}{3}\right)$.

(b) Since x, y, and z are nonnegative, and $x + y + z = 1$, we must have $0 \le x \le 1$, $0 \le y \le 1$, and $0 \le z \le 1$, where not all of x, y, and z are 0 and at most one of x, y, and z is 1. Then for all possible (x, y, z), $x^2 \le x$, $y^2 \le y$, $z^2 \le z$ and thus $x^2 + y^2 + z^2 \le x + y + z = 1$. We note that for $(x, y, z) = (0, 0, 1)$, $x^2 + y^2 + z^2 = 1$. Hence $x^2 + y^2 + z^2$ can be as large as 1.

31 Let x, y, and z be the coordinates of the corner of the box in the first octant (assuming that the box is placed with its surfaces parallel to the coordinate planes and its centroid at the origin). $V = 8xyz$ and $x^2 + y^2 + z^2 = 1$, so $V = 8xy\sqrt{1 - x^2 - y^2}$ and $V^2 = 64x^2y^2(1 - x^2 - y^2)$, which is simpler to maximize. Then $(V^2)_x = 128xy^2 - 256x^3y^2 - 128xy^4$, $(V^2)_y = 128x^2y - 128x^4y - 256x^2y^3$, $(V^2)_{xx} = 128y^2 - 768x^2y^2 - 128y^4$, $(V^2)_{yy} = 128x^2 - 128x^4 - 768x^2y^2$, and $(V^2)_{xy} = 256xy - 512x^3y - 512xy^3$. We need $(V^2)_x = 0$ and $(V^2)_y = 0$. Thus $128xy^2(1 - 2x^2 - y^2) = 0$ and $128x^2y(1 - x^2 - 2y^2) = 0$. We can exclude the possibilities $x = 0$ and $y = 0$, since they give a box of volume 0. Thus $2x^2 + y^2 = 1$ and $x^2 + 2y^2 = 1$. Hence $x = \pm\sqrt{1/3}$ and $y = \pm\sqrt{1/3}$. Since we need consider

only positive values, we can take $x = \sqrt{1/3}$, $y = \sqrt{1/3}$, and $z = \sqrt{1/3}$, so each edge is $\frac{2\sqrt{3}}{3}$ and volume is $\frac{8\sqrt{3}}{9}$. To check that this is a maximum, note that $(V^2)_{xx} = (V^2)_{yy} = -\frac{512}{9}$, $(V^2)_{xy} = -\frac{256}{9}$, and $D = \frac{65{,}536}{27}$ at $\left(\sqrt{\frac{1}{3}}, \sqrt{\frac{1}{3}}, \sqrt{\frac{1}{3}}\right)$.

33 $f_x = 2kx + 5y$, $f_y = 5x + 8y$, $f_{xx} = 2k$, $f_{yy} = 8$, and $f_{xy} = 5$. $f_x(0, 0) = 0$ and $f_y(0, 0) = 0$, so $(0, 0)$ is a critical point. $D = 2k \cdot 8 - 25$. We need $D \geq 0$, so $16k \geq 25$. Therefore, $k \geq \frac{25}{16}$. If $k > \frac{25}{16}$ then $D > 0$ and $f_{xx} = 2k > 0$, so we do indeed have a relative minimum. If $k = \frac{25}{16}$ then

$$f(x, y) = \frac{25}{16}x^2 + 5xy + 4y^2 = \left(\frac{5}{4}x + 2y\right)^2 \geq 0.$$

So f is minimal at $(0, 0)$ for $k = \frac{25}{16}$ as well. All told, $k \geq \frac{25}{16}$.

35 (a) On the unit circle $x = \cos\theta$ and $y = \sin\theta$, so

$$f = (2x^2 + y^2)e^{-x^2-y^2} =$$

$$(2\cos^2\theta + \sin^2\theta)e^{-1} = \frac{1}{e}(\cos^2\theta + 1).$$

This has a maximum value of $\frac{2}{e}$, which occurs at $\theta = 0$ and π; its minimum value is $\frac{1}{e}$ and occurs at $\theta = \frac{\pi}{2}$ and $\frac{3\pi}{2}$. In rectangular coordinates, the function has a maximum of $\frac{2}{e}$ at $(1, 0)$ and $(-1, 0)$ and a minimum of $\frac{1}{e}$ at $(0, 1)$ and $(0, -1)$.

(b) On the circle $x^2 + y^2 = 4$ we have $x = 2\cos\theta$ and $y = 2\sin\theta$, so $f = 4e^{-4}(\cos^2\theta + 1)$. The maximum value is $8e^{-4}$ and occurs at $(\pm 2, 0)$; the minimum value is $4e^{-4}$ and occurs at $(0, \pm 2)$.

37 $f_x = y$, $f_y = x$, $f_{xx} = 0$, $f_{yy} = 0$, and $f_{xy} = 1$. The only critical point is $(0, 0)$, so f will have its minimum and maximum on the boundary of the triangle. In the triangular region $x \geq 0$ and $y \geq 0$, so $f(x, y) = xy \geq 0$. Therefore, it is clear that f is minimal on the line segments AB and AC, where $A = (0, 0)$, $B = (1, 0)$, and $C = (0, 1)$. The maximum must occur somewhere on BC. On BC we have $y = 1 - x$, so $f(x, 1 - x) = x - x^2$. Now $(x - x^2)' = 1 - 2x$, so f has a maximum at $\left(\frac{1}{2}, \frac{1}{2}\right)$; its value is $\frac{1}{4}$.

39 (a) Let $f(x, y) = x^2 - y^2 + 2xy$. Then $f_x = 2x + 2y$, $f_y = -2y + 2x$, $f_{xx} = 2$, $f_{yy} = -2$, and $f_{xy} = 2$. Setting $f_x = f_y = 0$ gives the critical point $(0, 0)$. At this point, $D = 2(-2) - 2^2 = -8 < 0$. Hence $(0, 0)$ is a saddle point, so $z = f(x, y)$ has no maximum or minimum, global or otherwise.

(b) On the unit circle, $x = \cos\theta$, $y = \sin\theta$, so $z = \cos^2\theta - \sin^2\theta + 2\cos\theta\sin\theta = \cos 2\theta + \sin 2\theta$. Then $z' =$

$-2\sin 2\theta + 2\cos 2\theta$, which equals 0 when $\tan 2\theta = 1$; this occurs for $\theta = \frac{\pi}{8}, \frac{5\pi}{8}, \frac{9\pi}{8}$, and $\frac{13\pi}{8}$. The corresponding extreme values of z are $\sqrt{2}$, $-\sqrt{2}$, $\sqrt{2}$, and $-\sqrt{2}$.

(c) From (a), the only critical point is the saddle point at (0, 0). Hence the extreme values of z lie on the boundary, as found in (b).

41 The constraint is $2x + y + 5z = 60$. Thus $y = 60 - 2x - 5z$ and $U = x^{1/2}(60 - 2x - 5z)^{1/3}z^{1/6}$. Hence $U_x = \left[x^{1/2}\cdot\frac{1}{3}(60 - 2x - 5z)^{-2/3}(-2) + \frac{1}{2}x^{-1/2}(60 - 2x - 5z)^{1/3}\right]z^{1/6} =$

$x^{-1/2}z^{1/6}(60 - 2x - 5z)^{-2/3}\left[-\frac{2}{3}x + \frac{1}{2}(60 - 2x - 5z)\right]$

and $U_z =$

$x^{1/2}z^{-5/6}(60 - 2x - 5z)^{-2/3}\left[\frac{1}{6}(60 - 2x - 5z) - \frac{5}{3}z\right]$.

Note that x, z, and $60 - 2x - 5z$ cannot be zero, so critical points will occur if

$-\frac{2}{3}x + \frac{1}{2}(60 - 2x - 5z) = 0$ and

$-\frac{5}{3}z + \frac{1}{6}(60 - 2x - 5z) = 0$. The only critical point occurs when $x = 15$ and $z = 2$. The corresponding y value at this maximum is 20.

43 (a) If (x_i, y_i) is one of the n points, then $d_i = |y_i - (mx_i + b)|$, since the ordinate of the line $y = mx + b$ is $mx_i + b$ for $x = x_i$. Thus $d_i^2 = [y_i - (mx_i + b)]^2$, so the line of regression minimizes $f(m, b) = \sum_{i=1}^{n}[y_i - (mx_i + b)]^2$ for all choices of m and b.

(b) $f_m(m, b) = \sum_{i=1}^{n} 2[y_i - (mx_i + b)](-x_i)$

$= 2\sum_{i=1}^{n}(mx_i^2 + bx_i - x_iy_i)$; $f_b(m, b)$

$= \sum_{i=1}^{n} 2[y_i - (mx_i + b)](-1)$

$= 2\sum_{i=1}^{n}(mx_i + b - y_i)$

(c) For $f_m = 0$, $0 = \sum_{i=1}^{n}(mx_i^2 + bx_i - x_iy_i)$

$= m\sum_{i=1}^{n}x_i^2 + b\sum_{i=1}^{n}x_i - \sum_{i=1}^{n}x_iy_i$, so

$m\sum_{i=1}^{n}x_i^2 + b\sum_{i=1}^{n}x_i = \sum_{i=1}^{n}x_iy_i$. For $f_b = 0$,

$0 = \sum_{i=1}^{n}(mx_i + b - y_i) =$

$m\sum_{i=1}^{n}x_i + nb - \sum_{i=1}^{n}y_i$, so $m\sum_{i=1}^{n}x_i + nb$

$= \sum_{i=1}^{n}y_i$.

(d) The determinant of the system of equations is $n\sum_{i=1}^{n}x_i^2 - \left(\sum_{i=1}^{n}x_i\right)^2$. Call this D. If $D \neq 0$, then the equations have a unique solution. If $D = 0$, then the two equations are equivalent to

$b = \dfrac{1}{n}\left(\sum_{i=1}^{n} y_i - m\sum_{i=1}^{n} x_i\right)$, which has infinitely many solutions. So the solution is unique if and only if $D \neq 0$. For $n = 2$ and $n = 3$, we find, respectively, that $D = 2(x_1^2 + x_2^2) - (x_1 + x_2)^2 = (x_1 - x_2)^2$, and $D = 3(x_1^2 + x_2^2 + x_3^2) - (x_1 + x_2 + x_3)^2 = 2x_1^2 + 2x_2^2 + 2x_3^3 - 2x_1x_2 - 2x_1x_3 - 2x_2x_3 = (x_1 - x_2)^2 + (x_1 - x_3)^2 + (x_2 - x_3)^2$. In each case $D = 0$ if and only if all x_i's are equal. More generally, $D = \sum_{i=1}^{n-1}\sum_{j=i+1}^{n}(x_i - x_j)^2$. (To see this, compute the coefficients of x_i^2 and x_ix_j on both sides of the equation; we find that both sides equal $(n-1)\sum_{i=1}^{n} x_i^2 - 2\sum_{i=1}^{n-1}\sum_{j=i+1}^{n} x_ix_j$.) Thus D is a sum of squares, so it equals 0 if and only if $x_i - x_j = 0$ for all $i < j$. Thus the line of regression is unique if and only if not all x_i's are equal.

(e) Let $(x_1, y_1) = (1, 1)$, $(x_2, y_2) = (2, 3)$, and $(x_3, y_3) = (3, 5)$. Here $\sum_{i=1}^{3} x_i = 6$, $\sum_{i=1}^{3} y_i = 9$, $\sum_{i=1}^{3} x_i^2 = 14$, and $\sum_{i=1}^{3} x_iy_i = 22$. By (c) we must solve the system of equations $14m + 6b = 22$ and $6m + 3b = 9$. (Since not all the x_i are equal, the solution is unique.) Dividing the first equation by 2 and subtracting the second from the result gives $m = 2$. Then $b = -1$, and the line of regression is $y = 2x - 1$. (In this case the results are so nice because the three points happen to lie on a line.)

45 (a) For any $x \neq 0$, $f(x, 0) = (0 - x^2)(0 - 2x^2) = 2x^4 > 0 = f(0, 0)$, so f does not have a local maximum at (0, 0). Also, for $x \neq 0$, $f\left(x, \frac{3}{2}x^2\right) = \left(\frac{3}{2}x^2 - x^2\right)\left(\frac{3}{2}x^2 - 2x^2\right) = -\frac{1}{4}x^4 < 0 = f(0, 0)$, so f does not have a local minimum at (0, 0).

(b) Let L be a line through (0, 0). If L is the y axis, then $f(x, y) = f(0, y) = y^2$ on L and has a minimum at $y = 0$. If L is the x axis, then $f(x, y) = f(x, 0) = 2x^4$ on L and has a minimum at $x = 0$. In either of these cases, f therefore has a local minimum at (0, 0). If L is neither of these axes, then L has the equation $y = ax$ for some $a \neq 0$. Hence $f(x, y) = f(x, ax) = (ax - x^2)(ax - 2x^2) = 2x^2(a - x)\left(\frac{a}{2} - x\right)$. This quantity is negative only when x is between $\frac{a}{2}$ and a. Hence, for $|x| < \left|\frac{a}{2}\right|$, $f(x, ax) \geq 0 = f(0, 0)$, so f, considered only on L, has a minimum at (0, 0).

47 Suppose the maximum occurs at a point $P = (p, q)$. If L is any chord of R passing through P, we parameterize L as $x = p + \alpha t$, $y = q + \beta t$, $t_0 \leq t \leq t_1$, where $t_0 \leq 0$ and $t_1 \geq 0$. On L, $f(x, y) = a(p + \alpha t) + b(q + \beta t) + c = (a\alpha + b\beta)t +$

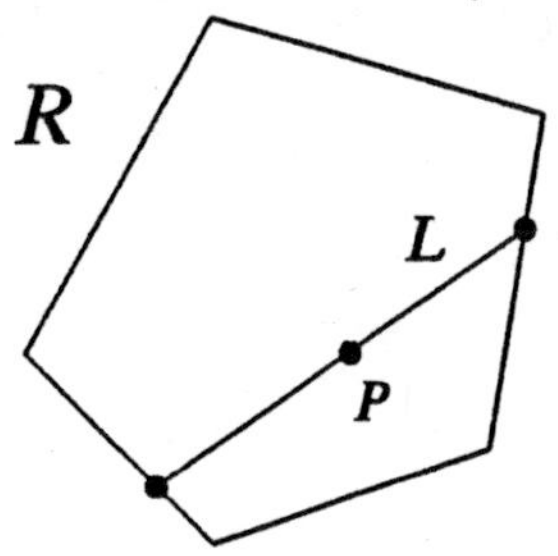

$(ap + bq + c)$ is a linear function. Linear functions are either constant or monotonically increasing or decreasing, so the value of $f(x, y)$ at one of the two endpoints E of L is greater than or equal to the value at P. Since f is maximal at P, the value at E must equal that at P. Thus the maximum of f occurs at some point E on the boundary of R. If E is not a vertex, then it lies on some edge. As above, the value of f at one endpoint A must be greater than or equal to the value at E. Since f is maximal at E, the value of A must equal that at E. Thus, the maximum occurs at some vertex A of R. (The proof for a minimum is similar.)

49 (a) Suppose that a, b, and c are nonnegative numbers and let $S = a + b + c$. If $S = 0$, then $a = b = c = 0$ and the assertion is true. Therefore assume that $S > 0$ and define $A = \frac{a}{S}$, $B = \frac{b}{S}$, and $C = \frac{c}{S}$. A, B, and C are also nonnegative and $A + B + C = \frac{a}{S} + \frac{b}{S} + \frac{c}{S} = \frac{a+b+c}{S} = \frac{S}{S} = 1$, so, by Exercise 48, $ABC \le \frac{1}{27}$; that is, $\frac{a}{S}\cdot\frac{b}{S}\cdot\frac{c}{S} \le \frac{1}{27}$, so $abc \le \frac{S^3}{27}$, and $\sqrt[3]{abc} \le \frac{S}{3} = \frac{a+b+c}{3}$.

(b) By the arguments used in Exercise 48, we can show that if x, y, z, and w are each nonnegative and $w + x + y + z = 1$, then $P = wxyz$ is maximized at $w = x = y = z = \frac{1}{4}$. Note that $P = (1 - x - y - z)xyz = xyz - x^2yz - xy^2z - xyz^2$, so $P_x = yz - 2xyz - y^2z - yz^2 = yz(1 - 2x - y - z)$, $P_y = xz(1 - x - 2y - z)$, and $P_z = xy(1 - x - y - 2z)$. If $P_x = P_y = P_z = 0$, then $0 = 1 - 2x - y - z = 1 - x - 2y - z = 1 - x - y - 2z$. From this system of equations we get $x = y = z = \frac{1}{4}$ and thus $w = 1 - \frac{1}{4} - \frac{1}{4} - \frac{1}{4} = \frac{1}{4}$. Therefore, the maximum occurs at $\left(\frac{1}{4}, \frac{1}{4}, \frac{1}{4}, \frac{1}{4}\right)$ and its value is $\frac{1}{4^2} = \frac{1}{256}$.

Now suppose that a, b, c, and d are nonnegative. By analogy with (a), let $S = a + b + c + d$ and define $A = \frac{a}{S}$, $B = \frac{b}{S}$, $C = \frac{c}{S}$, and $D = \frac{d}{S}$. Then A, B, C, and D are nonnegative and $A + B + C + D = 1$. As shown above, $ABCD \le \frac{1}{256}$ so $abcd \le \left(\frac{S}{4}\right)^4 = \left(\frac{a+b+c+d}{4}\right)^4$. Thus we have the desired corresponding result: For any four nonnegative numbers, a, b, c, and d, $\sqrt[4]{abcd}$

$\le \frac{a+b+c+d}{4}$. (In the case $S = 0$, we must have $a = b = c = d = 0$, so the desired result holds.)

51 Assume that $D < 0$; that is, using the result from the proof of case 1 in the text, assume $AC - B^2 < 0$, or $B^2 - AC > 0$. We shall exhibit two lines such that f achieves a relative maximum at (0, 0) along one and a relative minimum at (0, 0) along the other. Consider the line $y = Ex$, and note that $f(x, Ex) = Ax^2 + 2Bx(Ex) + C(Ex)^2 = (A + 2BE + CE^2)x^2$. If $C = 0$, choose $E_1 < -A/(2B)$ and $E_2 > -A/(2B)$. (Note that $B \neq 0$; otherwise, $B^2 - AC = 0$.) Then along the line $y = E_1x$, $f(x, E_1x) = (A + 2BE_1)x^2 < 0$ for all $x \neq 0$. Hence f has a relative maximum at (0, 0) in this case. Similarly, it can be shown that f has a relative minimum at (0, 0) along the line $y = E_2x$. Now suppose $C \neq 0$. Then the real roots of the quadratic equation in E, $A + 2BE + CE^2 = 0$, are given by $r = \frac{-2B \pm \sqrt{(2B)^2 - 4AC}}{2C} = -\frac{B}{C} \pm \frac{\sqrt{B^2 - AC}}{C}$. Since $B^2 - AC > 0$, there are indeed two real roots; call them r_1 and r_2. Let E_3 be between r_1 and r_2, and let E_4 be less than the smaller of r_1 and r_2. Hence $A + 2BE_3 + CE_3^2$ and $A + 2BE_4 + CE_4^2$ have opposite signs and, arguing as above, we see that f takes on a relative maximum at (0, 0) along one of the lines $y = E_3x$ and $y = E_4x$ and a relative minimum at (0, 0) along the other.

14.10 Lagrange Multipliers

1 Let $L(x, y, \lambda) = xy - \lambda(x^2 + y^2 - 4)$. Then $L_x = y - 2\lambda x$, $L_y = x - 2\lambda y$, $L_\lambda = -(x^2 + y^2 - 4)$. Setting these equal to 0 yields $y = 2\lambda x$, $x = 2\lambda y$, and $x^2 + y^2 = 4$. Hence $y = 2\lambda x = 2\lambda(2\lambda y) = (2\lambda)^2 y$, so either $y = 0$ or $\lambda = \pm\frac{1}{2}$. But if $y = 0$, then $x = 2\lambda y = 0$ and $x^2 + y^2 = 0 \neq 4$. Hence we must have $\lambda = \pm\frac{1}{2}$, so $x = \pm y$. Then $4 = x^2 + y^2 = 2x^2$, so $x^2 = 2$ and $x = \pm\sqrt{2}$, $y = \pm\sqrt{2}$. Computing xy for the four choices of signs shows that the maximum value is 2 and occurs at $(\sqrt{2}, \sqrt{2})$ and $(-\sqrt{2}, -\sqrt{2})$.

3 Let $L(x, y, \lambda) = 2x + 3y - \lambda(xy - 1)$. Then $L_x = 2 - \lambda y$, $L_y = 3 - \lambda x$, and $L_\lambda = -(xy - 1)$. Setting these equal to 0 results in $\lambda y = 2$ and $\lambda x = 3$, so $y = \frac{2}{\lambda} = \frac{2}{(3/x)} = \frac{2}{3}x$. Hence $1 = xy = \frac{2}{3}x^2$, so $x^2 = \frac{3}{2}$, $x = \sqrt{\frac{3}{2}}$, and $y = \frac{1}{x} = \sqrt{\frac{2}{3}}$. Therefore, $2x + 3y = 2\sqrt{\frac{3}{2}} + 3\sqrt{\frac{2}{3}} = 2\sqrt{6}$.

5 Let x and y be the lengths, in centimeters, of the sides of the rectangle. We wish to maximize the area xy subject to the constraint that $2x + 2y = 12$. The constraint is equivalent to $x + y - 6 = 0$, so let $L(x, y, \lambda) = xy - \lambda(x + y - 6)$. Then $L_x = y - \lambda$, $L_y = x - \lambda$, and $L_\lambda = -(x + y - 6)$. Setting these equal to 0 results in $y = \lambda = x$ and $6 = x + y = 2x$. Thus $x = y = 3$ and the maximal area is 9 cm^2.

7 As suggested, minimize the square of the distance to the origin, $x^2 + y^2 + z^2$, subject to the constraint $x + 2y + 3z = 6$. Let $L(x, y, z, \lambda) = x^2 + y^2 + z^2 - \lambda(x + 2y + 3z - 6)$; then $L_x = 2x - \lambda$, $L_y = 2y - 2\lambda$, $L_z = 2z - 3\lambda$, and $L_\lambda = -(x + 2y + 3z - 6)$. Equating these to 0 yields $\lambda = 2x$, $y = \lambda = 2x$, $z = \frac{3}{2}\lambda = 3x$, and $6 = x + 2y + 3z = x + 2\cdot 2x + 3\cdot 3x = 14x$. Hence $x = \frac{6}{14} = \frac{3}{7}$, $y = \frac{6}{7}$, and $z = \frac{9}{7}$. The point on the plane that is closest to the origin is $\left(\frac{3}{7}, \frac{6}{7}, \frac{9}{7}\right)$.

9 Minimize $(x - 1)^2 + (y - 3)^2 + (z - 2)^2$ subject to the constraint $2x + y + z = 5$. Let $L(x, y, z, \lambda) = (x - 1)^2 + (y - 3)^2 + (z - 2)^2 - \lambda(2x + y + z - 5)$. Then $L_x = 2(x - 1) - 2\lambda$, $L_y = 2(y - 3) - \lambda$, $L_z = 2(z - 2) - \lambda$, and $L_\lambda = -(2x + y + z - 5)$. Equating these to 0 produces $\lambda = 2(y - 3) = 2y - 6$, $x = \lambda + 1 = 2y - 5$, $z = \frac{1}{2}\lambda + 2 = y - 1$, and $5 = 2x + y + z = (4y - 10) + y + (y - 1) = 6y - 11$. Hence $y = \frac{8}{3}$, so $x = 2y - 5 = \frac{1}{3}$ and $z = y - 1 = \frac{5}{3}$. The minimum distance is $\sqrt{\left(\frac{1}{3} - 1\right)^2 + \left(\frac{8}{3} - 3\right)^2 + \left(\frac{5}{3} - 2\right)^2} = \sqrt{\left(-\frac{2}{3}\right)^2 + \left(-\frac{1}{3}\right)^2 + \left(-\frac{1}{3}\right)^2} = \sqrt{\frac{2}{3}}$ and it occurs at $\left(\frac{1}{3}, \frac{8}{3}, \frac{5}{3}\right)$.

11 Let $L(x, y, z, \lambda) = x^2y^2z^2 - \lambda(x^2 + y^2 + z^2 - 1)$. Then $L_x = 2xy^2z^2 - 2\lambda x$, $L_y = 2x^2yz^2 - 2\lambda y$, $L_z = 2x^2y^2z - 2\lambda z$, and $L_\lambda = -(x^2 + y^2 + z^2 - 1)$. Setting $L_x = 0$ yields the equation $2x(y^2z^2 - \lambda) = 0$, so either $x = 0$, or $y^2z^2 = \lambda$. If $x = 0$, then $x^2y^2z^2 = 0$, which is certainly not maximal. Hence $y^2z^2 = \lambda$. Similarly, $x^2z^2 = \lambda$ and $x^2y^2 = \lambda$, so $x^2 = y^2 = z^2$. Since $x^2 + y^2 + z^2 = 1$, we have $x^2 = y^2 = z^2 = \frac{1}{3}$ and $x^2y^2z^2 = \frac{1}{27}$. The maximum is thus $\frac{1}{27}$ and occurs at the eight points of the form $\left(\pm\frac{1}{\sqrt{3}}, \pm\frac{1}{\sqrt{3}}, \pm\frac{1}{\sqrt{3}}\right)$, where the signs of the coordinates are taken in all possible combinations.

13 $L(x, y, z, \lambda, \mu) = x^2 + y^2 + z^2 - \lambda(x + 2y + 3z) - \mu(2x + 3y + z - 4)$. $L_x = 2x - \lambda - 2\mu$, $L_y = 2y - 2\lambda - 3\mu$, $L_z = 2z - 3\lambda - \mu$, $L_\lambda = -(x + 2y + 3z)$, and $L_\mu = -(2x + 3y + z - 4)$. Setting these equal to 0 and solving for x, y, and z in terms of λ and μ results in $x = \frac{\lambda + 2\mu}{2}$, $y = \frac{2\lambda + 3\mu}{2}$, and $z = \frac{3\lambda + \mu}{2}$. Hence $0 = x + 2y + 3z = \frac{1}{2}[(\lambda + 2\mu) + 2(2\lambda + 3\mu) + 3(3\lambda + \mu)] = \frac{1}{2}(14\lambda + 11\mu)$, so $14\lambda + 11\mu = 0$. Also, $4 = 2x + 3y + z = \frac{1}{2}[2(\lambda + 2\mu) + 3(2\lambda + 3\mu) + (3\lambda + \mu)] = \frac{1}{2}(11\lambda + 14\mu)$, so $11\lambda + 14\mu = 8$. Subtracting

gives $3(\mu - \lambda) = 8$, so $\mu = \lambda + \frac{8}{3}$. Hence $0 = 14\lambda + 11\mu = 14\lambda + 11\left(\lambda + \frac{8}{3}\right) = 25\lambda + \frac{88}{3}$, so $\lambda = -\frac{88}{75}$ and $\mu = \lambda + \frac{8}{3} = \frac{112}{75}$. Then $x = \frac{\lambda + 2\mu}{2} = \frac{68}{75}$, $y = \frac{2\lambda + 3\mu}{2} = \frac{16}{15}$, and $z = \frac{3\lambda + \mu}{2} = -\frac{76}{75}$. The minimum is $x^2 + y^2 + z^2 = \frac{224}{75}$.

15 Let the base have dimensions x and y and let the height be z; then we minimize $xy + 2xz + 2yz$ subject to the constraint $xyz = 1$. Let $L(x, y, z, \lambda) = xy + 2xz + 2yz - \lambda(xyz - 1)$. Then $L_x = y + 2z - \lambda yz$, $L_y = x + 2z - \lambda xz$, $L_z = 2x + 2y - \lambda xy$, and $L_\lambda = -(xyz - 1)$. Setting $L_\lambda = 0$ gives $z = \frac{1}{xy}$. Equating L_x, L_y, and L_z to 0 and solving for λ results in $\lambda = \frac{y + 2z}{yz} = xy + \frac{2}{y}$, $\lambda = \frac{x + 2z}{xz} = xy + \frac{2}{x}$, and $\lambda = \frac{2x + 2y}{xy}$. From the first two equations we have $x = y$. Then $\lambda = \frac{2x + 2y}{xy} = \frac{4}{x}$ and $\lambda = xy + \frac{2}{y} = x^2 + \frac{2}{x}$. So $\frac{4}{x} = x^2 + \frac{2}{x}$, $x^3 - 2 = 0$, and $x = 2^{1/3} = y$. Also, $z = \frac{1}{xy} = 2^{-2/3}$. The results agree with Exercise 25 of Sec. 14.9.

17 Call the dimensions x, y, and z. We wish to maximize xyz subject to the constraint $2xy + 2xz + 2yz = 12$. Let $L(x, y, z, \lambda) = xyz - \lambda(2xy + 2xz + 2yz - 12)$. Then $L_x = yz - 2\lambda y - 2\lambda z$, $L_y = xz - 2\lambda x - 2\lambda z$, $L_z = xy - 2\lambda x - 2\lambda y$, and $L_\lambda = -(2xy + 2xz + 2yz - 12)$. Setting these equal to 0 and solving for λ yields $\lambda = \frac{yz}{2(y + z)} = \frac{xz}{2(x + z)} = \frac{xy}{2(x + y)}$. So $x = y = z$. Applying the equation $2xy + 2xz + 2yz - 12 = 0$, we have $6x^2 - 12 = 0$, or $x = y = z = \sqrt{2}$. The box is a cube.

19 (a) Let $L(t, u, v, \lambda) = \sin t \sin u \sin v - \lambda(t + u + v - \pi/2)$. Then $L_t = \cos t \sin u \sin v - \lambda$, $L_u = \sin t \cos u \sin v - \lambda$, $L_v = \sin t \sin u \cos v - \lambda$, and $L_\lambda = -(t + u + v - \pi/2)$. Equating the first three partial derivatives to 0 yields $\lambda = \cos t \sin u \sin v = \sin t \cos u \sin v = \sin t \sin u \cos v$. If any of t, u, or v are 0, then $\sin t \sin u \sin v$ is not maximal, so assume t, u, and v are positive. Since $t + u + v = \pi/2$, t, u, and v must be less than $\pi/2$ as well. From the above equations for λ, we have $\cos t \sin u = \sin t \cos u$ and $\cos u \sin v = \sin u \cos v$. So $\tan t = \tan u$ and $\tan u = \tan v$, and thus $\tan t = \tan u = \tan v$. Since $0 < t, u, v < \pi/2$, it follows that $t = u = v = \pi/6$. Hence the maximum value of $\sin t \sin u \sin v$ is $(\sin \pi/6)^3 = 1/8$.

(b) See Figs. 5–7 on p. 863.

(c) If $|\mathbf{W}|$ is the force of the wind, then the fraction of this force that pushes the boat north is $\dfrac{|\mathbf{W}| \sin t \sin u \sin v}{|\mathbf{W}|}$

$= \sin t \sin u \sin v = \dfrac{1}{8}$.

21 The problem is equivalent to minimizing the square of the distance from the origin to the line common to the two planes $x + 2y + 3z = 0$ and $2x + 3y + z = 4$. Let $f(x, y, z) = x + 2y + 3z$ and $g(x, y, z) = 2x + 3y + z$. Then $\nabla f \times \nabla g =$

$$(\mathbf{i} + 2\mathbf{j} + 3\mathbf{k}) \times (2\mathbf{i} + 3\mathbf{j} + \mathbf{k}) = \begin{vmatrix} \mathbf{i} & \mathbf{j} & \mathbf{k} \\ 1 & 2 & 3 \\ 2 & 3 & 1 \end{vmatrix} =$$

$-7\mathbf{i} + 5\mathbf{j} - \mathbf{k}$ is a vector parallel to the line common to $f(x, y, z) = 0$ and $g(x, y, z) = 4$. To find a point on the line, let $z = 0$. Then $x + 2y = 0$ and $2x + 3y = 4$. Solving for x and y gives $x = 8$, $y = -4$. Hence $Q = (8, -4, 0)$ lies on the line.

Let $P = (0, 0, 0)$. Then $\overrightarrow{QP} = -8\mathbf{i} + 4\mathbf{j}$, $\nabla f \times \nabla g = -7\mathbf{i} + 5\mathbf{j} - \mathbf{k}$, and the square of the distance is $\left\| \overrightarrow{QP} - \left(\overrightarrow{QP} \cdot \dfrac{\nabla f \times \nabla g}{|\nabla f \times \nabla g|} \right) \dfrac{\nabla f \times \nabla g}{\|\nabla f \times \nabla g\|} \right\|^2$

$$= \left\| -8\mathbf{i} + 4\mathbf{j} - \left(\frac{76}{\sqrt{75}} \right) \left(\frac{-7\mathbf{i} + 5\mathbf{j} - \mathbf{k}}{\sqrt{75}} \right) \right\|^2$$

$$= \left\| -8\mathbf{i} + 4\mathbf{j} + \frac{76}{75}(7\mathbf{i} - 5\mathbf{j} + \mathbf{k}) \right\|^2$$

$$= \left\| -\frac{68}{75}\mathbf{i} - \frac{80}{75}\mathbf{j} + \frac{76}{75}\mathbf{k} \right\|^2$$

$$= \left(-\frac{68}{75} \right)^2 + \left(-\frac{80}{75} \right)^2 + \left(\frac{76}{75} \right)^2 = \frac{224}{75}.$$ This checks with Exercise 13 of this section.

23 (a)

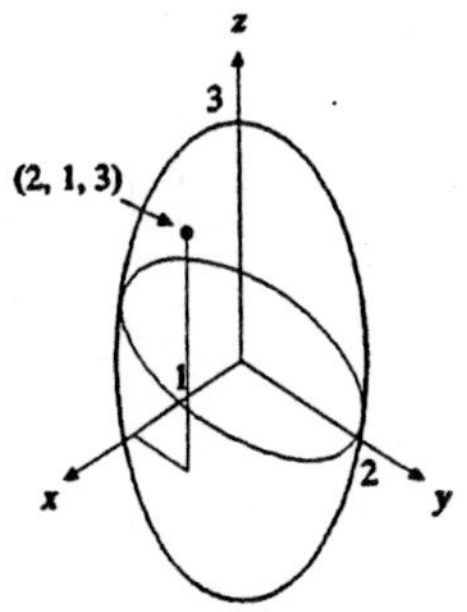

(b) We wish to minimize the distance from (2, 1, 3) to all points (x, y, z) subject to the constraint $x^2 + \dfrac{y^2}{4} + \dfrac{z^2}{9} = 1$. It is easier to minimize the square of the distance from (2, 1, 3) to (x, y, z), $(x - 2)^2 + (y - 1)^2 + (z - 3)^2$. Let $L(x, y, z, \lambda) = (x - 2)^2 + (y - 1)^2 + (z - 3)^2 - \lambda\left(x^2 + \dfrac{y^2}{4} + \dfrac{z^2}{9} - 1\right)$.

Then $L_x = 2(x - 2) - 2\lambda x$, $L_y = 2(y - 1) - \dfrac{\lambda y}{2}$, $L_z = 2(z - 3) - \dfrac{2\lambda z}{9}$, and $L_\lambda = -(x^2 + y^2/4 + z^2/9 - 1)$. Equating L_x, L_y, and L_z to 0 and solving for x, y, and z in terms of λ results in $x = \dfrac{2}{1 - \lambda}$, $y = \dfrac{4}{4 - \lambda}$, and $z = \dfrac{27}{9 - \lambda}$. Letting $L_\lambda = 0$ gives $x^2 + \dfrac{y^2}{4} + \dfrac{z^2}{9} = \left(\dfrac{2}{1 - \lambda}\right)^2 + \dfrac{1}{4}\left(\dfrac{4}{4 - \lambda}\right)^2 + \dfrac{1}{9}\left(\dfrac{27}{9 - \lambda}\right)^2 = 1$. This equation reduces to a sixth-order polynomial in λ: $\lambda^6 - 28\lambda^5 + 205\lambda^4 - 450\lambda^3 - 700\lambda + 4176\lambda - 5508 = 0$. This has two real roots,

$\lambda \approx -2.6080$ and $\lambda \approx 18.1550$. These give the points (0.5543, 0.6053, 2.3260) and (−0.1166, −0.2826, −2.9492), respectively. The first is the point on the ellipsoid nearest P and the second is the point farthest from P.

(c) The minimum distance is a perpendicular distance, so the angle between PQ and the tangent plane at Q is $\pi/2$.

25 Let $L(x, y, z, \lambda, \mu) = x^3 + y^3 + 2z^3 - \lambda(x^2 + y^2 + z^2 - 4) - \mu((x - 3)^2 + y^2 + z^2 - 4)$. Then $L_x = 3x^2 - 2\lambda x - 2\mu x + 6\mu$, $L_y = 3y^2 - 2\lambda y - 2\mu y$, and $L_z = 6z^2 - 2\lambda z - 2\mu z$. As before, $L_\lambda = 0$ and $L_\mu = 0$ yield the original constraints. Setting $L_y = 0$ gives either $y = 0$ or $\lambda + \mu = \frac{3}{2}y$. Setting $L_z = 0$ yields either $z = 0$ or $\lambda + \mu = 3z$. Hence either $y = 0$, or $z = 0$, or $\lambda + \mu = \frac{3}{2}y = 3z$.

Suppose $y = 0$. Then $x^2 + z^2 = 4$ and $(x - 3)^2 + z^2 = 4$. Subtracting the second equation from the first yields $6x - 9 = 0$, so $x = \frac{3}{2}$ and $z = \pm\sqrt{4 - \left(\frac{3}{2}\right)^2} = \pm\frac{\sqrt{7}}{2}$. If $z = 0$, similar reasoning results in $x = \frac{3}{2}$ and $y = \pm\frac{\sqrt{7}}{2}$. If $\lambda + \mu = \frac{3}{2}y = 3z$, then $y = 2z$ and thus $x = \frac{3}{2}$, $y = \pm\sqrt{\frac{7}{5}}$, and $z = \pm\sqrt{\frac{7}{20}}$. Computing $x^3 + y^3 + 2z^3$ at the six points $\left(\frac{3}{2}, 0, \pm\frac{\sqrt{7}}{2}\right)$, $\left(\frac{3}{2}, \pm\frac{\sqrt{7}}{2}, 0\right)$, and $\left(\frac{3}{2}, \pm\sqrt{\frac{7}{5}}, \pm\sqrt{\frac{7}{20}}\right)$, we find that the maximum occurs at $\left(\frac{3}{2}, 0, \frac{\sqrt{7}}{2}\right)$ and is equal to

$$\left(\frac{3}{2}\right)^3 + 2\left(\frac{\sqrt{7}}{2}\right)^3 = \frac{27 + 14\sqrt{7}}{8}.$$

27 We maximize $f(x, y) = \sin x \sin y \sin(\pi - x - y)$. Now $f_x = \cos x \sin y \sin(\pi - x - y) - \sin x \sin y \cos(\pi - x - y)$ and $f_y = \sin x \cos y \sin(\pi - x - y) - \sin x \sin y \cos(\pi - x - y)$. Equating f_x and f_y to 0 yields the critical point $(\pi/3, \pi/3)$ since $0 < x < \pi$ and $0 < y < \pi$. To see that a maximum occurs at $(\pi/3, \pi/3)$, note that $f_{xx} = -2 \sin x \sin y \sin(\pi - x - y) - 2 \cos x \sin y \cos(\pi - x - y)$, $f_{yy} = -2 \sin x \sin y \sin(\pi - x - y) - 2 \sin x \cos y \cos(\pi - x - y)$, and $f_{xy} = (\cos x \cos y - \sin x \sin y) \sin(\pi - x - y) - (\cos x \sin y + \sin x \cos y) \cos(\pi - x - y) = \cos(x + y) \sin(\pi - x - y) - \sin(x + y) \cos(\pi - x - y)$, so $f_{xx} = f_{yy} = -\sqrt{3}$ and $f_{xy} = -\sqrt{3}/2$ at $(\pi/3, \pi/3)$. Hence $D = (-\sqrt{3})(-\sqrt{3}) - \left(-\frac{\sqrt{3}}{2}\right)^2 = 3 - \frac{3}{4} = \frac{9}{4} > 0$ and $f_{xx} < 0$, so f has a relative maximum at $(\pi/3, \pi/3)$. This agrees with Exercise 26.

29 (a) $L(x_1, x_2, \cdots, x_n, \lambda) = x_1x_2 \cdots x_n - \lambda\left(\sum_{i=1}^{n} x_i - 1\right)$. $L_{x_i} = x_1x_2 \cdots x_{i-1}x_{i+1} \cdots x_n - \lambda$

and $L_\lambda = -\left(\sum_{i=1}^{n} x_i - 1\right)$. If any of the x_i are 0, then $x_1 \cdots x_n = 0$, which will be a minimum value, so we can assume that $x_i > 0$ for all i. Setting $L_{x_1} = 0$ and $L_{x_2} = 0$ gives $x_2x_3\cdots x_n = \lambda = x_1x_3\cdots x_n$, from which it follows that $x_1 = x_2$. Similarly $x_1 = x_2 = x_3 = \cdots = x_n$, so $x_i = \frac{1}{n}$ for all i and $x_1x_2 \cdots x_n = \frac{1}{n^n}$.

(b) Let $S = a_1 + a_2 + \cdots + a_n > 0$. If $S = 0$ then all a_i's are 0 so the inequality holds. If $S > 0$, then $\sum_{i=1}^{n} \frac{a_i}{S} = 1$. By part (a), $\frac{a_1}{S}\frac{a_2}{S} \cdots \frac{a_n}{S} \le \frac{1}{n^n}$. Hence $a_1a_2 \cdots a_n \le \left(\frac{S}{n}\right)^n$, so $\sqrt[n]{a_1a_2 \cdots a_n} \le \frac{S}{n} = \frac{a_1 + a_2 + \cdots + a_n}{n}$.

31 Let $L(x_1, \cdots, x_n, \lambda) = \sum_{i=1}^{n} a_ix_i - \lambda\left(\sum_{i=1}^{n} x_i^2 - 1\right)$. Then, for $1 \le i \le n$, $L_{x_i} = a_i - 2\lambda x_i$. Setting these equal to 0 yields $x_i = \frac{a_i}{2\lambda}$ for all values of i. But $\sum_{i=1}^{n} x_i^2 = 1$, so $\sum_{i=1}^{n}\left(\frac{a_i}{2\lambda}\right)^2 = 1$, $\sum_{i=1}^{n} a_i^2 = (2\lambda)^2$, and $2\lambda = \pm\left(\sum_{i=1}^{n} a_i^2\right)^{1/2}$ and $x_i = \frac{a_i}{2\lambda} = \pm a_i\left(\sum_{i=1}^{n} a_i^2\right)^{-1/2}$. Then $\sum_{i=1}^{n} a_ix_i = \pm\left(\sum_{i=1}^{n} a_i^2\right)^{1/2}$; we desire a maximum, so we choose the positive sign. Therefore the maximum is $\left(\sum_{i=1}^{n} a_i^2\right)^{1/2}$ and occurs when $x_i = a_i\left(\sum_{i=1}^{n} a_i^2\right)^{-1/2}$.

33 (a) Let $L(x_1, \cdots, x_n, \lambda) = u(x_1, \cdots, x_n) - \lambda\left(\sum_{i=1}^{n} p_ix_i - B\right)$. Then $\frac{\partial L}{\partial x_i} = \frac{\partial u}{\partial x_i} - \lambda p_i$. At the maximum, $\frac{\partial L}{\partial x_i} = 0$, so $\frac{\partial u}{\partial x_i} = \lambda p_i$ for all values of i. That is, $\frac{\partial u/\partial x_i}{p_i} = \lambda$. Since λ does not depend on i, the result follows.

(b) The quantity $\frac{\partial u}{\partial x_i}$ is the rate of change of utility with respect to x_i, the number of ith items. Since p_i is the price in dollars per item, $\frac{1}{p_i}\cdot\frac{\partial u}{\partial x_i}$ is the change in utility with respect to the number of dollars spent on the ith item. If this quantity is not equal for all values of i, it means that it is possible to get more "bang for the buck" by reallocating dollars from items with lesser utility impact to those with more. At maximum utility, however, this cannot be done, so it must be that utility impact is equal for all items.

35 Let $L(b_1, \cdots, b_k, \lambda) = \sum_{j=1}^{k} p_j 2^{B-b_j} - \lambda\left(\sum_{j=1}^{k} b_j - B\right)$. Then $L_{b_i} = p_i 2^{B-b_i}(-\ln 2) - \lambda$. So if $L_{b_i} = 0$, then $p_i 2^{B-b_i} = -\frac{\lambda}{\ln 2}$. Let $c = -\frac{\lambda}{\ln 2}$. At the minimum we have $p_i 2^{B-b_i} = c$ for $1 \le i \le k$, so $b_i = B - \log_2 \frac{c}{p_i} = (B - \log_2 c) + \log_2(p_i)$. But $\sum_{j=1}^{k} b_j = B$, so $B = \sum_{j=1}^{k} ((B - \log_2 c) + \log_2 p_j) = k(B - \log_2 c) + \sum_{j=1}^{k} \log_2(p_j)$. Hence $B - \log_2 c = \frac{1}{k}\left(B - \sum_{j=1}^{k} \log_2 p_j\right)$ and $b_i = \frac{1}{k}\left(B - \sum_{j=1}^{k} \log_2 p_j\right) + \log_2 p_i$.

14.11 The Chain Rule Revisited

1 Consider u as a function of y and z only: $u = f(h(y, z), y, z) = k$. To compute $\left(\frac{\partial u}{\partial z}\right)_y$, we refer to

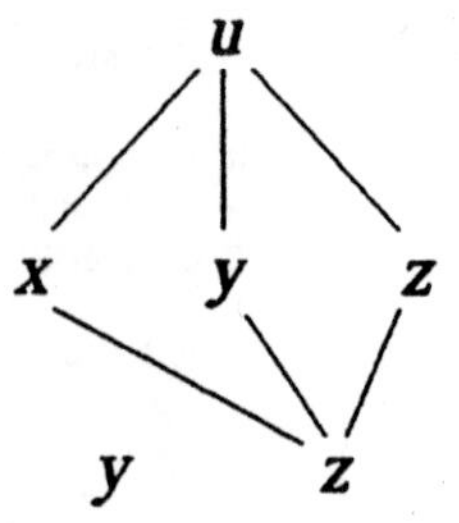

the paths in the figure. By the chain rule, $\left(\frac{\partial u}{\partial z}\right)_y = \left(\frac{\partial u}{\partial x}\right)_{y,z} \frac{\partial x}{\partial z} + \left(\frac{\partial u}{\partial y}\right)_{x,z} \frac{\partial y}{\partial z} + \left(\frac{\partial u}{\partial z}\right)_{x,y} \frac{\partial z}{\partial z}$, where $u = f(x, y, z)$ on the right-hand side of the equation. Now, since on the left-hand side of the equation, $u = f(h(y, z), y, z) = k$, $\left(\frac{\partial u}{\partial z}\right)_y = 0$. It is clear that $\frac{\partial z}{\partial z} = 1$ and $\frac{\partial y}{\partial z} = 0$. Finally, $x = h(y, z)$, so we may write $\frac{\partial x}{\partial z} = \left(\frac{\partial x}{\partial z}\right)_y$. Then the above equation reduces to $0 = \left(\frac{\partial u}{\partial x}\right)_{y,z}\left(\frac{\partial x}{\partial z}\right)_y + \left(\frac{\partial u}{\partial z}\right)_{x,y}$. Solving for $\left(\frac{\partial x}{\partial z}\right)_y$, we have $\left(\frac{\partial x}{\partial z}\right)_y = -\frac{(\partial u/\partial z)_{x,y}}{(\partial u/\partial x)_{y,z}}$.

3 Considering $u = f(x, y, z)$, we have $\left(\frac{\partial u}{\partial x}\right)_{y,z} = \frac{\partial f}{\partial x} = \frac{\partial}{\partial x}(x^2 + 2y^2 + z^2) = 2x$ and $\left(\frac{\partial u}{\partial z}\right)_{x,y} = \frac{\partial f}{\partial z} = 2z$. Then $\frac{-(\partial u/\partial x)_{y,z}}{(\partial u/\partial z)_{x,y}} = \frac{-2x}{2z} = -\frac{x}{z}$. Now consider $z = \sqrt{4 - x^2 - 2y^2}$. Hence $\left(\frac{\partial z}{\partial x}\right)_y = \frac{-2x}{2\sqrt{4 - x^2 - 2y^2}} = -\frac{x}{z} = -\frac{(\partial u/\partial x)_{y,z}}{(\partial u/\partial z)_{x,y}}$, and Theorem 1 is verified.

5 Rewriting $\left(\frac{\partial P}{\partial T}\right)_V = -\frac{(\partial V/\partial T)_P}{(\partial V/\partial P)_T}$ as

$$\left(\frac{\partial V}{\partial P}\right)_T\left(\frac{\partial P}{\partial T}\right)_V + \left(\frac{\partial V}{\partial T}\right)_P = 0$$ suggests that we

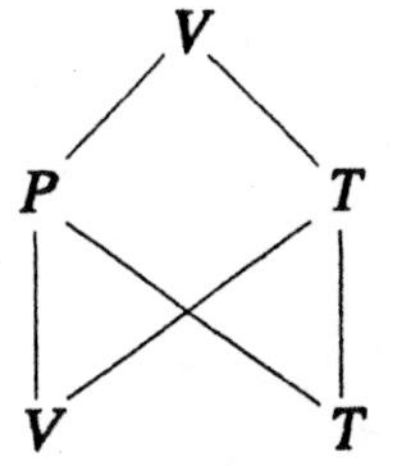

consider V as the top variable, P and T as middle variables, and V and T as bottom variables. In view of the diagram, let $V = f(V, T) = 0 \cdot T + V$ and from thermodynamics theory we know also that $V = g(P, T)$ for some function g. Then, by the chain rule, $\left(\frac{\partial f}{\partial T}\right)_V = \left(\frac{\partial g}{\partial P}\right)_T\left(\frac{\partial P}{\partial T}\right)_V + \left(\frac{\partial g}{\partial T}\right)_P\left(\frac{\partial T}{\partial T}\right)_V =$

$\left(\frac{\partial V}{\partial P}\right)_T\left(\frac{\partial P}{\partial T}\right)_V + \left(\frac{\partial V}{\partial T}\right)_P$. Now $\left(\frac{\partial f}{\partial T}\right)_V = 0$, so $0 =$

$\left(\frac{\partial V}{\partial P}\right)_T\left(\frac{\partial P}{\partial T}\right)_V + \left(\frac{\partial V}{\partial T}\right)_P$ and the desired equality

follows.

7 Consider P as the top variable, T and V as middle variables, and P and V as bottom variables. Then 1

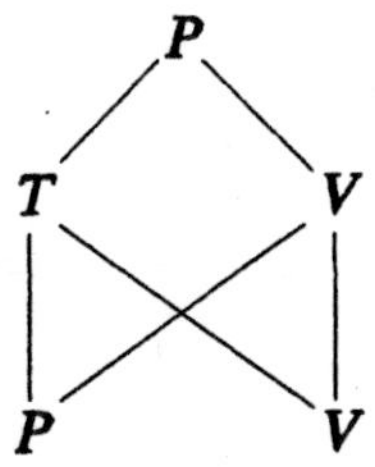

$= \left(\frac{\partial P}{\partial P}\right)_V = \left(\frac{\partial P}{\partial T}\right)_V\left(\frac{\partial T}{\partial P}\right)_V + \left(\frac{\partial P}{\partial V}\right)_T\left(\frac{\partial V}{\partial P}\right)_V =$

$\left(\frac{\partial P}{\partial T}\right)_V\left(\frac{\partial T}{\partial P}\right)_V + \left(\frac{\partial P}{\partial V}\right)_T \cdot 0 = \left(\frac{\partial P}{\partial T}\right)_V\left(\frac{\partial T}{\partial P}\right)_V.$

9 (a) The partial of z with respect to r with s and t held constant, or the partial derivative of z with respect to r with s and u held constant.

(b) $\left(\frac{\partial z}{\partial r}\right)_{s,t} = 2r$, $\left(\frac{\partial z}{\partial r}\right)_{s,u} = \frac{\partial}{\partial r}[r^2 + s^2 + (rsu)^2]$

$= 2r + 2rs^2u^2$

11 We are given $u = f(x, y)$ and $v = g(x, y)$ for some functions f and g. Let $S = F(u, v, x, y)$ and $T = G(u, v, x, y)$. Then S and T may also be thought of as functions of x and y; that is, $S = H(x, y) = F(f(x, y), g(x, y), x, y) = 0$ and $T = I(x, y) = G(f(x, y), g(x, y), x, y) = 0$ where $H(x, y) = I(x, y) = 0$ is given. We wish to solve for $\frac{\partial u}{\partial x}$, so we differentiate $S = H(x, y)$ and $T = I(x, y)$ with respect to x. By the chain rule, $\left(\frac{\partial S}{\partial x}\right)_y =$

$\left(\frac{\partial S}{\partial u}\right)_{x,y,v}\left(\frac{\partial u}{\partial x}\right)_y + \left(\frac{\partial S}{\partial v}\right)_{x,y,u}\left(\frac{\partial v}{\partial x}\right)_y + \left(\frac{\partial S}{\partial x}\right)_{y,u,v}\left(\frac{\partial x}{\partial x}\right)_y$

$+ \left(\frac{\partial S}{\partial y}\right)_{x,u,v}\left(\frac{\partial y}{\partial x}\right)_y$ and $\left(\frac{\partial T}{\partial x}\right)_y = \left(\frac{\partial T}{\partial u}\right)_{x,y,v}\left(\frac{\partial u}{\partial x}\right)_y +$

$\left(\frac{\partial T}{\partial v}\right)_{x,y,u}\left(\frac{\partial v}{\partial x}\right)_y + \left(\frac{\partial T}{\partial x}\right)_{y,u,v}\left(\frac{\partial x}{\partial x}\right)_y +$

$\left(\frac{\partial T}{\partial y}\right)_{x,u,v}\left(\frac{\partial y}{\partial x}\right)_y$. Since $\left(\frac{\partial S}{\partial x}\right)_y = \left(\frac{\partial T}{\partial x}\right)_y = \left(\frac{\partial y}{\partial x}\right)_y =$

0, $\left(\frac{\partial x}{\partial x}\right)_y = 1$, and $S = F(u, v, x, y)$ and $T =$

$G(u, v, x, y)$ on the right-hand sides of the above two equations, the equations become $0 = \frac{\partial F}{\partial u}\left(\frac{\partial u}{\partial x}\right)_y + \frac{\partial F}{\partial v}\left(\frac{\partial v}{\partial x}\right)_y + \frac{\partial F}{\partial x}$ and $0 = \frac{\partial G}{\partial u}\left(\frac{\partial u}{\partial x}\right)_y + \frac{\partial G}{\partial v}\left(\frac{\partial v}{\partial x}\right)_y + \frac{\partial G}{\partial x}$. Solving for $\left(\frac{\partial v}{\partial x}\right)_y$ in the second equation gives $\left(\frac{\partial v}{\partial x}\right)_y =$

$$\frac{-\left(\frac{\partial G}{\partial x} + \frac{\partial G}{\partial u}\left(\frac{\partial u}{\partial x}\right)_y\right)}{\partial G/\partial v}.$$

Substituting this expression for $\left(\frac{\partial v}{\partial x}\right)_y$ into the first equation yields $0 =$

$$\frac{\partial F}{\partial u}\left(\frac{\partial u}{\partial x}\right)_y - \frac{\partial F/\partial v}{\partial G/\partial v}\left(\frac{\partial G}{\partial x} + \frac{\partial G}{\partial u}\left(\frac{\partial u}{\partial x}\right)_y\right) + \frac{\partial F}{\partial x}.$$

Solving for $\left(\frac{\partial u}{\partial x}\right)_y$ in the above equation, we have

$$\left(\frac{\partial u}{\partial x}\right)_y = \frac{-\frac{\partial F}{\partial x} + \frac{\partial F/\partial v}{\partial G/\partial v}\left(\frac{\partial G}{\partial x}\right)}{\frac{\partial F}{\partial u} - \frac{\partial F/\partial v}{\partial G/\partial v}\left(\frac{\partial G}{\partial u}\right)}$$

$$= \frac{-\frac{\partial F}{\partial x}\frac{\partial G}{\partial v} + \frac{\partial F}{\partial v}\frac{\partial G}{\partial x}}{\frac{\partial F}{\partial u}\frac{\partial G}{\partial v} - \frac{\partial F}{\partial v}\frac{\partial G}{\partial u}}.$$

14.S Guide Quiz

2 (a) See Theorem 1 of Sec. 14.6.

(b) See Theorem 3 of Sec. 14.6.

3

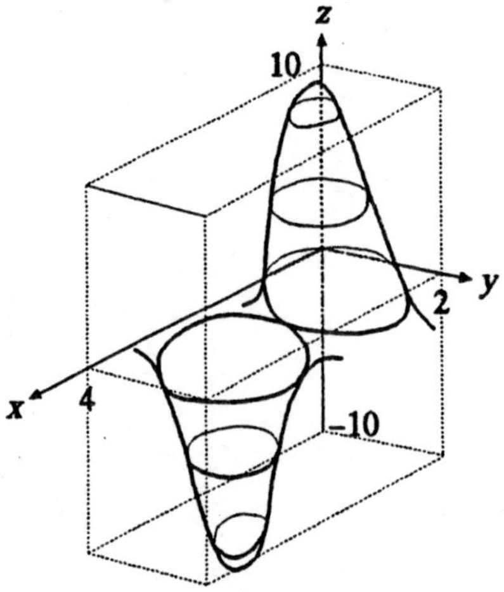

4 (a) $\frac{\partial f}{\partial x}(0, 0) \approx \frac{\Delta f}{\Delta x}$

$$= \frac{f(0 + 0.03, 0) - f(0, 0)}{0.03}$$

$$= \frac{5.02 - 5.00}{0.03} = \frac{2}{3}$$

(b) $\frac{\partial f}{\partial y}(0, 0) \approx \frac{\Delta f}{\Delta y}$

$$= \frac{f(0, 0 + 0.01) - f(0, 0)}{0.01}$$

$$= \frac{5.02 - 5.00}{0.01} = 2$$

(c) $\nabla f(0, 0) = f_x(0, 0)\mathbf{i} + f_y(0, 0)\mathbf{j} \approx \frac{2}{3}\mathbf{i} + 2\mathbf{j}$

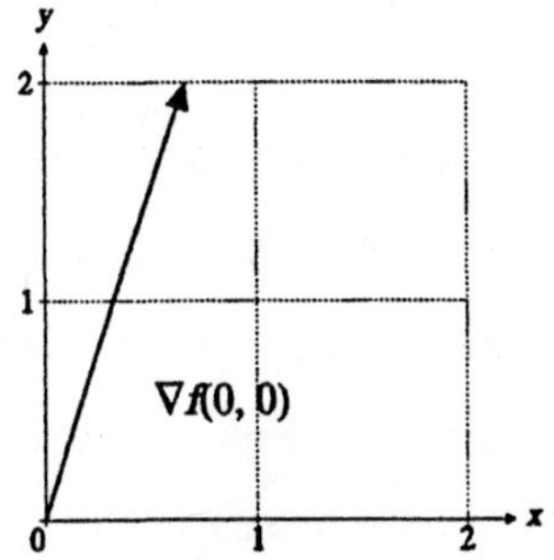

(d) The gradient is always perpendicular to the level curve, so the angle is 90°.

(e) $D_{\mathbf{u}}f = \nabla f(0, 0)\cdot\mathbf{u}$

$\approx \left(\frac{2}{3}\mathbf{i} + 2\mathbf{j}\right)\cdot(\cos 37°\ \mathbf{i} + \sin 37°\ \mathbf{j})$

$= \frac{2}{3}\cos 37° + 2\sin 37°$

5 (a) $f(x, y) = 2x + 3y$

(b) Since $\frac{\partial f}{\partial x} = 2$, it follows that $f(x, y) =$

$\int \frac{\partial f}{\partial x}\,dx + g(y) = 2x + C_1 + g(y)$. Then

$\frac{\partial f}{\partial y} = g'(y) = 3$, which implies that $g(y) =$

$3y + C_2$. So $f(x, y) = 2x + 3y + C$, where C $= C_1 + C_2$ is constant. Using the fact that $f(0, 0) = 0$, we see that $C = 0$. Therefore $f(x, y) = 2x + 3y$ is the only function with the given properties.

6 (a) $\frac{\partial}{\partial y}[\sec^3(x + 2y)\ln(1 + 2xy)] = 3\sec^2(x + 2y)$

$\times \sec(x + 2y)\tan(x + 2y)\,2\ln(1 + 2xy) +$

$\sec^3(x + 2y)\frac{2x}{1 + 2xy} = 2\sec^3(x + 2y) \times$

$\left[3\tan(x + 2y)\ln(1 + 2xy) + \frac{x}{1 + 2xy}\right]$

(b) $\frac{\partial}{\partial x}(x\tan^{-1} 3xy) = x\cdot\frac{1}{1 + (3xy)^2}\cdot 3y +$

$\tan^{-1} 3xy = \frac{3xy}{1 + 9x^2y^2} + \tan^{-1} 3xy$

(c) $\frac{\partial}{\partial x}(ye^{x^3y}) = ye^{x^3y}\cdot 3x^2y = 3x^2y^2e^{x^3y}$

(d) $\frac{\partial^2}{\partial x\,\partial y}[\cos(2x + 3y)] = \frac{\partial}{\partial x}\left[\frac{\partial}{\partial y}(\cos(2x + 3y))\right]$

$= \frac{\partial}{\partial x}[-3\sin(2x + 3y)] = -6\cos(2x + 3y)$

7 (a) By Theorem 3 of Sec. 14.6, we have $\frac{\partial w}{\partial x} =$

$\frac{\partial w}{\partial u}\cdot\frac{\partial u}{\partial x} + \frac{\partial w}{\partial v}\cdot\frac{\partial v}{\partial x} = \frac{\partial w}{\partial u} + \frac{\partial w}{\partial v}$ and $\frac{\partial w}{\partial y} =$

$\frac{\partial w}{\partial u}\cdot\frac{\partial u}{\partial y} + \frac{\partial w}{\partial v}\cdot\frac{\partial v}{\partial y} = \frac{\partial w}{\partial u} - \frac{\partial w}{\partial v}$. Thus

$\frac{\partial w}{\partial x}\cdot\frac{\partial w}{\partial y} = \left(\frac{\partial w}{\partial u} + \frac{\partial w}{\partial v}\right)\left(\frac{\partial w}{\partial u} - \frac{\partial w}{\partial v}\right) =$

$\left(\frac{\partial w}{\partial u}\right)^2 - \left(\frac{\partial w}{\partial v}\right)^2 = \left(\frac{\partial f}{\partial u}\right)^2 - \left(\frac{\partial f}{\partial v}\right)^2.$

(b) As above, $\frac{\partial w}{\partial y} = \frac{\partial w}{\partial u} - \frac{\partial w}{\partial v}$, so $\frac{\partial^2 w}{\partial x\,\partial y} =$

$\frac{\partial}{\partial x}\left(\frac{\partial w}{\partial u} - \frac{\partial w}{\partial v}\right) = \frac{\partial}{\partial u}\left(\frac{\partial w}{\partial u} - \frac{\partial w}{\partial v}\right)\frac{\partial u}{\partial x} +$

$\frac{\partial}{\partial v}\left(\frac{\partial w}{\partial u} - \frac{\partial w}{\partial v}\right)\frac{\partial v}{\partial x}$

$= \frac{\partial^2 w}{\partial u^2} - \frac{\partial^2 w}{\partial u\,\partial v} + \frac{\partial^2 w}{\partial v\,\partial u} - \frac{\partial^2 w}{\partial v^2}$

$= \dfrac{\partial^2 w}{\partial u^2} - \dfrac{\partial^2 w}{\partial v^2} = \dfrac{\partial^2 f}{\partial u^2} - \dfrac{\partial^2 f}{\partial v^2}$, where the cancellations of the mixed partial derivatives follow from the fact that the partial derivatives of f are continuous, so the mixed partials are equal.

8 See the figures in the answer section of the text.

9 (a) $y(x, t) = y_0 \sin(kx - kvt)$,

$\dfrac{\partial y}{\partial x} = ky_0 \cos(kx - kvt)$,

$\dfrac{\partial^2 y}{\partial x^2} = -k^2 y_0 \sin(kx - kvt)$,

$\dfrac{\partial y}{\partial t} = -kvy_0 \cos(kx - kvt)$, and

$\dfrac{\partial^2 y}{\partial t^2} = -k^2v^2y_0 \sin(kx - kvt) = v^2 \cdot \dfrac{\partial^2 y}{\partial x^2}$, so

y satisfies $\dfrac{\partial^2 y}{\partial x^2} = \dfrac{1}{v^2} \cdot \dfrac{\partial^2 y}{\partial t^2}$.

(b) Let $y = f(x + vt)$. Then $\dfrac{\partial y}{\partial x} = f'(x + vt)$,

$\dfrac{\partial^2 y}{\partial x^2} = f''(x + vt)$, $\dfrac{\partial y}{\partial t} = vf'(x + vt)$, and

$\dfrac{\partial^2 y}{\partial t^2} = v^2 f''(x + vt)$. Hence $\dfrac{\partial^2 y}{\partial x^2} = \dfrac{1}{v^2} \cdot \dfrac{\partial^2 y}{\partial t^2}$,

provided that $v \neq 0$. Applying this result with v replaced by $-v$, we see that if $y = f(x - vt)$, then $\dfrac{\partial^2 y}{\partial x^2} = \dfrac{1}{(-v)^2} \cdot \dfrac{\partial^2 y}{\partial t^2} = \dfrac{1}{v^2} \cdot \dfrac{\partial^2 y}{\partial t^2}$, so $f(x - vt)$ also satisfies the equation.

10 (a) By Theorem 2 of Sec. 14.9, if it is also true that $D(0, 0) > 0$, then $(0, 0)$ will be a local minimum. In particular, $f(x, y) = x^2 + y^2$ is such a function, since $f_{xx} = f_{yy} = 2$, $f_{xy} = 0$, and $D = 2 \cdot 2 - 0^2 = 4 > 0$.

(b) If it is also true that $D(0, 0) < 0$, then $(0, 0)$ is neither a local maximum nor minimum. For $f(x, y) = x^2 - 3xy + y^2$, $D = -5$.

11 Let the floor have dimensions x feet by y feet, and let the height be z feet. The heat admitted by the walls is $5(2xz + 2yz) = 10(xz + yz)$, by the floor is xy, and the roof is $3xy$. Thus the total rate at which heat enters is $R = 10(xz + yz) + 4xy$. Now $xyz = 10000 = k$, so $z = \dfrac{k}{xy}$. Hence, $R(x, y) = 10k\left(\dfrac{1}{y} + \dfrac{1}{x}\right) + 4xy$, $R_x = -\dfrac{10k}{x^2} + 4y$, and $R_y = -\dfrac{10k}{y^2} + 4x$. Setting $R_x = 0$ and $R_y = 0$ yields $4x^2y = 10k = 4xy^2$, from which we see that $x = y$ and therefore $x = y = \sqrt[3]{5k/2}$. This provides the only critical point. At the critical point, $R_{xx} = \dfrac{20k}{x^3} = \dfrac{20k}{5k/2} = 8$, $R_{yy} = \dfrac{20k}{y^3} = 8$, $R_{xy} = 4$, and $D = R_{xx}R_{yy} - (R_{xy})^2 = 48$. Since $D > 0$ and $R_{xx} > 0$, the critical point provides a minimum for R. The house should therefore have a square floor with sides of length $\sqrt[3]{5k/2} = \sqrt[3]{25{,}000} = 10\sqrt[3]{25}$ and height $z = \dfrac{k}{xy} = \dfrac{10000}{100(25)^{2/3}} = \dfrac{100}{25^{2/3}} = 4\sqrt[3]{25}$.

12 (a) $(1.1)^2\ln(1.2) \approx 0.2206$

(b) Let $x = y = 1$, $\Delta x = 0.1$, and $\Delta y = 0.2$. Then $(1.1)^2 \ln(1.2) = f(x + \Delta x, y + \Delta y)$ $\approx f(x, y) + f_x(x, y)\Delta x + f_y(x, y)\Delta y$ $= f(1, 1) + f_x(1, 1)(0.1) + f_y(1, 1)(0.2)$. But $f_x = 2x \ln y$ and $f_y = x^2/y$, so $f_x(1, 1) = 0$, $f_y(1, 1) = 1$, and thus $(1.1)^2 \ln(1.2) \approx 0 + 0(0.1) + 1(0.2) = 0.2$.

13 That the maximum errors in measuring m and v are 1% and 3%, respectively, means that $\left|\frac{\Delta m}{m}\right| \le 0.01$ and $\left|\frac{\Delta v}{v}\right| \le 0.03$. We wish to estimate $\left|\frac{\Delta K}{K}\right|$. But $\Delta K \approx dK = K_m \Delta m + K_v \Delta v$ $= \frac{1}{2}v^2 \Delta m + mv\Delta v$, so $\frac{\Delta K}{K} \approx \frac{dK}{K}$ $= \frac{\Delta m}{m} + 2\frac{\Delta v}{v}$, and $\left|\frac{\Delta K}{K}\right| \approx \left|\frac{\Delta m}{m} + 2\frac{\Delta v}{v}\right| \le$ $\left|\frac{\Delta m}{m}\right| + 2\left|\frac{\Delta v}{v}\right| \le 0.01 + 2(0.03) = 0.07$. The maximum error in measuring K is about 7%.

14 We are given that $\mathbf{r} = x(t)\mathbf{i} + y(t)\mathbf{j} = t\mathbf{i} + 2t^2\mathbf{j}$. We wish to find $\frac{dT}{dt}$; by Theorem 2 of Sec. 14.6, $\frac{dT}{dt} = \frac{\partial T}{\partial x}\cdot\frac{dx}{dt} + \frac{\partial T}{\partial y}\cdot\frac{dy}{dt}$ since $x = x(t)$ and $y = y(t)$. Now $\frac{\partial T}{\partial x}(1,2) \approx \frac{\Delta T}{\Delta x} =$ $\frac{T(1.01, 2) - T(1, 2)}{0.01} = \frac{0.03}{0.01} = 3$ and $\frac{\partial T}{\partial y}(1,2)$ $\approx \frac{\Delta T}{\Delta y} = \frac{T(1, 2.01) - T(1, 2)}{0.01} = \frac{-0.02}{0.01} =$ -2. Also, at $(1, 2)$ we have $t = 1$. Thus $\frac{dT}{dt}(1, 2)$ $= \frac{\partial T}{\partial x}(1, 2)\cdot\left.\frac{\partial x}{\partial t}\right|_{t=1} + \frac{\partial T}{\partial y}(1, 2)\cdot\left.\frac{\partial y}{\partial t}\right|_{t=1}$ $\approx 3\cdot 1 + (-2)\cdot 4 = -5$.

15 (b) Before trying to find the solution with Lagrange multipliers, consider the situation geometrically. The ellipsoid $x^2 + \frac{y^2}{4} + \frac{z^2}{9} = 1$ has axis intercepts $x = \pm 1$, $y = \pm 2$, and $z = \pm 3$. The surface is closest to the origin at the points $(1, 0, 0)$ and $(-1, 0, 0)$. Algebraically, $x^2 + y^2 + z^2 \ge x^2 + \frac{y^2}{4} + \frac{z^2}{9} = 1$, so the distance is always at least 1. At $(\pm 1, 0, 0)$, however, this minimum is obtained. Since the point $(1, 0, 0)$ also lies on the trace created by the plane $x + y + z = 1$, the point $(1, 0, 0)$ is our solution.

The Lagrange approach is more involved, and a computer algebra system is highly recommended. We minimize the square of the distance $x^2 + y^2 + z^2$; let $L(x, y, z, \lambda, \mu) = x^2 + y^2 + z^2 - \lambda(x^2 + y^2/4 + z^2/9 - 1) - \mu(x + y + z - 1)$. Then $L_x = 2x - 2\lambda x - \mu$, $L_y = 2y - \frac{1}{2}\lambda y - \mu$, and $L_z = 2z - \frac{2}{9}\lambda z - \mu$. Setting $L_x = L_y = L_z = 0$ yields $x = \frac{\mu}{2 - 2\lambda}$, $y = \frac{\mu}{2 - \frac{1}{2}\lambda} =$

$\frac{2\mu}{4 - \lambda}$, and $z = \frac{\mu}{2 - \frac{2}{9}\lambda} = \frac{9\mu}{18 - 2\lambda}$.

Substituting these values into the two constraints gives

$\left(\frac{\mu}{2 - 2\lambda}\right)^2 + \frac{1}{4}\left(\frac{2\mu}{4 - \lambda}\right)^2 + \frac{1}{9}\left(\frac{9\mu}{18 - 2\lambda}\right)^2 - 1$

$= 0$ and $\frac{\mu}{2 - 2\lambda} + \frac{2\mu}{4 - \lambda} + \frac{9\mu}{18 - 2\lambda} - 1 =$

0. Solving the second equation for μ results in

$\mu = \left(\frac{1}{2 - 2\lambda} + \frac{2}{4 - \lambda} + \frac{9}{18 - 2\lambda}\right)^{-1}$.

Substituting this value of μ into the first equation gives a 4th order polynomial in λ: $(\lambda - 1)(91\lambda^3 - 1183\lambda^2 + 4626\lambda - 4950) = 0$. The roots are $\lambda = 1$, $\lambda \approx 1.74226$, $\lambda \approx 4.94847$, and $\lambda \approx 6.30926$. These give $\mu = 0$, $\mu \approx 1.20155$, $\mu \approx -0.88921$, and $\mu \approx 1.40419$, respectively, giving the points $(1, 0, 0)$, $(-0.80938, 1.06439, 0.74500)$, $(0.11260, 1.87503, -0.98764)$, and $(-0.13224, -1.21614, 2.34838)$. The point nearest the origin is $(1, 0, 0)$.

16 (b) An equation for the tangent plane is $(1)(x - 1) + (-1)(y - 2) + (-1)(z - 3) = 0$ or $x - y - z + 4 = 0$.

(c) The line is given parametrically by $x = 4t$, $y = t$, and $z = 2t$. It intersects the plane when $4t - t - 2t + 4 = 0$, or when $t = -4$. The point of intersection is $(4t, t, 2t) = (-16, -4, -8)$.

17 The equation suggests that H is considered a function of T and P, while T is considered a function of H and P. Hence we construct the

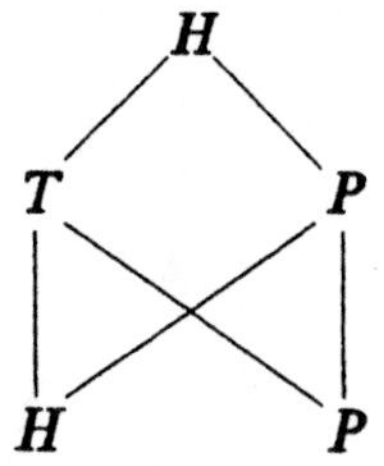

accompanying figure. By the chain rule, $\left(\frac{\partial H}{\partial P}\right)_H = \left(\frac{\partial H}{\partial T}\right)_P\left(\frac{\partial T}{\partial P}\right)_H + \left(\frac{\partial H}{\partial P}\right)_T\left(\frac{\partial P}{\partial P}\right)_H$. However, $\left(\frac{\partial H}{\partial P}\right)_H = 0$ and $\left(\frac{\partial P}{\partial P}\right)_H = 1$. Then the above equation becomes $0 = \left(\frac{\partial H}{\partial T}\right)_P\left(\frac{\partial T}{\partial P}\right)_H + \left(\frac{\partial H}{\partial P}\right)_T$ and thus the formula on the board is correct.

18 (a),(b)

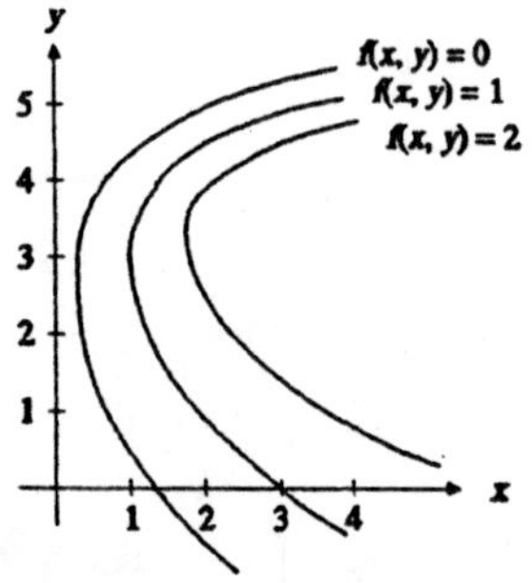

(c) The gradient vectors show the direction in which f is increasing. Since we are given the level curve $f(x, y) = 1$, the gradient vectors point toward the level curve $f(x, y) = 2$. The longer gradient vectors indicate a more rapid change in f, so the level curves are closer when the arrows are longer. Similar reasoning shows that the level curve $f(x, y) = 0$ lies on the other side of the $f(x, y) = 1$ curve.

14.S Review Exercises

1 (a) Let $P = (x, y, z)$ be an arbitrary point on the surface of revolution that comes from revolving a point Q on the curve $zy = 1$, $y > 0$. The x coordinate of Q is 0, while the y coordinate of Q is the same as that of P. From $\overline{PR} = \overline{QR}$, the z coordinate of Q is $\sqrt{x^2 + z^2}$, and $Q = \left(0, y, \sqrt{x^2 + z^2}\right)$. Now Q lies on the curve $zy = 1$, $y > 0$, so a formula for the surface of revolution is $y\sqrt{x^2 + z^2} = 1$.

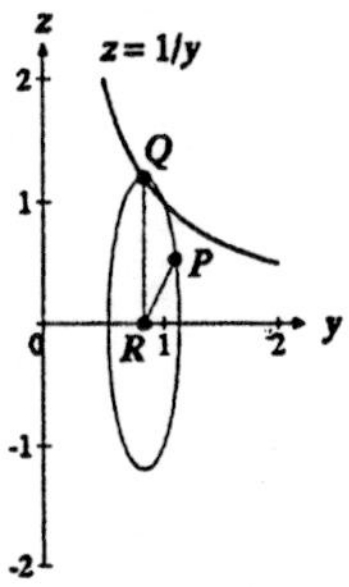

(b) Squaring both sides of $y\sqrt{x^2 + z^2} = 1$ gives $y^2(x^2 + z^2) = 1$. (The surface of revolution is given by this equation when $y > 0$.) Since this equation is not of the form (1) in Sec. 14.2, the surface is not a quadric.

3 (a) Let $f(x, y, z) = x^2 + 2y^2 - z$. Then $\nabla f(1, 1, 3) = 2\mathbf{i} + 4\mathbf{j} - \mathbf{k}$ is normal to the surface at (1, 1, 3).

(b) An equation for the tangent plane is $\frac{\partial f}{\partial x}(1, 1, 3)(x - 1) + \frac{\partial f}{\partial y}(1, 1, 3)(y - 1) + \frac{\partial f}{\partial z}(1, 1, 3)(z - 3) = 0$, or $2(x - 1) + 4(y - 1) + (-1)(z - 3) = 0$. Simplifying gives $2x + 4y - z - 3 = 0$. The distance from (1, 1, 1) to the tangent plane is $\frac{|2\cdot 1 + 4\cdot 1 + (-1)\cdot 1 + (-3)|}{\sqrt{2^2 + 4^2 + (-1)^2}} = \frac{2}{\sqrt{21}}$.

5 See the figures in the answer section of the text.

7 (a) Differentiate both sides of the equation $xyz + x + y + z^5 + 3 = 0$ with respect to x, holding y fixed: $xy\left(\frac{\partial z}{\partial x}\right)_y + yz + 1 + 0 + 5z^4\left(\frac{\partial z}{\partial x}\right)_y + 0 = 0$, $(xy + 5z^4)\left(\frac{\partial z}{\partial x}\right)_y + (1 + yz) = 0$; hence $\left(\frac{\partial z}{\partial x}\right)_y = -\frac{1 + yz}{xy + 5z^4}$.

(b) Differentiate with respect to y, holding z fixed: $(x_y yz + xz) + x_y + 1 + 0 + 0 = 0$, $(1 + yz)x_y + (1 + xz) = 0$; hence $\left(\frac{\partial x}{\partial y}\right)_z = x_y = -\frac{1 + xz}{1 + yz}$.

(c) Differentiate with respect to z, holding x fixed: $(xy_z z + xy) + 0 + y_z + 5z^4 + 0 = 0$, $(1 + xz)y_z + (xy + 5z^4) = 0$; hence $\left(\frac{\partial y}{\partial z}\right)_x = y_z = -\frac{xy + 5z^4}{1 + xz}$.

(d) Combining parts (a), (b), and (c), we obtain $\left(\frac{\partial z}{\partial x}\right)_y\left(\frac{\partial x}{\partial y}\right)_z\left(\frac{\partial y}{\partial z}\right)_x = (-1)^3\frac{1 + yz}{xy + 5z^4}\cdot\frac{1 + xz}{1 + yz}\cdot\frac{xy + 5z^4}{1 + xz} = -1$.

9 We may view f as a function of x and y, where $x = u\cos\theta - v\sin\theta$ and $y = u\sin\theta + v\cos\theta$ or as a function of u and v, specifically $g(u, v)$. By the chain rule, $\frac{\partial g}{\partial u} = \left(\frac{\partial f}{\partial u}\right)_v = \frac{\partial f}{\partial x}\cdot\frac{\partial x}{\partial u} + \frac{\partial f}{\partial y}\cdot\frac{\partial y}{\partial u}$

$= \cos\theta \dfrac{\partial f}{\partial x} + \sin\theta \dfrac{\partial f}{\partial y}$ and $\dfrac{\partial g}{\partial v} = \left(\dfrac{\partial f}{\partial v}\right)_u =$

$\dfrac{\partial f}{\partial x}\cdot\dfrac{\partial x}{\partial v} + \dfrac{\partial f}{\partial y}\cdot\dfrac{\partial y}{\partial v} = -\sin\theta \dfrac{\partial f}{\partial x} + \cos\theta \dfrac{\partial f}{\partial y}.$

Hence $\left(\dfrac{\partial g}{\partial u}\right)^2 + \left(\dfrac{\partial g}{\partial v}\right)^2 = \cos^2\theta \left(\dfrac{\partial f}{\partial x}\right)^2 +$

$2\sin\theta\cos\theta \dfrac{\partial f}{\partial x}\cdot\dfrac{\partial f}{\partial y} + \sin^2\theta\left(\dfrac{\partial f}{\partial y}\right)^2 +$

$\sin^2\theta\left(\dfrac{\partial f}{\partial x}\right)^2 - 2\sin\theta\cos\theta \dfrac{\partial f}{\partial x}\cdot\dfrac{\partial f}{\partial y} +$

$\cos^2\theta\left(\dfrac{\partial f}{\partial y}\right)^2 = \left(\dfrac{\partial f}{\partial x}\right)^2 + \left(\dfrac{\partial f}{\partial y}\right)^2.$

11 (a) See the figure in the answer section of the text.

(b) Consider the portion of the fence with y coordinate between y and $y + dy$. This is approximately a rectangle whose base is the line segment joining $(1 - y, y, 0)$ and $(1 - y - dy, y + dy, 0)$ and whose height is $z = y^2$. The length of the base is $[[(1 - y) - (1 - y - dy)]^2 + (y - (y + dy))^2]^{1/2} = \sqrt{2}\,dy$, so the local approximation of the area is $\sqrt{2}y^2\,dy$. By the informal approach, the fence's area is $\int_0^1 \sqrt{2}y^2\,dy = \dfrac{\sqrt{2}}{3}y^3\Big|_0^1 = \dfrac{\sqrt{2}}{3}.$

13 (a) See the figure in the answer section of the text.

(b) $\dfrac{\partial z}{\partial x}(2, 3) = 1$. The trace must be parallel to the yz plane (with x varying).

(c) $\dfrac{\partial z}{\partial y}(2, 3) = \dfrac{2}{3}$. The trace must be parallel to the xz plane (with y varying).

(d) $D_u z = \dfrac{1}{2}\cdot 1 + \dfrac{\sqrt{3}}{2}\cdot\dfrac{2}{3} = \dfrac{1}{2} + \dfrac{\sqrt{3}}{3}$

(e) Let $f(x, y, z) = \dfrac{x^2}{4} + \dfrac{y^2}{9} - z$. Then $\nabla f(2, 3, 2) = \mathbf{i} + \dfrac{2}{3}\mathbf{j} - \mathbf{k}$ is such a normal.

(f) Since $z = \dfrac{x^2}{4} + \dfrac{y^2}{9} \geq 0$ for all x and y, the trace in $z = -2$ is empty.

(g) The level curve is $\dfrac{x^2}{4} + \dfrac{y^2}{9} = 2.$

(h) A vector normal to $\dfrac{x^2}{4} + \dfrac{y^2}{9} = 2$ is $\nabla z(2, 3) = \mathbf{i} + \dfrac{2}{3}\mathbf{j}$. See the figure in the answer section of the text.

15 See Theorem 2 of Sec. 14.8.

17 (a) The bird's temperature at time t is $T = x^2 + y^2 + z^2 = t^2 + (t^2)^2 + (t^3)^2 = t^2 + t^4 + t^6$. Hence $\dfrac{dT}{dt} = 2t + 4t^3 + 6t^5$. At $(1, 1, 1)$, $t = 1$, so $\dfrac{dT}{dt} = 2 + 4 + 6 = 12.$

(b) Let s denote the arc length along the curve; then $\dfrac{ds}{dt} = \sqrt{\left(\dfrac{dx}{dt}\right)^2 + \left(\dfrac{dy}{dt}\right)^2 + \left(\dfrac{dz}{dt}\right)^2} = \sqrt{1^2 + (2t)^2 + (3t^2)^2} = \sqrt{1 + 4t^2 + 9t^4}$; at t

$= 1, \dfrac{ds}{dt} = \sqrt{1 + 4 + 9} = \sqrt{14}$. Using the result in (a), we have $\dfrac{dT}{ds} = \dfrac{dT}{dt}\cdot\dfrac{dt}{ds} = 12\cdot\dfrac{1}{\sqrt{14}} = \dfrac{12}{\sqrt{14}}$.

19 (a)

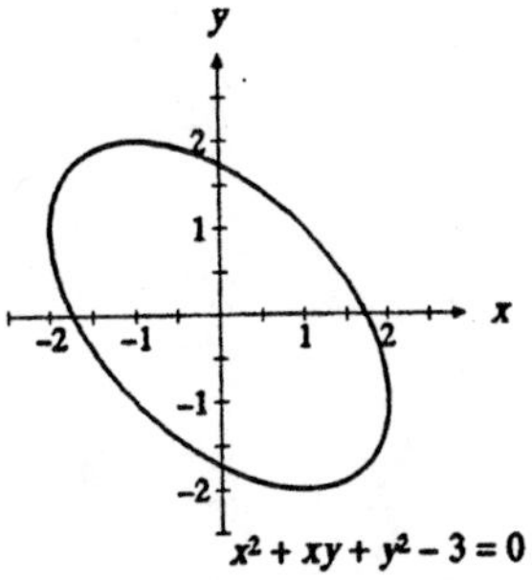

(b) We wish to minimize and maximize $x^2 + y^2$ subject to $x^2 + xy + y^2 = 3$. Let $L(x, y, \lambda) = x^2 + y^2 - \lambda(x^2 + xy + y^2 - 3)$. Then $L_x = 2x - 2\lambda x - \lambda y$, $L_y = 2y - 2\lambda y - \lambda x$, and $L_\lambda = -(x^2 + xy + y^2 - 3)$. Setting $L_x = L_y = 0$ results in $2x - 2\lambda x - \lambda y = 0$ and $2y - 2\lambda y - \lambda x = 0$. Solving for x in the first equation gives $x = \dfrac{\lambda y}{2 - 2\lambda}$. Substituting this value of x into the second equation gives $2y - 2\lambda y - \lambda\left(\dfrac{\lambda y}{2 - 2\lambda}\right) = y\left(2 - 2\lambda - \dfrac{\lambda^2}{2 - 2\lambda}\right) = 0$. So either $y = 0$ or $2 - 2\lambda - \dfrac{\lambda^2}{2 - 2\lambda} = 0$. If $y = 0$, then $x = 0$ and the constraint $x^2 + xy + y^2 = 3$ is not satisfied. So suppose $2 - 2\lambda - \dfrac{\lambda^2}{2 - 2\lambda} = 0$. Then $\lambda^2 - (2 - 2\lambda)^2 = 0$ and $[\lambda + (2 - 2\lambda)][\lambda - (2 - 2\lambda)] = (2 - \lambda)(3\lambda - 2) = 0$. Thus $\lambda = 2$ or $\lambda = 2/3$. When $\lambda = 2$, we have $x = -y$, in which case $(-y)^2 + (-y)y + y^2 = y^2 = 3$ and $(\pm\sqrt{3}, \mp\sqrt{3})$ are possible points of extrema. When $\lambda = 2/3$, we have $x = y$, $y^2 + y\cdot y + y^2 = 3y^2 = 3$, and $(\pm 1, \pm 1)$ are also possible points of extrema. Inspection reveals that $(\pm 1, \pm 1)$ are points closest to the origin while $(\pm\sqrt{3}, \mp\sqrt{3})$ are points farthest from the origin.

21 By the chain rule, $\left(\dfrac{\partial z}{\partial u}\right)_v = \left(\dfrac{\partial z}{\partial x}\right)_y\left(\dfrac{\partial x}{\partial u}\right)_v + \left(\dfrac{\partial z}{\partial y}\right)_x\left(\dfrac{\partial y}{\partial u}\right)_v$ or $\dfrac{\partial p}{\partial u} = \dfrac{\partial f}{\partial x}\cdot\dfrac{\partial g}{\partial u} + \dfrac{\partial f}{\partial y}\cdot\dfrac{\partial h}{\partial u}$. By the product rule (and the chain rule) $\dfrac{\partial^2 p}{\partial u^2} = \dfrac{\partial}{\partial u}\left(\dfrac{\partial p}{\partial u}\right) = \dfrac{\partial}{\partial u}\left(\dfrac{\partial f}{\partial x}\cdot\dfrac{\partial g}{\partial u}\right) + \dfrac{\partial}{\partial u}\left(\dfrac{\partial f}{\partial y}\cdot\dfrac{\partial h}{\partial u}\right) = \dfrac{\partial g}{\partial u}\cdot\dfrac{\partial}{\partial u}\left(\dfrac{\partial f}{\partial x}\right) + \dfrac{\partial f}{\partial x}\cdot\dfrac{\partial^2 g}{\partial u^2} + \dfrac{\partial h}{\partial u}\cdot\dfrac{\partial}{\partial u}\left(\dfrac{\partial f}{\partial y}\right) + \dfrac{\partial f}{\partial y}\cdot\dfrac{\partial^2 h}{\partial u^2} = \dfrac{\partial g}{\partial u}\left[\dfrac{\partial}{\partial x}\left(\dfrac{\partial f}{\partial x}\right)\dfrac{\partial g}{\partial u} + \dfrac{\partial}{\partial y}\left(\dfrac{\partial f}{\partial x}\right)\dfrac{\partial h}{\partial u}\right] + \dfrac{\partial f}{\partial x}\cdot\dfrac{\partial^2 g}{\partial u^2} + \dfrac{\partial h}{\partial u}\left[\dfrac{\partial}{\partial x}\left(\dfrac{\partial f}{\partial y}\right)\dfrac{\partial g}{\partial u} + \dfrac{\partial}{\partial y}\left(\dfrac{\partial f}{\partial y}\right)\dfrac{\partial h}{\partial u}\right] + \dfrac{\partial f}{\partial y}\cdot\dfrac{\partial^2 h}{\partial u^2} = \left(\dfrac{\partial g}{\partial u}\right)^2\dfrac{\partial^2 f}{\partial x^2} + 2\dfrac{\partial g}{\partial u}\cdot\dfrac{\partial h}{\partial u}\cdot\dfrac{\partial^2 f}{\partial y\,\partial x} + \dfrac{\partial f}{\partial x}\cdot\dfrac{\partial^2 g}{\partial u^2} +$

$\left(\frac{\partial h}{\partial u}\right)^2\frac{\partial^2 f}{\partial y^2} + \frac{\partial f}{\partial y}\cdot\frac{\partial^2 h}{\partial u^2}$, since $\frac{\partial^2 f}{\partial y\,\partial x} = \frac{\partial^2 f}{\partial x\,\partial y}$ by the assumption that f has continuous partial derivatives of all orders. Similarly, $\frac{\partial^2 p}{\partial v^2} =$

$\left(\frac{\partial g}{\partial v}\right)^2\frac{\partial^2 f}{\partial x^2} + 2\frac{\partial g}{\partial v}\cdot\frac{\partial h}{\partial v}\cdot\frac{\partial^2 f}{\partial y\,\partial x} + \frac{\partial f}{\partial x}\cdot\frac{\partial^2 g}{\partial v^2} +$

$\left(\frac{\partial h}{\partial v}\right)^2\frac{\partial^2 f}{\partial y^2} + \frac{\partial f}{\partial y}\cdot\frac{\partial^2 h}{\partial v^2}$. Adding, we have

$$\frac{\partial^2 p}{\partial u^2} + \frac{\partial^2 p}{\partial v^2} = \left(\frac{\partial^2 f}{\partial x^2} + \frac{\partial^2 f}{\partial y^2}\right)\left(\left(\frac{\partial g}{\partial u}\right)^2 + \left(\frac{\partial g}{\partial v}\right)^2\right) +$$

$$2\frac{\partial^2 f}{\partial x\,\partial y}\left[\frac{\partial g}{\partial u}\frac{\partial h}{\partial u} + \frac{\partial g}{\partial v}\frac{\partial h}{\partial v}\right] =$$

$\left(\frac{\partial^2 f}{\partial x^2} + \frac{\partial^2 f}{\partial y^2}\right)\left(\left(\frac{\partial g}{\partial u}\right)^2 + \left(\frac{\partial g}{\partial v}\right)^2\right)$, since $\frac{\partial g}{\partial u} = \frac{\partial h}{\partial v}$

and $\frac{\partial g}{\partial v} = -\frac{\partial h}{\partial u}$ implies that $\frac{\partial g}{\partial u}\frac{\partial h}{\partial u} + \frac{\partial g}{\partial v}\frac{\partial h}{\partial v} = 0$.

23 (a) Since $y(x, t) = f(x)g(t)$, we have $y_x = f'(x)g(t)$, $y_{xx} = f''(x)g(t)$, $y_t = f(x)g'(t)$, and $y_{tt} = f(x)g''(t)$. Hence $f(x)g''(t) = y_{tt} = a^2 y_{xx} = a^2 f''(x)g(t)$, as required.

(b) We have, for any x and t, $\frac{f''(x)}{f(x)} = \frac{g''(t)}{a^2 g(t)}$.

In particular, for any x, $\frac{f''(x)}{f(x)} = \frac{g''(0)}{a^2 g(0)} =$

k, where k is a constant (assuming $g(0) \neq 0$).

On the other hand, $\frac{g''(t)}{a^2 g(t)} = \frac{f''(x)}{f(x)} = k$, so

f''/f and $g''/(a^2 g)$ equal the same constant.

(c) We have $f'(x) = c_1\sqrt{k}e^{\sqrt{k}x} + c_2(-\sqrt{k})e^{-\sqrt{k}x} = \sqrt{k}\left(c_1 e^{\sqrt{k}x} - c_2 e^{-\sqrt{k}x}\right)$ and $f''(x) = \sqrt{k}\left(c_1\sqrt{k}e^{\sqrt{k}x} - c_2(-\sqrt{k})e^{-\sqrt{k}x}\right) = k\left(c_1 e^{\sqrt{k}x} + c_2 e^{-\sqrt{k}x}\right) = kf(x)$, so $\frac{f''(x)}{f(x)} = k$.

(d) Proceeding as in (c), we obtain $g'(t) = a\sqrt{k}\left(c_3 e^{a\sqrt{k}t} - c_4 e^{-a\sqrt{k}t}\right)$ and $g''(t) = a^2 k g(t)$,

so $\frac{g''(t)}{a^2 g(t)} = k$.

25 By substitution, $dz = \frac{\partial z}{\partial u}du + \frac{\partial z}{\partial v}dv =$

$$\frac{\partial z}{\partial u}\left(\frac{\partial u}{\partial x}dx + \frac{\partial u}{\partial y}dy\right) + \frac{\partial z}{\partial v}\left(\frac{\partial v}{\partial x}dx + \frac{\partial v}{\partial y}dy\right)$$

$$= \left(\frac{\partial z}{\partial u}\cdot\frac{\partial u}{\partial x} + \frac{\partial z}{\partial v}\cdot\frac{\partial v}{\partial x}\right)dx + \left(\frac{\partial z}{\partial u}\cdot\frac{\partial u}{\partial y} + \frac{\partial z}{\partial v}\cdot\frac{\partial v}{\partial y}\right)dy$$

$= \left(\frac{\partial z}{\partial x}\right)_y dx + \left(\frac{\partial z}{\partial y}\right)_x dy$, where the chain rule was

applied in the last step. Writing $\left(\frac{\partial z}{\partial x}\right)_y = \frac{\partial z}{\partial x}$ and

$\left(\frac{\partial z}{\partial y}\right)_x = \frac{\partial z}{\partial y}$, we see that the two expressions for

dz are indeed equal.

27 With regard to F, we are given $x = g_1(X, Y, Z)$ and $y = g_2(X, Y, Z)$, where g_1 and g_2 are differentiable. With regard to F^{-1}, we are given $X = h_1(x, y)$, $Y = h_2(x, y)$, and $Z = h_3(x, y)$, where h_1, h_2, and h_3 are differentiable. By the chain rule,

$\frac{\partial x}{\partial x} = \frac{\partial x}{\partial X}\cdot\frac{\partial X}{\partial x} + \frac{\partial x}{\partial Y}\cdot\frac{\partial Y}{\partial x} + \frac{\partial x}{\partial Z}\cdot\frac{\partial Z}{\partial x}$, $\frac{\partial y}{\partial y} =$

$\frac{\partial y}{\partial X}\cdot\frac{\partial X}{\partial y} + \frac{\partial y}{\partial Y}\cdot\frac{\partial Y}{\partial y} + \frac{\partial y}{\partial Z}\cdot\frac{\partial Z}{\partial y}$, $\frac{\partial X}{\partial X} =$

$\frac{\partial X}{\partial x}\cdot\frac{\partial x}{\partial X} + \frac{\partial X}{\partial y}\cdot\frac{\partial y}{\partial X}$, $\frac{\partial Y}{\partial Y} = \frac{\partial Y}{\partial x}\cdot\frac{\partial x}{\partial Y} + \frac{\partial Y}{\partial y}\cdot\frac{\partial y}{\partial Y}$,

and $\frac{\partial Z}{\partial Z} = \frac{\partial Z}{\partial x}\cdot\frac{\partial x}{\partial Z} + \frac{\partial Z}{\partial y}\cdot\frac{\partial y}{\partial Z}$. But we now see that $\frac{\partial x}{\partial x} + \frac{\partial y}{\partial y} = \frac{\partial X}{\partial X} + \frac{\partial Y}{\partial Y} + \frac{\partial Z}{\partial Z}$. However, since $\frac{\partial x}{\partial x} + \frac{\partial y}{\partial y} = 1 + 1 = 2$ and $\frac{\partial X}{\partial X} + \frac{\partial Y}{\partial Y} + \frac{\partial Z}{\partial Z} = 1 + 1 + 1 = 3$, these cannot actually be equal. Hence the functions cannot exist as stipulated.

29 (a) $\frac{\partial}{\partial V}(nRTV^{-1}) = -nRTV^{-2} = -\frac{nRT}{V^2}$

$\frac{\partial}{\partial T}(nRTV^{-1}) = nRV^{-1} = \frac{nR}{V}$

(b) Differentiate the equation $0 = f(P, V, T)$ with respect to V, holding P fixed: $0 =$

$\frac{\partial}{\partial V}[f(P, V, T)] =$

$\frac{\partial f}{\partial P}\left(\frac{\partial P}{\partial V}\right)_P + \frac{\partial f}{\partial V}\left(\frac{\partial V}{\partial V}\right)_P + \frac{\partial f}{\partial T}\left(\frac{\partial T}{\partial V}\right)_P = \frac{\partial f}{\partial P}\cdot 0$

$+ \frac{\partial f}{\partial V}\cdot 1 + \frac{\partial f}{\partial T}\left(\frac{\partial T}{\partial V}\right)_P = \frac{\partial f}{\partial V} + \frac{\partial f}{\partial T}\left(\frac{\partial T}{\partial V}\right)_P$,

so $\left(\frac{\partial T}{\partial V}\right)_P = -\frac{\partial f/\partial V}{\partial f/\partial T}$. Next, differentiate the equation with respect to V, holding T fixed: 0

$= \frac{\partial f}{\partial P}\left(\frac{\partial P}{\partial V}\right)_T + \frac{\partial f}{\partial V}\left(\frac{\partial V}{\partial V}\right)_T + \frac{\partial f}{\partial T}\left(\frac{\partial T}{\partial V}\right)_T =$

$\frac{\partial f}{\partial P}\left(\frac{\partial P}{\partial V}\right)_T + \frac{\partial f}{\partial V}$, so $\left(\frac{\partial P}{\partial V}\right)_T = -\frac{\partial f/\partial V}{\partial f/\partial P}$.

Finally, differentiate with respect to T, holding V fixed: $0 =$

$\frac{\partial f}{\partial P}\left(\frac{\partial P}{\partial T}\right)_V + \frac{\partial f}{\partial V}\left(\frac{\partial V}{\partial T}\right)_V + \frac{\partial f}{\partial T}\left(\frac{\partial T}{\partial T}\right)_V =$

$\frac{\partial f}{\partial P}\left(\frac{\partial P}{\partial T}\right)_V + \frac{\partial f}{\partial T}$, so $\left(\frac{\partial P}{\partial T}\right)_V = -\frac{\partial f/\partial T}{\partial f/\partial P}$.

Combining these results, we obtain

$-\frac{(\partial P/\partial V)_T}{(\partial P/\partial T)_V} = -\frac{-(\partial f/\partial V)/(\partial f/\partial P)}{-(\partial f/\partial T)/(\partial f/\partial P)} =$

$-\frac{\partial f/\partial V}{\partial f/\partial T} = \left(\frac{\partial T}{\partial V}\right)_P$, as required.

(c) We have $T = \frac{PV}{nR}$, so $\left(\frac{\partial T}{\partial V}\right)_P = \frac{P}{nR}$. From (a), $-\frac{(\partial P/\partial V)_T}{(\partial P/\partial T)_V} = -\frac{-nRT/V^2}{nR/V} = \frac{T}{V}$. Since $P = \frac{nRT}{V}$, $\frac{P}{nR} = \frac{T}{V}$, and the results agree.

31 $\Delta f = \ln\frac{(x + \Delta x)^2 + (y + \Delta y)^2}{(x + \Delta x)(y + \Delta y)} - \ln\frac{x^2 + y^2}{xy} =$

$\ln\frac{(1.01)^2 + (2.02)^2}{(1.01)(2.02)} - \ln\frac{5}{2} = \ln\frac{5}{2} - \ln\frac{5}{2} = 0$;

$df = \frac{xy}{x^2 + y^2}\left(\frac{(xy)\cdot 2x - (x^2 + y^2)\cdot y}{(xy)^2}\right)\Delta x +$

$\frac{xy}{x^2 + y^2}\left(\frac{(xy)\cdot 2y - (x^2 + y^2)\cdot x}{(xy)^2}\right)\Delta y =$

$\frac{2}{5}\left(\frac{2\cdot 2 - 5\cdot 2}{2^2}\right)(0.01) + \frac{2}{5}\left(\frac{2\cdot 4 - 5\cdot 1}{2^2}\right)(0.02) = 0$

33 $\Delta f = (x + \Delta x)\tan^{-1}[(x + \Delta x)(y + \Delta y)] - x\tan^{-1}xy$

$= 1.3\tan^{-1}[(1.3)(1.1)] - \tan^{-1}1 \approx 0.463$;

$$df = \left(x\left(\frac{1}{1+x^2y^2}\right)y + \tan^{-1}xy\right)\Delta x + x\left(\frac{1}{1+x^2y^2}\right)x\Delta y$$

$$= \left(\frac{1}{2} + \frac{\pi}{4}\right)(0.3) + \frac{1}{2}(0.1) \approx 0.436$$

35 $\Delta f = f(3 + 0.02, 4 + 0.03) - f(3, 4)$
$= f(3.02, 4.03) - f(3, 4) \approx 5.03600 - 5$
$= 0.036;$

$$df = \frac{1}{2}(x^2+y^2)^{-1/2}(2x)\Delta x + \frac{1}{2}(x^2+y^2)^{-1/2}(2y)\Delta y$$

$$= \frac{3}{5}(0.02) + \frac{4}{5}(0.03) = 0.012 + 0.024 = 0.036$$

37 $\Delta f = f(7 + 0.2, 9 - 0.3) - f(7, 9)$
$= f(7.2, 8.7) - f(7, 9) = -11.2 - (-13) = 1.8;$
$df = 3\Delta x - 4\Delta y = 3(0.2) - 4(-0.3) = 1.8$

39 $df = \frac{3x^2}{y^2}\Delta x - \frac{2x^3}{y^3}\Delta y$, $\left|\frac{\Delta x}{x}\right| \le 0.03$ and $\left|\frac{\Delta y}{y}\right| \le 0.04$, so $\left|\frac{df}{f}\right| = \left|\left(\frac{3x^2}{y^2}\Delta x - \frac{2x^3}{y^3}\Delta y\right)\frac{y^2}{x^3}\right| = \left|\frac{3\Delta x}{x} - \frac{2\Delta y}{y}\right| \le 3(0.03) + 2(0.04) = 0.17$
$= 17\%$.

41 $df = mx^{m-1}y^n\Delta x + nx^my^{n-1}\Delta y$, $\left|\frac{df}{f}\right| = \left|\frac{mx^{m-1}y^n\Delta x + nx^my^{n-1}\Delta y}{x^my^n}\right| = \left|\frac{m\Delta x}{x} + \frac{n\Delta y}{y}\right| \le \left|\frac{m\Delta x}{x}\right| + \left|\frac{n\Delta y}{y}\right| \le 3|m| + 4|n|$ percent

43 (a) Let $V(x, y, z) = xyz$. Then $V_x = yz$, $V_y = xz$, and $V_z = xy$, so $\Delta V \approx dV = yz\Delta x + xz\Delta y + xy\Delta z$. The desired volume is
$V(1.03, 2.02, 3.99)$
$= V(1 + 0.03, 2 + 0.02, 4 - 0.01)$
$= V(1, 2, 4) + \Delta V \approx V(1, 2, 4) + dV$
$= 8 + [2\cdot 4(0.03) + 1\cdot 4(0.02) + 1\cdot 2(-0.01)]$
$= 8.3$ cubic meters.

(b) Volume $= xyz = 1.03\cdot 2.02\cdot 3.99$
$= 8.301594 \text{ m}^3$

45 (a) Let a and b be two sides of the triangle and let C be the included angle. With b treated as the base of the triangle, we see that its corresponding height is $a \sin C$; thus the area of the triangle is $f(a, b, C) = \frac{1}{2}ab \sin C$.

Now $df = f_a\Delta a + f_b\Delta b + f_C\Delta C = \frac{\Delta a}{2}(b \sin C) + \frac{\Delta b}{2}(a \sin C) + \frac{\Delta C}{2}(ab \cos C)$. Here $a = 8$, $b = 11$, and $C = 45° = \pi/4$, and for maximal error $|\Delta a| = 0.2$ cm, $|\Delta b| = 0.2$ cm, and $|\Delta C| = 3° = \pi/60$. Hence $|df| =$

$$\left|\frac{\Delta a}{2}b \sin C + \frac{\Delta b}{2}a \sin C + \frac{\Delta C}{2}ab \cos C\right|$$

$$\le \frac{|\Delta a|}{2}b \sin C + \frac{|\Delta b|}{2}a \sin C + \frac{|\Delta C|}{2}ab \cos C$$

$$= \frac{\sqrt{2}}{4}\left(0.2\cdot 11 + 0.2\cdot 8 + \frac{\pi}{60}\cdot 8\cdot 11\right)$$

$\approx 2.9726 \text{ cm}^2$.

(b) With f defined as above, we have $f(8, 11, \pi/4) = 22\sqrt{2} \text{ cm}^2$. Considering f at the eight possible points of extreme error $(8 \pm 0.2, 11 \pm 0.2, \pi/4 \pm \pi/60)$, we see that Δf is maximal when $\Delta a = 0.2$, $\Delta b = 0.2$, and $\Delta C = -\pi/60$ and is about 3.01251 cm^2.

47 (b) Let (x, y, z) be a point on the intersection of the surfaces $z = x^2 + y^2$ and $z = x + y - 1$. Then $x^2 + y^2 = x + y - 1$, so it follows that $x^2 + y^2 - x - y + 1 = 0$. Completing the squares, we have $\left(x - \frac{1}{2}\right)^2 + \left(y - \frac{1}{2}\right)^2 = -\frac{1}{2}$, which is impossible, since the sum of two squares is never negative.

(c) By Theorem 5 of Sec. 12.4, the distance from the point (x, y, z) to the plane $x + y - z = 1$ is $f(x, y, z) = \frac{|x + y - z - 1|}{\sqrt{3}} = \pm\frac{1}{\sqrt{3}}(x + y - z - 1)$. Note that any extremum of $x + y - z - 1$ is also an extremum of $f(x, y, z)$. Since the distance between points on the two surfaces is unbounded, any extremum we find must be a minimum for the distance. Therefore let $L(x, y, z, \lambda) = x + y - z - 1 - \lambda(x^2 + y^2 - z)$; then $L_x = 1 - 2\lambda x$, $L_y = 1 - 2\lambda y$, $L_z = -1 + \lambda$, and $L_\lambda = -x^2 - y^2 + z$. Setting them equal to zero shows that $\lambda = 1$, $x = y = \frac{1}{2}$, and $z = x^2 + y^2 = \frac{1}{4} + \frac{1}{4} = \frac{1}{2}$. The nearest point is $\left(\frac{1}{2}, \frac{1}{2}, \frac{1}{2}\right)$ and its distance is

$$f\left(\frac{1}{2}, \frac{1}{2}, \frac{1}{2}\right) = \left|\frac{1}{\sqrt{3}}\left(\frac{1}{2} + \frac{1}{2} - \frac{1}{2} - 1\right)\right| = \frac{1}{2\sqrt{3}}.$$

49 Let $s(x, y) = \frac{x}{x^2 + y^2}$ and $t(x, y) = \frac{y}{x^2 + y^2}$.

Then $g(x, y) = f(s, t)$. By the chain rule, $g_x = f_s s_x + f_t t_x$, $g_{xx} = f_{ss}(s_x)^2 + f_s s_{xx} + f_{tt}(t_x)^2 + f_t t_{xx}$, $g_y =$

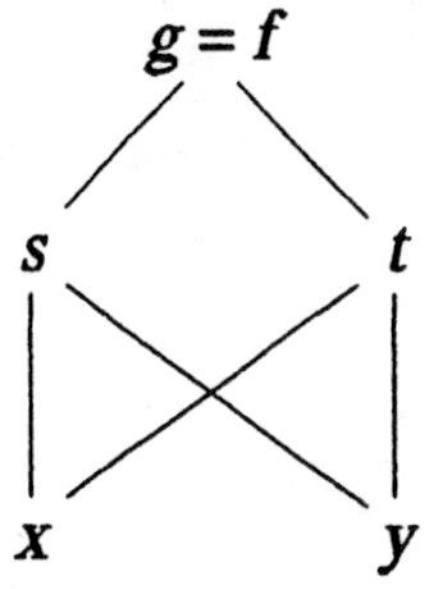

$f_s s_y + f_t t_y$, $g_{yy} = f_{ss}(s_y)^2 + f_s s_{yy} + f_{tt}(t_y)^2 + f_t t_{yy}$. We must show that $g_{xx} + g_{yy} = f_{ss}[(s_x)^2 + (s_y)^2] + f_{tt}[(t_x)^2 + (t_y)^2] + f_s(s_{xx} + s_{yy}) + f_t(t_{xx} + t_{yy}) = 0$.

Now $s_x = \frac{(x^2 + y^2)(1) - x(2x)}{(x^2 + y^2)^2} = \frac{y^2 - x^2}{(x^2 + y^2)^2}$,

$$s_{xx} = \frac{(x^2 + y^2)^2(-2x) - (y^2 - x^2)2(x^2 + y^2)(2x)}{(x^2 + y^2)^4} = \frac{2x(x^2 - 3y^2)}{(x^2 + y^2)^3},$$

$s_y = -x(x^2 + y^2)^{-2}(2y) = \frac{-2yx}{(x^2 + y^2)^2}$, $s_{yy} =$

$$\frac{(x^2 + y^2)^2(-2x) - (-2yx)2(x^2 + y^2)(2y)}{(x^2 + y^2)^4} = \frac{-2x(x^2 - 3y^2)}{(x^2 + y^2)^3} = -s_{xx}.$$

Similarly, $t_x = \frac{-2xy}{(x^2 + y^2)^2}$

$= s_y$, $t_{xx} = \dfrac{2y(3x^2 - y^2)}{(x^2 + y^2)^3}$, $t_y = \dfrac{x^2 - y^2}{(x^2 + y^2)^2} = -s_x$,

$t_{yy} = \dfrac{-2y(3x^2 - y^2)}{(x^2 + y^2)^3} = -t_{xx}$. It follows that $(s_x)^2 +$ $(s_y)^2 = (t_x)^2 + (t_y)^2$, $s_{xx} + s_{yy} = 0$, and $t_{xx} + t_{yy} =$ 0. We also know that $f(s, t)$ satisfies $f_{ss} + f_{tt} = 0$. Then $g_{xx} + g_{yy} = (f_{ss} + f_{tt})[(s_x)^2 + (s_y)^2] + f_s(s_{xx} +$ $s_{yy}) + f_t(t_{xx} + t_{yy}) = 0[(s_x)^2 + (s_y)^2] + f_s\cdot 0 + f_t\cdot 0$ $= 0$.

51 Let f be the function whose level curves are shown.

(a) $\nabla f \approx f_x(0.02, 0.05)\mathbf{i} + f_y(0.02, 0.05)\mathbf{j}$

$\approx \dfrac{f(0.03, 0.05) - f(0.02, 0.05)}{0.01}\mathbf{i} +$

$\dfrac{f(0.02, 0.06) - f(0.02, 0.05)}{0.01}\mathbf{j} \approx -\mathbf{i} + 0\mathbf{j}$

$= -\mathbf{i}$

(b) Since the level curves of f are closer together at P, the gradient is longer there.

(c) $f_x(0.02, 0.05) \approx -1$

(d) $D_{\mathbf{u}}f = \nabla f(0.02, 0.05)\cdot\mathbf{u} =$

$f_x(0.02, 0.05)\cos 20° + f_y(0.02, 0.05)\sin 20°$

$\approx -\cos 20° \approx -0.9$

53 (a) We wish to minimize the surface area of the tank, $S = 2\pi rh + 4\pi r^2$, subject to the volume constraint, $\pi r^2 h + \dfrac{4}{3}\pi r^3 = c$. So let

$L(r, h, \lambda) = 2\pi rh + 4\pi r^2 -$

$\lambda\left(\pi r^2 h + \dfrac{4}{3}\pi r^3 - c\right)$. Then $L_r = 2\pi h + 8\pi r$

$- 2\pi\lambda rh - 4\pi\lambda r^2$, $L_h = 2\pi r - \pi\lambda r^2$, and

$L_\lambda = -\left(\pi r^2 h + \dfrac{4}{3}\pi r^3 - c\right)$. Setting these equal to zero, we obtain $2\pi(h + 4r - \lambda rh -$ $2\lambda r^2) = 0$ and $\pi r(2 - \lambda r) = 0$. So $h + 4r -$ $\lambda rh - 2\lambda r^2 = 0$ and either $r = 0$ or $2 - \lambda r$ $= 0$. If $r = 0$, then the tank is nonexistent. Therefore $2 - \lambda r = 0$, so $r = 2/\lambda$. Substituting this value of r into the first equation gives $h + \dfrac{8}{\lambda} - 2h - \dfrac{8}{\lambda} = 0$; thus

$h = 0$ and $\dfrac{h}{r} = 0$. The tank of minimal surface area is a sphere with volume c ft^3 and radius $r = \left(\dfrac{3c}{4\pi}\right)^{1/3}$ ft.

(b) Again, we must minimize S. Differentiating S $= 2\pi rh + 4\pi r^2$ and $\pi r^2 h + \dfrac{4}{3}\pi r^3 = c$ with respect to r yields $\dfrac{dS}{dr} = 2\pi r\dfrac{dh}{dr} + 2\pi h +$ $8\pi r$ and $\pi r^2\dfrac{dh}{dr} + 2\pi rh + 4\pi r^2 = 0$. Setting $\dfrac{dS}{dr} = 0$, we see that $2\pi r\dfrac{dh}{dr} + 2\pi h + 8\pi r$ $= 0$, or $r\dfrac{dh}{dr} + h + 4r = 0$. Dividing the other differential equation by πr gives $r\dfrac{dh}{dr} + 2h + 4r = 0$. So $r\dfrac{dh}{dr} = -2h - 4r$ $= -h - 4r$, from which it follows that $h =$ 0. As expected, $h/r = 0$.

(c) We are given that $\pi r^2 h + \dfrac{4}{3}\pi r^3 = c$. Solving for h results in $h = \dfrac{c - \frac{4}{3}\pi r^3}{\pi r^2} =$

$\dfrac{c}{\pi r^2} - \dfrac{4r}{3}$. Substituting this value for h into S, we obtain $S = 2\pi rh + 4\pi r^2 =$

$$2\pi r\left(\frac{c}{\pi r^2} - \frac{4r}{3}\right) + 4\pi r^2 = \frac{2c}{r} - \frac{8\pi r^2}{3} + 4\pi r^2 = \frac{2c}{r} + \frac{4\pi r^2}{3}.$$

Then $\dfrac{dS}{dr} = -\dfrac{2c}{r^2} + \dfrac{8\pi r}{3} = 0$ when $-2c + \dfrac{8\pi r^3}{3} = 0$, or when $r^3 = \dfrac{3c}{4\pi}$. For this value of r, $h =$

$$\frac{c - \frac{4}{3}\pi \cdot \frac{3c}{4\pi}}{\pi\left(\frac{3c}{4\pi}\right)^{2/3}} = \frac{c - c}{\pi\left(\frac{3c}{4\pi}\right)^{2/3}} = 0 \text{ and } \frac{h}{r} = 0.$$

55 (a) The maximum and minimum of f on C are 3 and -2.

(b) The maximum occurs at $(1, 0)$; the minimum occurs at $(3, -2)$.

57 Rewriting $\int_{x+3}^{x+4} g(y, t)\,dt$ as $\int_{x+3}^{0} g(y, t)\,dt + \int_0^{x+4} g(y, t)\,dt$, we see from the first fundamental theorem of calculus (since y is treated as a constant) that $\dfrac{\partial f}{\partial x}$

$$= \frac{\partial}{\partial x}\left(\int_{x+3}^{x+4} g(y, t)\,dt\right)$$

$$= \frac{\partial}{\partial x}\left(-\int_0^{x+3} g(y, t)\,dt + \int_0^{x+4} g(y, t)\,dt\right)$$

$$= \frac{\partial}{\partial x}\left(-\int_0^{x+3} g(y, t)\,dt\right) + \frac{\partial}{\partial x}\left(\int_0^{x+4} g(y, t)\,dt\right)$$

$$= -g(y, x+3)(1) + g(y, x+4)(1)$$

$= g(y, x+4) - g(y, x+3)$. Note that the chain rule was used, where $\dfrac{\partial}{\partial x}(x+3) = \dfrac{\partial}{\partial x}(x+4) = 1$.

59 If $\dfrac{\partial f}{\partial x} = 0$, then $f(x, y) = g(y)$ for some function $g(y)$. Since $\dfrac{\partial f}{\partial y}(0, y) = 1$, it follows that $g'(y) = 1$, hence $g(y) = y + C$ for some constant C. Therefore $f(x, y) = y + C$. Now $f(0, 0) = 0$, so $0 + C = 0$ and $C = 0$. Thus $f(x, y) = y$ is the only function satisfying the given conditions.

61 $\psi_x = -Ak \sin(kx - \omega t) + Bk \cos(kx - \omega t)$,

$\psi_{xx} = -Ak^2 \cos(kx - \omega t) - Bk^2 \sin(kx - \omega t)$,

$\psi_t = A\omega \sin(kx - \omega t) - B\omega \cos(kx - \omega t)$, and

$\psi_{tt} = -A\omega^2 \cos(kx - \omega t) - B\omega^2 \sin(kx - \omega t)$. So

$\alpha\psi_{tt} = -A\alpha\omega^2 \cos(kx - \omega t) - B\alpha\omega^2 \sin(kx - \omega t)$

$= -Ak^2 \cos(kx - \omega t) - Bk^2 \sin(kx - \omega t) = \psi_{xx}$

since $\alpha\omega^2 = k^2$. Then $\psi_{xx} - \alpha\psi_{tt} = 0$, as claimed.

63 (a) See Theorem 2 of Sec. 14.8.

(b) See Theorem 1 of Sec. 14.8.

65 (a) Let $f(x, y) = 5x^2 + 3y^2$; $f(1, 1) = 8$, so $5x^2 + 3y^2 = 8$ is the level curve of f that passes through the point $(1, 1)$.

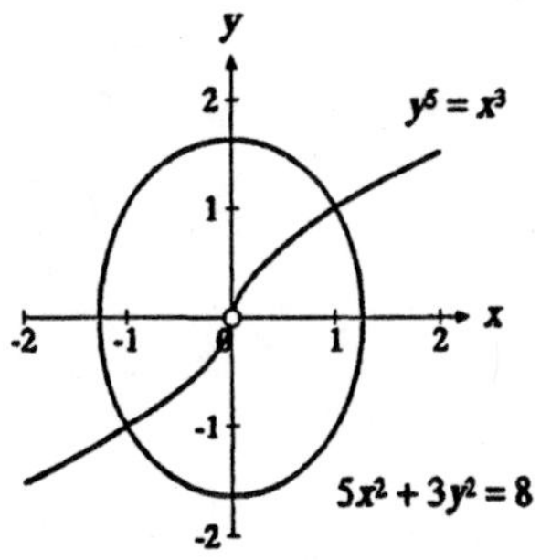

(b) Let $g(x, y) = y^5/x^3$; $g(1, 1) = 1$, so $y^5/x^3 = 1$ is the level curve of g that passes through

(1, 1). (It is undefined at (0, 0).)

(c) $\nabla f = 10x\mathbf{i} + 6y\mathbf{j}$ and $\nabla g = (-3y^5/x^4)\mathbf{i} + (5y^4/x^3)\mathbf{j}$; at (1, 1), $\nabla f = 10\mathbf{i} + 6\mathbf{j}$ and $\nabla g = -3\mathbf{i} + 5\mathbf{j}$, so $\nabla f \cdot \nabla g = 10(-3) + 6\cdot 5 = 0$. Thus the normals to the curves are perpendicular, so the curves must cross at right angles.

(d) $\nabla f \cdot \nabla g = (10x)(-3y^5/x^4) + (6y)(5y^4/x^3) = -30y^5x^{-3} + 30y^5x^{-3} = 0$, so, in general, the level curves of f and g cross at right angles.

67 See the figures in the answer section of the text.

69 The roots of $a + 2bx + cx^2 = 0$ are $\dfrac{-2b \pm \sqrt{4b^2 - 4ac}}{2c} = \dfrac{-b \pm \sqrt{b^2 - ac}}{c}$. Since $b^2 \neq ac$, there are indeed two roots r_1 and r_2. Now let $u = x + r_1y$ and $v = x + r_2y$. Then $z = f(u) + g(v) = h(x, y)$. By the chain rule, $z_x = f_uu_x + g_vv_x = f_u + g_v$, $z_{xx} = f_{uu}u_x + g_{vv}v_x = f_{uu} + g_{vv}$, $z_y = f_uu_y + g_vv_y = r_1f_u + r_2g_v$, $z_{yy} = r_1f_{uu}u_y + r_2g_{vv}v_y = r_1^2f_{uu} + r_2^2g_{vv}$ and $z_{yx} = r_1f_{uu}u_x + r_2g_{vv}v_x = r_1f_{uu} + r_2g_{vv}$. Then $az_{xx} + 2bz_{yx} + cz_{yy} = a(f_{uu} + g_{vv}) + 2b(r_1f_{uu} + r_2g_{vv}) + c(r_1^2f_{uu} + r_2^2g_{vv}) = (a + 2br_1 + cr_1^2)f_{uu} + (a + 2br_2 + cr_2^2)g_{vv} = 0\cdot f_{uu} + 0\cdot g_{vv} = 0$ since r_1 and r_2 are roots of $a + 2bx + cx^2 = 0$.

71 Let $s = y/x$ and $t = z/x$. Then $u = x^4f(s, t)$. By the chain rule, $\dfrac{\partial u}{\partial x} = \dfrac{\partial}{\partial x}(x^4)f + x^4\dfrac{\partial f}{\partial x} = 4x^3f + x^4(f_ss_x + f_tt_x)$, $\dfrac{\partial u}{\partial y} = x^4\dfrac{\partial f}{\partial y} = x^4(f_ss_y + f_tt_y)$, and $\dfrac{\partial u}{\partial z} = x^4\dfrac{\partial f}{\partial z} = x^4(f_ss_z + f_tt_z)$. Now $s_x = -\dfrac{y}{x^2}$, $s_y = \dfrac{1}{x}$, $s_z = 0$, $t_x = -\dfrac{z}{x^2}$, $t_y = 0$, and $t_z = \dfrac{1}{x}$. So $\dfrac{\partial u}{\partial x} = 4x^3f + x^4\left[f_s\left(-\dfrac{y}{x^2}\right) + f_t\left(-\dfrac{z}{x^2}\right)\right] = 4x^3f - yx^2f_s - zx^2f_t$, $\dfrac{\partial u}{\partial y} = x^4\left[f_s\left(\dfrac{1}{x}\right) + f_t\cdot 0\right] = x^3f_s$, and $\dfrac{\partial u}{\partial z} = x^4\left[f_s\cdot 0 + f_t\left(\dfrac{1}{x}\right)\right] = x^3f_t$.

Therefore, $x\dfrac{\partial u}{\partial x} + y\dfrac{\partial u}{\partial y} + z\dfrac{\partial u}{\partial z} = x(4x^3f - yx^2f_s - zx^2f_t) + y(x^3f_s) + z(x^3f_t) = 4x^4f - yx^3f_s - zx^3f_t + yx^3f_s + zx^3f_t = 4(x^4f) = 4u$.

73 We first seek critical points of f in the interior of the square region. Since $f_x = 8x + 2y$ and $f_y = -6y + 2x$, we find that $f_x = f_y = 0$ if and only if $x = y = 0$. Thus there are no such critical points. We now examine f on each edge of the square. If $x = 0$, then $f(x, y) = f(0, y) = -3y^2$. The maximum on this edge is $f(0, 0) = 0$ and the minimum is $f(0, 1) = -3$. If $y = 0$, then $f(x, y) = f(x, 0) = 4x^2$. The extrema are $f(1, 0) = 4$ and $f(0, 0) = 0$. If $x = 1$, then $f(x, y) = f(1, y) = 4 - 3y^2 + 2y$. Now $(4 - 3y^2 + 2y)' = -6y + 2 = 0$ when $y = \dfrac{1}{3}$, so the extrema on this edge are found among $f(1, 0) = 4$, $f(1, 1/3) = \dfrac{13}{3}$, and $f(1, 1) = 3$. Therefore the maximum is $f(1, 1/3) = \dfrac{13}{3}$ and the minimum is $f(1, 1) = 3$. If $y = 1$, then $f(x, y) = f(x, 1) = 4x^2 - 3 + 2x$. Now $(4x^2 - 3 + 2x)' =$

$8x + 2 = 0$ when $x = -\frac{1}{4}$, which is not in the desired interval. Checking the endpoints, we find that the extrema are $f(1, 1) = 3$ and $f(0, 1) = -3$. Comparing the value of f at the points found above, we find that the overall maximum of f is $f(1, 1/3) = \frac{13}{3}$ and the minimum is $f(0, 1) = -3$.

75 (a) Since $\mathbf{u}_r$ and $\mathbf{u}_\theta$ are perpendicular, $D_{\mathbf{u}_r}f = \nabla f\cdot\mathbf{u}_r = (A\mathbf{u}_r + B\mathbf{u}_\theta)\cdot\mathbf{u}_r = A$ and $D_{\mathbf{u}_\theta}f = \nabla f\cdot\mathbf{u}_\theta = (A\mathbf{u}_r + B\mathbf{u}_\theta)\cdot\mathbf{u}_\theta = B$.

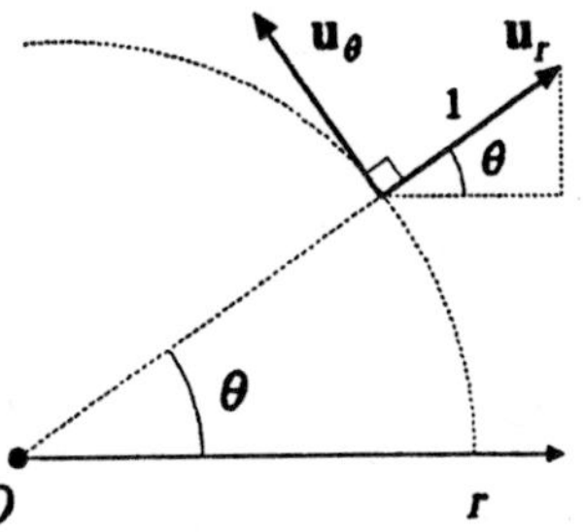

(b) Note that $\mathbf{u}_r = \cos\theta\,\mathbf{i} + \sin\theta\,\mathbf{j}$, as shown in the figure. Consider f as a function of x and y, which are functions of r and θ: $x = r\cos\theta$ and $y = r\sin\theta$. By the chain rule, $\frac{\partial f}{\partial r}$

$$= \frac{\partial f}{\partial x}\cdot\frac{\partial x}{\partial r} + \frac{\partial f}{\partial y}\cdot\frac{\partial y}{\partial r} = f_x\cos\theta + f_y\sin\theta$$

$$= (f_x\mathbf{i} + f_y\mathbf{j})\cdot(\cos\theta\,\mathbf{i} + \sin\theta\,\mathbf{j}) = \nabla f\cdot\mathbf{u}_r$$

$$= D_{\mathbf{u}_r}f.$$

(c) Note that $\mathbf{u}_\theta = -\sin\theta\,\mathbf{i} + \cos\theta\,\mathbf{j}$, as shown in the figure. As in (b), $\frac{\partial f}{\partial \theta} =$

$$\frac{\partial f}{\partial x}\cdot\frac{\partial x}{\partial \theta} + \frac{\partial f}{\partial y}\cdot\frac{\partial y}{\partial \theta}$$

$$= f_x(-r\sin\theta) + f_y(r\cos\theta)$$

$$= r(-f_x\sin\theta + f_y\cos\theta)$$

$$= r(f_x\mathbf{i} + f_y\mathbf{j})\cdot(-\sin\theta\,\mathbf{i} + \cos\theta\,\mathbf{j}) = r\nabla f\cdot\mathbf{u}_\theta$$

$$= rD_{\mathbf{u}_\theta}f, \text{ so } D_{\mathbf{u}_\theta}f = \frac{1}{r}\cdot\frac{\partial f}{\partial \theta}.$$

(d) From (a) and (b), $A = \frac{\partial f}{\partial r}$. From (a) and (c), $B = \frac{1}{r}\cdot\frac{\partial f}{\partial \theta}$. Thus from (a), $\nabla f = A\mathbf{u}_r + B\mathbf{u}_\theta$

$$= \frac{\partial f}{\partial r}\mathbf{u}_r + \frac{1}{r}\cdot\frac{\partial f}{\partial \theta}\mathbf{u}_\theta.$$

77 We already know (see Exercise 75) that $\mathbf{u}_r = \cos\theta\,\mathbf{i} + \sin\theta\,\mathbf{j}$ and $\mathbf{u}_\theta = -\sin\theta\,\mathbf{i} + \cos\theta\,\mathbf{j}$. Hence $(\sin\theta)\mathbf{u}_r = \sin\theta\cos\theta\,\mathbf{i} + \sin^2\theta\,\mathbf{j}$ and $(\cos\theta)\mathbf{u}_\theta = -\cos\theta\sin\theta\,\mathbf{i} + \cos^2\theta\,\mathbf{j}$. Adding these equations, we obtain $(\sin\theta)\mathbf{u}_r + (\cos\theta)\mathbf{u}_\theta = \mathbf{j}$. Similarly, $\mathbf{i} = (\cos\theta)\mathbf{u}_r - (\sin\theta)\mathbf{u}_\theta$. Since $r = \sqrt{x^2 + y^2}$ and $\theta = \tan^{-1}\frac{y}{x}$, we have $f(r, \theta) = g(x, y)$ and by the chain rule, $f_x = f_r r_x + f_\theta\theta_x$ and $f_y = f_r r_y + f_\theta\theta_y$. Now $r_x = \frac{2x}{2\sqrt{x^2 + y^2}} = \frac{r\cos\theta}{r}$

$$= \cos\theta,\ r_y = \frac{2y}{2\sqrt{x^2 + y^2}} = \frac{r\sin\theta}{r} = \sin\theta,$$

$$\theta_x = \frac{-y/x^2}{1 + (y/x)^2} = -\frac{y}{x^2 + y^2} = -\frac{\sin\theta}{r},$$

and $\theta_y = \frac{1/x}{1 + (y/x)^2} = \frac{x}{x^2 + y^2} = \frac{\cos\theta}{r}$, so

$f_x = \cos\theta\, f_r - \frac{\sin\theta}{r}f_\theta$ and $f_y =$

$\sin\theta\, f_r + \frac{\cos\theta}{r}f_\theta$. Hence $f_x\mathbf{i} + f_y\mathbf{j} =$

$$\left(\cos\theta\, f_r - \frac{\sin\theta}{r} f_\theta\right)(\cos\theta\, \mathbf{u}_r - \sin\theta\, \mathbf{u}_\theta) +$$

$$\left(\sin\theta\, f_r + \frac{\cos\theta}{r} f_\theta\right)(\sin\theta\, \mathbf{u}_r + \cos\theta\, \mathbf{u}_\theta)$$

$$= \left(\cos^2\theta\, f_r - \frac{\sin\theta\cos\theta}{r} f_\theta + \sin^2\theta\, f_r + \right.$$

$$\left.\frac{\sin\theta\cos\theta}{r} f_\theta\right)\mathbf{u}_r + \left(-\sin\theta\cos\theta\, f_r + \frac{\sin^2\theta}{r} f_\theta\right.$$

$$\left.+ \sin\theta\cos\theta\, f_r + \frac{\cos^2\theta}{r} f_\theta\right)\mathbf{u}_\theta = f_r\mathbf{u}_r + \frac{f_\theta}{r}\mathbf{u}_\theta,$$

as expected.

15 Definite Integrals over Plane and Solid Regions

15.1 The Definite Integral of a Function over a Region in the Plane

1 (a) Using lower left corners, the sampling points are $P_1 = (0, 1)$, $P_2 = (0, 0)$, $P_3 = (2, 1)$, and $P_4 = (2, 0)$. The resulting approximation for the volume is $(0^2 + 1^2)2 + (0^2 + 0^2)2 + (2^2 + 1^2)2 + (2^2 + 0^2)2 = 1\cdot2 + 0\cdot2 + 5\cdot2 + 4\cdot2 = 20$ cubic inches.

(b) Using upper right corners, the sampling points are $P_1 = (2, 2)$, $P_2 = (2, 1)$, $P_3 = (4, 2)$, and $P_4 = (4, 1)$. The resulting approximation is $(2^2 + 2^2)2 + (2^2 + 1^2)2 + (4^2 + 2^2)2 + (4^2 + 1^2)2 = 8\cdot2 + 5\cdot2 + 20\cdot2 + 17\cdot2 = 50\cdot2 = 100$ cubic inches.

(c) Using lower left corners provides an underestimate of the volume while upper right corners provides an overestimate. The volume is therefore between 20 and 100 cubic inches.

3 (a) $f(P) = 5$ for every point P in the region R. It follows that $\sum_{i=1}^{n} f(P_i)A_i = \sum_{i=1}^{n} 5A_i = \sum_{i=1}^{n} A_i$

$= 5A$.

(b) By (a), every approximating sum for $\int_R f(P)\, dA$ is equal to $5A$. Since the approximating sums are constant, their limit as the mesh approaches zero must equal the constant. Hence $\int_R f(P)\, dA = 5A$.

5 (a) By Problem 2, the approximate value of $\int_R f(P)\, dA$ is 50 grams. The region has area 8 cm^2, so the average value of the function is approximately $\frac{50}{8} = 6.25\ \text{g/cm}^2$.

(b) From Exercise 2, the value of the integral is between 32 and 80 grams. Therefore the average value of the function is between $\frac{32}{8} = 4\ \text{g/cm}^2$ and $\frac{80}{8} = 10\ \text{g/cm}^2$, as stated.

7

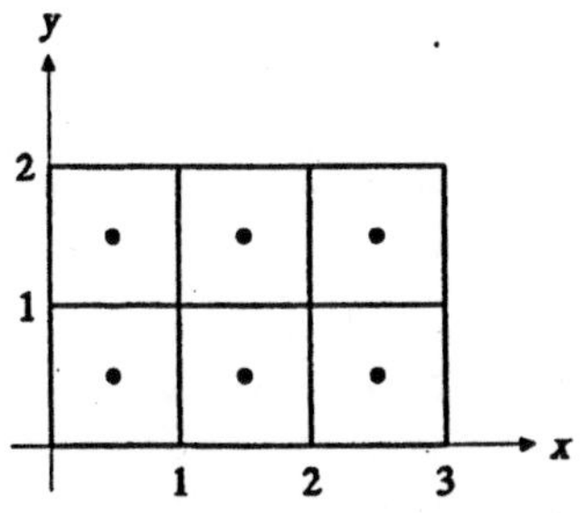

(a) As shown in the figure, each square has area 1. The sampling points are (1/2, 1/2), (1/2, 3/2), (1/2, 5/2), (3/2, 1/2), (3/2, 3/2), and (3/2, 5/2). The approximation of $\int_R \sqrt{x + y}\, dA$ obtained from this partition is therefore $\sqrt{\frac{1}{2}+\frac{1}{2}} + \sqrt{\frac{1}{2}+\frac{3}{2}} +$

$\sqrt{\frac{1}{2}+\frac{5}{2}} + \sqrt{\frac{3}{2}+\frac{1}{2}} + \sqrt{\frac{3}{2}+\frac{3}{2}} +$

$\sqrt{\frac{3}{2}+\frac{5}{2}} = \sqrt{1} + 2\sqrt{2} + 2\sqrt{3} + \sqrt{4}$

$= 3 + 2\sqrt{2} + 2\sqrt{3} \approx 9.29.$

(b) Since R has area 6, the average of f is about

$\frac{9.29}{6} \approx 1.55.$

9

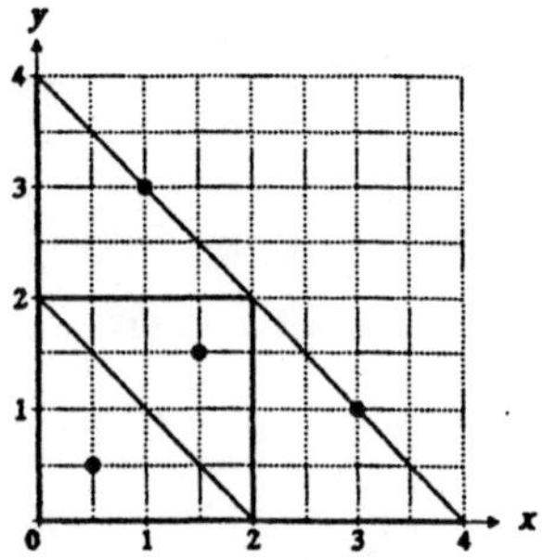

(a) As seen in the figure, the area of each triangle is 2, and the indicated sampling points are (1/2, 1/2), (3/2, 3/2), (1, 3) and (3, 1). Now $f(x, y) = x^2y$, so $\int_R f(P)\,dA \approx$

$2\left[\left(\frac{1}{2}\right)^2\frac{1}{2} + \left(\frac{3}{2}\right)^2\frac{3}{2} + 1^2(3) + 3^2(1)\right] =$

$2\left(\frac{1}{8} + \frac{27}{8} + 3 + 9\right) = 31.$

(b) The function $f(x, y) = x^2y$ is increasing in both x and y; that is, increasing either x or y results in an increase in the value of the function. Therefore the maximum of f must occur on the boundary $x + y = 4$; that is, $y = 4 - x$. We thus consider the function $f(x, y) = f(x, 4 - x) = x^2(4 - x) = 4x^2 - x^3$ for $0 \le x \le 4$. Now $(4x^2 - x^3)' = 8x - 3x^2 = x(8 - 3x)$; in the interval (0, 4), this equals 0 for $x = 8/3$, and we have $f\left(\frac{8}{3}, 4 - \frac{8}{3}\right) =$

$\left(\frac{8}{3}\right)^2\left(4 - \frac{8}{3}\right) = \frac{64}{9}\cdot\frac{4}{3} = \frac{256}{27}$. Since x varies between 0 and 4, we also check for an endpoint maximum: $f(0, 4) = f(4, 0) = 0$.

The maximum value of f is $\frac{256}{27}$ and occurs at $\left(\frac{8}{3}, \frac{4}{3}\right)$.

(c) $\int_R f(P)\,dA \le \frac{256}{27}(\text{Area of } R) = \frac{256}{27}\cdot 8$

$= \frac{2048}{27}$

11 (a) The integral $\int_R f(P)\,dA$, where $f(P)$ is the annual rainfall in inches at the point P, represents the total rainfall in cubic inches for the geographical region R, where the area of R is given in square inches.

(b) $\frac{\int_R f(P)\,dA}{\text{Area of } R}$ is the average annual rainfall, the average being taken over all points of R.

13 (a) Place the center of the unit disk at the pole of a polar coordinate system. Then $f(P) = r$ for $P = (r, \theta)$, and R is partitioned by $0 = r_0 < r_1 < r_2 < \cdots < r_m = 1$ and $0 = \theta_0 < \theta_1 < \theta_2 < \cdots < \theta_n = 2\pi$. The area of the region between r_i and r_{i+1} and θ_j and θ_{j+1} is

$\frac{1}{2}r_{i+1}^2(\theta_{j+1} - \theta_j) - \frac{1}{2}r_i^2(\theta_{j+1} - \theta_j) =$

$\frac{1}{2}\left(r_{i+1}^2 - r_i^2\right)(\theta_{j+1} - \theta_j)$. If we choose (r_i, θ_j)

for our sampling point, then $f(r_i, \theta_j) = r_i$, so the integral of f over the small region is approximated by $\frac{1}{2}r_i\left(r_{i+1}^2 - r_i^2\right)(\theta_{j+1} - \theta_j)$.

Hence $\int_R f(P)\, dA$ is approximated by

$\sum_{i=0}^{m-1}\sum_{j=0}^{n-1} \frac{1}{2}r_i\left(r_{i+1}^2 - r_i^2\right)(\theta_{j+1} - \theta_j)$. For fixed i,

$$\sum_{j=0}^{n-1} \frac{1}{2}r_i\left(r_{i+1}^2 - r_i^2\right)\left(\frac{1}{2}\theta_{j+1} - \theta_j\right)$$

$$= \frac{1}{2}r_i\left(r_{i+1}^2 - r_i^2\right)\sum_{j=0}^{n-1}(\theta_{j+1} - \theta_j)$$

$$= \frac{1}{2}r_i\left(r_{i+1}^2 - r_i^2\right)(\theta_n - \theta_0)$$

$$= \frac{1}{2}r_i\left(r_{i+1}^2 - r_i^2\right)(2\pi - 0) = \pi r_i\left(r_{i+1}^2 - r_i^2\right).$$

(Observe that the series telescoped.) This is an estimate of the integral over the circular ring between r_i and r_{i+1}. The value over the entire region is approximated by

$\sum_{i=0}^{n-1} \pi r_i\left(r_{i+1}^2 - r_i^2\right)$. In the manner of the informal approach, let $r_i = r$ and $r_{i+1} = r + dr$. Then our local approximation $\pi r_i\left(r_{i+1}^2 - r_i^2\right)$ is equal to $\pi r[(r + dr)^2 - r^2]$

$= \pi r(r^2 + 2r\,dr + dr^2 - r^2)$

$= \pi r(2r\,dr + dr^2) = 2\pi r^2\,dr + \pi r\,dr^2$. For small values of dr, the term containing dr^2 is negligible and may be dropped. Our local approximation is then $2\pi r^2\,dr$. Since r ranges from 0 to 1, the desired integral is equal to

$$\int_0^1 2\pi r^2\,dr = \left.\frac{2\pi r^3}{3}\right|_0^1 = \frac{2\pi}{3}.$$

(b) The area of R is π, so the average of f is

$$\frac{1}{\pi}\int_R f(P)\,dA = \frac{1}{\pi}\cdot\frac{2\pi}{3} = \frac{2}{3}.$$

15 You could divide R into pieces $R_1, R_2, \cdots, R_n$ and add up a series of terms such as $F_1A_1 + F_2A_2 + \cdots + F_nA_n$ where F_i is your estimate of the average value of f in R_i and A_i is the area of R_i. I did this in a rather loose fashion using trapezoids and triangles and arrived at an estimate of 3.25.

17 (a) The value of the integral is approximately 1.008. (Your result is unlikely to be accurate to as many decimal places because of the limited number of sampling points.)

(b) Multiply your result in (a) by π to obtain an approximation of $\int_R f(P)\,dA$. Its value is approximately 3.166.

(c) Since $-1 \le x, y \le 1$ it follows that $-1 \le x^2y \le 1$. Thus $1/e \le e^{x^2y} \le e$.

Therefore, $\int_R \frac{1}{e}\,dA \le \int_R e^{x^2y}\,dA$

$\le \int_R e\,dA$. Therefore $\pi/e \le \int_R e^{x^2y}\,dA$

$\le \pi e$.

19 Since the diameter of the square is $\sqrt{2}$, the mesh cannot exceed $\sqrt{2}$. Any partition containing a piece that encloses a diagonal of the square will have a mesh of $\sqrt{2}$.

15.2 Computing $\int_R f(P)\,dA$ using Rectangular Coordinates

1 The edge from (0, 0) to (2, 1) has equation $y = x/2$; that is, $x = 2y$.

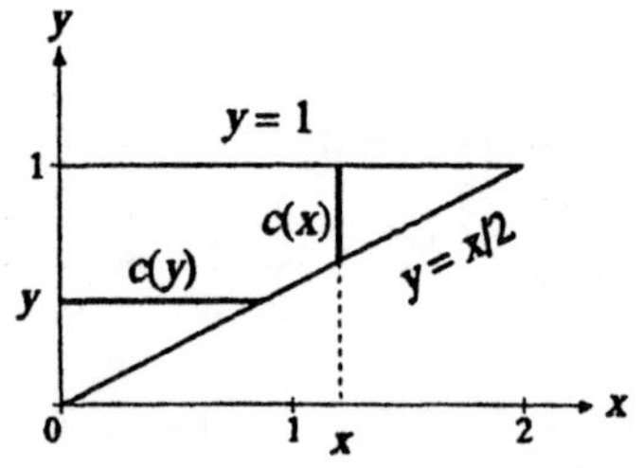

(a) For $0 \le x \le 2$, y ranges from $x/2$ to 1. The description is $0 \le x \le 2$, $x/2 \le y \le 1$.

(b) For $0 \le y \le 1$, x ranges from 0 to $2y$. The description is $0 \le y \le 1$, $0 \le x \le 2y$.

3

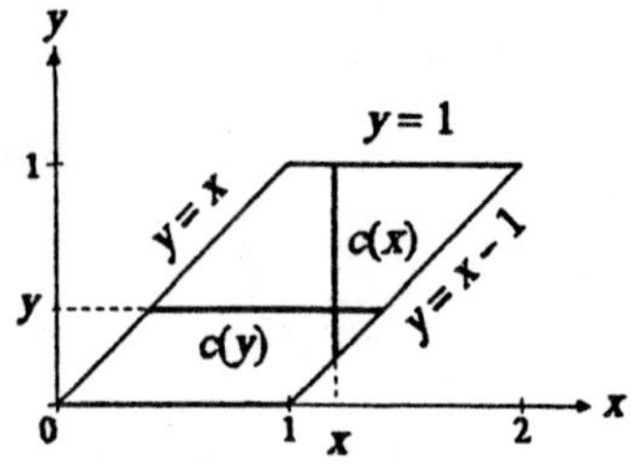

(a) For $0 \le x \le 1$, y goes from 0 to x; for $1 \le x \le 2$, y goes from $x - 1$ to 1. The description is $0 \le x \le 1$, $0 \le y \le x$ and $1 \le x \le 2$, $x - 1 \le y \le 1$.

(b) $0 \le y \le 1$, $y \le x \le y + 1$

5

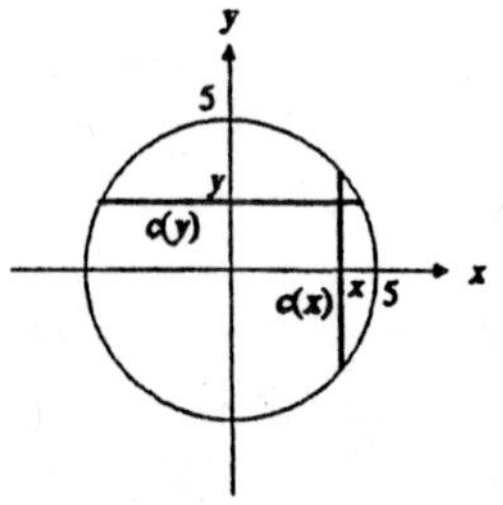

(a) Following Example 1, $-5 \le x \le 5$, $-\sqrt{25 - x^2} \le y \le \sqrt{25 - x^2}$.

(b) By the symmetry, the description is $-5 \le y \le 5$, $-\sqrt{25 - y^2} \le x \le \sqrt{25 - y^2}$.

7

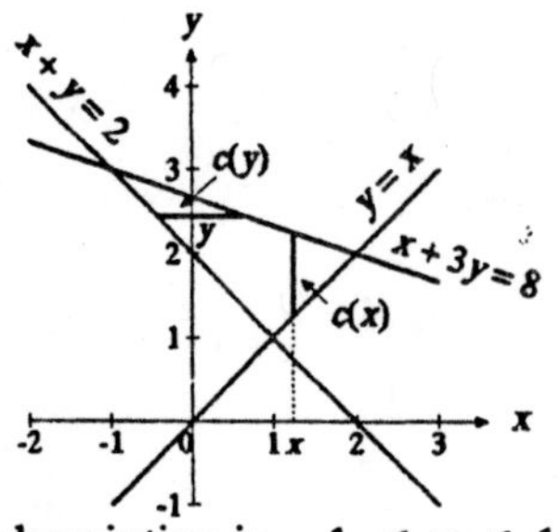

(a) The description is $-1 \le x \le 1$, $2 - x \le y \le \dfrac{8 - x}{3}$ and $1 \le x \le 2$, $x \le y \le \dfrac{8 - x}{3}$.

(b) $1 \le y \le 2$, $2 - y \le x \le y$ and $2 \le y \le 3$, $2 - y \le x \le 8 - 3y$.

9

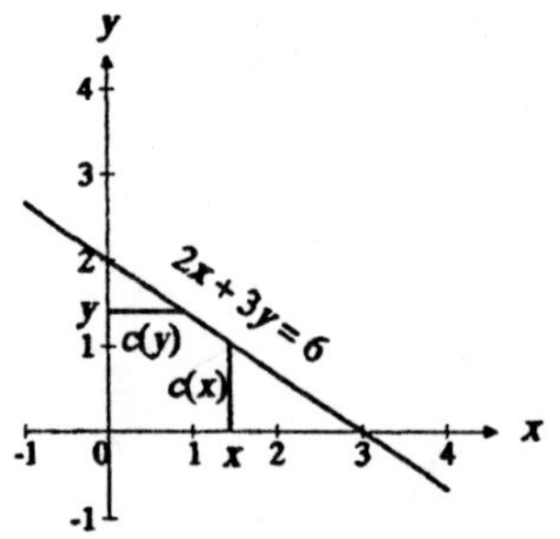

(a) $0 \le x \le 3$, $0 \le y \le \dfrac{6 - 2x}{3}$

(b) $0 \le y \le 2$, $0 \le x \le \dfrac{6 - 3y}{2}$

11

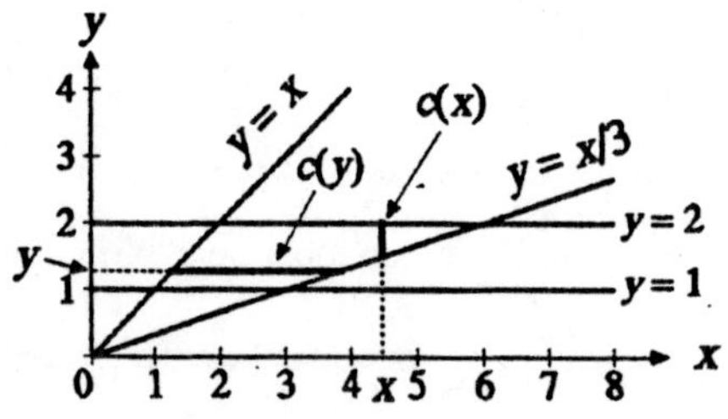

(a) From the figure, the description is $1 \le x \le 2$, $1 \le y \le x$ and $2 \le x \le 3$, $1 \le y \le 2$

and $3 \le x \le 6$, $x/3 \le y \le 2$.

(b) $1 \le y \le 2$, $y \le x \le 3y$

13

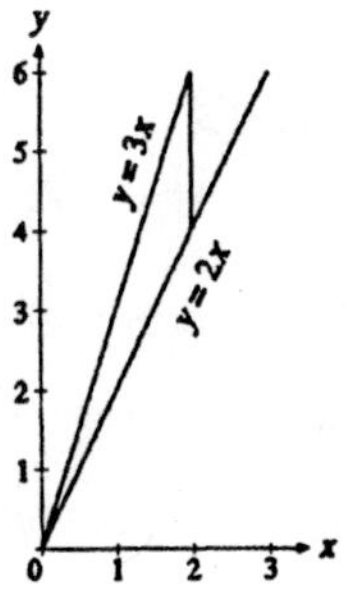

$0 \le y \le 4$, $y/3 \le x \le y/2$ and $4 \le y \le 6$, $y/3 \le x \le 2$

15

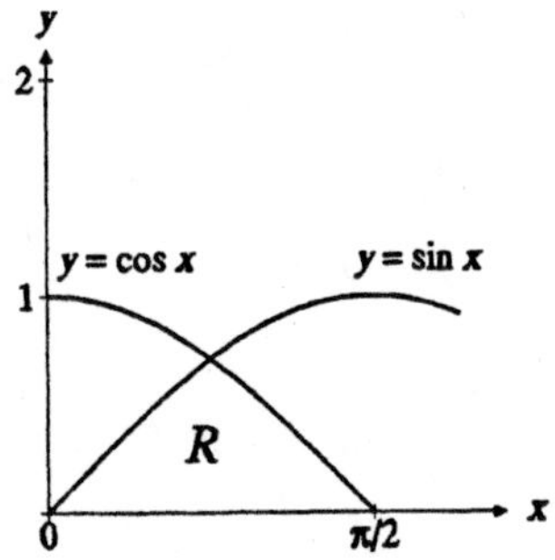

$0 \le y \le \dfrac{1}{\sqrt{2}}$, $\sin^{-1} y \le x \le \cos^{-1} y$

17 $\displaystyle\int_0^1 \int_0^x (x + 2y)\,dy\,dx = \int_0^1 (xy + y^2)\Big|_0^x\,dx$

$\displaystyle= \int_0^1 (x^2 + x^2)\,dx = \frac{2x^3}{3}\Big|_0^1 = \frac{2}{3}$

19 $\displaystyle\int_0^2 \int_0^{x^2} xy^2\,dy\,dx = \int_0^2 \frac{xy^3}{3}\Big|_0^{x^2}\,dx = \int_0^2 \frac{x^7}{3}\,dx$

$\displaystyle= \frac{x^8}{24}\Big|_0^2 = \frac{256}{24} = \frac{32}{3}$

21 $\displaystyle\int_1^2 \int_0^{\sqrt{y}} yx^2\,dx\,dy = \int_1^2 \frac{yx^3}{3}\Big|_0^{\sqrt{y}}\,dy$

$\displaystyle= \int_1^2 \frac{y^{5/2}}{3}\,dy = \frac{2}{21}y^{7/2}\Big|_1^2 = \frac{2}{21}(2^{7/2} - 1)$

23 (a)

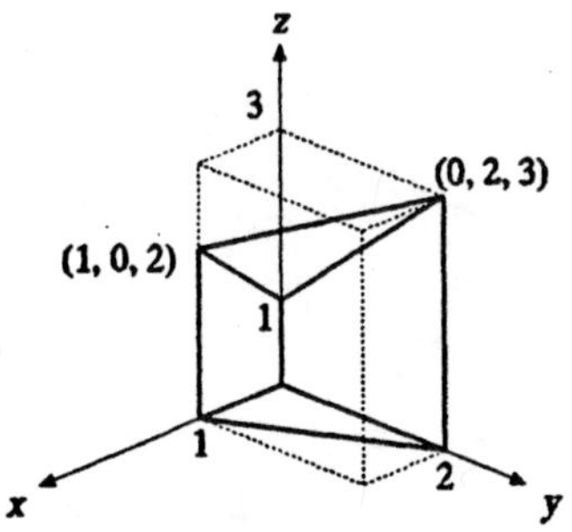

(b) $0 \le x \le 1$, $0 \le y \le 2 - 2x$

(c) $\displaystyle V = \int_R z\,dA = \int_0^1 \int_0^{2-2x} (1 + x + y)\,dy\,dx$

(d) $\displaystyle\int_0^1 \int_0^{2-2x} (1 + x + y)\,dy\,dx$

$\displaystyle= \int_0^1 \left(y + xy + \frac{1}{2}y^2\right)\Big|_0^{2-2x}\,dx$

$\displaystyle= \int_0^1 \left[2 - 2x + 2x - 2x^2 + \frac{1}{2}(2 - 2x)^2\right]dx$

$\displaystyle= \int_0^1 (4 - 4x)\,dx = 4x - 2x^2\big|_0^1 = 2$

(e) $\displaystyle V = \int_R z\,dA$

$\displaystyle= \int_0^2 \int_0^{1-y/2} (1 + x + y)\,dx\,dy$

$\displaystyle= \int_0^2 \left(x + \frac{1}{2}x^2 + yx\right)\Big|_0^{1-y/2}\,dy$

$\displaystyle= \int_0^2 \left[1 - \frac{y}{2} + \frac{1}{2}\left(1 - \frac{y}{2}\right)^2 + y\left(1 - \frac{y}{2}\right)\right]dy$

$\displaystyle= \int_0^2 \left(1 - \frac{y}{2} + \frac{1}{2} - \frac{y}{2} + \frac{y^2}{8} + y - \frac{y^2}{2}\right)dy$

$\displaystyle= \int_0^2 \left(-\frac{3}{8}y^2 + \frac{3}{2}\right)dy = -\frac{y^3}{8} + \frac{3y}{2}\Big|_0^2 = 2$

25 (a)

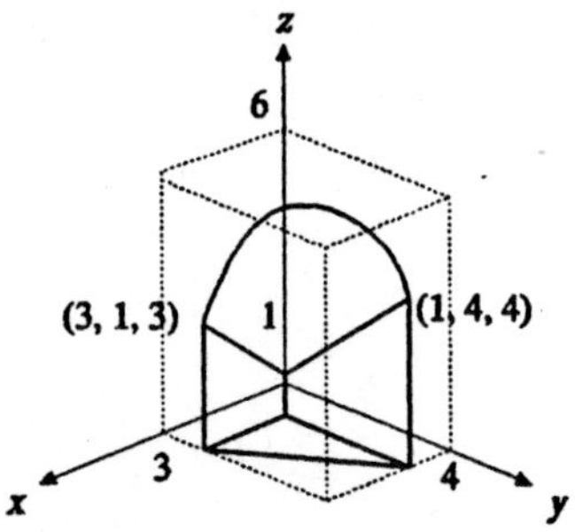

(b) Using vertical cross sections x varies from 1 to 3. For x in [1, 3] y varies from 1 to the value of y on the line through (1, 4) and (3, 1). This line has the equation $y = -\frac{3}{2}x + \frac{11}{2}$. Therefore the description of R by vertical cross sections is $1 \le x \le 3$, $1 \le y \le \frac{11}{2} - \frac{3}{2}x$.

(c) $V = \int_R z\,dA = \int_1^3 \int_1^{\frac{11}{2}-\frac{3}{2}x} xy\,dy\,dx$

(d) $$V = \int_1^3 \int_1^{\frac{11}{2}-\frac{3}{2}x} xy\,dy\,dx$$

$$= \int_1^3 \frac{1}{2}xy^2\Big|_1^{\frac{11}{2}-\frac{3}{2}x} dx$$

$$= \int_1^3 \frac{1}{2}x\left[\left(\frac{11}{2} - \frac{3}{2}x\right)^2 - 1\right] dx$$

$$= \int_1^3 \frac{1}{2}x\left(\frac{117}{4} - \frac{33}{2}x + \frac{9}{4}x^2\right) dx$$

$$= \int_1^3 \left(\frac{117}{8}x - \frac{33}{4}x^2 + \frac{9}{8}x^3\right) dx$$

$$= \frac{117}{16}x^2 - \frac{33}{12}x^3 + \frac{9}{32}x^4\Big|_1^3 = \frac{117\cdot 9}{16} - \frac{33\cdot 27}{12} + \frac{9\cdot 81}{32} - \frac{117}{16} + \frac{33}{12} - \frac{9}{32}$$

$$= \frac{117\cdot 8}{16} - \frac{33\cdot 26}{12} + \frac{9\cdot 80}{32}$$

$$= \frac{117}{2} - \frac{11\cdot 13}{2} + \frac{9\cdot 5}{2} = \frac{19}{2}.$$

(e) Using horizontal cross sections y varies from 1 to 4 and x varies from 1 to the line through (1, 4) and (3, 1). This line has the equation $x = \frac{-2}{3}y + \frac{11}{3}$. Therefore R is described by

$$1 \le y \le 4,\ 1 \le x \le \frac{11}{3} - \frac{2}{3}y.$$

$$V = \int_1^4 \int_1^{\frac{11}{3}-\frac{2}{3}y} xy\,dx\,dy$$

$$= \int_1^4 \frac{1}{2}x^2y\Big|_1^{\frac{11}{3}-\frac{2}{3}y} dy$$

$$= \int_1^4 \frac{1}{2}y\left[\left(\frac{11}{3} - \frac{2}{3}y\right)^2 - 1\right] dy$$

$$= \int_1^4 \frac{1}{2}y\left(\frac{121}{9} - \frac{44}{9}y + \frac{4}{9}y^2 - 1\right) dy$$

$$= \int_1^4 \left(\frac{56}{9}y - \frac{22}{9}y^2 + \frac{2}{9}y^3\right) dy$$

$$= \frac{28}{9}y^2 - \frac{22}{27}y^3 + \frac{1}{18}y^4\Big|_1^4 = \frac{28\cdot 16}{9} - \frac{22\cdot 64}{27} + \frac{256}{18} - \frac{28}{9} + \frac{22}{27} - \frac{1}{18}$$

$$= \frac{28\cdot 15}{9} - \frac{22\cdot 63}{27} + \frac{255}{18}$$

$$= \frac{28\cdot 5}{3} - \frac{22\cdot 7}{3} + \frac{85}{6} = \frac{19}{2}.$$

27 To describe the lamina by coordinates, we must find the points of intersection of its boundary curves $y = 2x^2$ and $y = 5x - 3$. The curves intersect whenever $2x^2 = 5x - 3$; this occurs when $2x^2 - 5x + 3 = (2x - 3)(x - 1) = 0$; that is, for $x = 1$ or $3/2$. Hence the description of the lamina is $1 \le x \le 3/2$, $2x^2 \le y \le 5x - 3$. Since the density function of the lamina is xy, we express the mass of the lamina as Mass $= \int_1^{3/2} \int_{2x^2}^{5x-3} xy\,dy\,dx$

$$= \int_1^{3/2} \frac{xy^2}{2}\bigg|_{2x^2}^{5x-3} dx$$

$$= \int_1^{3/2} \left[\frac{x}{2}(5x-3)^2 - \frac{x}{2}(2x^2)^2\right] dx$$

$$= \int_1^{3/2} \left(\frac{25x^3}{2} - 15x^2 + \frac{9}{2}x - 2x^5\right) dx$$

$$= \frac{25x^4}{8} - 5x^3 + \frac{9}{4}x^2 - \frac{x^6}{3}\bigg|_1^{3/2}$$

$$= \frac{25}{8}\cdot\frac{81}{16} - 5\cdot\frac{27}{8} + \frac{81}{16} - \frac{243}{64} - \frac{25}{8} +$$

$$5 - \frac{9}{4} + \frac{1}{3} = \frac{65}{384}.$$

29 The average temperature in the triangle R is $\dfrac{\int_R T(x,y)\,dA}{\text{Area of } R}$. The description of the region R

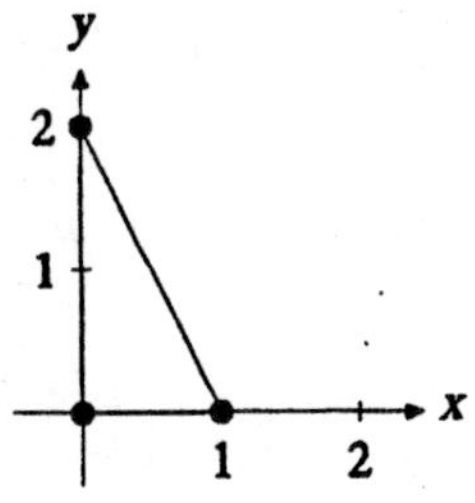

using vertical cross sections is: $0 \le x \le 1$, $0 \le y \le 2 - 2x$. Therefore $\int_R T(x,y)\,dA$

$$= \int_0^1 \int_0^{2-2x} \cos(x + 2y)\,dy\,dx$$

$$= \int_0^1 \frac{1}{2}\sin(x + 2y)\bigg|_0^{2-2x} dx$$

$$= \int_0^1 \left(\frac{1}{2}\sin(x + 2(2-2x)) - \frac{1}{2}\sin x\right) dx$$

$$= \int_0^1 \left[\frac{1}{2}\sin(4 - 3x) - \frac{1}{2}\sin x\right] dx$$

$$= \frac{1}{2}\left[\frac{1}{3}\cos(4 - 3x) + \cos x\right]\bigg|_0^1$$

$$= \frac{1}{2}\left[\frac{1}{3}\cos 1 + \cos 1 - \frac{1}{3}\cos 4 - \cos 0\right]$$

$$= \frac{1}{2}\left(\frac{4}{3}\cos 1 - \frac{1}{3}\cos 4 - 1\right) \approx -0.03086.$$

Since the area of R is 1, the average temperature is about -0.0309.

31

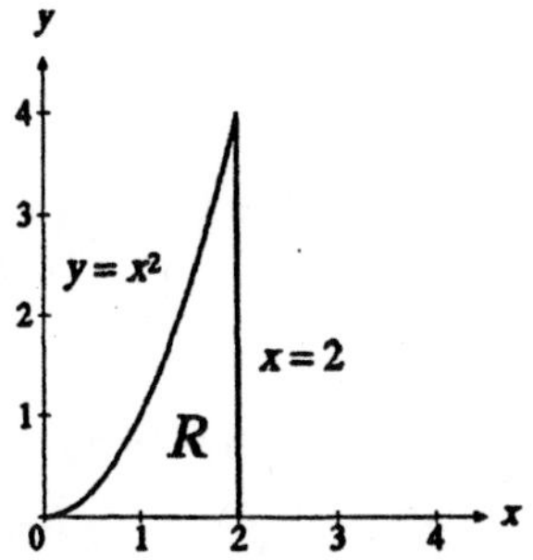

The region of integration is described using horizontal cross sections as $0 \le y \le 4$, $\sqrt{y} \le x \le 2$. Hence the integral equals $\int_0^4 \int_{\sqrt{y}}^2 x^3 y\,dx\,dy$.

33 The region of integration is the triangle with vertices (0, 0), (1, 1), (2, 1). Using horizontal

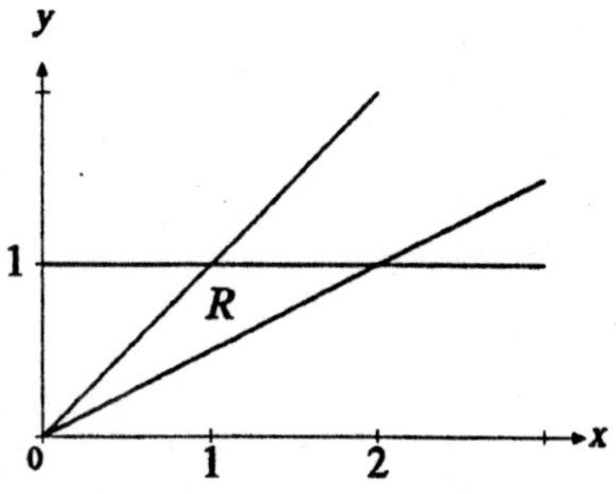

cross sections, its description is $0 \le y \le 1$, $y \le x \le 2y$. So the integral equals $\int_0^1 \int_y^{2y} xy\, dx\, dy$.

35 Since $\sin y^2$ does not have an elementary antiderivative, reverse the order of integration. The

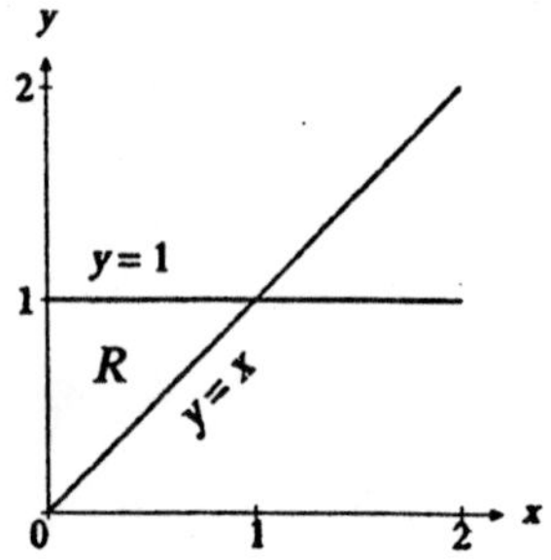

region can be described by $0 \le y \le 1$, $0 \le x \le y$, so the integral equals $\int_0^1 \int_0^y \sin y^2\, dx\, dy =$

$$\int_0^1 y \sin y^2\, dy = -\frac{1}{2} \cos y^2 \Big|_0^1 = \frac{1}{2}(1 - \cos 1).$$

37 The region of integration is described by $0 \le y \le$

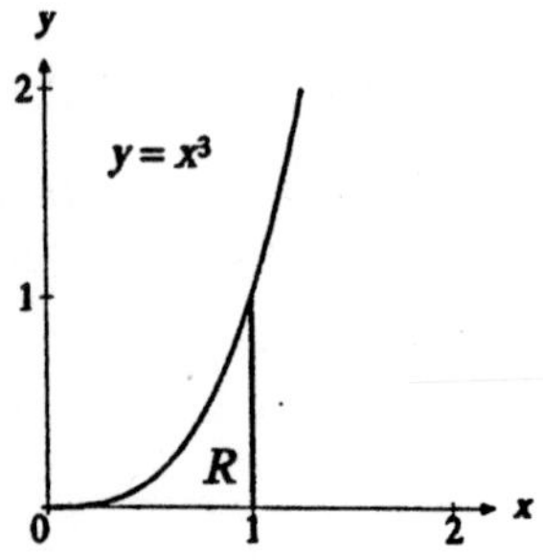

1, $\sqrt[3]{y} \le x \le 1$. This is equivalent to $0 \le x \le 1$, $0 \le y \le x^3$. Hence the integral equals

$$\int_0^1 \int_0^{x^3} \sqrt{1 + x^4}\, dy\, dx = \int_0^1 x^3\sqrt{1 + x^4}\, dx =$$

$$\frac{1}{4} \cdot \frac{(1 + x^4)^{3/2}}{3/2}\Big|_0^1 = \frac{1}{6}(2^{3/2} - 1^{3/2}) = \frac{1}{6}(\sqrt{8} - 1).$$

39 The description of R by vertical cross sections is in two parts: $0 \le x \le a$, $x/2 \le y \le x$, and $a \le x$

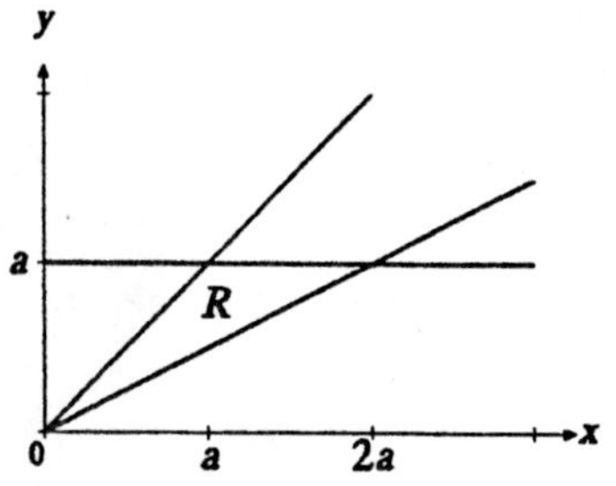

$\le 2a$, $x/2 \le y \le a$. For horizontal cross sections, we have $0 \le y \le a$, $y \le x \le 2y$.

(a) By the above information, we have

$$\int_R f(P)\, dA =$$

$$\int_0^a \int_{x/2}^x y^2 e^{y^2}\, dy\, dx + \int_a^{2a} \int_{x/2}^a y^2 e^{y^2}\, dy\, dx,$$

or $\int_R f(P)\, dA = \int_0^a \int_y^{2y} y^2 e^{y^2}\, dx\, dy$.

(b) The second integral is easier:

$$\int_0^a \int_y^{2y} y^2 e^{y^2}\, dx\, dy = \int_0^a y^2 e^{y^2} x \Big|_y^{2y} dy$$

$= \int_0^a y^3 e^{y^2}\, dy$. Use integration by parts with

$u = y^2$, $dv = ye^{y^2}\, dy$, $du = 2y\, dy$, and $v = \frac{1}{2}e^{y^2}$ to obtain $\frac{1}{2}y^2 e^{y^2}\Big|_0^a - \int_0^a ye^{y^2}\, dy =$

$$\frac{1}{2}a^2e^{a^2} - \frac{1}{2}e^{y^2}\Big|_0^a = \frac{1}{2}a^2e^{a^2} - \frac{1}{2}e^{a^2} + \frac{1}{2}.$$

41 The region R lies below the curve $y = f(x)$ and above the interval $[a, b]$; its description is therefore $a \le x \le b$, $0 \le y \le f(x)$. The density of the region is 1, so the moment about the x axis is M_x $= \int_R y\sigma\, dA = \int_a^b \int_0^{f(x)} y\, dy\, dx =$

$$\int_a^b \frac{y^2}{2}\Big|_0^{f(x)} dx = \int_a^b \frac{1}{2}(f(x))^2\, dx,$$ which agrees with the formula from Sec. 8.6.

43 (a) Some obvious examples of such functions are $f(x, y) = xy$, $f(x, y) = x^3y$, and $f(x, y) =$ $\sin xy$.

(b) If R is symmetric about the y axis, then $\int_R f(x, y)\, dA = 0$. To see this, consider any partition and sampling points with are symmetric with respect to the y axis. The approximating sum is 0, since each term is cancelled by the corresponding term from the other side of the y axis.

45 For fixed x, $g(x) = \int_R f(P)\, dA =$

$$\int_3^x \int_0^5 f(t, y)\, dy\, dt.$$ Let $h(x, y) = \int_0^5 f(x, u)\, du$.

Then $g(x) = \int_3^x h(t, y)\, dt$, and $\frac{dg}{dx} = h(x, y)$ by the first fundamental theorem of calculus.

15.3 Moments and Centers of Mass

1

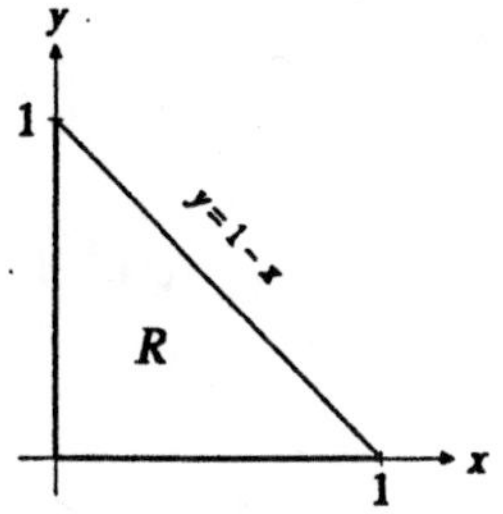

Using the definition, $\bar{x}$

$$= \frac{\int_0^1 \int_0^{1-x} x(x + y)\, dy\, dx}{\int_0^1 \int_0^{1-x} (x + y)\, dy\, dx}$$

$$= \frac{\int_0^1 \left(x^2y + \frac{1}{2}xy^2\right)\Big|_0^{1-x} dx}{\int_0^1 \left(xy + \frac{1}{2}y^2\right)\Big|_0^{1-x} dx}$$

$$= \frac{\int_0^1 \left(x^2(1 - x) + \frac{1}{2}x(1 - x)^2\right) dx}{\int_0^1 \left(x(1 - x) + \frac{1}{2}(1 - x)^2\right) dx}$$

$$= \frac{\int_0^1 \left(x^2 - x^3 + \frac{1}{2}x - x^2 + \frac{1}{2}x^3\right) dx}{\int_0^1 \left(x - x^2 + \frac{1}{2} - x + \frac{1}{2}x^2\right) dx}$$

$$= \frac{\int_0^1 \left(\frac{1}{2}x - \frac{1}{2}x^3\right) dx}{\int_0^1 \left(\frac{1}{2} - \frac{1}{2}x^2\right) dx} = \frac{\frac{1}{4}x^2 - \frac{1}{8}x^4\Big|_0^1}{\frac{1}{2}x - \frac{1}{6}x^3\Big|_0^1}$$

$$= \frac{1/4 - 1/8}{1/2 - 1/6} = \frac{1/8}{1/3} = \frac{3}{8}.$$ By symmetry $\bar{y} =$ 3/8. Therefore the center of mass is (3/8, 3/8).

3 The mass of the lamina is $\int_0^1 \int_0^1 y \tan^{-1} x \, dy \, dx$

$= \int_0^1 \frac{y^2}{2} \tan^{-1} x \Big|_0^1 dx = \int_0^1 \frac{1}{2} \tan^{-1} x \, dx$

$= \frac{1}{2}\left[x \tan^{-1} x - \frac{1}{2} \ln(1 + x^2)\right]\Big|_0^1$

$= \frac{1}{2}\left(\tan^{-1} 1 - \frac{1}{2} \ln 2\right) - \frac{1}{2}\left(0 - \frac{1}{2} \ln 1\right)$

$= \frac{\pi}{8} - \frac{1}{4} \ln 2$. To compute the center of mass, we now need to calculate the moments with respect to the x and y axes: $M_x = \int_R y\sigma \, dA$

$= \int_0^1 \int_0^1 y^2 \tan^{-1} x \, dy \, dx = \int_0^1 (\tan^{-1} x)\frac{y^3}{3}\Big|_0^1 dx$

$= \frac{1}{3} \int_0^1 \tan^{-1} x \, dx = \frac{1}{3}\left(\frac{\pi}{4} - \frac{1}{2} \ln 2\right)$

$= \frac{\pi}{12} - \frac{1}{6} \ln 2$; $M_y = \int_R x\sigma \, dA$

$= \int_0^1 \int_0^1 xy \tan^{-1} x \, dy \, dx$

$= \int_0^1 (x \tan^{-1} x)\frac{y^2}{2}\Big|_0^1 dx = \frac{1}{2} \int_0^1 x \tan^{-1} x \, dx$

$= \frac{1}{2}\left(\frac{x^2}{2} \tan^{-1} x - \frac{x}{2} + \frac{1}{2} \tan^{-1} x\right)\Big|_0^1$

$= \frac{1}{2}\left(\frac{1}{2}\cdot\frac{\pi}{4} - \frac{1}{2} + \frac{1}{2}\cdot\frac{\pi}{4}\right) = \frac{\pi}{8} - \frac{1}{4}$. (We used integration by parts with $u = \tan^{-1} x$ and $dv = x \, dx$.) Now $\bar{x} = \frac{M_y}{\text{Mass}} = \frac{\pi/8 - 1/4}{\pi/8 - 1/4 \ln 2} = \frac{\pi - 2}{\pi - 2 \ln 2}$, and $\bar{y} = \frac{M_x}{\text{Mass}} = \frac{\frac{\pi}{12} - \frac{\ln 2}{6}}{\frac{\pi}{8} - \frac{\ln 2}{4}}$

$= \frac{2\pi - 4 \ln 2}{3\pi - 6 \ln 2} = \frac{2}{3}$.

5 The mass of the lamina is $\int_0^1 \int_{2x}^{3x} xy \, dy \, dx$

$= \int_0^1 \frac{xy^2}{2}\Big|_{2x}^{3x} dx = \int_0^1 \frac{x}{2}(9x^2 - 4x^2) \, dx$

$= \int_0^1 \frac{5}{2}x^3 \, dx = \frac{5}{8}x^4\Big|_0^1 = \frac{5}{8}$. It follows that $\bar{x}$

$= \frac{M_y}{\text{Mass}} = \frac{8}{5}M_y = \frac{8}{5} \int_0^1 \int_{2x}^{3x} x(xy) \, dy \, dx$

$= \frac{8}{5} \int_0^1 \frac{x^2y^2}{2}\Big|_{2x}^{3x} dx = \frac{8}{5} \int_0^1 \frac{1}{2}x^2(9x^2 - 4x^2) \, dx$

$= \frac{8}{5} \int_0^1 \frac{5}{2}x^4 \, dx = \frac{8}{5}\cdot\frac{x^5}{2}\Big|_0^1 = \frac{8}{5}\cdot\frac{1}{2} = \frac{4}{5}$.

Finally, $\bar{y} = \frac{M_x}{\text{Mass}} = \frac{8}{5} \int_0^1 \int_{2x}^{3x} y(xy) \, dy \, dx =$

$\frac{8}{5} \int_0^1 \frac{xy^3}{3}\Big|_{2x}^{3x} dx = \frac{8}{5} \int_0^1 x\left(9x^3 - \frac{8x^3}{3}\right) dx =$

$\frac{8}{5} \int_0^1 \frac{19x^4}{3} \, dx = \frac{8}{5}\cdot\frac{19x^5}{15}\Big|_0^1 = \frac{152}{75}$.

7 The mass of the lamina is $\int_0^3 \int_{x^2}^{x+6} 2x \, dy \, dx =$

$\int_0^3 2xy\Big|_{x^2}^{x+6} dx = \int_0^3 (2x^2 + 12x - 2x^3) \, dx$

$$= \frac{2x^3}{3} + 6x^2 - \frac{x^4}{2}\Big|_0^3 = \frac{63}{2}.\text{ Then } \bar{x} = \frac{M_y}{\text{Mass}}$$

$$= \frac{2}{63}\int_0^3\int_{x^2}^{x+6} 2x^2\,dy\,dx = \frac{2}{63}\int_0^3 2x^2y\Big|_{x^2}^{x+6} dx$$

$$= \frac{2}{63}\int_0^3 (2x^3 + 12x^2 - 2x^4)\,dx$$

$$= \frac{2}{63}\left(\frac{x^4}{2} + 4x^3 - \frac{2}{5}x^5\right)\Bigg|_0^3 = \frac{2}{63}\cdot\frac{513}{10} = \frac{57}{35}$$

and $\bar{y} = \frac{M_x}{\text{Mass}} = \frac{2}{63}\int_0^3\int_{x^2}^{x+6} 2xy\,dy\,dx$

$$= \frac{2}{63}\int_0^3 xy^2\Big|_{x^2}^{x+6} dx$$

$$= \frac{2}{63}\int_0^3 (x^3 + 12x^2 + 36x - x^5)\,dx$$

$$= \frac{2}{63}\left(\frac{x^4}{4} + 4x^3 + 18x^2 - \frac{x^6}{6}\right)\Bigg|_0^3 = \frac{2}{63}\cdot\frac{675}{4}$$

$$= \frac{75}{14}.$$

9 Since the triangle is homogeneous, its density at any point is $\sigma(P) = M/\text{Area} = \frac{M}{\sqrt{3}a^2/4} = \frac{4M}{\sqrt{3}a^2}$.

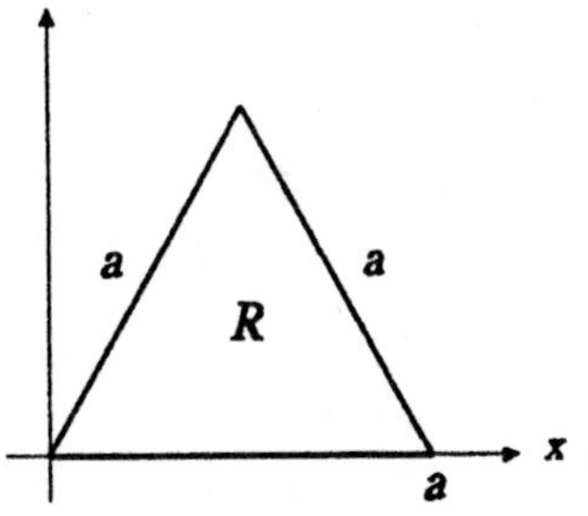

Let the line L pass through (0, 0). Then the moment of inertia about L is $\int_R r^2\sigma(P)\,dA = \int_R (x^2 + y^2)\sigma\,dA = \frac{4M}{\sqrt{3}a^2}\int_R (x^2 + y^2)\,dA$.

Describe R using horizontal cross sections. The line joining (0, 0) and $\left(\frac{a}{2}, \frac{\sqrt{3}a}{2}\right)$ has the equation $y = \sqrt{3}x$ or $x = y/\sqrt{3}$. The line joining $\left(\frac{a}{2}, \frac{\sqrt{3}a}{2}\right)$ and $(a, 0)$ has the equation $y = -\sqrt{3}x + \sqrt{3}a$ or $x = a - y/\sqrt{3}$. Hence R has the description $0 \le y \le \frac{\sqrt{3}a}{2}$, $\frac{y}{\sqrt{3}} \le x \le a - \frac{y}{\sqrt{3}}$. Therefore the moment of inertia is $\frac{4M}{\sqrt{3}a^2}\int_0^{\sqrt{3}a/2}\int_{y/\sqrt{3}}^{a-y/\sqrt{3}} (x^2 + y^2)\,dx\,dy$

$$= \frac{4M}{\sqrt{3}a^2}\int_0^{\sqrt{3}a/2} \frac{1}{3}x^3 + y^2x\Bigg|_{y/\sqrt{3}}^{a-y/\sqrt{3}} dy =$$

$$\frac{4M}{\sqrt{3}a^2}\int_0^{\sqrt{3}a/2}\left[\frac{1}{3}\left(a - \frac{y}{\sqrt{3}}\right)^3 + y^2\left(a - \frac{y}{\sqrt{3}}\right) - \frac{1}{3}\left(\frac{y}{\sqrt{3}}\right)^3 - y^2\left(\frac{y}{\sqrt{3}}\right)\right] dy$$

$$= \frac{4M}{\sqrt{3}a^2}\int_0^{\sqrt{3}a/2}\left[\frac{1}{3}\left(a^3 - \frac{3a^2y}{\sqrt{3}} + \frac{3ay^2}{3} - \frac{y^3}{3\sqrt{3}}\right) + ay^2 - \frac{y^3}{\sqrt{3}} - \frac{y^3}{9\sqrt{3}} - \frac{y^3}{\sqrt{3}}\right] dy$$

$$= \frac{4M}{\sqrt{3}a^2}\int_0^{\sqrt{3}a/2}\left(\frac{a^3}{3} + \frac{a^2y}{\sqrt{3}} + \frac{4ay^2}{3} - \frac{20y^3}{9\sqrt{3}}\right) dy$$

$$= \frac{4M}{\sqrt{3}a^2}\left(\frac{a^3y}{3} - \frac{a^2y^2}{2\sqrt{3}} + \frac{4ay^3}{9} - \frac{5y^4}{9\sqrt{3}}\right)\Bigg|_0^{\sqrt{3}a/2}$$

$$= \frac{4M}{\sqrt{3}a^2}\left(\frac{\sqrt{3}a^4}{6} - \frac{a^2}{2\sqrt{3}}\,\frac{3a^2}{4} + \frac{4a}{9}\,\frac{3\sqrt{3}a^3}{8} - \frac{5}{9\sqrt{3}}\,\frac{9a^4}{16}\right)$$

$$= \frac{4M}{\sqrt{3}a^2}\left(\frac{\sqrt{3}a^4}{6} - \frac{3a^4}{8\sqrt{3}} + \frac{\sqrt{3}a^4}{6} - \frac{5a^4}{16\sqrt{3}}\right)$$

$$= 4Ma^2\left(\frac{1}{6} - \frac{1}{8} + \frac{1}{6} - \frac{5}{48}\right) = \frac{5Ma^2}{12}$$

11 Let the line about which the square rotates pass through (0, 0). Since the square is homogenous,

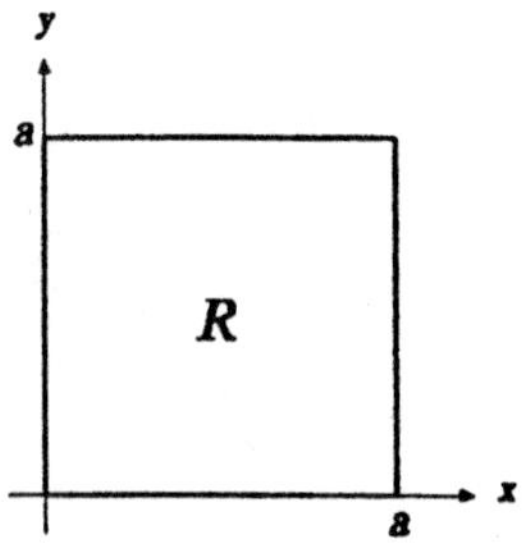

$\sigma(P) = M/a^2$. Thus the moment of inertia about L is $\int_R r^2\sigma(P)\,dA = \frac{M}{a^2}\int_R (x^2 + y^2)\,dA$

$$= \frac{M}{a^2}\int_0^a\int_0^a (x^2 + y^2)\,dy\,dx$$

$$= \frac{M}{a^2}\int_0^a \left(x^2y + \frac{y^3}{3}\right)\Bigg|_0^a dx$$

$$= \frac{M}{a^2}\int_0^a \left(ax^2 + \frac{a^3}{3}\right) dx = \frac{M}{a^2}\left(\frac{ax^3}{3} + \frac{a^3x}{3}\right)\Bigg|_0^a$$

$$= \frac{M}{3a^2}(a^4 + a^4) = \frac{2Ma^2}{3}.$$

13 $I = \int_R r^2\sigma(P)\,dA$. First describe R. It is easier to use horizontal cross sections. Clearly y ranges from

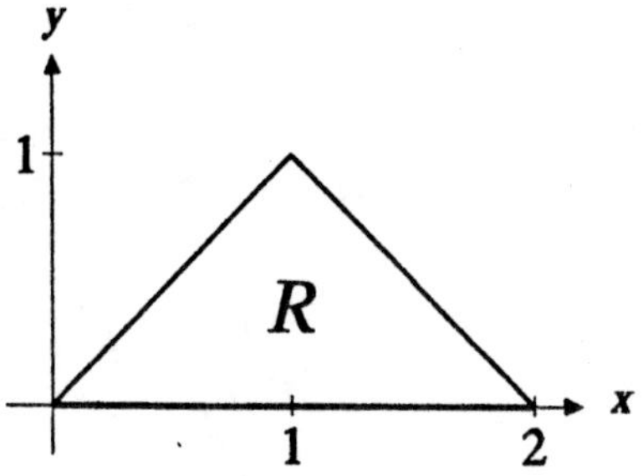

0 to 1. The line through (0, 0) and (1, 1) has the equation $y = x$ and the line through (1, 1) and (2, 0) has the equation $y = -x + 2$ or $x = 2 - y$. Then the description of R is $0 \le y \le 1$, $y \le x \le 2 - y$. We may now calculate $I =$

$$\int_0^1\int_y^{2-y} (x^2 + y^2)(xy)\,dx\,dy$$

$$= \int_0^1\int_y^{2-y} (x^3y + xy^3)\,dx\,dy$$

$$= \int_0^1 \frac{x^4y}{4} + \frac{x^2y^3}{2}\Bigg|_y^{2-y} dy = \int_0^1 \left(\frac{y^5}{4} - 2y^4 + 6y^3 - 8y^2 + 4y + 2y^3 - 2y^4 + \frac{y^5}{2} - \frac{y^5}{4} - \frac{y^5}{2}\right) dy = \int_0^1 (-4y^4 + 8y^3 - 8y^2 + 4y)\,dy$$

$$= \frac{-4y^5}{5} + \frac{8y^4}{4} - \frac{8y^3}{3} + \frac{4y^2}{2}\Bigg|_0^1$$

$$= \frac{-4}{5} + 2 - \frac{8}{3} + 2 = \frac{8}{15}.$$

15 (a) The general form for I remains the same, that is $I = \int_R r^2\sigma(P)\,dA$. However, r is now the distance from a point (x, y) in R to the line L. If L has the equation $ax + by + c = 0$, then by Theorem 2 of Sec. 12.4, $r = \dfrac{|ax + by + c|}{\sqrt{a^2 + b^2}}$ and thus $I = \dfrac{1}{a^2 + b^2}\int_R (ax + by + c)^2\sigma(x, y)\,dA$. This expression could get messy so choose your axes so that L is one of them. Then if L is the x axis, $I_x = \int_R y^2\sigma(x, y)\,dA$ and if L is the y axis, $I_y = \int_R x^2\sigma(x, y)\,dA$.

(b) From above it is clear that $I_z = I_x + I_y$.

17 Just use the definitions. $\bar{x}_1 = \dfrac{\int_{R_1} x\sigma(P)dA}{M_1}$, $\bar{x}_2 = \dfrac{\int_{R_2} x\sigma(P)dA}{M_2}$, and $\bar{x} = \dfrac{\int_R x\sigma(P)dA}{M} = \dfrac{\int_{R_1} x\sigma(P)dA + \int_{R_2} x\sigma(P)dA}{M_1 + M_2} = \dfrac{M_1\bar{x}_1 + M_2\bar{x}_2}{M_1 + M_2}$.

Similarly, $\bar{y} = \dfrac{M_1\bar{y}_1 + M_2\bar{y}_2}{M_1 + M_2}$.

19 (a) By definition, $M_L = \int_R (y - \bar{y})\,\sigma(x, y)\,dA = \int_R y\sigma(x, y)\,dA - \bar{y}\int_R \sigma(P)\,dA$. Let $M = \int_R \sigma(P)\,dA$. Then the above expression becomes $M_x - \bar{y}M = \bar{y}M - \bar{y}M = 0$. Thus $M_L = 0$.

(b) The line L passes through the center of mass, so it is a "balancing line" for the region. Thus is makes sense that the moment about L is 0.

15.4 Computing $\int_R f(P)\,dA$ using Polar Coordinates

1 $0 \le \theta \le 2\pi$, $0 \le r \le 3 + \cos\theta$

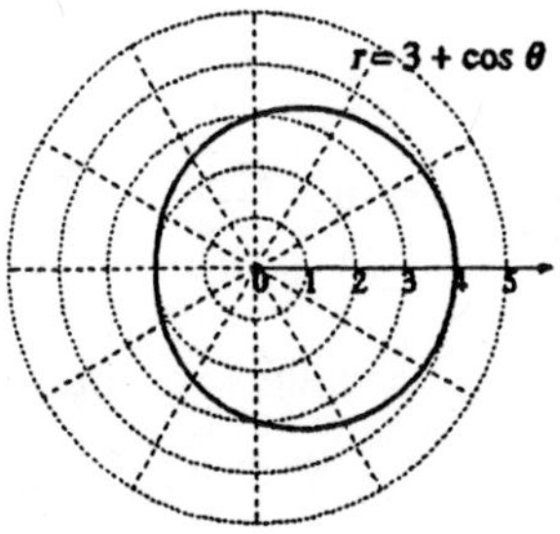

3 The edges through $(0, 0)$ are $\theta = \tan^{-1} 1 = \pi/4$ and $\theta = \tan^{-1}\sqrt{3} = \pi/3$. The vertical edge is $x = 1$; that is, $r\cos\theta = 1$. So the description is $\pi/4 \le \theta \le \pi/3$, $0 \le r \le \sec\theta$.

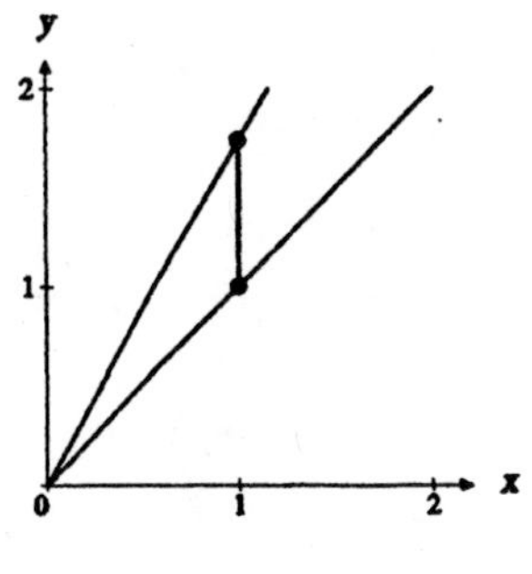

5 $\dfrac{\pi}{3} \le \theta \le \dfrac{5\pi}{3}$, $0 \le r \le 5$

7 (a)

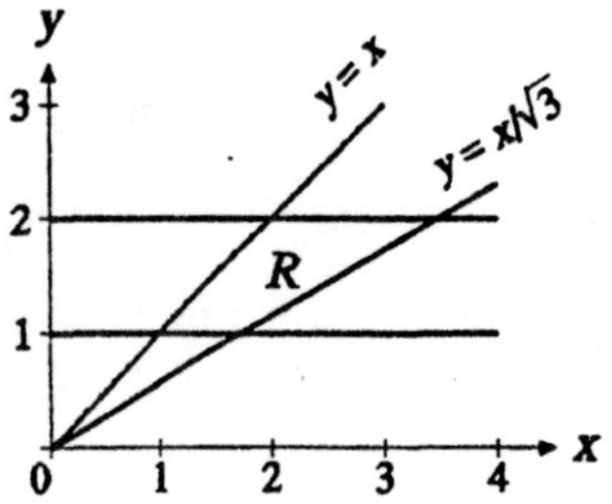

(b) $1 \le y \le 2,\ y \le x \le \sqrt{3}y$

(c) $1 \le x \le \sqrt{3},\ 1 \le y \le x$, and $\sqrt{3} \le x \le 2$, $\dfrac{x}{\sqrt{3}} \le y \le x$, and $2 \le x \le 2\sqrt{3}$, $\dfrac{x}{\sqrt{3}} \le y \le 2$.

(d) θ ranges from $\tan^{-1}\dfrac{1}{\sqrt{3}} = \dfrac{\pi}{6}$ to $\tan^{-1} 1 = \pi/4$. The horizontal sides are $y = 1$ and $y = 2$; that is, $r = \csc\theta$ and $r = 2\csc\theta$. So the description is $\pi/6 \le \theta \le \pi/4$, $\csc\theta \le r \le 2\csc\theta$.

9

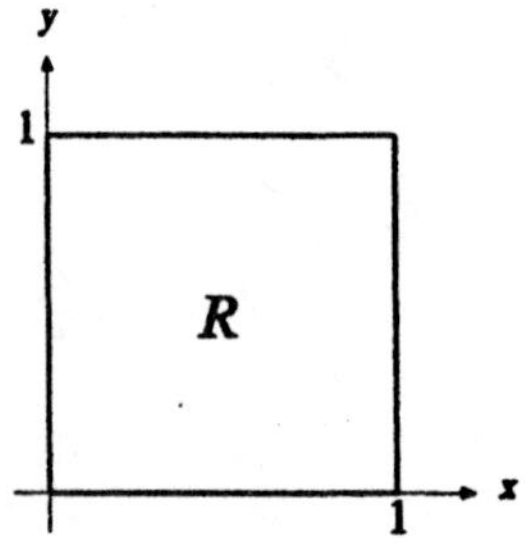

The right side is described by $x = 1$; that is, $r = \sec\theta$. The top is described by $y = 1$; that is, $r = \csc\theta$. The top right corner has polar coordinates $\left(\sqrt{2}, \dfrac{\pi}{4}\right)$. Hence, $0 \le \theta \le \pi/4$, $0 \le r \le \sec\theta$ and $\pi/4 \le \theta \le \pi/2$, $0 \le r \le \csc\theta$.

11

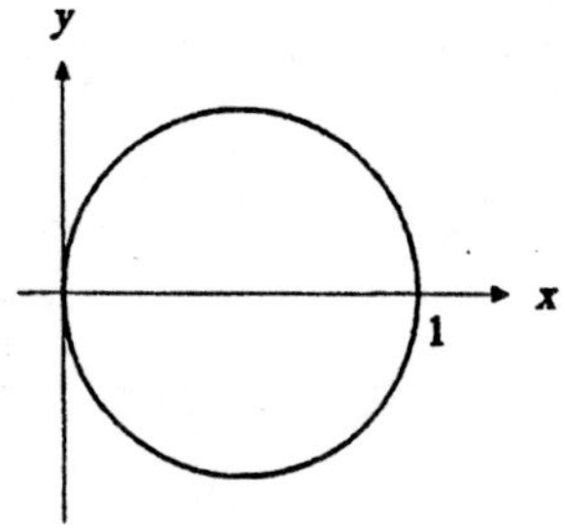

$$\int_R r^2\, dA = \int_{-\pi/2}^{\pi/2}\int_0^{\cos\theta} r^3\, dr\, d\theta$$

$$= \int_{-\pi/2}^{\pi/2} \frac{r^4}{4}\bigg|_0^{\cos\theta} d\theta = \frac{1}{4}\int_{-\pi/2}^{\pi/2} \cos^4\theta\, d\theta$$

$$= \frac{2}{4}\int_0^{\pi/2} \cos^4\theta\, d\theta = \frac{1}{2}\cdot\frac{3\pi}{16} = \frac{3\pi}{32}.$$ (Note the use of symmetry and Formula 73 from the table of integrals to evaluate the integral.)

13

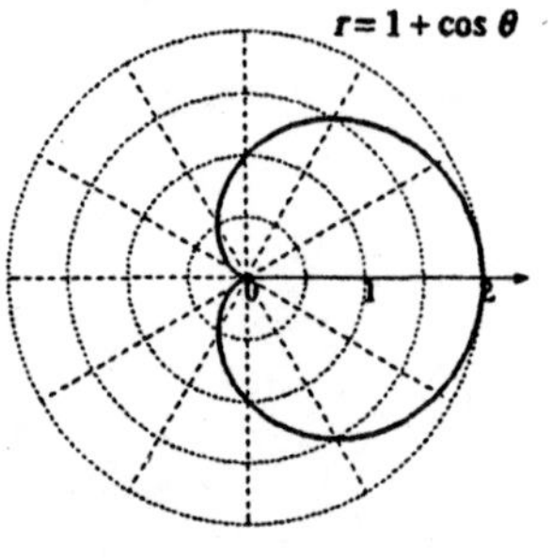

$$\int_R r^2\, dA = \int_0^{2\pi}\int_0^{1+\cos\theta} r^3\, dr\, d\theta$$

$$= \int_0^{2\pi} \frac{r^4}{4}\bigg|_0^{1+\cos\theta} d\theta = \frac{1}{4}\int_0^{2\pi} (1 + \cos\theta)^4\, d\theta$$

$$= \frac{1}{4}\int_0^{2\pi} (1 + 4\cos\theta + 6\cos^2\theta + 4\cos^3\theta + \cos^4\theta)\, d\theta.$$ The second and fourth terms of the integrand are odd powers of $\cos\theta$; by symmetry, their value is zero over $[0, 2\pi]$. Also by symmetry, the third and fifth terms can be evaluated with the

help of Formula 73 from the table of integrals. We therefore have $\frac{1}{4}\left[2\pi + 0 + 6\cdot 4\cdot\frac{\pi}{4} + 0 + 4\cdot\frac{3\pi}{16}\right]$

$$= \frac{1}{4}\left(\frac{35\pi}{4}\right) = \frac{35\pi}{16}.$$

15 $\int_R y^2\,dA = \int_0^{2\pi}\int_0^a (r\sin\theta)^2\, r\,dr\,d\theta$

$$= \int_0^{2\pi} \frac{r^4}{4}\sin^2\theta\Big|_{r=0}^{r=a} d\theta = \frac{a^4}{4}\int_0^{2\pi}\sin^2\theta\,d\theta$$

$$= \frac{4a^2}{4}\int_0^{\pi/2}\sin^2\theta\,d\theta = a^2\cdot\frac{\pi}{4} = \frac{\pi a^2}{4}.$$

17 The region is described by $0 \le \theta \le 2\pi$, $0 \le r \le 1 + \sin\theta$. Since $y = r\sin\theta$, the integral is

$$\int_0^{2\pi}\int_0^{1+\sin\theta}(r\sin\theta)^2\, r\,dr\,d\theta$$

$$= \int_0^{2\pi}\sin^2\theta\int_0^{1+\sin\theta} r^3\,dr\,d\theta$$

$$= \int_0^{2\pi}\sin^2\theta\,\frac{1}{4}(1+\sin\theta)^4\,d\theta = \frac{1}{4}\int_0^{2\pi}(\sin^2\theta$$

$$+ 4\sin^3\theta + 6\sin^4\theta + 4\sin^5\theta + \sin^6\theta)\,d\theta.$$

By symmetry the odd powers of $\sin\theta$ contribute nothing, while the even powers contribute four times the amounts given by Formula 73 from the table of integrals. Hence the desired integral equals

$$\int_0^{\pi/2}(\sin^2\theta + 6\sin^4\theta + \sin^6\theta)\,d\theta =$$

$$\frac{1}{2}\cdot\frac{\pi}{2} + 6\cdot\frac{1\cdot 3}{2\cdot 4}\cdot\frac{\pi}{2} + \frac{1\cdot 3\cdot 5}{2\cdot 4\cdot 6}\cdot\frac{\pi}{2} = \frac{49\pi}{32}.$$

19 The region is described by $0 \le \theta \le 2\pi$, $0 \le r \le 1 + \cos\theta$. Its area is $\int_0^{2\pi}\int_0^{1+\cos\theta} r\,dr\,d\theta$

$$= \int_0^{2\pi}\frac{1}{2}(1+\cos\theta)^2\,d\theta$$

$$= \frac{1}{2}\int_0^{2\pi}(1 + 2\cos\theta + \cos^2\theta)\,d\theta$$

$= \frac{1}{2}(2\pi + 0 + \pi) = \frac{3\pi}{2}$. Also $M_y =$

$$\int_0^{2\pi}\int_0^{1+\cos\theta} xr\,dr\,d\theta$$

$$= \int_0^{2\pi}\int_0^{1+\cos\theta} r^2\cos\theta\,dr\,d\theta$$

$$= \int_0^{2\pi}\cos\theta\left(\frac{r^3}{3}\right)\Big|_0^{1+\cos\theta} d\theta$$

$$= \frac{1}{3}\int_0^{2\pi}\cos\theta\,(1+\cos\theta)^3\,d\theta =$$

$$\frac{1}{3}\int_0^{2\pi}(\cos\theta + 3\cos^2\theta + 3\cos^3\theta + \cos^4\theta)\,d\theta$$

$$= \frac{1}{3}\left(0 + 3\cdot 4\cdot\frac{1}{2}\cdot\frac{\pi}{2} + 0 + 4\cdot\frac{1}{2}\cdot\frac{3}{4}\cdot\frac{\pi}{2}\right) = \frac{5\pi}{4}.$$

Hence $\bar{x} = \frac{5\pi/4}{3\pi/2} = 5/6$. By symmetry, $\bar{y} = 0$, so the centroid is $(5/6, 0)$.

21 The region is described by $0 \le \theta \le \pi/3$, $0 \le r \le \sin 3\theta$. Its area is $\int_0^{\pi/3}\int_0^{\sin 3\theta} r\,dr\,d\theta =$

$\int_0^{\pi/3}\frac{1}{2}\sin^2 3\theta\,d\theta = \frac{\pi}{12}$. Also, $\int_R f(P)\,dA =$

$$\int_0^{\pi/3}\int_0^{\sin 3\theta} r^2\,dr\,d\theta = \int_0^{\pi/3}\frac{1}{3}\sin^3 3\theta\,d\theta$$

$$= \frac{1}{9}\int_0^{\pi}\sin^3 u\,du = \frac{2}{9}\int_0^{\pi/2}\sin^3 u\,du = \frac{2}{9}\cdot\frac{2}{3}$$

$= \frac{4}{27}$, by Formula 73. Hence the average is $\frac{4/27}{\pi/12}$

$$= \frac{16}{9\pi}.$$

23 Place the polar axis so that r is the interior of $r = 2a\cos\theta$. Then r is described by $-\frac{\pi}{2} \le \theta \le \frac{\pi}{2}$, $0 \le r \le 2a\cos\theta$, the area is πa^2, and the function is $f(P) = r$. Thus $\int_R f(P)\, dA =$

$$\int_{-\pi/2}^{\pi/2} \int_0^{2a\cos\theta} r^2\, dr\, d\theta = 2\int_0^{\pi/2} \frac{1}{3}(2a\cos\theta)^3\, d\theta$$

$$= \frac{16a^3}{3}\int_0^{\pi/2} \cos^3\theta\, d\theta = \frac{16a^3}{3}\cdot\frac{2}{3} = \frac{32a^3}{9}.$$

Hence the average is $\frac{32a^3}{9}\cdot\frac{1}{\pi a^2} = \frac{32a}{9\pi}$.

25 The region is described by $0 \le \theta \le \pi/4$, $0 \le r \le \sec\theta$. Since $\sqrt{x^2 + y^2} = r$, the integral equals

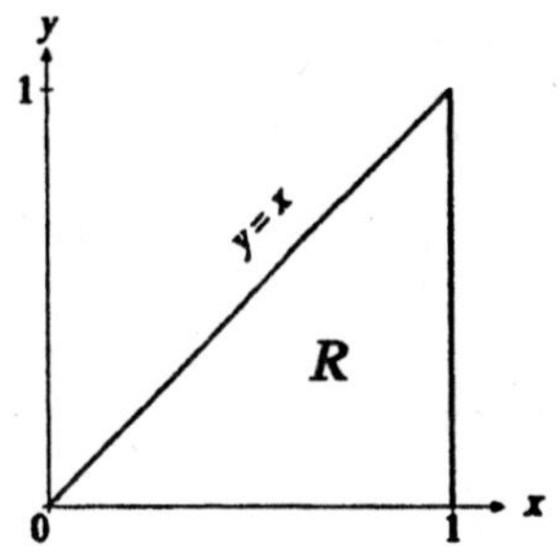

$$\int_0^{\pi/4} \int_0^{\sec\theta} r^2\, dr\, d\theta = \int_0^{\pi/4} \frac{1}{3}\sec^3\theta\, d\theta$$

$$= \frac{1}{6}(\sec\theta\tan\theta + \ln|\sec\theta + \tan\theta|)\Big|_0^{\pi/4}$$

$$= \frac{1}{6}(\sqrt{2} + \ln(1 + \sqrt{2})).$$

27 The region is described in polar coordinates by $\frac{\pi}{4} \le \theta \le \frac{\pi}{2}$, $0 \le r \le 1$. Hence

$$\int_0^{1/\sqrt{2}} \int_x^{\sqrt{1-x^2}} xy\, dy\, dx$$

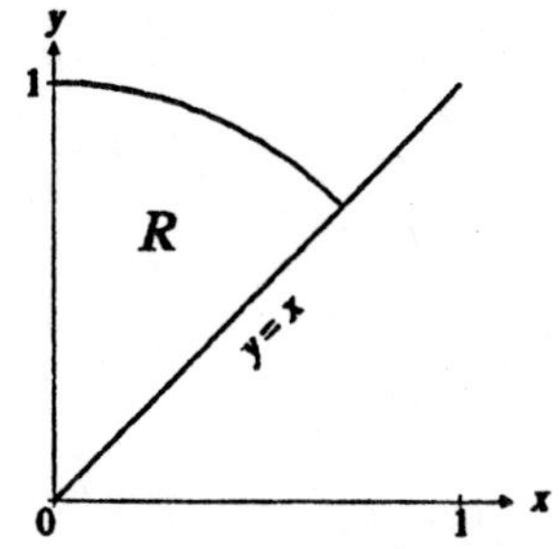

$$= \int_{\pi/4}^{\pi/2} \int_0^1 (r\cos\theta)(r\sin\theta)\, r\, dr\, d\theta$$

$$= \int_{\pi/4}^{\pi/2} \int_0^1 r^3\cos\theta\sin\theta\, dr\, d\theta$$

$$= \int_{\pi/4}^{\pi/2} \frac{1}{4}\cos\theta\sin\theta\, d\theta = \frac{1}{8}\int_{\pi/4}^{\pi/2} \sin 2\theta\, d\theta$$

$$= -\frac{1}{16}\cos 2\theta\Big|_{\pi/4}^{\pi/2} = \frac{1}{16}.$$

29 (a) The region is described by $0 \le \theta \le \pi/2$, $0 \le r \le a$, so the integral equals

$$\int_0^{\pi/2} \int_0^a \cos(r^2)\, r\, dr\, d\theta$$

$$= \int_0^{\pi/2} \frac{1}{2}\sin(r^2)\Big|_0^a d\theta = \int_0^{\pi/2} \frac{1}{2}\sin a^2\, d\theta$$

$$= \frac{1}{2}\sin a^2 \int_0^{\pi/2} d\theta = \frac{\pi}{4}\sin a^2.$$

(b) The region is described by $0 \le \theta \le \frac{\pi}{4}$, $0 \le r \le 2\sec\theta$. Hence the integral equals

$$\int_0^{\pi/4} \int_0^{2\sec\theta} r\cdot r\, dr\, d\theta$$

$$= \int_0^{\pi/4} \frac{1}{3}(2\sec\theta)^3\, d\theta = \frac{8}{3}\int_0^{\pi/4} \sec^3\theta\, d\theta$$

$$= \frac{8}{3}\cdot\frac{1}{2}(\sec\theta\tan\theta + \ln|\sec\theta + \tan\theta|)\Big|_0^{\pi/4}$$

$$= \frac{4}{3}(\sqrt{2} + \ln(1 + \sqrt{2})).$$

31 The equation $r = 2a \sin\theta$ describes a circle of radius a whose center has polar coordinates $(a, \pi/2)$, which is on the y axis. Since the region and the density function $\sigma = \sin\theta$ are both symmetric with respect to the y axis, we see that $\bar{x} = 0$. To find $\bar{y}$ we must compute mass and M_x:

Mass $= \int_R \sigma(P)\, dA = \int_0^\pi \int_0^{2a\sin\theta} \sin\theta \, r\, dr\, d\theta$

$= \int_0^\pi \sin\theta \left.\frac{r^2}{2}\right|_0^{2a\sin\theta} d\theta = \frac{1}{2}\int_0^\pi 4a^2 \sin^3\theta\, d\theta$

$= 4a^2 \int_0^{\pi/2} \sin^3\theta\, d\theta = 4a^2 \cdot \frac{2}{3} = \frac{8a^2}{3}$. Also, M_x

$= \int_0^\pi \int_0^{2a\sin\theta} (r\sin\theta)(\sin\theta)\, r\, dr\, d\theta$

$= \int_0^\pi \int_0^{2a\sin\theta} r^2 \sin^2\theta\, dr\, d\theta$

$= \int_0^\pi \frac{r^3}{3} \sin^2\theta \Big|_{r=0}^{r=2a\sin\theta} dr = \frac{1}{3}\int_0^\pi 8a^3 \sin^5\theta\, d\theta$

$= \frac{8a^3}{3} \cdot 2 \int_0^{\pi/2} \sin^5\theta\, d\theta = \frac{16a^3}{3}\cdot\frac{8}{15}$, so $\bar{y} =$

$\frac{M_x}{\text{Mass}} = \frac{16a^3}{3}\cdot\frac{8}{15}\cdot\frac{3}{8a^2} = \frac{16a}{15}$.

33 The definition of the moment of inertia is independent of any coordinate system so $I = \int_R r^2 \sigma(P)\, dA$. Since the lamina is homogeneous, $\sigma(P) = \frac{M}{\pi a^2}$ and $I = \frac{M}{\pi a^2}\int_0^{2\pi}\int_0^a r^3\, dr\, d\theta =$

$\frac{M}{\pi a^2}\int_0^{2\pi} \frac{a^4}{4}\, d\theta = \left(\frac{M}{\pi a^2}\right)\left(\frac{a^4}{4}\right)(2\pi) = \frac{Ma^2}{2}$.

35 Choose coordinate axes so that the x and y axes are diameters. Then the disk has the description $0 \le \theta \le 2\pi$, $0 \le r \le a$ and the distance of an element of mass to the axis of rotation (in this case the x axis) is $r\sin\theta$. Thus $I =$

$\frac{M}{\pi a^2}\int_0^{2\pi}\int_0^a r^3 \sin^2\theta\, dr\, d\theta$

$= \frac{M}{\pi a^2}\int_0^{2\pi} \frac{r^4}{4}\sin^2\theta \Big|_0^a d\theta$

$= \frac{Ma^2}{4\pi}\int_0^{2\pi} \sin^2\theta\, d\theta = \frac{Ma^2}{4\pi}\cdot 4\cdot\frac{1}{2}\cdot\frac{\pi}{2} = \frac{Ma^2}{4}$.

Another way to work this problem is to remember that $I_z = I_x + I_y = 2I_x$ by symmetry. By Exercise 33, $I_z = \frac{Ma^2}{2}$, so $I_x = \frac{Ma^2}{4}$.

37 (a) Consider two circles, one of radius r_0 and the other of radius $r_0 + \Delta r$. Then the area of the ring is $\pi(r_0 + \Delta r)^2 - \pi r_0^2 =$

$\pi(2r_0\Delta r + (\Delta r)^2)$.

(b) If $\Delta\theta = \pi$ then the fraction is 1/2. In general, the fractional area is $\frac{\Delta\theta}{2\pi}$.

(c) $A = \left(\frac{\Delta\theta}{2\pi}\right)\pi(2r_0\Delta r + (\Delta r)^2) =$

$r_0\Delta r\Delta\theta + \frac{(\Delta r)^2\Delta\theta}{2} = \left(r_0 + \frac{\Delta r}{2}\right)\Delta r\,\Delta\theta$.

39 (a) R_1 is described by $0 \le \theta \le \pi/2$, $0 \le r \le a$. Thus $\int_{R_1} f(P)\, dA = \int_0^{\pi/2}\int_0^a e^{-r^2} r\, dr\, d\theta$

$= \int_0^{\pi/2} -\frac{1}{2}e^{-r^2}\Big|_0^a d\theta$

$= \int_0^{\pi/2}\left(\frac{1}{2} - \frac{1}{2}e^{-a^2}\right) d\theta = \frac{\pi}{2}\left(\frac{1}{2} - \frac{1}{2}e^{-a^2}\right)$

$= \frac{\pi}{4}(1 - e^{-a^2})$. R_3 is described by $0 \le \theta \le \pi/2$, $0 \le r \le a\sqrt{2}$. Thus $\int_{R_3} f(P)\, dA$

$= \int_0^{\pi/2} \int_0^{a\sqrt{2}} e^{-r^2} r\, dr\, d\theta$

$= \int_0^{\pi/2} -\frac{1}{2}e^{-r^2}\Big|_0^{a\sqrt{2}}\, d\theta$

$= \int_0^{\pi/2} \left(\frac{1}{2} - \frac{1}{2}e^{-2a^2}\right) d\theta = \frac{\pi}{4}(1 - e^{-2a^2})$.

(b) Using rectangular coordinates, we have $f(P) = e^{-(x^2+y^2)}$. Thus $\int_{R_2} f(P)\, dA$

$= \int_0^a \int_0^a e^{-(x^2+y^2)}\, dy\, dx$

$= \int_0^a \int_0^a e^{-x^2} e^{-y^2}\, dy\, dx$

$= \int_0^a e^{-x^2} \left(\int_0^a e^{-y^2}\, dy\right) dx$

$= \left(\int_0^a e^{-y^2}\, dy\right) \int_0^a e^{-x^2}\, dx$

$= \left(\int_0^a e^{-x^2}\, dx\right)^2$, since y is a dummy variable in the first definite integral. Also, from the figure, R_1 is contained in R_2, which in turn is contained in R_3. Since R_2 does not equal R_1 or R_3 (and $f(P) > 0$ for all P), we have

$\int_{R_1} f(P)\, dA < \int_{R_2} f(P)\, dA < \int_{R_3} f(P)\, dA$.

Applying these results together with those from (a), the given inequalities follow.

(c) As $a \to \infty$, the first inequality in (b) becomes $\frac{\pi}{4} \le \left(\int_0^\infty e^{-x^2}\, dx\right)^2$ since $\lim_{a\to\infty} (1 - e^{-a^2}) = 1$. (Recall that an inequality can become an equality in taking a limit.) Similarly, the second inequality in (b) becomes $\left(\int_0^\infty e^{-x^2}\, dx\right)^2 \le \frac{\pi}{4}$. The only way both inequalities can be true is if $\left(\int_0^\infty e^{-x^2}\, dx\right)^2 = \frac{\pi}{4}$, so we have $\int_0^\infty e^{-x^2}\, dx = \frac{\sqrt{\pi}}{2}$, since $e^{-x^2} > 0$ for all x.

41 Introduce the coordinate system given in the figure. By symmetry, $\bar{x} = 0$. The area of the region is

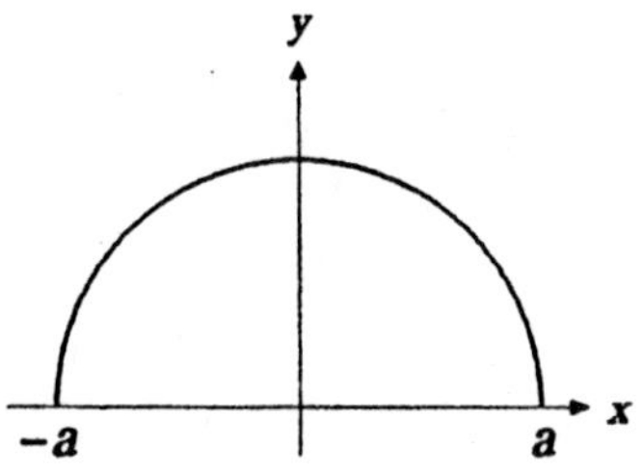

$\frac{1}{2}\pi a^2$. The moment about the x axis is

$\int_{-a}^{a} \int_0^{\sqrt{a^2-x^2}} y\, dy\, dx = \int_{-a}^{a} \frac{1}{2}(a^2 - x^2)\, dx =$

$\left(\frac{1}{2}a^2x - \frac{1}{6}x^3\right)\Big|_{-a}^{a} = \frac{2}{3}a^3$. Hence $\bar{y} = \frac{2a^3/3}{\pi a^2/2} = \frac{4a}{3\pi}$, so the center of mass is $\left(0, \frac{4a}{3\pi}\right)$.

43 (a) Suppose that dA is a small patch of area containing the point P. If the population density is δ, then the population of the patch is $\delta\, dA$ and the probability of contagion is $f(P)\, \delta\, dA$. The probability over the whole region is therefore $\delta \int_R f(P)\, dA$.

(b) When Q is the center of town, agreeing to place Q at the origin, we have $\int_R f(P)\,dA =$

$$\int_0^{2\pi}\int_0^1 (2 - r)r\,dr\,d\theta = \frac{4\pi}{3} \approx 4.2.$$

When Q is on the edge of town, agreeing to place the center of town at $(x, y) = (0, 1)$ and Q at the origin, we have $\int_R f(P)\,dA =$

$$\int_0^{\pi}\int_0^{2\sin\theta} (2 - r)\,r\,dr\,d\theta = 2\pi - \frac{32}{9}$$

$\approx 2.7.$

(c) The edge of town is safer.

45 Let R be the regular hexagon placed as shown in a rectangular coordinate system. Let S be the

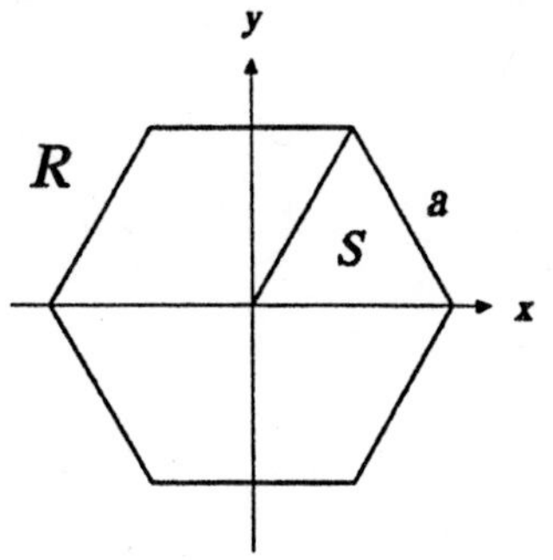

indicated sixth of the hexagon, a triangle of side a. Since the hexagon has area A and the triangle has area $\frac{A}{6} = \frac{\sqrt{3}a^2}{4}$, we have $a = \frac{\sqrt{2A}}{3^{3/4}}$. Note that the line bounded the right edge of S has slope $-\sqrt{3}$ and passes through $(a, 0)$, so the line has equation $y = -\sqrt{3}(x - a)$. Hence $\sqrt{3}x + y = a\sqrt{3}$, which in polar coordinates becomes $\sqrt{3}r\cos\theta + r\sin\theta$ $= a\sqrt{3}$, so $r\left(\frac{\sqrt{3}}{2}\cos\theta + \frac{1}{2}\sin\theta\right) = \frac{a\sqrt{3}}{2}$ and $r\left(\sin\frac{\pi}{3}\cos\theta + \cos\frac{\pi}{3}\sin\theta\right) = r\sin\left(\theta + \frac{\pi}{3}\right) = \frac{a\sqrt{3}}{2}$; that is, $r = \frac{a\sqrt{3}}{2}\csc\left(\theta + \frac{\pi}{3}\right)$. We now see that S has polar description $0 \le \theta \le \pi/3$, $0 \le r \le \frac{a\sqrt{3}}{2}\csc\left(\theta + \frac{\pi}{3}\right)$. The distance to the pole is just r, so the average distance is $\frac{1}{A}\int_R r\,dA$

$$= \frac{6}{A}\int_S r\,dA = \frac{6}{A}\int_0^{\pi/3}\int_0^{\frac{a\sqrt{3}}{2}\csc(\theta+\pi/3)} r^2\,dr\,d\theta$$

$$= \frac{6}{A}\int_0^{\pi/3}\frac{1}{3}\left(\frac{a\sqrt{3}}{2}\right)^3\csc^3\left(\theta + \frac{\pi}{3}\right)d\theta$$

$$= \frac{6}{A}\cdot\frac{a^3 3^{3/2}}{2^3}\cdot\frac{1}{3}\left[-\frac{1}{2}\csc\left(\theta + \frac{\pi}{3}\right)\cot\left(\theta + \frac{\pi}{3}\right) + \frac{1}{2}\ln\left|\csc\left(\theta + \frac{\pi}{3}\right) - \cot\left(\theta + \frac{\pi}{3}\right)\right|\right]\Bigg|_0^{\pi/3}$$

$$= \frac{3^{3/2}}{4A}\left(\frac{\sqrt{2A}}{3^{3/4}}\right)^3\left[-\frac{1}{2}\csc\frac{2\pi}{3}\cot\frac{2\pi}{3} + \frac{1}{2}\ln\left|\csc\frac{2\pi}{3} - \cos\frac{2\pi}{3}\right| + \frac{1}{2}\csc\frac{\pi}{3}\cot\frac{\pi}{3} - \frac{1}{2}\ln\left|\csc\frac{\pi}{3} - \cot\frac{\pi}{3}\right|\right]$$

$$= \frac{3^{3/2}}{4A}\cdot\frac{2\sqrt{2}A^{3/2}}{3^{9/4}}\cdot\frac{1}{2}\left[-\frac{2}{\sqrt{3}}\left(-\frac{1}{\sqrt{3}}\right) + \ln\left|\frac{2}{\sqrt{3}} + \frac{1}{\sqrt{3}}\right| + \frac{2}{\sqrt{3}}\frac{1}{\sqrt{3}} - \ln\left|\frac{2}{\sqrt{3}} - \frac{1}{\sqrt{3}}\right|\right]$$

$$= \frac{\sqrt{2A}}{4\cdot 3^{3/4}}\left[\frac{2}{3} + \ln\sqrt{3} + \frac{2}{3} - \ln\frac{1}{\sqrt{3}}\right]$$

$= \frac{\sqrt{2A}}{4\cdot 3^{3/4}}\left(\frac{4}{3} + \ln 3\right) = \frac{\sqrt{2A}}{3^{3/4}}\left(\frac{1}{3} + \frac{1}{4}\ln 3\right)$, as claimed.

47 Place an equilateral triangle of side a with its centroid at the origin so that one side is vertical

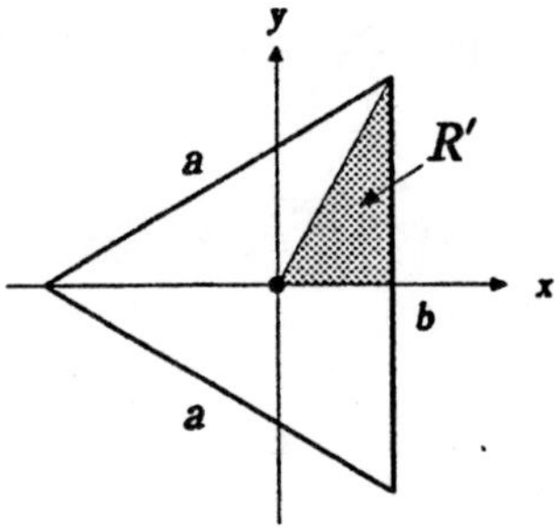

with equation $x = b$, where $b = a\sqrt{3}/6$. The area of the triangle is $A = \frac{\sqrt{3}}{4}a^2$, so $a = \sqrt{\frac{4A}{\sqrt{3}}} = \frac{2\sqrt{A}}{3^{1/4}}$. If R is the entire triangle and R' is the shaded sixth in the figure, then symmetry lets us say that $\frac{1}{A}\int_R r\,dA = \frac{6}{A}\int_{R'} r\,dA$, where r, as usual, is the distance to the pole (origin). R' can be described by $0 \le \theta \le \pi/3$, $0 \le r \le b\sec\theta$, so

$$\frac{6}{A}\int_{R'} r\,dA = \frac{24}{a^2\sqrt{3}}\int_0^{\pi/3}\int_0^{b\sec\theta} r\,r\,dr\,d\theta$$

$$= \frac{24}{a^2\sqrt{3}}\int_0^{\pi/3} \frac{r^3}{3}\bigg|_0^{b\sec\theta} d\theta$$

$$= \frac{8}{a^2\sqrt{3}}\int_0^{\pi/3} b^3\sec^3\theta\,d\theta =$$

$$\frac{8}{a^2\sqrt{3}}\left(\frac{a\sqrt{3}}{6}\right)^3\left(\frac{1}{2}\sec\theta\tan\theta + \frac{1}{2}\ln|\sec\theta + \tan\theta|\right)\bigg|_0^{\pi/3}$$

$$= \frac{a}{9}\left(\frac{1}{2}\cdot 2\sqrt{3} + \frac{1}{2}\ln(2 + \sqrt{3})\right)$$

$$= \frac{a}{18}(2\sqrt{3} + \ln(2 + \sqrt{3}))$$

$$= \frac{\sqrt{A}}{3^{9/4}}(2\sqrt{3} + \ln(2 + \sqrt{3})) \approx 0.404\sqrt{A},$$ as claimed.

49 Place the square R with its center at the origin and its sides parallel to the coordinate axes. The description of R is therefore $-a/2 \le x \le a/2$, $-a/2 \le y \le a/2$, where $a = \sqrt{A}$. Then the average metropolitan distance of points in R from the center is $\frac{1}{A}\int_R (|x| + |y|)\,dA$

$$= \frac{4}{A}\int_0^{a/2}\int_0^{a/2} (x + y)\,dy\,dx$$

$$= \frac{4}{a^2}\int_0^{a/2}\left(xy + \frac{y^2}{2}\right)\bigg|_0^{a/2} dx$$

$$= \frac{4}{a^2}\int_0^{a/2}\left(\frac{a}{2}x + \frac{a^2}{8}\right)dx = \frac{4}{a^2}\left(\frac{ax^2}{4} + \frac{a^2x}{8}\right)\bigg|_0^{a/2}$$

$$= \frac{4}{a^2}\left(\frac{a^3}{16} + \frac{a^3}{16}\right) = \frac{4}{a^2}\frac{a^3}{8} = \frac{a}{2} = \frac{\sqrt{A}}{2},$$ as claimed.

15.5 The Definite Integral of a Function over a Region in Space

1 Label the eight cubes as shown in the illustration. Each cube has volume 8, so a lower bound on the whole cube will be obtained by summing the minimum densities of the cubes and multiplying by

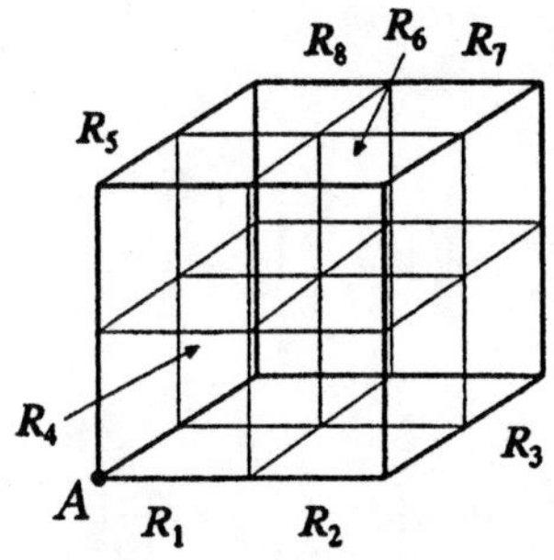

8. For example, the minimum density in cube R_1 is 0, while the minimum for R_2, R_4, or R_5 is $2^2 = 4$. Similarly, the minimum for R_3, R_6, or R_8 is $2^2 + 2^2 = 8$, and that for R_7 is $2^2 + 2^2 + 2^2 = 12$. Therefore the lower bound is $8(0 + 4 + 8 + 4 + 4 + 8 + 12 + 8) = 384$. The upper bound is similarly $8(12 + 24 + 36 + 24 + 24 + 36 + 48 + 36) = 8 \cdot 240 = 1920$. Since the dimensions of the cube are in centimeters and the density is given in grams per cubic centimeter, the mass of the cube must lie between 384 and 1920 grams.

3 Each approximating sum $\sum_{i=1}^{n} f(P_i)\, V_i$ equals $\sum_{i=1}^{n} 5 \cdot V_i = 5 \sum_{i=1}^{n} V_i$, which is just 5 times the volume of R, or $\frac{20\pi r^3}{3}$. Every approximating sum for $\int_R f(P)\, dV$ is equal to $\frac{20\pi r^3}{3}$, so this is the value of the integral.

5 Label the eight cubes as in Exercise 1. Note that the density at the center of R_1 is $1^2 + 1^2 + 1^2 = 3$, while the density at the center of R_2 is $3^2 + 1^2 + 1^2 = 11$, which is also the density at the center of R_4 or R_5. In R_3, R_6, and R_8, the density at the center is $3^2 + 3^2 + 1^2 = 19$, while in R_7 it is $3^2 + 3^2 + 3^2 = 27$. As in Exercise 1, we sum the densities in order and multiply by 8 to obtain the approximate mass: $8(3 + 11 + 19 + 11 + 11 + 19 + 27 + 19) = 8 \cdot 120 = 960$ grams.

7 The cube was partitioned into four boxes of dimension $2 \times 2 \times 4$. The diameter of such a box is $\sqrt{2^2 + 2^2 + 4^2} = \sqrt{24} = 2\sqrt{6}$; this is the mesh.

9 See the figure in the answer section of the text.

11 See the figure in the answer section of the text.

13 See the figure in the answer section of the text.

15 $$\int_0^1 \int_0^2 \int_0^x z\, dz\, dy\, dx = \int_0^1 \int_0^2 \frac{z^2}{2}\bigg|_0^x\, dy\, dx$$

$$= \int_0^1 \int_0^2 \frac{x^2}{2}\, dy\, dx = \int_0^1 \frac{1}{2}x^2 y \bigg|_{y=0}^{y=2} dx$$

$$= \int_0^1 x^2\, dx = \frac{x^3}{3}\bigg|_0^1 = \frac{1}{3}$$

17 $$\int_2^3 \int_x^{2x} \int_0^1 (x + z)\, dz\, dy\, dx$$

$$= \int_2^3 \int_x^{2x} \left(xz + \frac{z^2}{2}\right)\bigg|_{z=0}^{z=1} dy\, dx$$

$$= \int_2^3 \int_x^{2x} \left(x + \frac{1}{2}\right) dy\, dx = \int_2^3 \left(x + \frac{1}{2}\right) y \bigg|_{y=x}^{y=2x} dx$$

$$= \int_2^3 \left(x + \frac{1}{2}\right)(2x - x)\, dx = \int_2^3 \left(x^2 + \frac{1}{2}x\right) dx$$

$$= \left(\frac{x^3}{3} + \frac{x^2}{4}\right)\bigg|_2^3 = \left(9 + \frac{9}{4}\right) - \left(\frac{8}{3} + 1\right) = \frac{91}{12}$$

19 (a) $-a \le x \le a$, $-\sqrt{a^2 - x^2} \le y \le \sqrt{a^2 - x^2}$, $0 \le z \le h$

(b) $-a \le x \le a$, $0 \le z \le h$, $-\sqrt{a^2 - x^2} \le y \le \sqrt{a^2 - x^2}$

21 (a) The projection on the xy plane is the triangle with vertices (1, 1), (1, 0), and (2, 0). Its description is $1 \le x \le 2$, $0 \le y \le 2 - x$. The top plane of the tetrahedron is $z = 2$. The bottom plane contains the points (1, 1, 1), (1, 0, 2), and (2, 0, 2), so its equation is $y + z = 2$. Hence the tetrahedron is described by $1 \le x \le 2$, $0 \le y \le 2 - x$, $2 - y \le z \le 2$.

(b) The projection on the xz plane is the triangle with vertices (1, 1), (1, 2), and (2, 2). Its description is $1 \le x \le 2$, $x \le z \le 2$. The bounds on y are determined by two planes. The first passes through (1, 1, 1), (1, 0, 2), and (2, 0, 2); its equation is $y + z = 2$. The second passes through (1, 1, 1), (1, 1, 2), and (2, 0, 2); its equation is $x + y = 2$. The tetrahedron is described by $1 \le x \le 2$, $x \le z \le 2$, $2 - z \le y \le 2 - x$.

23 (a)

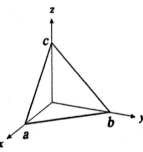

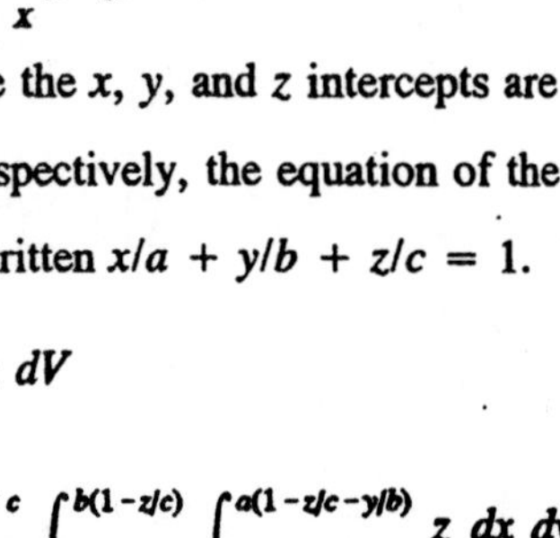

(b) Since the x, y, and z intercepts are a, b, and c, respectively, the equation of the plane may be written $x/a + y/b + z/c = 1$.

(c) $\int_R z\, dV$

$$= \int_0^c \int_0^{b(1-z/c)} \int_0^{a(1-z/c-y/b)} z\, dx\, dy\, dz$$

$$= \int_0^c \int_0^{b(1-z/c)} zx \Big|_0^{a(1-z/c-y/b)} dy\, dz$$

$$= \int_0^c \int_0^{b(1-z/c)} az\left(1 - \frac{z}{c} - \frac{y}{b}\right) dy\, dz$$

$$= \int_0^c az\left[\left(1 - \frac{z}{c}\right)y - \frac{y^2}{2b}\right]\Bigg|_0^{b(1-z/c)} dz$$

$$= \int_0^c az\left[b\left(1 - \frac{z}{c}\right) - \frac{b^2}{2b}\left(1 - \frac{z}{c}\right)^2\right] dz$$

$$= \int_0^c az\left[\left(b - \frac{b^2}{2b}\right)\left(1 - \frac{z}{c}\right)^2\right] dz$$

$$= \int_0^c \frac{ab}{2}\left[z\left(1 - \frac{z}{c}\right)^2\right] dz$$

$$= \frac{ab}{2}\int_0^c \left(z - \frac{2z^2}{c} + \frac{z^3}{c^2}\right) dz$$

$$= \frac{ab}{2}\left(\frac{z^2}{2} - \frac{2z^3}{3c} + \frac{z^4}{4c^2}\right)\Bigg|_0^c$$

$$= \frac{ab}{2}\left(\frac{c^2}{2} - \frac{2c^2}{3} + \frac{c^2}{4}\right) = \frac{abc^2}{24}$$

(d) The volume of the tetrahedron is $\frac{1}{3}$(Height)(Area of base) $= \frac{1}{3}c\left(\frac{1}{2}ab\right) = \frac{1}{6}abc$. Thus $\bar{z} = \dfrac{\int_R z\, dV}{\text{Volume of } R} = \dfrac{abc^2/24}{abc/6} = \dfrac{c}{4}$.

25 Place the cube in an xyz coordinate system so that the corner A is at the origin and x, y, and z all vary between 0 and 4. The density function $f(P)$ is the square of the distance of the point $P = (x, y, z)$ from A; that is, $f(P) = \overline{AP}^2 = x^2 + y^2 + z^2$. Thus

Mass $= \int_R f(P)\, dV$

$= \int_0^4 \int_0^4 \int_0^4 (x^2 + y^2 + z^2)\, dz\, dy\, dx$

$= \int_0^4 \int_0^4 \left(x^2z + y^2z + \dfrac{z^3}{3}\right)\Bigg|_{z=0}^{z=4} dy\, dx$

$= \int_0^4 \int_0^4 \left(4x^2 + 4y^2 + \dfrac{64}{3}\right) dy\, dx$

$= \int_0^4 \left(4x^2y + \dfrac{4y^3}{3} + \dfrac{64}{3}y\right)\Bigg|_0^4 dx$

$= \int_0^4 \left(16x^2 + \dfrac{512}{3}\right) dx = \left(\dfrac{16x^3}{3} + \dfrac{512x}{3}\right)\Bigg|_0^4$

$= \dfrac{1024}{3} + \dfrac{2048}{3} = \dfrac{3072}{3} = 1024$. The cube weighs 1024 grams.

27 The center of the cube is at $(a/2, a/2, a/2)$. The square of the distance from a point P_1 in the cube

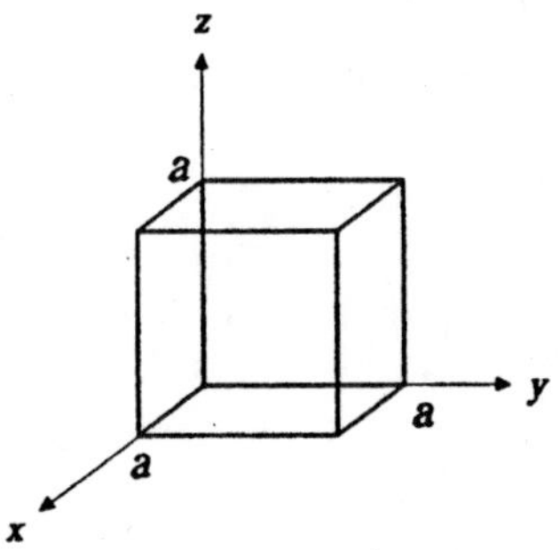

to the center is $(x - a/2)^2 + (y - a/2)^2 + (z - a/2)^2$. Therefore the average value of the squared distance is $\dfrac{1}{a^3} \int_0^a \int_0^a \int_0^a \left[\left(x - \dfrac{a}{2}\right)^2 + \left(y - \dfrac{a}{2}\right)^2 + \left(z - \dfrac{a}{2}\right)^2\right] dx\, dy\, dz =$

$\dfrac{1}{a^3} \int_0^a \int_0^a \int_0^a \left(x^2 + y^2 + z^2 - ax - ay - az + \dfrac{3a^2}{4}\right) dx\, dy\, dz = \dfrac{1}{a^3} \int_0^a \int_0^a \left(\dfrac{x^3}{3} + y^2x + z^2x - \dfrac{ax^2}{2} - ayx - azx + \dfrac{3a^2x}{4}\right)\Bigg|_0^a dy\, dz$

$= \dfrac{1}{a^3} \int_0^a \int_0^a \left(\dfrac{a^3}{3} + ay^2 + az^2 - \dfrac{a^3}{2} - a^2y - a^2z + \dfrac{3a^3}{4}\right) dy\, dx$

$= \dfrac{1}{a^2} \int_0^a \int_0^a \left(\dfrac{7a^2}{12} + y^2 + z^2 - ay - az\right) dy\, dx$

$= \dfrac{1}{a^2} \int_0^a \left(\dfrac{7a^2y}{12} + \dfrac{y^3}{3} + z^2y - \dfrac{ay^2}{2} - azy\right)\Bigg|_0^a dz$

$= \dfrac{1}{a^2} \int_0^1 \left(\dfrac{7}{12}a^3 + \dfrac{1}{3}a^3 + az^2 - \dfrac{1}{2}a^3 - a^2z\right) dz$

$= \dfrac{1}{a} \int_0^a \left(\dfrac{5}{12}a^2 + z^2 - az\right) dz$

$= \dfrac{1}{a}\left(\dfrac{5}{12}a^2z + \dfrac{1}{3}z^3 - \dfrac{1}{2}az^2\right)\Bigg|_0^a$

$= \dfrac{1}{a}\left(\dfrac{5}{12}a^3 + \dfrac{1}{3}a^3 - \dfrac{1}{2}a^3\right) = \dfrac{1}{4}a^2$

29 $\int_R xy\, dV = \int_0^1 \int_0^{1-x} \int_0^{1-x-y} xy\, dz\, dy\, dx$

$= \int_0^1 \int_0^{1-x} xy(1 - x - y)\, dy\, dx$

$= \int_0^1 \int_0^{1-x} (xy - x^2y - xy^2)\, dy\, dx$

$$= \int_0^1 \left(\frac{xy^2}{2} - \frac{x^2y^2}{2} - \frac{xy^3}{3} \right) \Bigg|_{y=0}^{y=1-x} dx$$

$$= \int_0^1 \left[\frac{1}{2}(x - x^2)(1 - x)^2 - \frac{x}{3}(1 - x)^3 \right] dx$$

$$= \int_0^1 \left(\frac{x}{2} - \frac{x}{3} \right)(1 - x)^3 \, dx = \int_0^1 \frac{x}{6}(1 - x)^3 \, dx$$

$$= \frac{1}{6} \int_0^1 (x - 3x^2 + 3x^3 - x^4) \, dx$$

$$= \frac{1}{6}\left(\frac{x^2}{2} - x^3 + \frac{3x^4}{4} - \frac{x^5}{5} \right) \Bigg|_0^1$$

$$= \frac{1}{6}\left(\frac{1}{2} - 1 + \frac{3}{4} - \frac{1}{5} \right) = \frac{1}{6} \cdot \frac{1}{20} = \frac{1}{120}$$

31 The top plane passes through (1, 1, 0), (0, 0, 2), and (1, 0, 0), so its equation is $z = 2 - 2x$. Hence

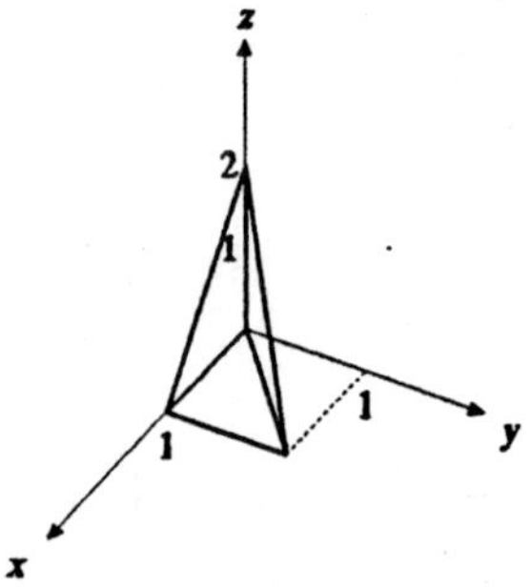

the region is given by $0 \le x \le 1$, $0 \le y \le x$, $0 \le z \le 2 - 2x$. The volume is $V =$

$$\int_0^1 \int_0^x \int_0^{2-2x} dz \, dy \, dx = \int_0^1 \int_0^x (2 - 2x) \, dy \, dx$$

$$= \int_0^1 (2x - 2x^2) \, dx = \left(x^2 - \frac{2}{3}x^3 \right) \Bigg|_0^1 = \frac{1}{3}.$$

Hence the average temperature is

$$\frac{1}{V} \int_0^1 \int_0^x \int_0^{2-2x} e^{-x-y-z} \, dz \, dy \, dx$$

$$= 3 \int_1^0 \int_0^x \left(-e^{-x-y-z} \right) \Big|_0^{2-2x} dy \, dx$$

$$= 3 \int_0^1 \int_0^x (e^{-x-y} - e^{x-2-y}) \, dy \, dx$$

$$= 3 \int_0^1 \left(-e^{-x-y} + e^{x-2-y} \right) \Big|_0^x dx$$

$$= 3 \int_0^1 (-e^{-2x} + e^{-2} + e^{-x} - e^{x-2}) \, dx$$

$$= 3\left(\frac{1}{2}e^{-2x} + e^{-2}x - e^{-x} - e^{x-2} \right) \Bigg|_0^1$$

$$= \frac{3}{2}(1 - 4e^{-1} + 5e^{-2}).$$

33 The box is homogeneous, so its density is the constant $\frac{M}{abc}$. Place the box in an xyz coordinate system so that its description is $0 \le x \le b$, $0 \le y \le c$, $0 \le z \le a$. The desired moment of inertia is

$$I_z = \int_R (x^2 + y^2)\, \sigma(P) \, dA$$

$$= \frac{M}{abc} \int_0^c \int_0^b \int_0^a (x^2 + y^2) \, dz \, dx \, dy$$

$$= \frac{M}{abc} \int_0^c \int_0^b (x^2 + y^2)z \Big|_{z=0}^{z=a} dx \, dy$$

$$= \frac{M}{bc} \int_0^c \int_0^b (x^2 + y^2) \, dx \, dy$$

$$= \frac{M}{bc} \int_0^c \left(\frac{x^3}{3} + xy^2 \right) \Bigg|_{y=0}^{y=b} dy$$

$$= \frac{M}{bc} \int_0^c \left(\frac{b^3}{3} + by^2 \right) dy = \frac{M}{c}\left(\frac{b^2}{3}y + \frac{y^3}{3} \right) \Bigg|_0^c$$

$$= \frac{M}{c}\left(\frac{b^2}{3}c + \frac{c^3}{3} \right) = \frac{M}{3}(b^2 + c^2).$$

35 (a) The greatest distance that a point of the cone can be from the cone's axis is a, the radius of

the cone's base. Suppose the entire mass M were concentrated at a point a distance a from the axis. The moment of inertia of the point mass would be (Mass)(Distance)^2 = Ma^2. Since most of the cone's mass is much closer to the axis, Ma^2 provides an upper bound on the cone's moment of inertia.

(b) Place the cone so that its vertex is at the origin and its axis coincides with the z axis, as

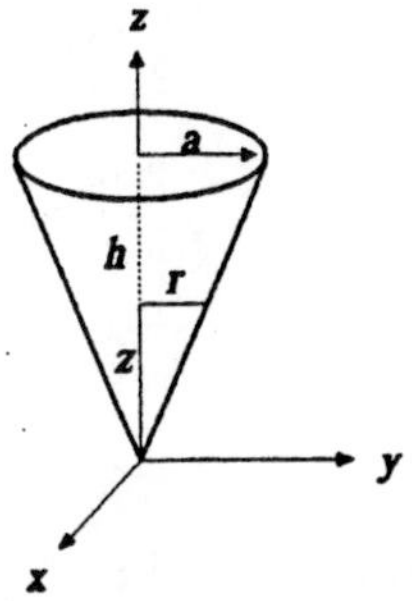

shown. A horizontal cross section of the cone produces a disk whose circumference has the equation $x^2 + y^2 = r^2$, where r depends on z. By similar triangles, we have $r/z = a/h$, so $r = \frac{a}{h}z$. The equation of the cone is therefore

$x^2 + y^2 = \frac{a^2}{h^2}z^2$, so the region R has

description $0 \le z \le h$, $-\frac{az}{h} \le x \le \frac{az}{h}$,

$-\sqrt{\frac{a^2z^2}{h^2} - x^2} \le y \le \sqrt{\frac{a^2z^2}{h^2} - x^2}$. The

volume of the cone is $\frac{1}{3}\pi a^2 h$ and the mass is

M, so the constant density is given by $\delta = \frac{3M}{\pi a^2 h}$. To simplify the calculations, we compute I_z for the region R', the quarter of the cone lying above the first quadrant, and multiply by 4: $\int_R (x^2 + y^2)\,\delta\,dV$

$$= 4\int_{R'} (x^2 + y^2)\,\delta\,dV =$$

$$4\delta \int_0^h \int_0^{az/h} \int_0^{\sqrt{a^2z^2/h^2 - x^2}} (x^2 + y^2)\,dy\,dx\,dz$$

$$= 4\delta \int_0^h \int_0^{az/h} \left(x^2y + \frac{y^3}{3}\right)\Bigg|_0^{\sqrt{a^2z^2/h^2 - x^2}} dx\,dz$$

$$= 4\delta \int_0^h \int_0^{az/h} \left(x^2 + \frac{1}{3}\left(\frac{a^2z^2}{h^2} - x^2\right)\right)\sqrt{\frac{a^2z^2}{h^2} - x^2}\,dx\,dz$$

$$= 4\delta \int_0^h \int_0^{az/h} \frac{1}{3}\left(\frac{a^2z^2}{h^2} + 2x^2\right)\sqrt{\frac{a^2z^2}{h^2} - x^2}\,dx\,dz;$$

now let $x = \frac{az}{h}u$, so that the inner integral becomes

$$\frac{1}{3}\int_0^1 \left(\frac{a^2z^2}{h^2} + 2\frac{a^2z^2u^2}{h^2}\right)\left(\frac{a^2z^2}{h^2} - \frac{a^2z^2u^2}{h^2}\right)^{1/2}\frac{az}{h}\,du$$

$$= \frac{1}{3}\left(\frac{az}{h}\right)^4 \int_0^1 (1 + 2u^2)(1 - u^2)^{1/2}\,du =$$

$$\frac{1}{3}\left(\frac{az}{h}\right)^4\left[\int_0^1 \sqrt{1 - u^2}\,du + 2\int_0^1 u^2\sqrt{1 - u^2}\,du\right] =$$

$$\frac{1}{3}\left(\frac{az}{h}\right)^4\left[\frac{\pi}{4} + 2\left(-\frac{u}{4}(1 - u^2)^{3/2} + \frac{u}{8}\sqrt{1 - u^2} + \frac{1}{8}\sin^{-1}u\right)\right]\Bigg|_0^1$$

$$= \frac{1}{3}\left(\frac{az}{h}\right)^4\left[\frac{\pi}{4} + 2\left(\frac{1}{8}\sin^{-1}1\right)\right]$$

$$= \frac{1}{3}\left(\frac{az}{h}\right)^4\left[\frac{\pi}{4} + \frac{\pi}{8}\right] = \frac{\pi}{8}\left(\frac{az}{h}\right)^4.$$ Plugging

this result back in, we have $4\delta \int_0^h \frac{\pi}{8}\left(\frac{az}{h}\right)^4 dz$

$$= 4\left(\frac{3M}{\pi a^2 h}\right)\cdot\frac{\pi}{8}\cdot\frac{a^4}{h^4}\int_0^h z^4\,dz = \frac{3Ma^2}{2h^5}\cdot\frac{z^5}{5}\Big|_0^h$$

$$= \frac{3Ma^2}{10}.$$

37 (a) Place the axes so that the line L is the line parallel to the x axis that passes through the point $(0, k, 0)$. Then $I_L =$

$\int_R ((y - k)^2 + z^2)\,\delta(P)\,dV =$

$\int_R (y^2 + z^2)\,\delta(P)\,dV - 2k\int_R y\,\delta(P)\,dV +$

$k^2\int_R \delta(P)\,dV = I - 2kM_{xz} + k^2M$

$= I - 2k\cdot 0 + k^2M = I + k^2M.$

(b) Since $I_L = I + k^2M$, I_L is the least when $k = 0$, that is, when L is the x axis.

39 Total work $= \int_M h(P)g(P)\,dV$, where M is the mountain.

15.6 Computing $\int_R f(P)\,dV$ using Cylindrical Coordinates

1

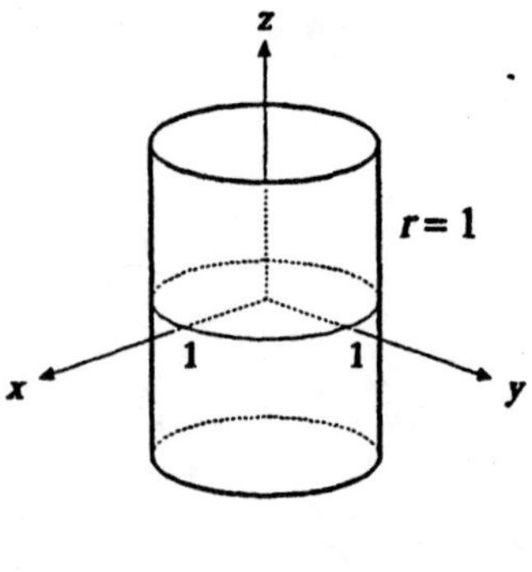

3

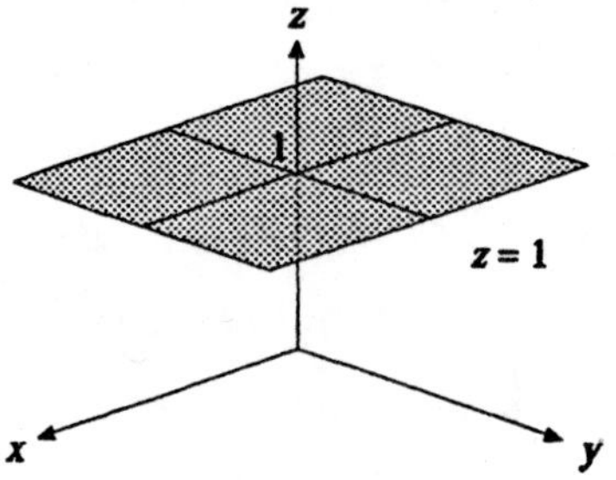

5

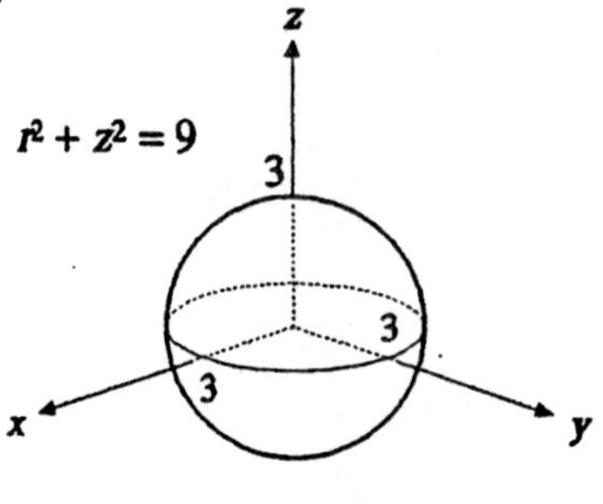

7 The region is a ball, so the cross section is a half-disk in the θ direction.

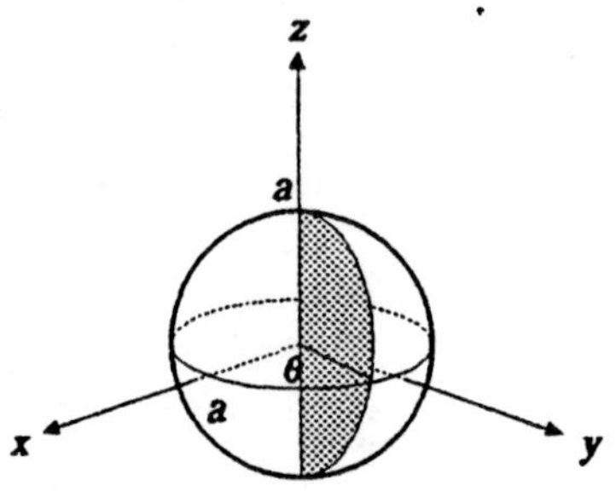

9 The equation in cylindrical coordinates of a sphere of radius a is $r^2 + z^2 = a^2$. Therefore $z = \pm\sqrt{a^2 - r^2}$, where the plus sign gives us the upper hemisphere, and the minus sign corresponds to the lower hemisphere. The description of the solid ball is thus $0 \le \theta \le 2\pi$, $0 \le r \le a$, $-\sqrt{a^2 - r^2} \le z \le \sqrt{a^2 - r^2}$.

11 (a) In cylindrical coordinates, r is the distance to the z axis, so $r = \sqrt{x^2 + y^2}$. Since (x, y) is

assumed to lie in the first quadrant, we have $\tan \theta = y/x$, so $\theta = \tan^{-1}(y/x)$. The z coordinate in cylindrical coordinates is the same as the z coordinate in rectangular coordinates.

(b) As in polar coordinates, $x = r \cos \theta$ and $y = r \sin \theta$; the z coordinate is the same.

13 (a) The equation of the xy plane is the same in cylindrical coordinates as in rectangular: $z = 0$.

(b) The plane is described by the equation $\theta = \pi/4$ with the restriction $r \geq 0$.

(c) The equation $r = 0$ is satisfied by all points at distance zero from the z axis; this is just the z axis itself, as required.

15 $0 \leq \theta \leq 2\pi,\ 0 \leq r \leq 1,\ r \leq z \leq 1$

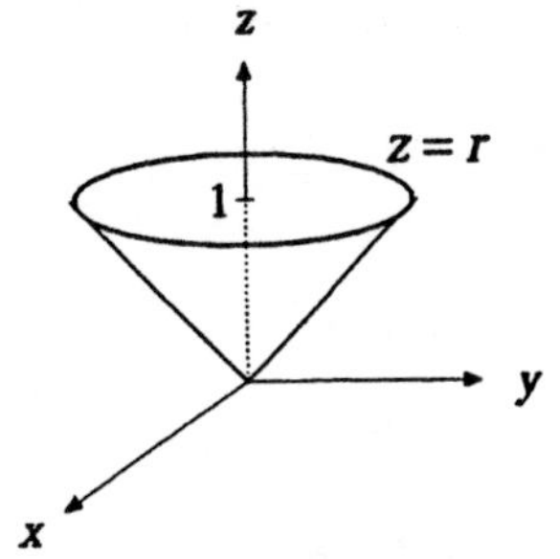

17 $0 \leq \theta \leq \pi/2,\ 0 \leq r \leq 1,\ 0 \leq z \leq r \cos \theta$

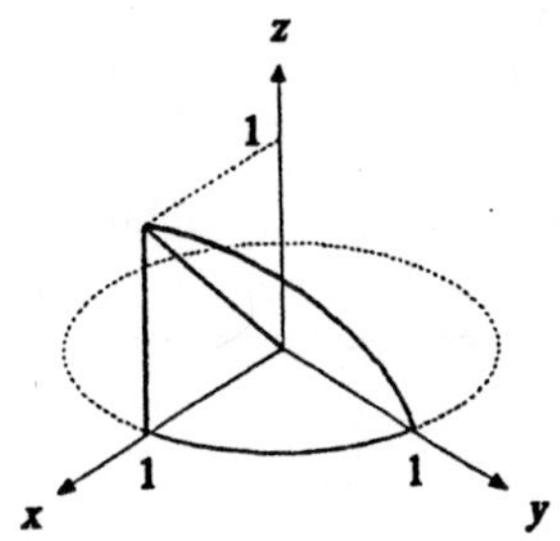

19 $0 \leq \theta \leq \pi/2,\ 1 \leq r \leq 1 + \sin \theta,\ r \leq z \leq 2$

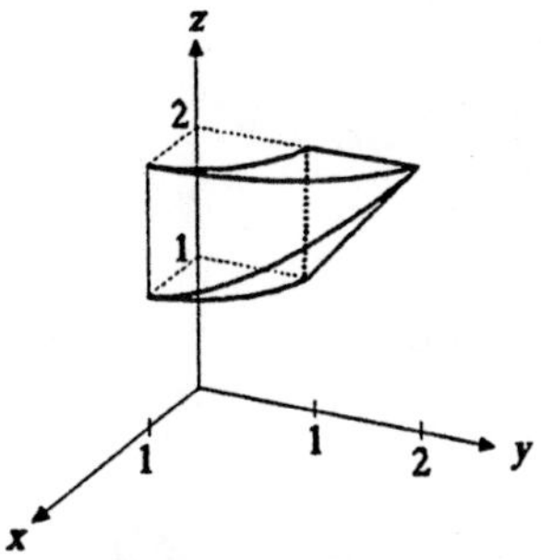

21 The disk is described by $0 \leq \theta \leq \pi$, $0 \leq r \leq 2 \sin \theta$. Hence R is described by $0 \leq \theta \leq \pi$, $0 \leq r \leq 2 \sin \theta$, $0 \leq z \leq r^2$.

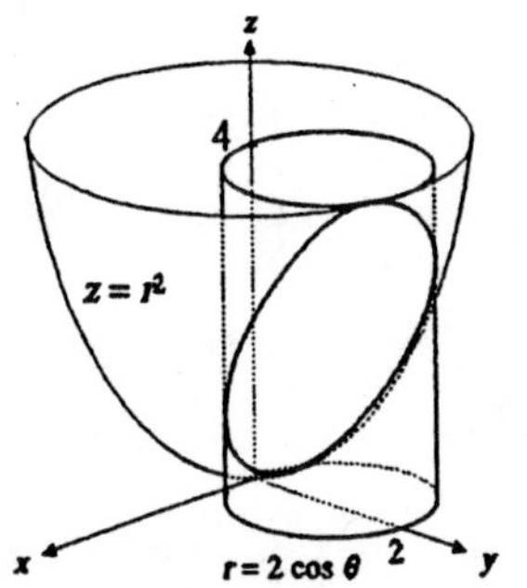

23 The projection on the xy plane is the square $0 \leq x \leq 1$, $0 \leq y \leq 1$. In polar coordinates, this square

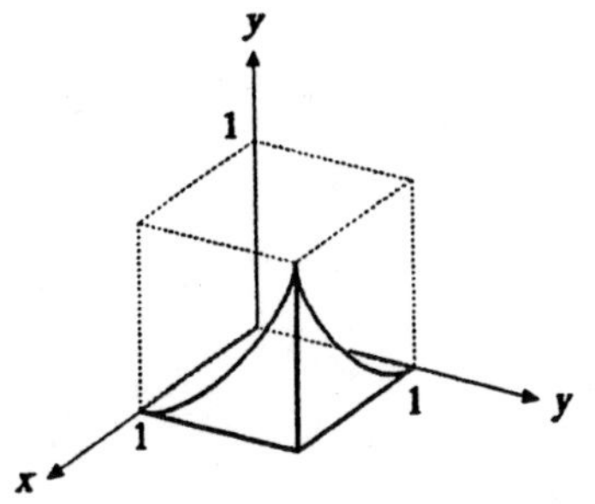

is described by $0 \leq \theta \leq \pi/4$, $0 \leq r \leq \sec \theta$ and $\pi/4 \leq \theta \leq \pi/2$, $0 \leq r \leq \csc \theta$. For each point in the square, z ranges from 0 to xy. Since $xy = (r \cos \theta)(r \sin \theta) = r^2 \sin \theta \cos \theta$, the region R is described by $0 \leq \theta \leq \pi/4$, $0 \leq r \leq \sec \theta$, $0 \leq z \leq r^2 \sin \theta \cos \theta$ and $\pi/4 \leq \theta \leq \pi/2$, $0 \leq r \leq \csc \theta$, $0 \leq z \leq r^2 \sin \theta \cos \theta$.

25 $\int_0^{2\pi} \int_0^1 \int_r^1 zr^3 \cos^2\theta \, dz \, dr \, d\theta$

$= \int_0^{2\pi} \int_0^1 r^3 \cos^2\theta \left.\frac{z^2}{2}\right|_r^1 dr \, d\theta$

$= \frac{1}{2} \int_0^{2\pi} \int_0^1 \cos^2\theta \, (r^3 - r^5) \, dr \, d\theta$

$= \frac{1}{2} \int_0^{2\pi} \cos^2\theta \left.\left(\frac{r^4}{4} - \frac{r^6}{6}\right)\right|_0^1 d\theta$

$= \frac{1}{24} \int_0^{2\pi} \cos^2\theta \, d\theta = \frac{1}{24}\cdot 4 \cdot \frac{1}{2} \cdot \frac{\pi}{2} = \frac{\pi}{24}$

27 The sphere $x^2 + y^2 + z^2 = 1$ has the equation $r^2 + z^2 = 1$ in cylindrical coordinates. The cone $z = \sqrt{x^2 + y^2}$ similarly simplifies to $z = r$. Observe that the sphere and cone intersect when $r^2 + r^2 = 1$; therefore, $r = 1/\sqrt{2}$. Hence the solid region R between the cone and the sphere can be described by $0 \le \theta \le 2\pi$, $0 \le r \le 1/\sqrt{2}$, $r \le z \le \sqrt{1 - r^2}$. The density is equal to z, so Mass $=$

$\int_R z \, dV = \int_0^{2\pi} \int_0^{1/\sqrt{2}} \int_r^{\sqrt{1-r^2}} zr \, dz \, dr \, d\theta$

$= \int_0^{2\pi} \int_0^{1/\sqrt{2}} \left.\frac{z^2 r}{2}\right|_r^{\sqrt{1-r^2}} dr \, d\theta$

$= \frac{1}{2} \int_0^{2\pi} \int_0^{1/\sqrt{2}} [(1 - r^2)r - r^3] \, dr \, d\theta$

$= \frac{1}{2} \int_0^{2\pi} \int_0^{1/\sqrt{2}} (r - 2r^3) \, dr \, d\theta$

$= \frac{1}{2} \int_0^{2\pi} \left.\left(\frac{r^2}{2} - \frac{r^4}{2}\right)\right|_0^{1/\sqrt{2}} d\theta$

$= \frac{1}{2} \int_0^{2\pi} \left(\frac{1}{4} - \frac{1}{8}\right) d\theta = \frac{1}{2}\cdot\frac{1}{8}\cdot 2\pi = \frac{\pi}{8}.$

29 First describe the region in cylindrical coordinates. Clearly $0 \le \theta \le 2\pi$. Also from the description 0

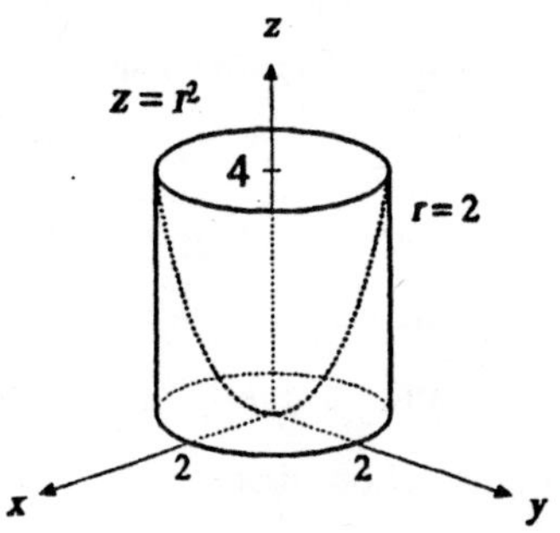

$\le r \le 2$. Now the region is "capped" on top by the circle $z = x^2 + y^2 = r^2$. Thus $0 \le z \le r^2$, and $\bar{z}$

$= \dfrac{\int_0^{2\pi} \int_0^2 \int_0^{r^2} rz \, dz \, dr \, d\theta}{\int_0^{2\pi} \int_0^2 \int_0^{r^2} r \, dz \, dr \, \theta} =$

$\dfrac{\frac{1}{2} \int_0^{2\pi} \int_0^2 rz^2\Big|_0^{r^2} dr \, d\theta}{\int_0^{2\pi} \int_0^2 rz\Big|_0^{r^2} dr \, d\theta} = \dfrac{\frac{1}{2} \int_0^{2\pi} \int_0^2 r^5 \, dr \, d\theta}{\int_0^{2\pi} \int_0^2 r^3 \, dr \, d\theta}$

$= \dfrac{\frac{1}{12} \int_0^{2\pi} r^6\Big|_0^2 d\theta}{\frac{1}{4} \int_0^{2\pi} r^4\Big|_0^2 d\theta} = \dfrac{\frac{64}{12} \int_0^{2\pi} d\theta}{\frac{16}{4} \int_0^{2\pi} d\theta} = \frac{4}{3}.$

31 To simplify calculations, choose axes so that the z axis is the axis of the cylinder and its base is in the xy plane. Then $I = \int_R r^2 \sigma(P) \, dV$

$= \frac{M}{\pi a^2 h} \int_0^{2\pi} \int_0^a \int_0^h r^3 \, dz \, dr \, d\theta$

$= \frac{M}{\pi a^2 h} \int_0^{2\pi} \int_0^a z\Big|_0^h r^3 \, dr \, d\theta$

$$= \frac{M}{4\pi a^2}\int_0^{2\pi} r^4\Big|_0^a\, d\theta = \frac{Ma^2}{4\pi}\int_0^{2\pi} d\theta = \frac{Ma^2}{2}.$$

33 The trick, as in most moment of inertia calculations, is to choose coordinate axes wisely. In this case choose them as in Exercise 31. Then the distance from any point in R to the axis of rotation (in this case the y axis) is $\sqrt{z^2 + x^2}$. Thus $I =$

$$\frac{M}{\pi a^2 h}\int_0^{2\pi}\int_0^a\int_0^h (z^2 + x^2)\, r\, dz\, dr\, d\theta$$

$$= \frac{M}{\pi a^2 h}\int_0^{2\pi}\int_0^a\int_0^h (z^2 + r^2\cos^2\theta)\, r\, dz\, dr\, d\theta$$

$$= \frac{M}{\pi a^2 h}\int_0^{2\pi}\int_0^a \left(\frac{z^3 r}{3} + zr^3\cos^2\theta\right)\Bigg|_{z=0}^{z=h} dr\, d\theta$$

$$= \frac{M}{\pi a^2 h}\int_0^{2\pi}\int_0^a \left(\frac{h^3 r}{3} + hr^3\cos^2\theta\right) dr\, d\theta$$

$$= \frac{M}{\pi a^2}\int_0^{2\pi}\left(\frac{h^2r^2}{6} + \frac{r^4\cos^2\theta}{4}\right)\Bigg|_{r=0}^{r=a} d\theta$$

$$= \frac{M}{\pi a^2}\int_0^{2\pi}\left(\frac{h^2a^2}{6} + \frac{a^4\cos^2\theta}{4}\right) d\theta$$

$$= \frac{M}{\pi}\left(\frac{h^2\theta}{6} + \frac{a^2}{4}\left(\frac{\theta}{2} + \frac{\sin 2\theta}{4}\right)\right)\Bigg|_0^{2\pi}$$

$$= \frac{M}{\pi}\left(\frac{h^2\pi}{3} + \frac{a^2\pi}{4}\right) = M\left(\frac{h^2}{3} + \frac{a^2}{4}\right).$$

35 Example 3 on page 920 provides a description of the region R. Let the z axis coincide with the diameter about which the moment of inertia is to be calculated. Since the ball is homogeneous $\sigma(P) = \frac{\text{Mass}}{\text{Volume}} = \frac{3M}{4\pi a^3}$. The distance from any point in the sphere to the z axis is simply r. Then $I_z =$

$$\int_R r^2\sigma(P)\, dV$$

$$= \frac{3M}{4\pi a^3}\int_0^{2\pi}\int_0^a\int_{-\sqrt{a^2-r^2}}^{\sqrt{a^2-r^2}} r^3\, dz\, dr\, d\theta$$

$$= \frac{3M}{4\pi a^3}\int_0^{2\pi}\int_0^a r^3 z\Bigg|_{-\sqrt{a^2-r^2}}^{\sqrt{a^2-r^2}} dr\, d\theta$$

$$= \frac{3M}{4\pi a^3}\int_0^{2\pi}\int_0^a 2r^3\sqrt{a^2 - r^2}\, dr\, d\theta$$

$$= \frac{3M}{2\pi a^3}\int_0^{2\pi}\left[-\left(\frac{r^2}{5} + \frac{2a^2}{15}\right)(a^2 - r^2)^{3/2}\right]\Bigg|_0^a d\theta$$

$$= \frac{3M}{2\pi a^3}\int_0^{2\pi}\frac{2a^5}{15}\, d\theta = \frac{3Ma^2}{15\pi}(2\pi) = \frac{2Ma^2}{5}.$$

37 By Exercise 35, the moment of inertia of the ball about a line through its center of mass is $I_1 = \frac{2Ma^2}{5}$. By Exercise 37 of Sec. 15.5, the moment of inertia about a parallel line displaced by a distance k is $I_2 = I_1 + k^2M$, where M is the mass. A line tangent to the ball is a distance a from a line through the center, so $I_2 = \frac{2Ma^2}{5} + a^2M =$

$$\frac{7Ma^2}{5}.$$

39 Call the center of mass of the solid C. Let C be the origin of a rectangular coordinate system with the line L_1 being the x axis. Now L_2 is a line parallel to L_1 and a distance r from it. This line intersects the yz plane at the point $P = (0, y_0, z_0)$. The distance

from the element of mass in the solid with coordinates (x, y, z) to this line is $\sqrt{(y - y_0)^2 + (z - z_0)^2}$. Therefore $I_2 =$

$\int_R \left[(y - y_0)^2 + (z - z_0)^2\right] \delta(P)\, dV =$

$\int_R (y^2 + z^2)\, \delta(P)\, dV - 2y_0 \int_R y\, \delta(P)\, dV -$

$2z_0 \int_R z\, \delta(P)\, dV + \left(y_0^2 + z_0^2\right) \int_R \delta(P)\, dV$. The first integral is I_1, the moment of inertia about L_1, since the distance from the mass element to L_1 is $\sqrt{y^2 + z^2}$. So we have (using the definition of the center of mass) $I_2 =$

$I_1 - 2y_0 M\bar{y} - 2z_0 M\bar{z} + r^2 M$. Since the center of mass is at (0, 0, 0), $I_2 = I_1 + r^2 M$.

15.7 Computing $\int_R f(P)\, dV$ using Spherical Coordinates

1 (a) Rectangular coordinates describe a point by specifying three planes on which it lies.

(b) Spherical coordinates describe a point by specifying a sphere, a half-plane, and a cone on which the point lies.

(c) Cylindrical coordinates describe a point by specifying a cylinder, a half-plane, and a plane on which it lies.

3 See the figure in the answer section of the text.

5 See the figure in the answer section of the text.

7 See the figure in the answer section of the text.

9 From the figure we see that $z = \rho \cos \phi$ and $r = \rho \sin \phi$. The angle θ is the same in cylindrical coordinates as it is in spherical.

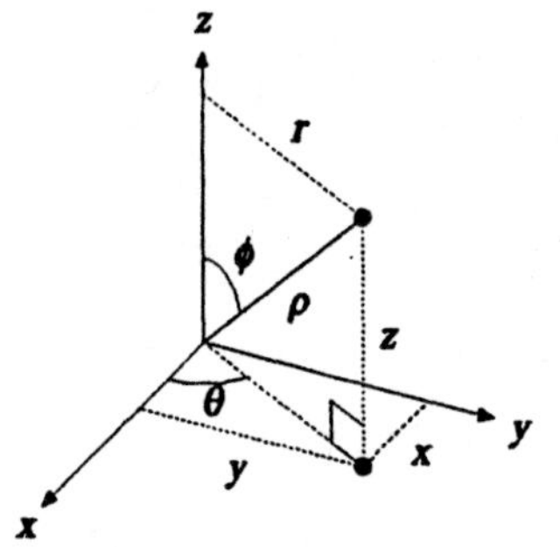

11 A sphere of radius a has equation $\rho = a$ in spherical coordinates. The ball of radius a has the description $0 \le \rho \le a$, $0 \le \phi \le \pi$, $0 \le \theta \le 2\pi$.

13 All three variables are bounded by constant limits: $0 \le \theta \le 2\pi$, $0 \le \phi \le \pi/6$, $0 \le \rho \le a$.

15 While θ and ϕ are as before, ρ now varies between 0 and the plane $z = \frac{3a}{5}$. Since $z = \rho \cos \phi$, we have $\rho \cos \phi = \frac{3a}{5}$ or $\rho = \frac{3a}{5} \sec \phi$. Thus the description of the region is $0 \le \theta \le 2\pi$, $0 \le \phi \le \pi/6$, $0 \le \rho \le \frac{3a}{5} \sec \phi$.

17 The region has the shape of a hollowed-out cantaloupe half. See the figure in the answer section of the text.

19 See the figure in the answer section of the text.

21 See the graph accompanying Exercise 31. By the Pythagorean theorem, $\rho^2 = r^2 + z^2$, so $\rho = \sqrt{r^2 + z^2}$. The angle θ is the same in both coordinate systems, while $\cos \phi = z/\rho$, so $\phi =$

$\cos^{-1} \frac{z}{\rho} = \cos^{-1} \frac{z}{\sqrt{r^2 + z^2}}$.

23 See the figure in the answer section of the text.

25 $\int_R z\, dV =$

$$\int_0^{2\pi}\int_0^{\tan^{-1}a/h}\int_0^{h/\cos\phi} \rho\cos\phi\,(\rho^2\sin\phi)\,d\rho\,d\phi\,d\theta$$

$$= \int_0^{2\pi}\int_0^{\tan^{-1}a/h}\int_0^{h/\cos\phi} \cos\phi\sin\phi\,\rho^3\,d\rho\,d\phi\,d\theta$$

$$= \int_0^{2\pi}\int_0^{\tan^{-1}a/h} \cos\phi\sin\phi\,\frac{\rho^4}{4}\bigg|_0^{h/\cos\phi} d\phi\,d\theta$$

$$= \int_0^{2\pi}\int_0^{\tan^{-1}a/h} \frac{h^4\sin\phi}{4\cos^3\phi}\,d\phi\,d\theta$$

$$= \frac{h^4}{4}\int_0^{2\pi} \frac{1}{2\cos^2\phi}\bigg|_0^{\tan^{-1}a/h} d\theta$$

$$= \frac{h^4}{8}\int_0^{2\pi}\left(\frac{1}{\cos^2(\tan^{-1}a/h)} - \frac{1}{2\cos^2 0}\right)d\theta$$

$$= \frac{h^4}{8}\int_0^{2\pi}\left(\frac{a^2+h^2}{h^2} - 1\right)d\theta$$

$$= \frac{h^4}{8}\left(\frac{a^2+h^2}{h^2} - 1\right)(2\pi) = \left(\frac{a^2h^2}{8}\right)(2\pi)$$

$$= \frac{\pi a^2h^2}{4}$$

27 The region is described by $0 \le \theta \le 2\pi$, $0 \le r \le 3$, $0 \le z \le 9 - r^2$. Its volume is $V =$

$$\int_0^{2\pi}\int_0^3\int_0^{9-r^2} r\,dz\,dr\,d\theta = 2\pi\int_0^3\int_0^{9-r^2} r\,dz\,dr$$

$$= 2\pi\int_0^3 (9r - r^3)\,dr = 2\pi\left(\frac{9}{2}r^2 - \frac{1}{4}r^4\right)\bigg|_0^3$$

$= 2\pi\cdot\frac{81}{4} = \frac{81\pi}{2}$. (Note: The outer integral was done first to simplify the expression. This was valid only because no inner integrals depended on θ).

29 Since in spherical coordinates $dV = \rho^2\sin\phi\,d\rho\,d\phi\,d\theta$, $f(P) = \rho\sin^2\theta$

31 (a) Observe that $x^2 + y^2 + z^2 = 1$ and $z = \sqrt{x^2+y^2}$ intersect when $x^2 + y^2 + (x^2 + y^2) = 1$, so $x^2 + y^2 = \frac{1}{2}$. Hence R can be described by $-\frac{1}{\sqrt2} \le x \le \frac{1}{\sqrt2}$, $-\sqrt{\frac{1}{2} - x^2} \le y \le \sqrt{\frac{1}{2} - x^2}$, $\sqrt{x^2+y^2} \le z \le \sqrt{1 - x^2 - y^2}$. The density is equal to z, so we have Mass $= \int_R z\,dV =$

$$\int_{-1/\sqrt2}^{1/\sqrt2}\int_{-\sqrt{1/2-x^2}}^{\sqrt{1/2-x^2}}\int_{\sqrt{x^2+y^2}}^{\sqrt{1-x^2-y^2}} z\,dz\,dy\,dx.$$

(b) See Exercise 27 of Sec. 15.6.

(c) In spherical coordinates, the sphere has the equation $\rho = 1$ and the cone is given by $\phi = \pi/4$. Hence the region R has the description $0 \le \theta \le 2\pi$, $0 \le \phi \le \pi/4$, $0 \le \rho \le 1$. The density is $z = \rho\cos\phi$, so we have Mass $=$

$$\int_0^{2\pi}\int_0^{\pi/4}\int_0^1 \rho\cos\phi\,\rho^2\sin\phi\,d\rho\,d\phi\,d\theta =$$

$$\int_0^{2\pi}\int_0^{\pi/4}\int_0^1 \rho^3\sin\phi\cos\phi\,d\rho\,d\phi\,d\theta.$$

(d) $\int_0^{2\pi}\int_0^{\pi/4}\int_0^1 \rho^3\sin\phi\cos\phi\,d\rho\,d\phi\,d\theta$

$$= \int_0^{2\pi}\int_0^{\pi/4} \frac{1}{4}\sin\phi\cos\phi\,d\phi\,d\theta$$

$$= \int_0^{2\pi}\frac{1}{8}\sin^2\phi\bigg|_0^{\pi/4} d\theta = \int_0^{2\pi}\frac{1}{16}\,d\theta$$

$$= \frac{2\pi}{16} = \frac{\pi}{8}$$

33 Place the ball with its center at the origin. The function to be averaged over the region is the distance from the origin. In rectangular coordinates this is $\sqrt{x^2 + y^2 + z^2}$; in cylindrical, $\sqrt{r^2 + z^2}$; in spherical, ρ. The volume of the ball is $V = \frac{4}{3}\pi a^3$, so $\frac{1}{V}\int_R f(P)\, dV =$

$$\frac{3}{4\pi a^3}\int_{-a}^{a}\int_{-\sqrt{a^2-x^2}}^{\sqrt{a^2-x^2}}\int_{-\sqrt{a^2-x^2-y^2}}^{\sqrt{a^2-x^2-y^2}} \sqrt{x^2+y^2+z^2}\, dz\, dy\, dx$$

$$= \frac{3}{4\pi a^3}\int_0^{2\pi}\int_{-a}^{a}\int_0^{\sqrt{a^2-z^2}} \sqrt{r^2 + z^2}\, r\, dr\, dz\, d\theta$$

$$= \frac{3}{4\pi a^3}\int_0^{2\pi}\int_0^{\pi}\int_0^{a} \rho\cdot\rho^2 \sin\phi\, d\rho\, d\phi\, d\theta.$$ The integrals in cylindrical and spherical coordinates are easier to compute. Using cylindrical coordinates, we have

$$\frac{6}{4\pi a^2}\int_0^{2\pi}\int_0^{a} \frac{1}{2}\cdot\frac{(r^2 + z^2)^{3/2}}{3/2}\Bigg|_{r=0}^{r=\sqrt{a^2-z^2}} dz\, d\theta$$

$$= \frac{1}{2\pi a^3}\int_0^{2\pi}\int_0^{a} (a^3 - z^3)\, dz\, d\theta$$

$$= \frac{1}{2\pi a^3}\int_0^{2\pi}\left(a^3 z - \frac{z^4}{4}\right)\Bigg|_0^a d\theta$$

$$= \frac{1}{2\pi a^3}\cdot\frac{3a^4}{4}\int_0^{2\pi} d\theta = \frac{3a}{8\pi}\cdot 2\pi = \frac{3}{4}a.$$ (Note that we chose to integrate z from 0 to a and double the result. Had we tried to go from $-a$ to a, we would have run into trouble when simplifying $(0 + z^2)^{3/2} = (z^2)^{3/2} = |z|^3$. It would not have been correct to write z^3 because z^3 is not always positive. Using symmetry is simpler.) In spherical coordinates, we have

$$\frac{3}{4\pi a^3}\int_0^{2\pi}\int_0^{\pi}\int_0^{a} \rho^3 \sin\phi\, d\rho\, d\phi\, d\theta$$

$$= \frac{3}{4\pi a^3}\cdot\frac{a^4}{4}\int_0^{2\pi}\int_0^{\pi} \sin\phi\, d\phi\, d\theta$$

$$= \frac{3a}{16\pi}\int_0^{2\pi} (-\cos\phi)\Big|_0^{\pi} d\theta = \frac{3a}{16\pi}\cdot 2\int_0^{2\pi} d\theta$$

$$= \frac{3a}{8\pi}\cdot 2\pi = \frac{3}{4}a.$$

35 The region is bounded above by a sphere of radius a and below by a cone of vertex angle $\pi/6$. The

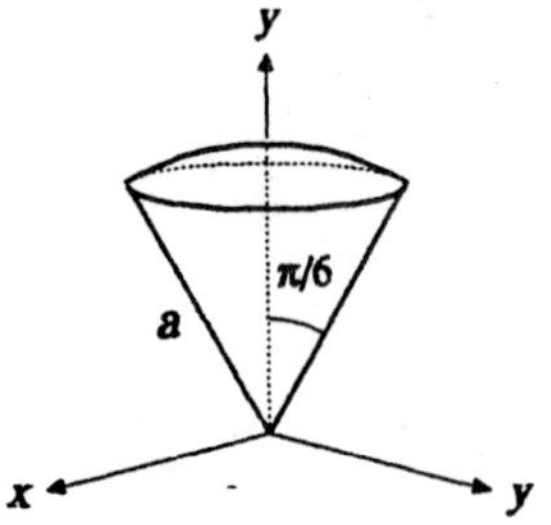

equation of the sphere is simply $x^2 + y^2 + z^2 = a^2$, $r^2 + z^2 = a^2$, or $\rho = a$. In spherical coordinates, the equation of the cone is just $\phi = \pi/6$. Since $r/z = \tan\phi$, we have $r = z \tan \pi/6 = z\cdot\frac{1}{\sqrt{3}}$ or $z = \sqrt{3}r$. It follows that in rectangular coordinates we have $z = \sqrt{3}\sqrt{x^2 + y^2}$. Note that r attains its greatest value at the intersection of the sphere and the cone; that is, when $\rho = a$ and $\phi = \pi/6$. Since $r = \rho \sin\phi$, we have $r = a \sin \pi/6 = a\cdot 1/2 = a/2$. Therefore r will vary from 0 to $a/2$. It also follows that x and y will vary within the circle

$x^2 + y^2 = a^2/4$. Thus $\int_R z\,dV$

$$= \int_{-a/2}^{a/2} \int_{-\sqrt{a^2/4-x^2}}^{\sqrt{a^2/4-x^2}} \int_{\sqrt{3}\sqrt{x^2+y^2}}^{\sqrt{a^2-x^2-y^2}} z\,dz\,dy\,dx$$

$$= \int_0^{2\pi} \int_0^{a/2} \int_{\sqrt{3}r}^{\sqrt{a^2-r^2}} zr\,dz\,dr\,d\theta$$

$$= \int_0^{2\pi} \int_0^{\pi/6} \int_0^a \rho\cos\phi\,\rho^2\sin\phi\,d\rho\,d\phi\,d\theta.$$

Cylindrical and spherical are both straightforward:

$$\int_0^{2\pi} \int_0^{a/2} \frac{z^2}{2}\bigg|_{\sqrt{3}r}^{\sqrt{a^2-r^2}} r\,dr\,d\theta$$

$$= \frac{1}{2}\int_0^{2\pi} \int_0^{a/2} (a^2 - r^2 - 3r^2)\,r\,dr\,d\theta$$

$$= \frac{1}{2}\int_0^{2\pi} \int_0^{a/2} (a^2 r - 4r^3)\,dr\,d\theta$$

$$= \frac{1}{2}\int_0^{2\pi} \left(\frac{a^2r^2}{2} - r^4\right)\bigg|_0^{a/2} d\theta$$

$$= \frac{1}{2}\left(\frac{a^4}{8} - \frac{a^4}{16}\right)\int_0^{2\pi} d\theta = \frac{a^4}{32}\cdot 2\pi = \frac{\pi a^4}{16} \text{ and}$$

$$\int_0^{2\pi} \int_0^{\pi/6} \int_0^a \rho^3\cos\phi\sin\phi\,d\rho\,d\phi\,d\theta$$

$$= \int_0^{2\pi} \int_0^{\pi/6} \frac{a^4}{4}\cos\phi\sin\phi\,d\phi\,d\theta$$

$$= \frac{a^4}{4}\int_0^{2\pi} \frac{\sin^2\phi}{2}\bigg|_0^{\pi/6} d\theta = \frac{a^4}{4}\cdot\frac{1}{8}\cdot\theta\bigg|_0^{2\pi}$$

$$= \frac{a^4}{32}\cdot 2\pi = \frac{\pi a^4}{16}.$$

37 (a) Consider the hemisphere whose description is $0 \le \theta \le \pi$, $-a \le z \le a$, $0 \le r \le \sqrt{a^2 - z^2}$. (It is just the half of the ball of radius a for which $y \le 0$.) I_z is then the moment of inertia of the hemisphere about a diameter of its circular base. The mass of the hemisphere is M and its volume is $\frac{2}{3}\pi a^3$, so the constant density is $\delta = \frac{3M}{2\pi a^3}$. Thus $I_z =$

$$\delta \int_0^{\pi} \int_{-a}^{a} \int_0^{\sqrt{a^2-z^2}} r^2\cdot r\,dr\,dz\,d\theta$$

$$= \delta\pi \int_{-a}^{a} \frac{r^4}{4}\bigg|_0^{\sqrt{a^2-z^2}} dz$$

$$= \frac{\delta\pi}{4}\int_{-a}^{a} (a^2 - z^2)^2\,dz$$

$$= \frac{3M}{2\pi a^3}\cdot\frac{\pi}{4}\int_{-a}^{a} (a^4 - 2a^2z^2 + z^4)\,dz$$

$$= \frac{3M}{8a^3}\cdot 2\left(a^4 z - \frac{2}{3}a^2z^3 + \frac{z^5}{5}\right)\bigg|_0^a$$

$$= \frac{3M}{4a^3}\left(a^5 - \frac{2}{3}a^5 + \frac{1}{5}a^5\right) = \frac{2Ma^2}{5}.$$

(b) The hemisphere is placed as in (a). Its description is $0 \le \theta \le \pi$, $0 \le \phi \le \pi$, $0 \le \rho \le a$. The integrand r^2 is equal to $\rho^2\sin^2\phi$, so $I_z =$

$$\delta \int_0^{\pi} \int_0^{\pi} \int_0^a \rho^2\sin^2\phi\,\rho^2\sin\phi\,d\rho\,d\phi\,d\theta$$

$$= \frac{\delta\pi a^5}{5}\int_0^{\pi} \sin^3\phi\,d\phi =$$

$$\frac{3M}{2\pi a^3}\cdot\pi\cdot\frac{a^5}{5}\left(2\cdot\frac{2}{3}\right) = \frac{3Ma^2}{10}\cdot\frac{4}{3} = \frac{2Ma^2}{5},$$ in agreement with (a).

39 By Example 5, $q = \sqrt{H^2 + \rho^2 - 2\rho H\cos\phi}$, so the average of q is

$$\frac{3}{4\pi a^3}\int_0^{2\pi}\int_0^a\int_0^\pi \sqrt{H^2+\rho^2-2\rho H\cos\phi}\;\rho^2\sin\phi\; d\phi\; d\rho\; d\theta$$

$$=\frac{3}{4\pi a^3}\cdot 2\pi\int_0^a\int_0^\pi \frac{\rho}{2H}\sqrt{H^2+\rho^2-2\rho H\cos\phi}\;2\rho H\sin\phi\; d\phi\; d\rho$$

$$=\frac{3}{4a^3H}\int_0^a \rho\cdot\frac{2}{3}(H^2+\rho^2-2\rho H\cos\phi)^{3/2}\Big|_0^\pi\; d\rho =$$

$$\frac{1}{2a^3H}\int_0^a \rho[H^2+\rho^2+2\rho H)^{3/2}-(H^2+\rho^2-2\rho H)^{3/2}]\; d\rho$$

$$=\frac{1}{2a^3H}\int_0^a \rho[(H+\rho)^3-(H-\rho)^3]\; d\rho$$

$$=\frac{1}{2a^3H}\int_0^a \rho[6H^2\rho+2\rho^3]\; d\rho$$

$$=\frac{1}{2a^3H}\left[2H^2\rho^3+\frac{2\rho^5}{5}\right]\Bigg|_0^a$$

$$=\frac{1}{2a^3H}\left[2H^2a^3+\frac{2a^5}{5}\right]=H+\frac{a^2}{5H}.$$

41 (a) By symmetry, $\int_R x^2\, dV = \int_R y^2\, dV$

$= \int_R z^2\, dV$, so $\int_R (x^2+y^2+z^2)\, dV$

$= 3\int_R x^2\, dV.$

(b) $\int_R (x^2+y^2+z^2)\, dV$

$$=\int_0^{2\pi}\int_0^\pi\int_0^a \rho^2\cdot\rho^2\sin\phi\; d\rho\; d\phi\; d\theta$$

$$=2\pi\int_0^\pi \frac{a^5}{5}\sin\phi\; d\phi = \frac{2\pi a^5}{5}[-\cos\phi]\Big|_0^\pi$$

$$=\frac{2\pi a^5}{5}\cdot 2=\frac{4\pi a^5}{5}$$

(c) $\int_R x^2\, dV = \frac{1}{3}\int_R (x^2+y^2+z^2)\, dV$

$$=\frac{1}{3}\cdot\frac{4\pi a^5}{5}=\frac{4\pi a^5}{15}$$

43 Let the ball have radius a and suppose its center coincides with the origin. Let the point A lie on the z axis a distance H from the center of the ball. As in Example 5, let $q = \overline{AP}$, the distance of a point P from the point A. We wish to find the average of $1/q$ over the given ball. It was shown in Example 5 that $\int_R \frac{1}{q}\, dV = \frac{4\pi a^3}{3H}$. The average of $1/q$ is

therefore $\frac{1}{\text{Volume of } R}\int_R \frac{1}{q}\, dV = \frac{3}{4\pi a^3}\cdot\frac{4\pi a^3}{3H}$

$= \frac{1}{H}$, the reciprocal of the distance from A to the center, as claimed.

45 Mass $= \int_0^a\int_0^\pi\int_0^{2\pi} g(\rho)\,\rho^2\sin\phi\; d\theta\; d\phi\; d\rho$

$= 4\pi\int_0^a g(\rho)\,\rho^2\, d\rho$

47 The fourth equation is incorrect:

$$\int_{-\pi/2}^{\pi/2}\left[-\frac{2}{3}(a^2-r^2)^{3/2}\right]\Bigg|_0^{a\cos\theta} d\theta =$$

$$\int_{-\pi/2}^{\pi/2}\left[-\frac{2}{3}(a^2-(a\cos\theta)^2)^{3/2}+\frac{2}{3}(a^2-0^2)^{3/2}\right] d\theta$$

$= \int_{-\pi/2}^{\pi/2}\frac{2a^3}{3}[1-(\sin^2\theta)^{3/2}]\, d\theta$. However, it is *not* always true that $(\sin^2\theta)^{3/2} = \sin^3\theta$. In fact, for $-\frac{\pi}{2} \le \theta \le 0$, $(\sin^2\theta)^{3/2}$ is positive while $\sin^3\theta$ is negative; for such θ, $(\sin^2\theta)^{3/2} = -\sin^3\theta$.

49 Exploration problems are in the *Instructor's Manual.*

15.S Guide Quiz

1 (a) The region is a half-disk of radius a: $-a \le x \le a$, $0 \le y \le \sqrt{a^2 - x^2}$.

(b) The region is a quadrilateral, as illustrated in the figure. The line $y = x/\sqrt{3}$ has slope $1/\sqrt{3}$ and angle of inclination $\theta = \tan^{-1}(1/\sqrt{3}) = \pi/6$. Similarly the line $y = x$ has angle of inclination $\pi/4$. The line $x = 1$ can be written as $r \cos\theta = 1$ or $r = \sec\theta$, and $x = 2$ is similarly equivalent to $r = 2\sec\theta$. The description is therefore $\pi/6 \le \theta \le \pi/4$, $\sec\theta \le r \le 2\sec\theta$.

2 (a) The region's description is $0 \le x \le 2$, $0 \le y \le x$. The moment of the region about the x axis is thus $M_x = \int_R y\, dA = \int_0^2 \int_0^x y\, dy\, dx$

$$= \int_0^2 \frac{x^2}{2}\, dx = \left.\frac{x^3}{6}\right|_0^2 = \frac{4}{3}.$$

(b) In polar coordinates the region is $0 \le \theta \le \pi/4$, $0 \le r \le 2\sec\theta$, and $y = r\sin\theta$, so M_x

$$= \int_R y\, dA = \int_0^{\pi/4} \int_0^{2\sec\theta} r\sin\theta\, r\, dr\, d\theta$$

$$= \int_0^{\pi/4} \frac{8\sec^3\theta}{3} \sin\theta\, d\theta$$

$$= \frac{8}{3}\int_0^{\pi/4} \tan\theta \sec^2\theta\, d\theta$$

$$= \frac{8}{3}\cdot\left.\frac{\tan^2\theta}{2}\right|_0^{\pi/4} = \frac{4}{3}.$$

3 (a) Place the center of the disk at the origin of the plane. The distance from a point to the center is thus equal to r. The average value of r over the disk of radius a is $\frac{1}{\text{Area of } R}\int_R r\, dA =$

$$\frac{1}{\pi a^2}\int_0^{2\pi}\int_0^a r^2\, dr\, d\theta = \frac{1}{\pi a^2}\int_0^{2\pi}\frac{a^3}{3}\, d\theta$$

$$= \frac{a}{3\pi}\cdot 2\pi = \frac{2a}{3}.$$

(b) Fewer than half the points of the disk are within a distance $a/2$ of the origin. The area of the disk of radius $a/2$ is $\pi a^2/4$, while the area outside the disk of radius $a/2$ and within the disk of radius a is $(\pi a^2) - (\pi a^2/4) = 3\pi a^2/4$. Since three-quarters of the points of the disk of radius a are at a greater distance than $a/2$ from the center, it is not surprising that the average of the distance exceeds $a/2$.

4 The region is the quarter of the disk of radius a that lies in the first quadrant: $0 \le \theta \le \pi/2$, $0 \le r \le a$. The integrand is r^3, so

$$\int_0^a \int_0^{\sqrt{a^2-x^2}} (x^2 + y^2)^{3/2}\, dy\, dx =$$

$$\int_0^{\pi/2}\int_0^a r^3 \cdot r\, dr\, d\theta = \frac{a^5}{5}\int_0^{\pi/2} d\theta = \frac{\pi a^5}{10}.$$

5 (a) The curve $r = \sin 2\theta$ describes one loop for θ in $[0, \pi/2]$. The density is 1, so the moment of inertia about a line through the pole and perpendicular to the plane is

$$\int_0^{\pi/2}\int_0^{\sin 2\theta} r^2 \cdot r\, dr\, d\theta = \int_0^{\pi/2} \frac{\sin^4 2\theta}{4}\, d\theta$$

$$= \frac{1}{8}\int_0^{\pi/2} \sin^4 2\theta\, (2\, d\theta) = \frac{1}{8}\int_0^{\pi} \sin^4 u\, du$$

$$= \frac{1}{8}\cdot 2\left(\frac{1\cdot 3}{2\cdot 4}\cdot\frac{\pi}{2}\right) = \frac{3\pi}{64}.$$ To evaluate the integral we made use of Formula 73.

(b) The region is as before. The density function is r^2, so we have Mass $= \int_R r^2\, dA$. This is precisely the integral evaluated in (a), so it follows that Mass $= \frac{3\pi}{64}$.

(c) The cross-sectional height of the solid is r^2, so the volume is $\int_R r^2\, dA$. By the result of (a), the volume is $\frac{3\pi}{64}$.

(d) The area of the region is $\int_0^{\pi/2} \int_0^{\sin 2\theta} r\, dr\, d\theta$

$= \int_0^{\pi/2} \frac{1}{2} \sin^2 2\theta\, d\theta = \frac{1}{4} \int_0^{\pi} \sin^2 u\, du =$

$\frac{1}{4} \cdot \frac{\pi}{2} = \frac{\pi}{8}$. Using (a), the average temperature is $\frac{1}{\text{Area of } R} \int_R r^2\, dA =$

$\frac{8}{\pi} \cdot \frac{3\pi}{64} = \frac{3}{8}$ degrees.

6 Let R_k be the disk of radius k feet centered at the sprinkler. The volume of water provided in one hour within R_k is $\int_{R_k} e^{-r}\, dA =$

$\int_0^{2\pi} \int_0^k e^{-r} r\, dr\, d\theta = \int_0^{2\pi} [-(r+1)e^{-r}]\big|_0^k\, d\theta =$

$2\pi[1 - (k+1)e^{-k}]$ ft^3.

(a) When $k = 100$, the water flow is $2\pi(1 - 101e^{-100})$ ft^3/hr.

(b) When $k = 50$, the water flow is $2\pi(1 - 51e^{-50})$ ft^3/hr.

7 Place the cylinder R so that its description is $0 \le \theta \le 2\pi$, $0 \le r \le a$, $-h/2 \le z \le h/2$. The density is $\delta = \frac{M}{\pi a^2 h}$. By symmetry, we may compute I_x for the upper half of R and double the result: $I_x =$

$\delta \int_R (y^2 + z^2)\, dV$

$= 2\delta \int_0^{2\pi} \int_0^a \int_0^{h/2} (r^2 \sin^2\theta + z^2)\, r\, dz\, dr\, d\theta$

$= 2\delta \int_0^{2\pi} \int_0^a \left(r^3 \sin^2\theta\, z + \frac{z^3 r}{3} \right)\Big|_0^{h/2} dr\, d\theta$

$= 2\delta \int_0^{2\pi} \int_0^a \left(\frac{1}{2} h r^3 \sin^2\theta + \frac{h^3 r}{24} \right) dr\, d\theta$

$= \delta h \int_0^{2\pi} \left(\frac{r^4}{4} \sin^2\theta + \frac{h^2}{12} \cdot \frac{r^2}{2} \right)\Big|_0^a d\theta$

$= \frac{M}{\pi a^2 h} \cdot h \cdot \frac{a^2}{4} \int_0^{2\pi} \left(a^2 \sin^2\theta + \frac{h^2}{6} \right) d\theta$

$= \frac{M}{4\pi}\left(a^2\pi + \frac{h^2}{6} \cdot 2\pi \right) = \frac{Ma^2}{4} + \frac{Mh^2}{12}$. (Note the use of Formula 73 integrate $\sin^2\theta$.)

8 In spherical coordinates we can describe the region R by $0 \le \theta \le 2\pi$, $0 \le \phi \le \pi/2$, $a \le \rho \le b$. By symmetry, the centroid lies on the z axis. The volume of R is $\frac{2}{3}\pi b^3 - \frac{2}{3}\pi a^3 = \frac{2}{3}\pi(b^3 - a^3)$.

Furthermore, $\int_R z\, dV =$

$\int_0^{2\pi} \int_0^{\pi/2} \int_a^b (\rho \cos\phi)\, \rho^2 \sin\phi\, d\rho\, d\phi\, d\theta$

$= \int_0^{2\pi} \int_0^{\pi/2} \frac{1}{4}(b^4 - a^4) \cos\phi \sin\phi\, d\phi\, d\theta$

$= \int_0^{2\pi} \frac{1}{4}(b^4 - a^4) \frac{\sin^2\phi}{2}\Big|_0^{\pi/2} d\theta$

$= 2\pi \cdot \frac{1}{4}(b^4 - a^4)\cdot\frac{1}{2} = \frac{\pi}{4}(b^4 - a^4)$, so $\bar{z} =$

$$\frac{\frac{\pi}{4}(b^4 - a^4)}{\frac{2}{3}\pi(b^3 - a^3)} = \frac{3(b^4 - a^4)}{8(b^3 - a^3)}.$$

9 (a) See p. 883 of Sec. 15.1.

(b) Each approximating sum $f(P_1)V_1 + f(P_2)V_2 + \cdots + f(P_n)V_n$ lies between $2(V_1 + V_2 + \cdots + V_n) = 2$(Volume of R) and $3(V_1 + V_2 + \cdots + V_n) = 3$(Volume of R). The limit of these sums, $\int_R f(P)\, dV$, must lie between the same bounds.

10 (a) The rectangular coordinates are $(3, 4, -3)$. Therefore $r = \sqrt{x^2 + y^2} = \sqrt{3^2 + 4^2} = 5$. Since $(3, 4)$ is in the first quadrant, $\theta = \tan^{-1}(y/x) = \tan^{-1}(4/3)$. Finally, $z = -3$, as in rectangular coordinates, so the cylindrical coordinates are $(5, \tan^{-1}(4/3), -3)$.

(b) The spherical coordinates are $(3, \pi/2, 2\pi/3)$. We have $x = \rho \sin\phi \cos\theta = 3 \sin \frac{2\pi}{3} \cos \pi/2 = 3\cdot\frac{\sqrt{3}}{2}\cdot 0 = 0$, $y = \rho \sin\phi \sin\theta = 3 \sin \frac{2\pi}{3} \sin \pi/2 = 3\cdot\frac{\sqrt{3}}{2}\cdot 1 = \frac{3\sqrt{3}}{2}$, and $z = \rho\cos\phi = 3\cos\frac{2\pi}{3} = 3\cdot\left(-\frac{1}{2}\right) = -\frac{3}{2}$. The rectangular coordinates are thus

$$\left(0, \frac{3\sqrt{3}}{2}, -\frac{3}{2}\right).$$

(c) The cylindrical coordinates are $(2, \pi/4, 2)$. Now $\rho = \sqrt{r^2 + z^2} = \sqrt{2^2 + 2^2} = \sqrt{8} = 2\sqrt{2}$. Since $z > 0$, $\phi = \tan^{-1}(r/z) = \tan^{-1}(2/2) = \tan^{-1}(1) = \pi/4$. Finally, θ is the same in spherical coordinates as in cylindrical, so the spherical coordinates are

$(2\sqrt{2}, \pi/4, \pi/4)$.

11 See the answer section of the text for the accompanying figures.

(a) The cross section is a disk lying in the plane $x = 1/2$ with center at $(1/2, 0, 0)$ and radius $\sqrt{3/4} = \sqrt{3}/2$.

(b) The cross section is part of a cone with vertex angle $\pi/3$.

(c) The cross section is a sphere of radius $1/2$.

(d) The cross section is a half disk.

(e) The cross section is a disk of radius $\sqrt{3/4} = \frac{\sqrt{3}}{2}$ and center $(0, 0, -\frac{1}{2})$.

12 (a) In cylindrical coordinates the factor r must be included in the integrand. In sphérical coordinates the factor $\rho^2 \sin\phi$ is required.

(b) See explanations in Secs. 15.6 and 15.7.

13 See pp. 924–925 of the text for a full discussion.

14 (a) Place the cylinder so its axis coincides with the z axis and its base lies in the xy plane. Then the square of the distance to the z axis is r^2, or $x^2 + y^2$. The volume of the cylinder is $\pi a^2 h$, so the average of r^2 is $\frac{1}{\pi a^2 h}\int_R r^2\, dV$

$$= \frac{1}{\pi a^2 h}\int_0^{2\pi}\int_0^a\int_0^h r^2\cdot r\, dz\, dr\, d\theta =$$

$$\frac{1}{\pi a^2 h}\int_{-a}^{a}\int_{-\sqrt{a^2-x^2}}^{\sqrt{a^2-x^2}}\int_0^h (x^2+y^2)\,dz\,dy\,dz.$$

Complications arise in spherical coordinates because the cylinder must be described in two pieces. First note that the top of the cylinder is described by $z = h$; since $z = \rho\cos\phi$, we also have $\rho\cos\phi = h$, or $\rho = h\sec\phi$. The vertical side of the cylinder has the equation $r = a$, where $r = \rho\sin\phi$, so $\rho = a\csc\phi$. Finally, let $\alpha = \tan^{-1}(a/h)$; α is the ϕ coordinate where the top of the cylinder meets the side. For $0 \le \phi \le \alpha$, ρ goes from 0 to the upper surface $\rho = h\sec\phi$, while for $\alpha \le \phi \le \pi/2$, ρ goes from 0 to the lateral surface $\rho = a\csc\phi$. If you insist, you can use this information to set up the integral (in two parts) in spherical coordinates.

(b) The integral in cylindrical coordinates is easy:

$$\frac{1}{\pi a^2 h}\int_0^{2\pi}\int_0^{a}\int_0^{h} r^3\,dz\,dr\,d\theta$$

$$= \frac{2\pi}{\pi a^2 h}\int_0^a hr^3\,dr = \frac{2}{a^2}\cdot\frac{a^4}{4} = \frac{a^2}{2}.$$

15 (a) $\int_0^1\int_0^1\int_0^x ye^{x^2}\,dz\,dy\,dx =$

$$\int_0^1\int_0^1 xye^{x^2}\,dy\,dx = \int_0^1 xe^{x^2}\frac{y^2}{2}\Big|_0^1\,dx =$$

$$\frac{1}{2}\int_0^1 xe^{x^2}\,dx = \frac{1}{2}\cdot\frac{1}{2}e^{x^2}\Big|_0^1$$

$$= \frac{1}{4}(e^1 - e^0)$$

$$= \frac{e-1}{4}$$

(b) Observe that the base of the region is the square in the xy plane where x and y vary

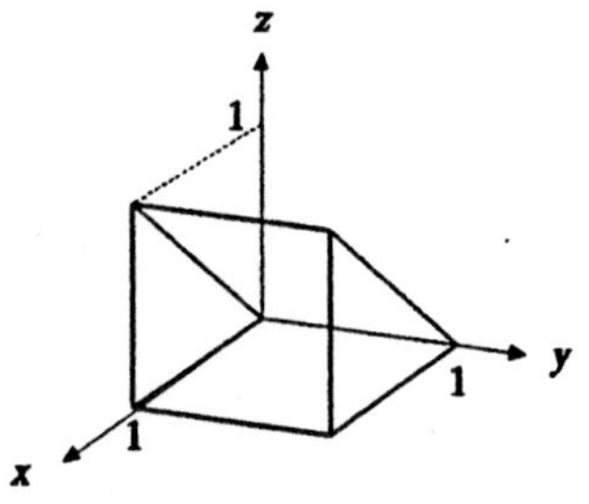

between 0 and 1. z goes from 0 to the plane $z = x$.

16 (a) If we place the cone in the customary position, with its vertex at the origin and its axis coinciding with the z axis, the moment of inertia about a line through its vertex and parallel to its base is equal to I_x or I_y. Since the region is homogeneous, the desired moment of inertia is just density multiplied by the integral of the square of the distance from the indicated axis.

(b) The rectangular equation for the cone was discussed in Exercise 35(b) of Sec. 15.5. The equation in cylindrical coordinates was discussed in Exercise 34 of Sec. 15.6. The equation in spherical coordinates was treated in Exercise 38 of Sec. 15.7. With this information, and the fact that $\delta = \dfrac{3M}{\pi a^2 h}$, we have $I_x = \delta\int_R (y^2+z^2)\,dV =$

$$\delta\int_{-a}^{a}\int_{-\sqrt{a^2-x^2}}^{\sqrt{a^2-x^2}}\int_{\frac{h}{a}\sqrt{x^2+y^2}}^{h}(y^2+z^2)\,dz\,dy\,dx$$

$$= \delta\int_0^{2\pi}\int_0^{h}\int_0^{az/h}(r^2\sin^2\theta + z^2)\,r\,dr\,dz\,d\theta$$

$= \delta \int_0^{2\pi} \int_0^{\alpha} \int_0^{h \sec \phi} (\rho^2 \sin^2 \phi \sin^2 \theta +$ $\rho^2 \cos^2 \phi)\, \rho^2 \sin \phi \, d\rho \, d\phi \, d\theta$. (Recall that $\alpha = \tan^{-1}(a/h)$.)

(c) The integral in cylindrical coordinates if relatively easy: $I_x =$

$$\delta \int_0^{2\pi} \int_0^{h} \int_0^{az/h} (r^3 \sin^2 \theta + rz^2)\, dr\, dz\, d\theta$$

$$= \delta \int_0^{2\pi} \int_0^{h} \left(\frac{r^4}{4} \sin^2 \theta + \frac{r^2 z^2}{2} \right)\Bigg|_0^{az/h} dz\, d\theta$$

$$= \int_0^{2\pi} \int_0^{h} \left(\frac{a^4 z^4}{4h^4} \sin^2 \theta + \frac{a^2 z^4}{2h^4} \right) dz\, d\theta$$

$$= \delta \int_0^{2\pi} \left(\frac{a^4}{4h^4} \sin^2 \theta + \frac{a^2}{2h^2} \right) \frac{z^5}{5}\Bigg|_0^{h} d\theta$$

$$= \frac{3M}{\pi a^2 h} \cdot \frac{h^5}{5} \left[\frac{a^4}{4h^4} \cdot \pi + \frac{a^2}{2h^2} \cdot 2\pi \right]$$

$$= \frac{3M}{20}(a^2 + 4h^2).$$

17 (a) $z = \rho \cos \phi$ in spherical coordinates, so $z = 3$ is equivalent to $3 = \rho \cos \phi$ or $\rho = 3 \sec \phi$.

(b) The surface $r = 2$ is equivalent to $x^2 + y^2 = 4$.

(c) $r = \rho \sin \phi$ in spherical coordinates, so $r = 2$ is equivalent to $\rho \sin \phi = 2$, or $\rho = 2 \csc \phi$.

(d) The surface $\rho = 3$ is a sphere of radius 3; in cylindrical coordinates, the equation is $r^2 + z^2 = 9$.

18 (a) The integrand $y^7 e^{-y^2}$ is an odd function and the interval of integration $[-1, 1]$ is symmetric with respect to 0. Hence the value of the integrand over $[-1, 0]$ will cancel out the value of the integrand over $[0, 1]$. Therefore $\int_{-1}^{1} y^7 e^{-y^2}\, dy = 0$.

(b) The integrand $\sin \theta$ is an odd function and the region R is symmetric with respect to the x axis; that is, $\sin \theta$ is negative for the part of R below the x axis and positive for the part of R above the x axis. The two parts will cancel out, so $\int_R \sin \theta \, dA = 0$.

(c) By symmetry over the spherical region R, each term of the integrand $2x + 3y + 4z$ will be negative as often as positive, so that $\int_R (2x + 3y + 4z)\, dV = 0$.

15.S Review Exercises

1 (a) The curves $y = x^2$ and $y = 4$ intersect at $(\pm 2, 4)$. As x varies from -2 to 2, each vertical cross section goes from $y = x^2$ up to $y = 4$. The description is $-2 \le x \le 2$, $x^2 \le y \le 4$.

(b) As y varies from 0 up to 4, x ranges from the curve $x = -\sqrt{y}$ to $x = \sqrt{y}$. The description is $0 \le y \le 4$, $-\sqrt{y} \le x \le \sqrt{y}$.

3 (a) $\int_1^{x^2} (x + y)\, dy = \left(xy + \frac{y^2}{2} \right)\Bigg|_{y=1}^{y=x^2}$

$$= \left(x^3 + \frac{x^4}{2} \right) - \left(x + \frac{1}{2} \right)$$

$$= \frac{1}{2}x^4 + x^3 - x - \frac{1}{2}$$

(b) $\int_y^{y^2} (x + y)\,dx = \left(\frac{x^2}{2} + xy\right)\Bigg|_{x=y}^{x=y^2}$

$= \left(\frac{y^4}{2} + y^3\right) - \left(\frac{y^2}{2} + y^2\right)$

$= \frac{1}{2}y^4 + y^3 - \frac{3}{2}y^2$

5 (a) The circle that forms one of the boundaries of the region has radius $a\sqrt{2}$, so its equation is $x^2 + y^2 = 2a^2$. With vertical cross sections, the description is thus $0 \le x \le a$, $0 \le y \le \sqrt{2a^2 - x^2}$.

(b) The description by horizontal cross sections has two parts: $0 \le y \le a$, $0 \le x \le a$; $a \le y \le a\sqrt{2}$, $0 \le x \le \sqrt{2a^2 - y^2}$.

(c) The line $x = a$ corresponds to the equation $r \cos\theta = a$ or $r = a \sec\theta$. The first part of the region's description in polar coordinates is therefore $0 \le \theta \le \pi/4$, $0 \le r \le a \sec\theta$. The rest of the region is a sector of a circle of radius $a\sqrt{2}$: $\frac{\pi}{4} \le \theta \le \frac{\pi}{2}$, $0 \le r \le a\sqrt{2}$.

7 As x goes from 0 to $1/\sqrt{2}$, y goes from the line $y = x$ to the semicircle $y = \sqrt{1 - x^2}$. The region is a sector of the unit disk: $\frac{\pi}{4} \le \theta \le \frac{\pi}{2}$, $0 \le r \le 1$. The integrand equals r, so

$\int_0^{1/\sqrt{2}} \int_x^{\sqrt{1-x^2}} \sqrt{x^2 + y^2}\,dy\,dx$

$= \int_{\pi/4}^{\pi/2} \int_0^1 r \cdot r\,dr\,d\theta = \int_{\pi/4}^{\pi/2} \frac{1}{3}\,d\theta = \frac{1}{3}\theta\Big|_{\pi/4}^{\pi/2} =$

$\frac{1}{3} \cdot \frac{\pi}{4} = \frac{\pi}{12}.$

9 (a) The limits on r and θ are all constant, so the region of integration is a sector of a disk.

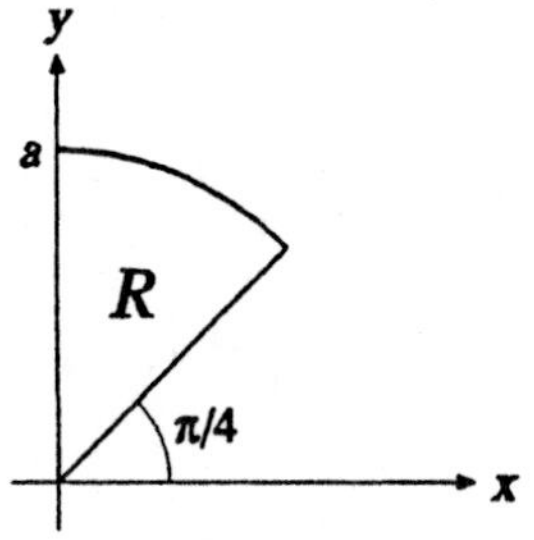

After removing the factor r that is part of dA, the integrand is $r \sin\theta$, which equals y. That is, $f(P) = y$.

(b) With vertical cross sections, we see that

$\int_{\pi/4}^{\pi/2} \int_0^a r^2 \sin\theta\,dr\,d\theta =$

$\int_0^{a/\sqrt{2}} \int_x^{\sqrt{a^2 - x^2}} y\,dy\,dx.$

(c) Both integrals in (b) are straightforward:

$\int_{\pi/4}^{\pi/2} \int_0^a r^2 \sin\theta\,dr\,d\theta = \int_{\pi/4}^{\pi/2} \frac{a^3}{3} \sin\theta\,d\theta$

$= \frac{a^3}{3}(-\cos\theta)\Big|_{\pi/4}^{\pi/2} = \frac{a^3}{3} \cdot \frac{1}{\sqrt{2}} = \frac{a^3}{3\sqrt{2}};$

$\int_0^{a/\sqrt{2}} \int_x^{\sqrt{a^2 - x^2}} y\,dy\,dx$

$= \int_0^{a/\sqrt{2}} \frac{y^2}{2}\Bigg|_x^{\sqrt{a^2 - x^2}} dx$

$= \frac{1}{2} \int_0^{a/\sqrt{2}} [(a^2 - x^2) - x^2]\,dx$

$= \frac{1}{2} \int_0^{a/\sqrt{2}} (a^2 - 2x^2)\,dx$

$$= \frac{1}{2}\left(a^2x - \frac{2}{3}x^3\right)\Bigg|_0^{a/\sqrt{2}} = \frac{1}{2}\left(\frac{a^3}{\sqrt{2}} - \frac{2a^3}{6\sqrt{2}}\right)$$

$$= \frac{a^3}{3\sqrt{2}}.$$

11 In rectangular coordinates, the region is $0 \le y \le a/\sqrt{2}$, $y \le x \le \sqrt{a^2 - y^2}$. Factoring out the r that belongs to dA, we see that the integrand is $r^2 = x^2 + y^2$. Therefore $\int_0^{\pi/4}\int_0^a r^3\,dr\,d\theta$

$$= \int_0^{a/\sqrt{2}}\int_y^{\sqrt{a^2-y^2}} (x^2 + y^2)\,dx\,dy$$

$$= \int_0^{a/\sqrt{2}} \left(\frac{x^3}{3} + xy^2\right)\Bigg|_y^{\sqrt{a^2-y^2}} dy$$

$$= \int_0^{a/\sqrt{2}} \left(\frac{a^2}{3}\sqrt{a^2-y^2} + \frac{2y^2}{3}\sqrt{a^2-y^2} - \frac{4y^3}{3}\right) dy$$

$$= \left[\frac{a^2}{6}\left(y\sqrt{a^2-y^2} + a^2\sin^{-1}\frac{y}{a}\right) + \right.$$

$$\frac{2}{3}\left(-\frac{y}{4}(a^2-y^2)^{3/2} + \frac{a^2y}{8}\sqrt{a^2-y^2} + \right.$$

$$\left.\left.\frac{a^4}{8}\sin^{-1}\frac{y}{a}\right) - \frac{y^4}{3}\right]\Bigg|_0^{a/\sqrt{2}} = \frac{a^2}{6}\left(\frac{a^2}{2} + a^2\cdot\frac{\pi}{4}\right) +$$

$$\frac{2}{3}\left(-\frac{a^4}{16} + \frac{a^4}{16} + \frac{a^4}{8}\cdot\frac{\pi}{4}\right) - \frac{a^4}{12} = \frac{\pi a^4}{16}.$$

13 (a) Place R in the first quadrant so that x and y vary between 0 and a. The density is $\delta = M/a^2$. Compute the moment of inertia about the x axis: $I_x = \int_R \delta y^2\,dA$

$$= \delta\int_0^a\int_0^a y^2\,dy\,dx = \frac{M}{a^2}\int_0^a \frac{a^3}{3}\,dx$$

$$= \frac{Ma}{3}\cdot a = \frac{Ma^2}{3}.$$

(b) Place R with a vertex at the origin and its diagonal coinciding with the x axis. By

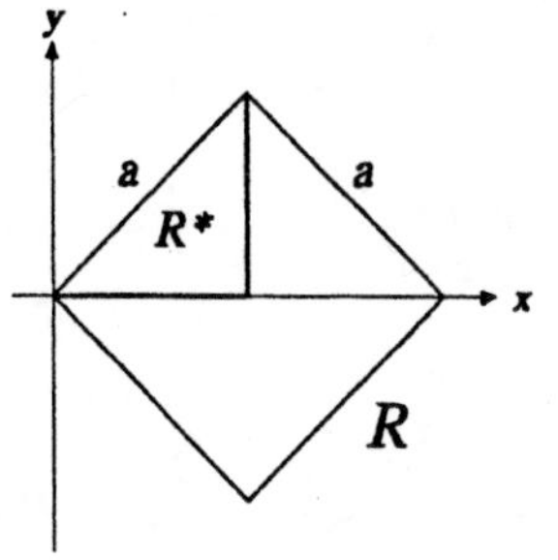

symmetry, the region R^* has one fourth of the moment of inertia about the x axis of R. Note that R^* has the simple description $0 \le x \le a/\sqrt{2}$, $0 \le y \le x$. We may now write $I_x = \int_R \delta y^2\,dA = 4\int_{R^*}\delta y^2\,dA$

$$= 4\delta\int_0^{a/\sqrt{2}}\int_0^x y^2\,dy\,dx = 4\delta\int_0^{a/\sqrt{2}}\frac{x^3}{3}\,dx$$

$$= \frac{4M}{a^2}\cdot\frac{x^4}{12}\Bigg|_0^{a/\sqrt{2}} = \frac{4M}{a^2}\cdot\frac{a^4}{48} = \frac{Ma^2}{12}.$$

15 The region has area $\int_0^2 (8 - x^3)\,dx =$ $\left(8x - \frac{x^4}{4}\right)\Big|_0^2 = 16 - 4 = 12$, so its density is $M/12$.

(a) $I_x = \int_R \delta y^2\,dA = \frac{M}{12}\int_0^2\int_{x^3}^8 y^2\,dy\,dx$

$$= \frac{M}{12}\int_0^2 \frac{y^3}{3}\Bigg|_{x^3}^8 dy = \frac{M}{12}\int_0^2\left(\frac{512}{3} - \frac{x^9}{3}\right)dx$$

$$= \frac{M}{36}\left(512x - \frac{x^{10}}{10}\right)\bigg|_0^2 = \frac{M}{36}\left(1024 - \frac{1024}{10}\right)$$

$$= \frac{128M}{5}.$$

(b) $$I_y = \int_R \delta x^2\, dA = \frac{M}{12}\int_0^2 \int_{x^3}^8 x^2\, dy\, dx$$

$$= \frac{M}{12}\int_0^2 x^2(8 - x^3)\, dx$$

$$= \frac{M}{12}\int_0^2 (8x^2 - x^5)\, dx = \frac{M}{12}\left(\frac{8x^3}{3} - \frac{x^6}{6}\right)\bigg|_0^2$$

$$= \frac{M}{12}\left(\frac{64}{3} - \frac{64}{6}\right) = \frac{8M}{9}$$

(c) $$I_z = \int_R \delta(x^2 + y^2)\, dA$$

$$= \int_R \delta x^2\, dA + \int_R \delta y^2\, dA = I_y + I_x$$

$$= \frac{8M}{9} + \frac{128M}{5} = \frac{40M + 1152M}{45}$$

$$= \frac{1192M}{45}$$

17 $$\int_R f(P)\, dA = \int_0^{\pi/2}\int_0^{\sin x} xy\, dy\, dx$$

$$= \int_0^{\pi/2} \frac{xy^2}{2}\bigg|_0^{\sin x} dx = \int_0^{\pi/2} \frac{1}{2}x \sin^2 x\, dx$$

$$= \frac{1}{2}\left(\frac{x^2}{4} - \frac{x \sin 2x}{4} - \frac{\cos 2x}{8}\right)\bigg|_0^{\pi/2} = \frac{1}{2}\left(\frac{\pi^2}{16} + \frac{1}{4}\right)$$

$$= \frac{\pi^2 + 4}{32}$$

19 $$\int_R xy\, dA = \int_0^{\pi/4}\int_0^{\cos 2\theta} (r\cos\theta)(r\sin\theta)\, r\, dr\, d\theta$$

$$= \int_0^{\pi/4}\int_0^{\cos 2\theta} r^3 \cos\theta \sin\theta\, dr\, d\theta$$

$$= \int_0^{\pi/4} \frac{\cos^4 2\theta}{4} \cos\theta \sin\theta\, d\theta$$

$$= \frac{1}{8}\int_0^{\pi/4} \cos^4 2\theta \sin 2\theta\, d\theta = -\frac{1}{16}\cdot\frac{\cos^5 2\theta}{5}\bigg|_0^{\pi/4}$$

$$= \frac{1}{80}$$

21 The depth of water at the point (r, θ) is 2^{-r}, so the volume of water is $\int_R 2^{-r}\, dA =$

$$\int_0^{2\pi}\int_0^a 2^{-r} r\, dr\, d\theta = \int_0^{2\pi} -\left[\frac{r \ln 2 + 1}{(\ln 2)^2}\cdot 2^{-r}\right]\bigg|_0^a d\theta$$

$$= \frac{2\pi}{(\ln 2)^2}[-(a \ln 2 + 1)\cdot 2^{-a} + 1]$$

$$= \frac{2\pi}{(\ln 2)^2}[1 - 2^{-a}(1 + a \ln 2)].$$

23 Label the vertices of R as shown; the triangle is described by $0 \le y \le b$, $\frac{a}{b}y \le x \le \frac{c}{b}y$. The

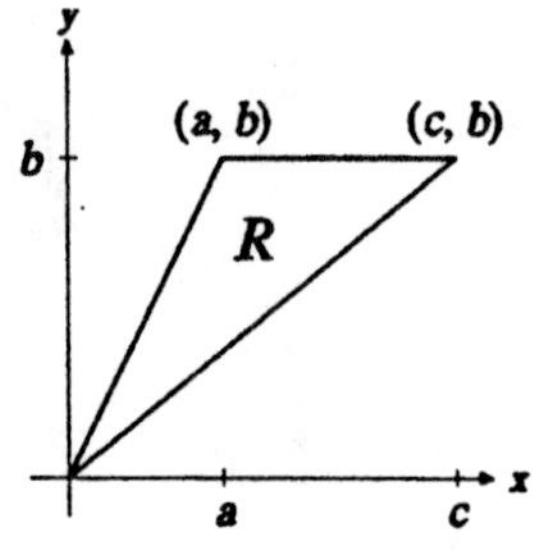

area of R is $\frac{1}{2}b(c - a)$, so the y coordinate of the centroid is $\bar{y} = \frac{1}{\text{Area}}\int_R y\, dA$

$$= \frac{2}{b(c - a)}\int_0^b \int_{ay/b}^{cy/b} y\, dx\, dy$$

$$= \frac{2}{b(c-a)} \int_0^b xy \Big|_{ay/b}^{cy/b} dy$$

$$= \frac{2}{b(c-a)} \int_0^b \frac{c-a}{b} \cdot y^2\, dy = \frac{2}{b^2} \cdot \frac{y^3}{3}\Big|_0^b$$

$$= \frac{2}{b^2} \cdot \frac{b^3}{3} = \frac{2b}{3}.$$

25

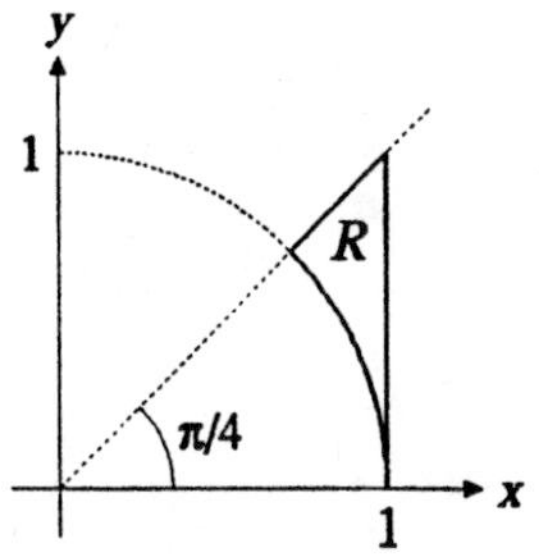

(a) The region R is shown in the figure. It is bounded by a circular arc and two straight lines. Note that the boundary line $x = 1$ is the same as $r \cos\theta = 1$ or $r = \sec\theta$. The line $y = x$ has angle of inclination $\pi/4$.

(b) From (a) and the figure, we see that in polar coordinates R has the description $0 \le \theta \le \pi/4$, $1 \le r \le \sec\theta$.

(c) The integrand of the double integral in rectangular coordinates is equal to $1/r$, so we have $\int_0^{\pi/4} \int_1^{\sec\theta} \frac{1}{r} \cdot r\, dr\, d\theta =$

$$\int_0^{\pi/4} \int_1^{\sec\theta} dr\, d\theta.$$

(d) Evaluating the integral in (c), we obtain

$$\int_0^{\pi/4} (\sec\theta - 1)\, d\theta$$

$$= (\ln|\sec\theta + \tan\theta| - \theta)\Big|_0^{\pi/4}$$

$$= \ln(\sqrt{2} + 1) - \frac{\pi}{4}.$$

27 (a) The cylindrical shell can be described by $0 \le \theta \le 2\pi$, $a \le r \le b$, $-\frac{h}{2} \le z \le \frac{h}{2}$. Its volume is $\pi(b^2 - a^2)h$. The moment of inertia about its axis is $I_z = \delta \int_R r^2\, dV$

$$= \delta \int_0^{2\pi} \int_a^b \int_{-h/2}^{h/2} r^3\, dz\, dr\, d\theta$$

$$= 2\pi\delta \int_a^b hr^3\, dr = \frac{2\pi Mh}{\pi(b^2 - a^2)h} \cdot \frac{r^4}{4}\Big|_a^b$$

$$= \frac{2M}{b^2 - a^2} \cdot \frac{b^4 - a^4}{4} = \frac{M(b^2 + a^2)}{2}.$$

(b) The desired moment of inertia is equal to I_x. In cylindrical coordinates, the integrand $y^2 + z^2$ is $r^2 \sin^2\theta + z^2$: $I_x =$

$$\delta \int_0^{2\pi} \int_a^b \int_{-h/2}^{h/2} (r^3 \sin^2\theta + rz^2)\, dz\, dr\, d\theta$$

$$= \delta \int_0^{2\pi} \int_a^b \left[(r^3 \sin^2\theta) z + \frac{rz^3}{3} \right]\Big|_{-h/2}^{h/2} dr\, d\theta$$

$$= \delta \int_0^{2\pi} \int_a^b \left(hr^3 \sin^2\theta + \frac{rh^3}{12} \right) dr\, d\theta$$

$$= \delta \int_0^{2\pi} \left(\frac{hr^4}{4} \sin^2\theta + \frac{r^2h^3}{24} \right)\Big|_a^b d\theta =$$

$$\frac{M}{\pi(b^2 - a^2)h} \int_0^{2\pi} \left[\frac{h}{4}(b^4 - a^4) \sin^2\theta + \frac{h^3}{24}(b^2 - a^2) \right] d\theta$$

$$= \frac{M}{\pi} \int_0^{2\pi} \left[\frac{1}{4}(b^2 + a^2) \sin^2\theta + \frac{h^2}{24} \right] d\theta$$

$$= \frac{M}{\pi}\left[\frac{1}{4}(b^2 + a^2)\cdot\pi + \frac{h^2}{24}\cdot 2\pi\right]$$

$$= \frac{M}{4}\left(b^2 + a^2 + \frac{h^2}{3}\right).$$

29 Place the disk R with its center at the pole of a polar coordinate system. Set the point A at $(H, 0)$

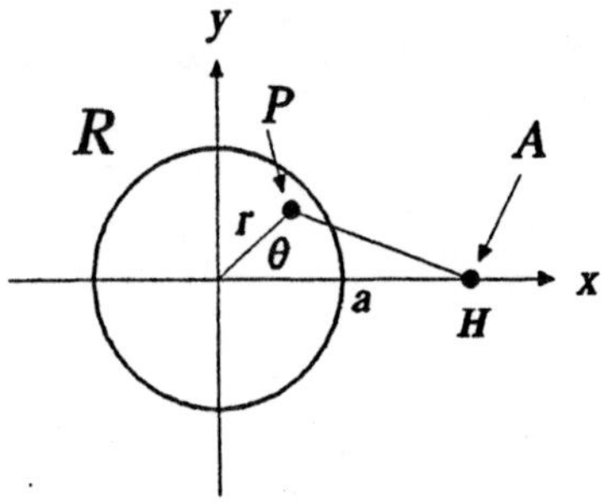

on the polar axis. By the law of cosines, the distance $\overline{PA}$ is given by $\overline{PA}^2 = H^2 + r^2 - 2Hr\cos\theta$. The area of R is πa^2, so the average of $\overline{PA}^2$ is given by $\frac{1}{\text{Area}}\int_R \overline{PA}^2\, dA$

$$= \frac{1}{\pi a^2}\int_0^{2\pi}\int_0^a (H^2 + r^2 - 2Hr\cos\theta)\, r\, dr\, d\theta$$

$$= \frac{1}{\pi a^2}\int_0^{2\pi}\int_0^a (H^2 r + r^3 - 2Hr^2\cos\theta)\, dr\, d\theta$$

$$= \frac{1}{\pi a^2}\int_0^{2\pi}\left(\frac{H^2a^2}{2} + \frac{a^4}{4} - \frac{2Ha^3}{3}\cos\theta\right) d\theta$$

$$= \frac{1}{\pi}\left(\frac{H^2}{2} + \frac{a^2}{4}\right)\cdot 2\pi - \frac{2Ha}{3}(\sin\theta)\Big|_0^{2\pi}$$

$$= H^2 + \frac{a^2}{2}.$$

31 Describe R by coordinates so that R has the description $0 \le x \le b$, $0 \le y \le a$. The density at

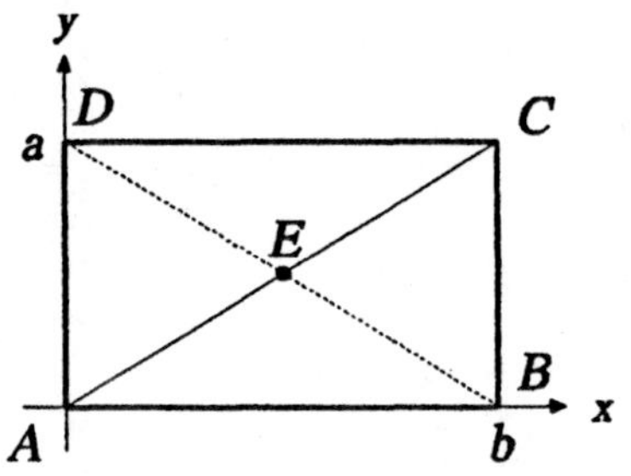

every point is $\frac{M}{ab}$. Hence, for any line L, the moment of inertia about L is $I_L =$ $\frac{M}{ab}\int_0^b\int_0^a r^2\, dy\, dx$, where r is the distance from (x, y) to L.

(a) $I_x = \frac{M}{ab}\int_0^b\int_0^a y^2\, dy\, dx = \frac{M}{ab}\int_0^b \frac{a^3}{3}\, dx$

$$= \frac{M}{ab}\cdot\frac{a^3}{3}\cdot b = \frac{Ma^2}{3}$$

(b) Interchanging a and b in I_x yields $I_y = \frac{Mb^2}{3}$.

(c) The line AC has equation $ax - by = 0$, so the distance from an arbitrary point (x, y) to AC is $\frac{|ax - by|}{\sqrt{a^2 + b^2}}$. (See Theorem 2 in Sec. 12.4.)

Hence $I_L = \frac{M}{ab}\int_0^b\int_0^a \frac{(ax - by)^2}{a^2 + b^2}\, dy\, dx =$

$$\frac{M}{ab(a^2 + b^2)}\int_0^b\int_0^a (b^2y^2 - 2abxy + a^2x^2)\, dy\, dx$$

$$= \frac{M}{ab(a^2+b^2)}\int_0^b \left(b^2\cdot\frac{a^3}{3} - 2abx\cdot\frac{a^2}{2} + a^2x^2a\right) dx$$

$$= \frac{Ma^3}{ab(a^2 + b^2)}\int_0^b \left(\frac{b^2}{3} - bx + x^2\right) dx$$

$$= \frac{Ma^2}{b(a^2+b^2)}\left(\frac{b^2}{3}\cdot b - b\cdot\frac{b^2}{2} + \frac{b^3}{3}\right)$$

$$= \frac{Ma^2b^2}{6(a^2+b^2)}.$$

(d) In this case $r = \left|y - \frac{a}{2}\right|$, so $I_L =$

$$\frac{M}{ab}\int_0^b\int_0^a\left(y-\frac{a}{2}\right)^2 dy\,dx =$$

$$\frac{M}{ab}\int_0^b \frac{1}{3}\left(y-\frac{a}{2}\right)^3\Big|_0^a dx = \frac{M}{ab}\int_0^b \frac{a^3}{12}\,dx =$$

$$\frac{Ma^2}{12}.$$

(e) $r = \sqrt{\left(x-\frac{b}{2}\right)^2 + \left(y-\frac{a}{2}\right)^2}$, so $I_L =$

$$\frac{M}{ab}\int_0^b\int_0^a\left[\left(x-\frac{b}{2}\right)^2+\left(y-\frac{a}{2}\right)^2\right]dy\,dx =$$

$$\frac{M}{ab}\int_0^b\int_0^a\left(x-\frac{b}{2}\right)^2 dy\,dx +$$

$\frac{M}{ab}\int_0^b\int_0^a\left(y-\frac{a}{2}\right)^2 dy\,dx$. By part (d), the last integral is $\frac{Ma^2}{12}$. By interchanging a and b in the result of (d), we get $\frac{Mb^2}{12}$ for the first integral. Hence $I_L = \frac{M(a^2+b^2)}{12}$.

33 The homogeneous disk has mass M and area πa^2, so its density is $\sigma = \frac{M}{\pi a^2}$.

(a) Place the disk with its center at the origin of a coordinate system so that its description in polar coordinates is $0 \le \theta \le 2\pi$, $0 \le r \le a$. The moment of inertia about the line perpendicular to the disk and passing through its center is $I_z = \int_R r^2\sigma\,dA$

$$= \sigma\int_0^{2\pi}\int_0^a r^3\,dr\,d\theta = \sigma\int_0^{2\pi}\frac{a^4}{4}\,d\theta$$

$$= \sigma(2\pi)\frac{a^4}{4} = \frac{M}{\pi a^2}\frac{\pi a^4}{2} = \frac{Ma^2}{2}.$$

(b) Place the disk so that its polar equation is $r = 2a\cos\theta$. The moment of inertia about a line perpendicular to the disk and passing through its border is $I_z = \int_R r^2\sigma\,dA$

$$= \sigma\int_{-\pi/2}^{\pi/2}\int_0^{2a\cos\theta} r^3\,dr\,d\theta$$

$$= 2\sigma\int_0^{\pi/2}\frac{1}{4}(16a^4\cos^4\theta)\,d\theta = 8a^4\sigma\frac{3\pi}{16}$$

$$= \frac{3\pi a^4}{2}\frac{M}{\pi a^2} = \frac{3Ma^2}{2}.$$

(c) Let the center of the disk be at the origin. We can compute either I_x or I_y for the moment of inertia about a diameter. $I_y = \int_R x^2\sigma\,dA$

$$= \sigma\int_0^{2\pi}\int_0^a r^2\cos^2\theta\,r\,dr\,d\theta$$

$$= \sigma\int_0^{2\pi}\frac{a^4}{4}\cos^2\theta\,d\theta = a^4\sigma\frac{\pi}{4}$$

$$= \frac{\pi a^4}{4}\frac{M}{\pi a^2} = \frac{Ma^2}{4}.$$

(d) Place the disk as in (b). Then the moment of inertia about a tangent line is equal to I_y. We have $I_y = \int_R x^2 \sigma \, dA$

$= \sigma \int_{-\pi/2}^{\pi/2} \int_0^{2a\cos\theta} r^2 \cos^2\theta \, r \, dr \, d\theta$

$= 2\sigma \int_0^{\pi/2} \frac{1}{4}(16a^4 \cos^4\theta) \cos^2\theta \, d\theta$

$= 8a^4\sigma \int_0^{\pi/2} \cos^6\theta \, d\theta = 8a^4 \frac{M}{\pi a^2} \frac{1\cdot 3\cdot 5}{2\cdot 4\cdot 6} \frac{\pi}{2}$

$= \frac{5Ma^2}{4}.$

35 The average value of $f(t)$ over $[0, 1]$ is $\frac{1}{1-0}\int_0^1 f(t)\,dt$, so Average $= \int_0^1 f(t)\,dt = \int_0^1 \int_t^1 e^{x^2}\,dx\,dt$. The inner integral is nonelementary, so we switch the order of integration. The region $0 \le t \le 1$, $t \le x \le 1$ is a triangle in the xt plane which can also be described by $0 \le x \le 1$, $0 \le t \le x$. Thus Average $=$

$\int_0^1 \int_0^x e^{x^2}\,dt\,dx = \int_0^1 e^{x^2} x\,dx = \frac{1}{2}e^{x^2}\Big|_0^1 =$

$\frac{1}{2}(e^1 - e^0) = \frac{1}{2}(e - 1).$

37 (a) The integrand is a square, so the integral must be nonnegative.

(b) By (a), $\int_R [f(x)g(y) - f(y)g(x)]^2\,dA \ge 0.$

The region R is described by $a \le x \le b$, $a \le y \le b$, so $\int_R [f(x)g(y) - f(y)g(x)]^2\,dA$

$= \int_a^b \int_a^b [f(x)g(y) - f(y)g(x)]^2\,dy\,dx$

$= \int_a^b \int_a^b [(f(x))^2(g(y))^2 - 2f(x)g(y)f(y)g(x)$

$+ (f(y))^2(g(x))^2]\,dy\,dx$

$= \int_a^b \int_a^b (f(x))^2(g(y))^2\,dy\,dx -$

$2\int_a^b \int_a^b f(x)g(y)f(y)g(x)\,dy\,dx +$

$\int_a^b \int_a^b (f(y))^2(g(x))^2\,dy\,dx$

$= \int_a^b (f(x))^2\,dx \int_a^b (g(y))^2\,dy -$

$2\int_a^b f(x)g(x)\,dx \int_a^b g(y)f(y)\,dy +$

$\int_a^b (g(x))^2\,dx \int_a^b (f(y))^2\,dy$

$= \int_a^b (f(x))^2\,dx \int_a^b (g(x))^2\,dx -$

$2\int_a^b f(x)g(x)\,dx \int_a^b g(x)f(x)\,dx +$

$\int_a^b (g(x))^2\,dx \int_a^b (f(x))^2\,dx$

$= 2\int_a^b (f(x))^2\,dx \int_a^b (g(x))^2\,dx -$

$2\left(\int_a^b f(x)g(x)\,dx\right) \ge 0.$ Therefore,

$\int_a^b (f(x))^2\,dx \int_a^b (g(x))^2\,dx \ge$

$\left(\int_a^b f(x)g(x)\,dx\right)^2$, as claimed.

39 Let R be the triangle in the xy plane whose vertices are (1, 1), (2, 1), and (2, 2); its description is $1 \le x \le 2$, $1 \le y \le x$. Therefore the volume above R and between the planes $z = 3x + y$ and $z = 4x + 2y$ is $\int_1^2 \int_1^x [(4x + 2y) - (3x + y)]\,dy\,dx$

$= \int_1^2 \int_1^x (x + y)\,dy\,dx = \int_1^2 \left(xy + \frac{y^2}{2}\right)\Big|_1^x\,dx$

$$= \int_1^2 \left(x^2 + \frac{x^2}{2} - x - \frac{1}{2}\right) dx$$

$$= \int_1^2 \left(\frac{3x^2}{2} - x - \frac{1}{2}\right) dx = \left(\frac{x^3}{2} - \frac{x^2}{2} - \frac{x}{2}\right)\Bigg|_1^2$$

$$= 4 - 2 - 1 - \frac{1}{2} + \frac{1}{2} + \frac{1}{2} = \frac{3}{2}.$$

41 R is described by $0 \le \theta \le 2\pi$, $0 \le r \le 1$, $0 \le z \le g(r)$. Therefore $\int_R z\, dV =$

$$\int_0^{2\pi} \int_0^1 \int_0^{g(r)} zr\, dz\, dr\, d\theta = 2\pi \int_0^1 \frac{r}{2}[g(r)]^2\, dr$$

$$= \int_0^1 \pi r[g(r)]^2\, dr = \int_0^1 \pi y[g(y)]^2\, dy,$$ which is one of the desired results. Since g is a decreasing function, it must be one-to-one and g^{-1} is therefore well-defined. R thus has the alternative description $0 \le \theta \le 2\pi$, $0 \le z \le g(0)$, $0 \le r \le g^{-1}(z)$.

Hence $\int_R z\, dV = \int_0^{2\pi} \int_0^{g(0)} \int_0^{g^{-1}(z)} zr\, dr\, dz\, d\theta$

$$= 2\pi \int_0^{g(0)} z \cdot \frac{1}{2}[g^{-1}(z)]^2\, dz$$

$$= \int_0^{g(0)} \pi[g^{-1}(z)]^2\, z\, dz.$$

43 The volume of the pyramid is $\frac{1}{3}abh$, so its density is $\delta = \frac{3M}{abh}$. The most obvious way to place the

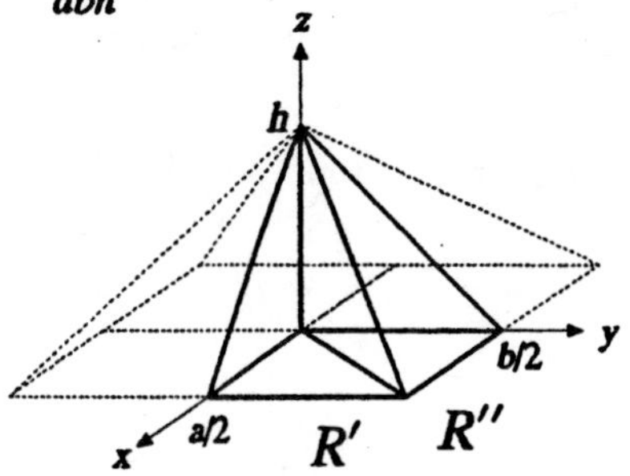

pyramid in an xyz coordinate system is with its base in the xy plane, sides parallel to the axes, and the vertex at the point $(0, 0, h)$. Describing the region R is then difficult because it is bounded by four different planes. However, symmetry can be used to simplify the problem. Consider instead R' and R'', subregions of R whose descriptions are, respectively, $0 \le x \le a/2$, $0 \le y \le \frac{b}{a}x$, $0 \le z \le h\left(1 - \frac{2}{a}x\right)$, and $0 \le y \le b/2$, $0 \le x \le \frac{a}{b}y$, $0 \le z \le h\left(1 - \frac{2}{b}y\right)$. (Observe that the two descriptions vary only in that x and y, and a and b, change places.) Let I' and I'' be the moments of inertia of R' and R'', respectively, about the z axis. By symmetry, $I' + I''$ is one fourth of the moment of inertia of R. We need, in fact, compute only I':

$$I' = \delta \int_{R'} (x^2 + y^2)\, dA$$

$$= \delta \int_0^{a/2} \int_0^{bx/a} \int_0^{h(1-2x/a)} (x^2 + y^2)\, dz\, dy\, dx$$

$$= h\delta \int_0^{a/2} \int_0^{bx/a} (x^2 + y^2)\left(1 - \frac{2}{a}x\right) dy\, dx$$

$$= h\delta \int_0^{a/2} \left(x^2y + \frac{y^3}{3}\right)\left(\frac{a - 2x}{a}\right)\Bigg|_0^{bx/a} dx$$

$$= \frac{h\delta}{a} \int_0^{a/2} \left(\frac{bx^3}{a} + \frac{b^3x^3}{3a^3}\right)(a - 2x)\, dx$$

$$= \frac{h}{a} \cdot \frac{3M}{abh} \cdot \frac{3a^2b + b^3}{3a^3}\left(\frac{ax^4}{4} - \frac{2x^5}{5}\right)\Bigg|_0^{a/2}$$

$$= \frac{M}{a^5}(3a^2 + b^2)\left(\frac{a^5}{64} - \frac{a^5}{80}\right)$$

$= M(3a^2 + b^2)\left(\frac{1}{64} - \frac{1}{80}\right) = \frac{M(3a^2 + b^2)}{320}$. It follows by interchanging a and b that $I'' = \frac{M(3b^2 + a^2)}{320}$. The moment of inertia of R about the z axis is therefore $4(I' + I'')$

$$= 4\left[\frac{M(3a^2 + b^2)}{320} + \frac{M(3b^2 + a^2)}{320}\right]$$

$$= \frac{4M}{320}[4a^2 + 4b^2] = \frac{M}{20}(a^2 + b^2).$$

45 The volume of the torus is $2\pi^2a^2b$. Identifying L and L' with the z and x axes, respectively, we can describe the torus by $0 \le \theta \le 2\pi$, $-a \le z \le a$, $b - \sqrt{a^2 - z^2} \le r \le b + \sqrt{a^2 - z^2}$. Therefore I_L

$$= \frac{M}{2\pi^2a^2b}\int_0^{2\pi}\int_{-a}^{a}\int_{b-\sqrt{a^2-z^2}}^{b+\sqrt{a^2-z^2}} r^2\cdot r\, dr\, dz\, d\theta = M\left(b^2 + \frac{3a^2}{4}\right).$$

47 We take advantage of the parallel axis theorem from Exercise 39 of Sec. 15.6, which gives an equation connecting the moment of inertia about an axis through the centroid of an object with the moment of inertia about any line parallel to that axis. We have $I_2 = I_1 + k^2M$, where I_1 is the moment of inertia about the axis through the centroid, M is the mass, k is the distance between the axis line and the parallel line, and I_2 is the moment of inertia about the parallel line. Place the cone R in a cylindrical coordinate system so that its description is $0 \le \theta \le 2\pi$, $0 \le z \le h$, $0 \le r \le az/h$. The density is $\delta = \frac{3M}{\pi a^2h}$. Thus $I_2 = I_x =$

$$\delta\int_R (y^2 + z^2)\, dV$$

$$= \delta\int_0^{2\pi}\int_0^h\int_0^{az/h}(r^2\sin^2\theta + z^2)\, r\, dr\, dz\, d\theta$$

$$= \delta\int_0^{2\pi}\int_0^h\left(\frac{r^4}{4}\sin^2\theta + z^2\frac{r^2}{2}\right)\Bigg|_0^{az/h} dz\, d\theta$$

$$= \delta\int_0^{2\pi}\int_0^h\left(\frac{a^4z^4}{4h^4}\sin^2\theta + z^2\frac{a^2z^2}{2h^2}\right) dz\, d\theta$$

$$= \frac{3M}{\pi a^2h}\int_0^{2\pi}\left(\frac{a^4}{4h^4}\frac{h^5}{5}\sin^2\theta + \frac{a^2}{2h^2}\frac{h^5}{5}\right) d\theta$$

$$= \frac{3M}{\pi a^2h}\left(\frac{a^4h}{20}\pi + \frac{a^2h^3}{10}2\pi\right) = \frac{3Ma^2}{20} + \frac{3Mh^2}{5}.$$

Now the centroid of R is at a height of $3h/4$, which is the distance between the x axis and a parallel line through the centroid. Hence $k = 3h/4$ and $I_1 =$

$$I_2 - k^2M = \frac{3Ma^2}{20} + \frac{3Mh^2}{5} - \frac{9h^2}{16}M$$

$$= \frac{3Ma^2}{20} + \frac{3Mh^2}{80} = \frac{3M}{20}\left(a^2 + \frac{h^2}{4}\right), \text{ as claimed.}$$

49 Place the bored-out cylinder atop the xy plane with its axis along the z axis. Then the moment of inertia is I_z

$$= \frac{M}{\pi(a^2 - b^2)h}\int_0^{2\pi}\int_b^a\int_0^h r^2\cdot r\, dz\, dr\, d\theta$$

$= \frac{1}{2}M(a^2 + b^2)$, where h, the height of the cylinder, turned out not to matter.

51 (a) Let $u = 2x$, then $\int_0^\infty e^{-4x^2}\, dx = \int_0^\infty e^{-u^2}\frac{du}{2}$

$$= \frac{1}{2}\frac{\sqrt{\pi}}{2} = \frac{\sqrt{\pi}}{4}.$$

(b) Let $u = \sqrt{x}$; then $x = u^2$ and $du = \dfrac{dx}{2\sqrt{x}}$, so

$$\int_0^\infty \frac{e^{-x}}{\sqrt{x}}\,dx = \int_0^\infty e^{-u^2}\,2\,du = 2\cdot\frac{\sqrt{\pi}}{2}$$

$= \sqrt{\pi}$.

(c) Integrate by parts, with $u = x$ and $dv = xe^{-x^2}\,dx$; then $du = dx$, $v = -\frac{1}{2}e^{-x^2}$, and

$$\int_0^\infty x^2e^{-x^2}\,dx$$

$$= x\left(-\frac{1}{2}e^{-x^2}\right)\Big|_0^\infty + \frac{1}{2}\int_0^\infty e^{-x^2}\,dx$$

$$= 0 + \frac{1}{2}\cdot\frac{\sqrt{\pi}}{2} = \frac{\sqrt{\pi}}{4}.$$

(d) Let $u = \sqrt{x}$; then $x = u^2$ and $dx = 2u\,du$, so

$$\int_0^\infty \sqrt{x}e^{-x}\,dx = \int_0^\infty ue^{-u^2}\,2u\,du$$

$$= 2\int_0^\infty u^2e^{-u^2}\,du = 2\cdot\frac{\sqrt{\pi}}{4} = \frac{\sqrt{\pi}}{2},$$ by (c).

(e) Let $u = \ln(1/x) = -\ln x$. Then $du = -dx/x$, so $dx = -x\,du = -e^{-u}\,du$ and $\displaystyle\int_0^1 \frac{dx}{\sqrt{\ln(1/x)}}$

$$= \int_\infty^0 \frac{-e^{-u}\,du}{\sqrt{u}} = \int_0^\infty \frac{e^{-u}}{\sqrt{u}}\,du = \sqrt{\pi},$$ from (b).

(f) Using the substitution from (e),

$$\int_0^1 \sqrt{\ln(1/x)}\,dx = \int_\infty^0 \sqrt{u}(-e^{-u})\,du =$$

$$\int_0^\infty \sqrt{u}e^{-u}\,du = \frac{\sqrt{\pi}}{2},$$ as shown in (d).

53 Since $\mathbf{u}_\rho$, $\mathbf{u}_\phi$, and $\mathbf{u}_\theta$ are perpendicular unit vectors, we have $\nabla f = A\mathbf{u}_\rho + B\mathbf{u}_\phi + C\mathbf{u}_\theta$, where $A = \nabla f\cdot\mathbf{u}_\rho = D_{\mathbf{u}_\rho}f$, $B = \nabla f\cdot\mathbf{u}_\phi = D_{\mathbf{u}_\phi}f$, and $C = \nabla f\cdot\mathbf{u}_\theta = D_{\mathbf{u}_\theta}f$. We must express the directional derivatives of f with respect to ρ, ϕ, and θ. First consider $\dfrac{\partial f}{\partial\rho}$. By definition, this is

$$\lim_{d\rho\to 0}\frac{f(\rho + d\rho, \phi, \theta) - f(\rho, \phi, \theta)}{d\rho}.$$ The point $(\rho + d\rho, \phi, \theta)$ is a distance $d\rho$ from (ρ, ϕ, θ) in the direction $\mathbf{u}_\rho$. Thus, for small values of $d\rho$, $f(\rho + d\rho, \phi, \theta) \approx f(\rho, \phi, \theta) + d\rho\, D_{\mathbf{u}_\rho}f$. Taking the limit as $d\rho \to 0$, we conclude that $\dfrac{\partial f}{\partial\rho} = D_{\mathbf{u}_\rho}f = A$. For $\dfrac{\partial f}{\partial\phi}$, the situation is slightly messier. For small $d\phi$, the point $(\rho, \phi + d\phi, \theta)$ is approximately a distance $\rho\,d\phi$ from (ρ, ϕ, θ), approximately in the direction $\mathbf{u}_\phi$. As $d\phi \to 0$, these approximations become more accurate, so $f(\rho, \phi + d\phi, \theta) \approx f(\rho, \phi, \theta) + \rho\,d\phi\,D_{\mathbf{u}_\phi}f$. We conclude that $\dfrac{\partial f}{\partial\phi} =$

$$\lim_{d\phi\to 0}\frac{f(\rho, \phi + d\phi, \theta) - f(\rho, \phi, \theta)}{d\rho} = \rho\,D_{\mathbf{u}_\phi}f.$$

Similarly, $(\rho, \phi, \theta + d\theta)$ is approximately a distance $\rho\sin\phi\,d\theta$ from (ρ, ϕ, θ), approximately in the direction $\mathbf{u}_\theta$. Therefore $f(\rho, \phi, \theta + d\theta) \approx f(\rho, \phi, \theta) + \rho\sin\phi\,d\theta\,D_{\mathbf{u}_\theta}f$ and $\dfrac{\partial f}{\partial\theta} = \rho\sin\phi\,D_{\mathbf{u}_\theta}f$. Putting these results together gives

$$\nabla f = \left(D_{\mathbf{u}_\rho}f\right)\mathbf{u}_\rho + \left(D_{\mathbf{u}_\phi}f\right)\mathbf{u}_\phi + \left(D_{\mathbf{u}_\theta}f\right)\mathbf{u}_\theta$$

$$= \frac{\partial f}{\partial\rho}\mathbf{u}_\rho + \frac{1}{\rho}\frac{\partial f}{\partial\phi}\mathbf{u}_\phi + \frac{1}{\rho\sin\phi}\frac{\partial f}{\partial\theta}\mathbf{u}_\theta.$$

16 Green's Theorem

16.1 Vector and Scalar Fields

1 $\mathbf{F} = x^2y\mathbf{i} + \sin xy\mathbf{j}$; $\nabla\cdot\mathbf{F} = \frac{\partial}{\partial x}(x^2y) + \frac{\partial}{\partial y}(\sin xy)$

$= 2xy + x\cos xy$

3 $\mathbf{r} = x\mathbf{i} + y\mathbf{j}$; $\nabla\cdot\mathbf{r} = \frac{\partial}{\partial x}(x) + \frac{\partial}{\partial y}(y) = 1 + 1 = 2$

5 $\mathbf{F} = e^{xy}\mathbf{i} + x\tan 2y\,\mathbf{j} + xz^2\mathbf{k}$;

$\nabla\cdot\mathbf{F} = \frac{\partial}{\partial x}(e^{xy}) + \frac{\partial}{\partial y}(x\tan 2y) + \frac{\partial}{\partial z}(xz^2)$

$= ye^{xy} + 2x\sec^2 2y + 2xz$

7 $\mathbf{F} = \frac{\hat{\mathbf{r}}}{|\mathbf{r}|} = \frac{\mathbf{r}}{|\mathbf{r}|^2} = \frac{x}{x^2+y^2}\mathbf{i} + \frac{y}{x^2+y^2}\mathbf{j}$, so

$\nabla\cdot\mathbf{F} = \frac{\partial}{\partial x}\left(\frac{x}{x^2+y^2}\right) + \frac{\partial}{\partial y}\left(\frac{y}{x^2+y^2}\right)$

$= \frac{(x^2+y^2)(1) - x(2x)}{(x^2+y^2)^2} + \frac{(x^2+y^2)(1) - y(2y)}{(x^2+y^2)^2}$

$= \frac{y^2 - x^2}{(x^2+y^2)^2} + \frac{x^2 - y^2}{(x^2+y^2)^2} = 0.$

9 $\mathbf{F} = (3x + 2y + 5z)\mathbf{i} + (2x - 3y + 4z)\mathbf{j} + (x + 6y + 7z)\mathbf{k}$;

$$\nabla\times\mathbf{F} = \begin{vmatrix} \mathbf{i} & \mathbf{j} & \mathbf{k} \\ \frac{\partial}{\partial x} & \frac{\partial}{\partial y} & \frac{\partial}{\partial z} \\ 3x+2y+5z & 2x-3y+4z & x+6y+7z \end{vmatrix}$$

$= (6 - 4)\mathbf{i} + (5 - 1)\mathbf{j} + (2 - 2)\mathbf{k} = 2\mathbf{i} + 4\mathbf{j}$

11 Let $\mathbf{F} = \frac{\hat{\mathbf{r}}}{|\mathbf{r}|} = \frac{\mathbf{r}}{|\mathbf{r}|^2}$. Define $r = |\mathbf{r}| = \sqrt{x^2 + y^2 + z^2}$. Then $\mathbf{F} = \frac{x}{r^2}\mathbf{i} + \frac{y}{r^2}\mathbf{j} + \frac{z}{r^2}\mathbf{k}$ and

$\nabla\times\mathbf{F} = \left[\frac{\partial}{\partial y}\left(\frac{z}{r^2}\right) - \frac{\partial}{\partial z}\left(\frac{y}{r^2}\right)\right]\mathbf{i} + \left[\frac{\partial}{\partial z}\left(\frac{x}{r^2}\right) - \frac{\partial}{\partial x}\left(\frac{z}{r^2}\right)\right]\mathbf{j} + \left[\frac{\partial}{\partial x}\left(\frac{y}{r^2}\right) - \frac{\partial}{\partial y}\left(\frac{x}{r^2}\right)\right]\mathbf{k}.$

Now $\frac{\partial}{\partial x}\left(\frac{1}{r^2}\right) = -(x^2 + y^2 + z^2)^{-2}\cdot 2x = \frac{-2x}{(x^2+y^2+z^2)^2} = -\frac{2x}{r^4}$. Similarly, $\frac{\partial}{\partial y}\left(\frac{1}{r^2}\right) = -\frac{2y}{r^4}$ and $\frac{\partial}{\partial z}\left(\frac{1}{r^2}\right) = -\frac{2z}{r^4}$. Thus $\nabla\times\mathbf{F} =$

$\left[z\left(-\frac{2y}{r^4}\right) - y\left(-\frac{2z}{r^4}\right)\right]\mathbf{i} + \left[x\left(-\frac{2z}{r^4}\right) - z\left(-\frac{2x}{r^4}\right)\right]\mathbf{j} + \left[y\left(-\frac{2x}{r^4}\right) - x\left(-\frac{2y}{r^4}\right)\right]\mathbf{k} = \mathbf{0}.$

13 Let $\mathbf{F} = \frac{\hat{\mathbf{r}}}{|\mathbf{r}|^3} = \frac{\mathbf{r}}{|\mathbf{r}|^4}$. Define $r = |\mathbf{r}| = \sqrt{x^2 + y^2 + z^2}$. Then $\mathbf{F} = \frac{x}{r^4}\mathbf{i} + \frac{y}{r^4}\mathbf{j} + \frac{z}{r^4}\mathbf{k}$ and

$\nabla\times\mathbf{F} = \left[\frac{\partial}{\partial y}\left(\frac{z}{r^4}\right) - \frac{\partial}{\partial z}\left(\frac{y}{r^4}\right)\right]\mathbf{i} +$

$\left[\frac{\partial}{\partial z}\left(\frac{x}{r^4}\right) - \frac{\partial}{\partial x}\left(\frac{z}{r^4}\right)\right]\mathbf{j} + \left[\frac{\partial}{\partial x}\left(\frac{y}{r^4}\right) - \frac{\partial}{\partial y}\left(\frac{x}{r^4}\right)\right]\mathbf{k}.$

Now $\frac{\partial}{\partial x}\left(\frac{1}{r^4}\right) = \frac{\partial}{\partial x}[(x^2 + y^2 + z^2)^{-2}] =$

$-2(x^2 + y^2 + z^2)^{-3}\cdot(2x) = -\frac{4x}{r^6}$. Similarly,

$\frac{\partial}{\partial y}\left(\frac{1}{r^4}\right) = -\frac{4y}{r^6}$ and $\frac{\partial}{\partial z}\left(\frac{1}{r^4}\right) = -\frac{4z}{r^6}$. Thus $\nabla \times \mathbf{F}$

$= \left[z\left(-\frac{4y}{r^6}\right) - y\left(-\frac{4z}{r^6}\right)\right]\mathbf{i} + \left[x\left(-\frac{4z}{r^6}\right) - z\left(-\frac{4x}{r^6}\right)\right]\mathbf{j} +$

$\left[y\left(-\frac{4x}{r^6}\right) - x\left(-\frac{4y}{r^6}\right)\right]\mathbf{k} = \mathbf{0}.$

15 The field $f\mathbf{F}$ is a vector field.

17 (a) $\nabla\times\nabla f = \begin{vmatrix} \mathbf{i} & \mathbf{j} & \mathbf{k} \\ \frac{\partial}{\partial x} & \frac{\partial}{\partial y} & \frac{\partial}{\partial z} \\ f_x & f_y & 0 \end{vmatrix}$

$= (0 - 0)\mathbf{i} + (0 - 0)\mathbf{j} + (f_{yx} - f_{xy})\mathbf{k} = \mathbf{0}$

(b) $\nabla\times\nabla f = \begin{vmatrix} \mathbf{i} & \mathbf{j} & \mathbf{k} \\ \frac{\partial}{\partial x} & \frac{\partial}{\partial y} & \frac{\partial}{\partial z} \\ f_x & f_y & f_z \end{vmatrix}$

$= (f_{zy} - f_{yz})\mathbf{i} - (f_{zx} - f_{xz})\mathbf{j} + (f_{yx} - f_{xy})\mathbf{k}$
$= 0\mathbf{i} - 0\mathbf{j} + 0\mathbf{k} = \mathbf{0}$

19 (a) curl **F** is a vector field.

(b) $\|\mathbf{F}\|$ is a scalar field.

(c) $\mathbf{F}\cdot\mathbf{F}$ is a scalar field.

(d) div **F** is a scalar field.

(e) $\nabla\cdot\mathbf{F}$ is a scalar field.

(f) $\mathbf{F} \times \mathbf{i}$ is a vector field.

(g) $\nabla \times \mathbf{F}$ is a vector field.

21 $f(x, y) = \ln(x^2 + y^2)$; $f_x = \frac{2x}{x^2 + y^2}$ and $f_y = \frac{2y}{x^2 + y^2}$. From Exercise 7 of this section, $f_{xx} = \frac{\partial}{\partial x}(f_x) = \frac{2(y^2 - x^2)}{(x^2 + y^2)^2}$ and $f_{yy} = \frac{2(x^2 - y^2)}{(x^2 + y^2)^2}$. Thus

$\nabla^2 f = f_{xx} + f_{yy} = \frac{2y^2 - 2x^2}{(x^2 + y^2)^2} + \frac{2x^2 - 2y^2}{(x^2 + y^2)^2} = 0$

23 Let $f = \frac{1}{|\mathbf{r}|} = \frac{1}{\sqrt{x^2 + y^2 + z^2}} = \frac{1}{r}$, where $r = |\mathbf{r}|$. Then $\nabla^2 f = f_{xx} + f_{yy} + f_{zz}$. Now $\frac{\partial}{\partial x}\left(\frac{1}{r}\right) =$

$-\frac{1}{2}\cdot(x^2 + y^2 + z^2)^{-3/2}\cdot 2x = -\frac{x}{r^3}$ and $\frac{\partial}{\partial x}\left(-\frac{x}{r^3}\right)$

$= -\frac{r^3\cdot\frac{\partial x}{\partial x} - x\cdot\frac{\partial r^3}{\partial x}}{r^6} = \frac{x\cdot 3r^2\cdot\frac{x}{r} - r^3}{r^6} =$

$\frac{3x^2}{r^5} - \frac{1}{r^3}$. Similarly, $\frac{\partial^2}{\partial y^2}\left(\frac{1}{r}\right) = \frac{3y^2}{r^5} - \frac{1}{r^3}$ and

$\frac{\partial^2}{\partial z^2}\left(\frac{1}{r}\right) = \frac{3z^2}{r^5} - \frac{1}{r^3}$. Thus $\nabla^2 f =$

$\left(\frac{3x^2}{r^5} - \frac{1}{r^3}\right) + \left(\frac{3y^2}{r^5} - \frac{1}{r^3}\right) + \left(\frac{3z^2}{r^5} - \frac{1}{r^3}\right) =$

$\frac{3(x^2 + y^2 + z^2)}{r^5} - \frac{3}{r^3} = \frac{3r^2}{r^5} - \frac{3}{r^3} = 0.$

25 (a) Suppose $\mathbf{F} = a\mathbf{i} + b\mathbf{j} + c\mathbf{k}$, where a, b, and c are constants. Then $\nabla\cdot\mathbf{F} =$

$\frac{\partial}{\partial x}(a) + \frac{\partial}{\partial y}(b) + \frac{\partial}{\partial z}(c) = 0 + 0 + 0 = 0.$

(b) Suppose $\mathbf{F} = f\mathbf{i} + g\mathbf{j} + h\mathbf{k}$. Then $\nabla\cdot\mathbf{F} = 0$ whenever $f_x + g_y + h_z = 0$. One example is $\mathbf{F} = x\mathbf{i} + y\mathbf{j} - 2z\mathbf{k}$: $\nabla\cdot\mathbf{F} = 1 + 1 - 2 = 0$.

27 $\nabla(fg) = (fg)_x\mathbf{i} + (fg)_y\mathbf{j} + (fg)_z\mathbf{k}$

$= (fg_x + f_xg)\mathbf{i} + (fg_y + f_yg)\mathbf{j} + (fg_z + f_zg)\mathbf{k}$

$= (fg_x)\mathbf{i} + (fg_y)\mathbf{j} + (fg_z)\mathbf{k} + (f_xg)\mathbf{i} + (f_yg)\mathbf{j} + (f_zg)\mathbf{k}$

$= f(g_x\mathbf{i} + g_y\mathbf{j} + g_z\mathbf{k}) + g(f_x\mathbf{i} + f_y\mathbf{j} + f_z\mathbf{k})$

$= f\nabla g + g\nabla f$

29 $\mathbf{F} = P\mathbf{i} + Q\mathbf{j} + R\mathbf{k}$ and $\mathbf{G} = L\mathbf{i} + M\mathbf{j} + N\mathbf{k}$, so $\mathbf{F} \times \mathbf{G} = (QN - RM)\mathbf{i} - (PN - RL)\mathbf{j} + (PM - QL)\mathbf{k}$. Therefore, $\nabla\cdot(\mathbf{F} \times \mathbf{G}) = (QN - RM)_x - (PN - RL)_y + (PM - QL)_z$

$= QN_x + Q_xN - RM_x - R_xM - PN_y - P_yN + RL_y + R_yL + PM_z + P_zM - QL_z - Q_zL$

$= L(R_y - Q_z) + M(P_z - R_x) + N(Q_x - P_y) - P(N_y - M_z) - Q(L_z - N_x) - R(M_x - L_y)$

$= \mathbf{G}\cdot(\text{curl } \mathbf{F}) - \mathbf{F}\cdot(\text{curl } \mathbf{G})$, as claimed.

31 Let $r = |\mathbf{r}| = \sqrt{x^2 + y^2 + z^2}$. Then $\mathbf{F} = \dfrac{\hat{\mathbf{r}}}{|\mathbf{r}|^k}$

$= \dfrac{\mathbf{r}}{|\mathbf{r}|^{k+1}} = \dfrac{\mathbf{r}}{r^{k+1}}$. In computing the divergence of

$\mathbf{F}$ we will need $(r^{k+1})_x = \dfrac{\partial}{\partial x}(r^{k+1}) = (k+1)r^k\dfrac{\partial r}{\partial x}$

$= (k+1)r^k\cdot\dfrac{1}{2}(x^2 + y^2 + z^2)^{-1/2}(2x)$

$= (k+1)r^k\cdot\dfrac{x}{r} = (k+1)xr^{k-1}$. Similarly, $(r^{k+1})_y$

$= (k+1)yr^{k-1}$ and $(r^{k+1})_z = (k+1)zr^{k-1}$. We

have $\mathbf{F} = \dfrac{1}{r^{k+1}}(x\mathbf{i} + y\mathbf{j} + z\mathbf{k})$; so $\nabla\cdot\mathbf{F}$

$$= \frac{\partial}{\partial x}\left(\frac{x}{r^{k+1}}\right) + \frac{\partial}{\partial y}\left(\frac{y}{r^{k+1}}\right) + \frac{\partial}{\partial z}\left(\frac{z}{r^{k+1}}\right)$$

$$= \frac{r^{k+1} - (k+1)x^2r^{k-1}}{r^{2(k+1)}} + \frac{r^{k+1} - (k+1)y^2r^{k-1}}{r^{2(k+1)}} + \frac{r^{k+1} - (k+1)z^2r^{k-1}}{r^{2(k+1)}}$$

$$= \frac{3r^{k+1} - (k+1)r^{k-1}(x^2 + y^2 + z^2)}{r^{2(k+1)}}$$

$= \dfrac{3r^{k+1} - (k+1)r^{k+1}}{r^{2(k+1)}} = \dfrac{3 - (k+1)}{r^{k+1}}$. Since $k \neq 2$, it follows that $k + 1 \neq 3$ and $\nabla\cdot\mathbf{F} \neq 0$.

33 Since $\mathbf{F}$ is a central force field, its magnitude is the same for a given distance from the origin. In addition, it is directed toward or away from the origin by definition. So $|\mathbf{F}(5, 0)| = |7\mathbf{i}| = 7$ implies that $|\mathbf{F}(4, 3)| = |\mathbf{F}(3, 4)| = |\mathbf{F}(0, 5)| = |\mathbf{F}(-5, 0)| = |\mathbf{F}(0, -5)| = 7$. Since $\mathbf{F}(5, 0)$ is directed away from $(0, 0)$, all other vectors are as well.

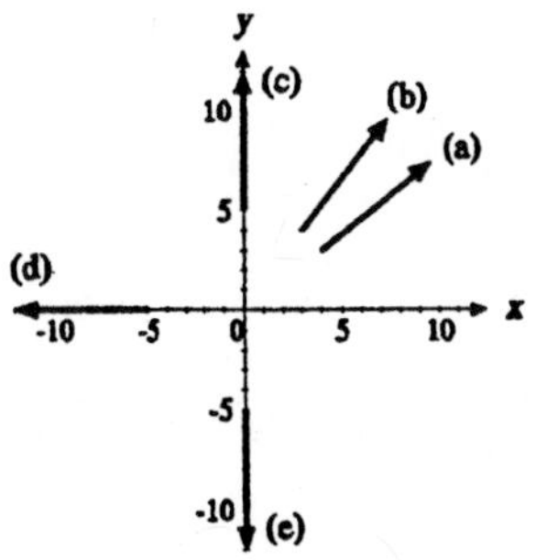

35 Since there is no z component contributing to the analysis, we have, given $\mathbf{F} = f(r)\hat{\mathbf{r}}$, that $\nabla\cdot\mathbf{F} = 0$ implies $f(r) = kr^{-1}$. (See Exercise 34(b).)

37 (a) Since $\mathbf{F} = P\mathbf{i} + Q\mathbf{j}$ is equal to $\nabla f = f_x\mathbf{i} + f_y\mathbf{j}$, we have $P = f_x$ and $Q = f_y$. Then $P_y = (f_x)_y = f_{xy}$ and $Q_x = (f_y)_x = f_{yx} = f_{xy} = P_y$; that is,

$P_y = Q_x$, as was to be shown.

(b) $\mathbf{F} = x^2y\mathbf{i} + x^2y^3\mathbf{j}$, so $P = x^2y$ and $Q = x^2y^3$. Since $P_y = x^2$ and $Q_x = 2xy^3$, which are not equal, there is no function such that $\mathbf{F} = \nabla f$.

16.2 Line Integrals

1 From $\mathbf{G}(t) = t\mathbf{i} + t^2\mathbf{j}$, the curve is described parametrically by $x = t$, $y = t^2$. So, by substitution, the curve described by $\mathbf{G}(t)$ for t in $[0, 1]$ is part of the parabola $y = x^2$. Its start occurs when $t = 0$, or at $(0, 0)$; its finish occurs when $t = 1$, or at $(1, 1)$.

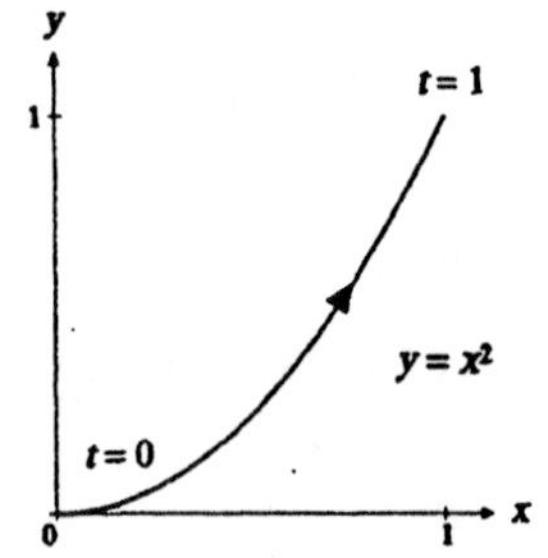

3 The curve is given parametrically as $x = 2t + 1$, $y = 3t$. Solving for t in the first equation gives $t = \dfrac{x - 1}{2}$. Substituting this value for t into the second equation yields $y = 3\left(\dfrac{x - 1}{2}\right)$; thus $\mathbf{G}(t) = (2t + 1)\mathbf{i} + 3t\mathbf{j}$ describes a portion of the line $y = \dfrac{3}{2}x - \dfrac{3}{2}$ when t is in $[0, 2]$. It starts when $t = 0$ at $(2\cdot0 + 1, 3\cdot0) = (1, 0)$; it finishes when $t = 2$ at $(2\cdot2 + 1, 3\cdot2) = (5, 6)$.

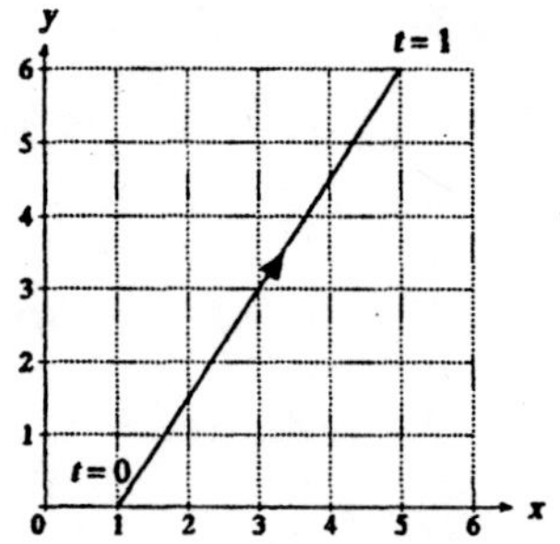

5 On the first part of the path, the horizontal segment, y is 0 and x increases from 0 to 1. Hence this part can be parameterized by $x = t$ and $y = 0$ for t in $[0, 1]$. For the vertical part, x is 1, y increases from 0 to 1, and we wish t to increase from 1 to some number. If we use the parameterization $x = 1$ and $y = t - 1$ for t in $[1, 2]$, then all these constraints are satisfied. Therefore the entire parameterization is $(x, y) = (t, 0)$ for $0 \le t \le 1$ and $(x, y) = (1, t - 1)$ for $1 \le t \le 2$.

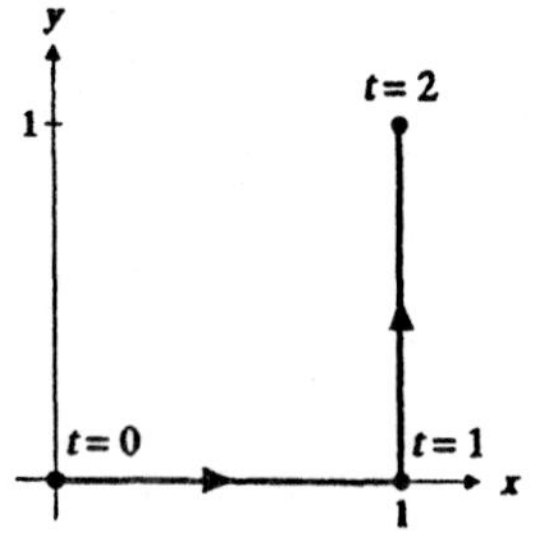

7 Note that the path sweeps out the upper half of a circle of radius 2 centered at the origin. The curve's orientation is counterclockwise, suggesting use of the parameterization of the circle given by $x = 2\cos t$, $y = 2\sin t$. Here the start point $(2, 0)$ occurs when $t = 0$. The finish point $(-2, 0)$ occurs when $t = \pi$. The entire parameterization is $(x, y) = (2\cos t, 2\sin t)$ for $0 \le t \le \pi$.

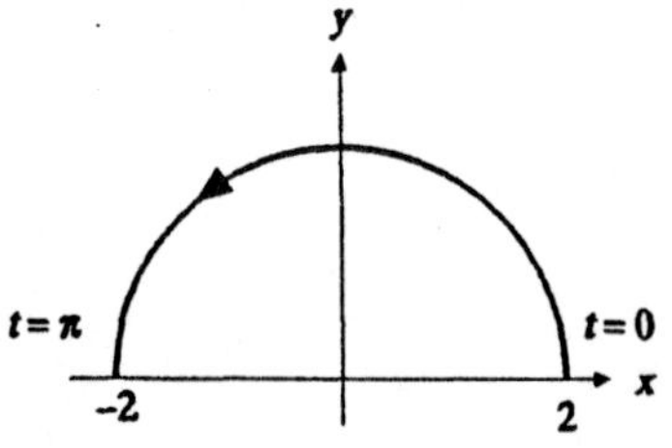

9 In this case, the curve C describes a portion of the line $y = x$. Then C can be parameterized by $x = t$,

$y = t$ for $1 \le t \le 3$, since the initial point is $(1, 1)$ and the terminal point is $(3, 3)$. So $\int_C xy\,dx$

$$= \int_1^3 xy \frac{dx}{dt}\,dt = \int_1^3 t \cdot t \cdot 1\,dt = \left.\frac{t^3}{3}\right|_1^3 =$$

$$\frac{27}{3} - \frac{1}{3} = \frac{26}{3}.$$

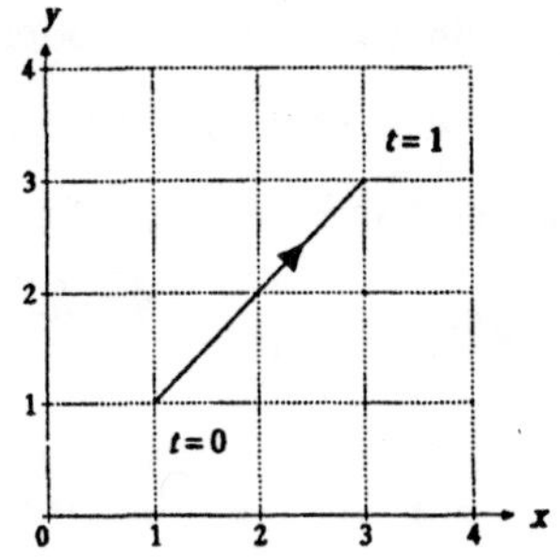

11 For this curve, y is 2 for all t and x varies from 3 to 7. So choose the parameterization $x = t$, $y = 2$, $3 \le t \le 7$ for C. Then $\int_C x^2\,dy = \int_3^7 x^2 \frac{dy}{dt}\,dt$

$$= \int_3^7 t^2 \cdot 0\,dt = 0.$$

13 We must first parameterize C as follows:

C_1: $x = t$, $y = 0$; $0 \le t \le a$

C_2: $x = a$, $y = t$; $0 \le t \le b$

C_3: $x = a - t$, $y = b$; $0 \le t \le a$

C_4: $x = 0$, $y = b - t$; $0 \le t \le b$.

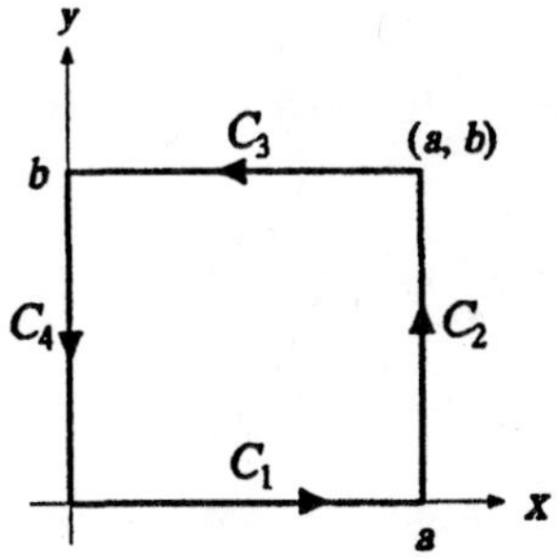

Now $\oint_C y\,dx = \int_{C_1} y\,dx + \int_{C_2} y\,dx + \int_{C_3} y\,dx$

$$+ \int_{C_4} y\,dx = \int_0^a y \frac{dx}{dt}\,dt + \int_0^b y \frac{dx}{dt}\,dt +$$

$$\int_0^a y \frac{dx}{dt}\,dt + \int_0^b y \frac{dx}{dt}\,dt = \int_0^a 0 \cdot 1\,dt +$$

$$\int_0^b t \cdot 0\,dt + \int_0^a b(-1)\,dt + \int_0^b (b - t) \cdot 0\,dt =$$

$-ab$, which is the negative of the area of the rectangle corresponding to R, as expected.

15 Let C be the curve given and introduce coordinate axes as in the figure. Denote the uppermost point

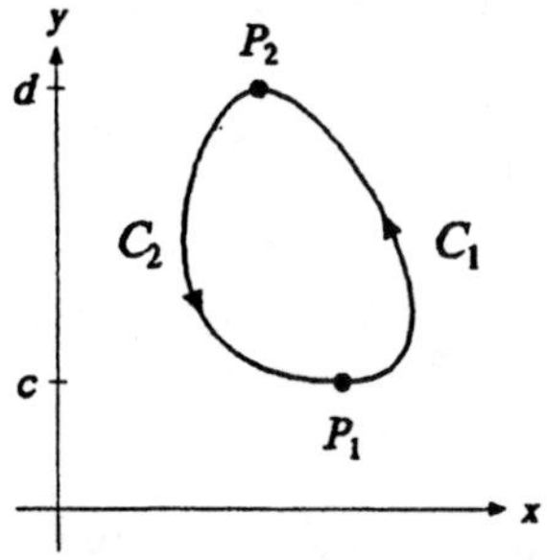

on C by P_2, and label its y coordinate d. Similarly, denote the lowermost point on C by P_1, and label its y coordinate c. Let the path from P_1 to P_2 be C_1, and let the path from P_2 to P_1 be C_2. (Note that if a horizontal line were to cross C in more than two points, we could divide C into several curves such that each curve will not be intersected more than twice by a horizontal line using the cancellation principle.) Then $\int_{C_1} x\,dy$ = (Area to the left of C_1 and to the right of $[c, d]$) $= A_1$ and $\int_{C_2} x\,dy$ = −(Area to the left of C_2 and to the right of $[c, d]$) $= -A_2$. Hence $\oint_C x\,dy = \int_{C_1} x\,dy + \int_{C_2} x\,dy =$ $A_1 - A_2$ = (Area bounded by C).

17 Here $\oint_C y\,dx = -(\text{Area bounded by } C) = -5.$

So $\oint_C 3y\,dx = 3\oint_C y\,dx = 3(-5) = -15.$

19 Recall that $\oint_C x\,dx = \oint_C y\,dy = 0$ and $\oint_C x\,dy$

$= (\text{Area bounded by } C) = 5.$ Then

$\oint_C [2x\,dx + (x + y)\,dy] =$

$2\oint_C x\,dx + \oint_C x\,dy + \oint_C y\,dy = 2\cdot 0 + 5 + 0$

$= 5.$

21 Any curve from (1, 3) to (2, 7) can be parameterized by $x = f(t)$, $y = g(t)$, $0 \le t \le 1$, where $f(0) = 1$, $f(1) = 2$, $g(0) = 3$, and $g(1) = 7$.

Then $\int_C x^3\,dx = \int_0^1 [f(t)]^3 f'(t)\,dt = \frac{1}{4}[f(t)]^4\Big|_0^1$

$= \frac{1}{4}[f(1)]^4 - \frac{1}{4}[f(0)]^4 = \frac{16}{4} - \frac{1}{4} = \frac{15}{4}.$

23 (a) The line through the points (1, 1) and (2, 4) is given by $y = 3x - 2$. Hence C_1 is parameterized by $x = t$, $y = 3t - 2$, $1 \le t \le 2$. So $\int_{C_1} (xy\,dx + x\,dy)$

$= \int_1^2 \left(xy\frac{dx}{dt} + x\frac{dy}{dt}\right) dt$

$= \int_1^2 [t(3t - 2)\cdot 1 + t\cdot 3]\,dt$

$= \int_1^2 (3t^2 + t)\,dt = t^3 + \frac{t^2}{2}\Big|_1^2 = \frac{17}{2}.$

(b) Along the parabola $y = x^2$, the path C_2 is parameterized by $x = t$, $y = t^2$, $1 \le t \le 2$.

Then $\int_{C_2} (xy\,dx + x\,dy)$

$= \int_1^2 \left(xy\frac{dx}{dt} + x\frac{dy}{dt}\right) dt$

$= \int_1^2 (t\cdot t^2\cdot 1 + t\cdot 2t)\,dt = \int_1^2 (t^3 + 2t^2)\,dt$

$= \frac{t^4}{4} + \frac{2t^3}{3}\Big|_1^2 = \frac{101}{12}.$

25 In order to apply the cancellation principle, we must assure that all curves are oriented counterclockwise (including the curve C). Since C_3 is the only curve whose orientation is not counterclockwise, consider $\oint_{-C_3} f\,dy = -\oint_{C_3} f\,dy$

$= -(-4) = 4.$ Then $\oint_C f\,dy =$

$\oint_{C_1} f\,dy + \oint_{C_2} f\,dy + \oint_{-C_3} f\,dy = 2 + 5 + 4 =$

11 by the cancellation principle.

27 (a) If $x = f(t)$ and $y = g(t)$ is a parameterization of a path from (0, 0) to (1, 2), then $dx = f'(t)\,dt$ and $dy = g'(t)\,dt$. Suppose that t ranges from a to b; then $f(a) = g(a) = 0$, $f(b) = 1$, and $g(b) = 2$. Hence we have

$\int_C y\,dx + x\,dy$

$= \int_a^b g(t)f'(t)\,dt + f(t)g'(t)\,dt$

$= \int_a^b (g(t)f'(t) + f(t)g'(t))\,dt$

$= \int_a^b (g(t)f(t))'\,dt = g(t)f(t)\big|_a^b$

$= g(b)f(b) - g(a)f(a) = 2\cdot 1 - 0\cdot 0 = 2.$

(b) The result is immediate from part (a). The value of the integral depended only on the values of f and g at the endpoints of the interval of integration. Since these were simply the coordinates of the endpoints of C, the result follows.

29 Label the two parts of C as shown. Then on C_1 we have $x = t$, $y = t$, and $z = 0$ as t goes from 0 to

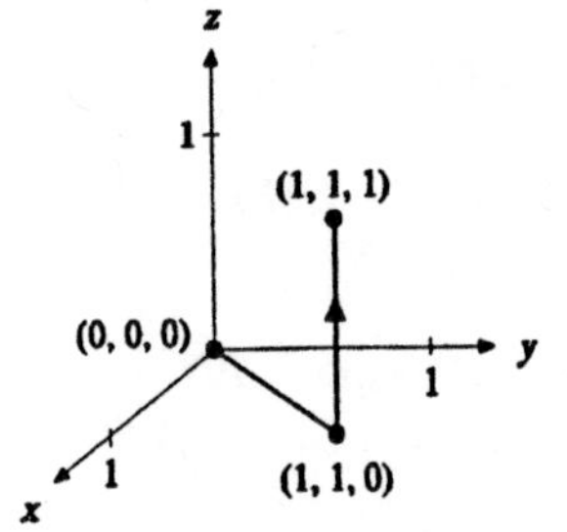

1; note also that $dx = dy = dt$ and $dz = 0$. On C_2 a suitable parameterization is $x = y = 1$ and $z = t$ as t goes from 0 to 1; in this case $dx = dy = 0$ and $dz = dt$. We may now write

$$\int_{C_1} e^x y\,dx + x\sin\pi y\,dy + xy\tan^{-1}z\,dz$$

$$= \int_0^1 (e^t t + t\sin\pi t)\,dt$$

$$= (t-1)e^t\Big|_0^1 + \left(\frac{1}{\pi^2}\sin\pi t - \frac{t}{\pi}\cos\pi t\right)\Bigg|_0^1$$

$$= 1 + \frac{1}{\pi}; \text{ and}$$

$$\int_{C_2} e^x y\,dx + x\sin\pi y\,dy + xy\tan^{-1}z\,dz =$$

$$\int_0^1 \tan^{-1}t\,dt = \left[t\tan^{-1}t - \frac{1}{2}\ln(1+t^2)\right]\Bigg|_0^1$$

$$= \frac{\pi}{4} - \frac{1}{2}\ln 2. \text{ Therefore}$$

$$\int_C e^x y\,dx + x\sin\pi y\,dy + xy\tan^{-1}z\,dz$$

$$= 1 + \frac{1}{\pi} + \frac{\pi}{4} - \frac{1}{2}\ln 2.$$

31 (a) We have $dx = 0$, $dy = dt$, and $0 \le t \le 1$, so $\int_C \frac{-y\,dx}{x^2+y^2} + \frac{x\,dy}{x^2+y^2} = \int_0^1 \frac{dt}{1+t^2} =$

$$\tan^{-1}t\Big|_0^1 = \frac{\pi}{4}.$$

(b) For the first path, we have $dx = -2\pi\sin 2\pi t\,dt$, $dy = 2\pi\cos 2\pi t\,dt$, $0 \le t \le \frac{1}{4}$; for the second, $dx = dt$, $dy = 0$, $0 \le t \le 1$. Thus $\int_C \frac{-y\,dx}{x^2+y^2} + \frac{x\,dy}{x^2+y^2} =$

$$\int_0^{1/4} [(-\sin 2\pi t)(-2\pi\sin 2\pi t) + (\cos 2\pi t)(2\pi\cos 2\pi t)]\,dt + \int_0^1 \frac{-dt}{t^2+1}$$

$$= \int_0^{1/4} 2\pi\,dt - \int_0^1 \frac{dt}{t^2+1} = \frac{\pi}{2} - \frac{\pi}{4}$$

$$= \frac{\pi}{4}.$$

33 $ds = \sqrt{1 + \left(\frac{dy}{dx}\right)^2}\,dx = \sqrt{1+x^2}\,dx$, so Mass $=$

$$\int_C x\,ds = \int_0^1 x\sqrt{1+x^2}\,dx = \frac{1}{3}(1+x^2)^{3/2}\Big|_0^1$$

$$= \frac{1}{3}(2\sqrt{2} - 1).$$

35 $ds = \sqrt{r^2 + (r')^2}\,d\theta = \sqrt{e^{2\theta} + e^{2\theta}}\,d\theta$

$= \sqrt{2}\,e^\theta\,d\theta$. $\int_C r\,ds = \int_0^{2\pi} e^\theta\sqrt{2}e^\theta\,d\theta$

$= \frac{1}{\sqrt{2}}e^{2\theta}\Big|_0^{2\pi} = \frac{1}{\sqrt{2}}(e^{4\pi} - 1)$ and $\int_C ds =$

$\int_0^{2\pi} \sqrt{2}e^\theta\, d\theta = \sqrt{2}(e^{2\pi} - 1)$. Thus the average temperature is $\dfrac{(e^{4\pi} - 1)/\sqrt{2}}{\sqrt{2}(e^{2\pi} - 1)} = \dfrac{e^{2\pi} + 1}{2}$.

37 (a) $dx = dy = dt$, $0 \le t \le 1$;

$$\int_C xy\, dx + x^2\, dy = \int_0^1 2t^2\, dt = \frac{2}{3}$$

(b) $dx = 2t\, dt$, $dy = 2t\, dt$, $0 \le t \le 1$;

$$\int_C xy\, dx + x^2\, dy = \int_0^1 (2t^5 + 2t^5)\, dt = \frac{2}{3}$$

(c) $dx = dt$, $dy = 2t\, dt$, $0 \le t \le 1$;

$$\int_C xy\, dx + x^2\, dy = \int_0^1 (t^3 + 2t^3)\, dt = \frac{3}{4}$$

(d) C_1: $x = 0$, $y = t$, $0 \le t \le 1$;

$$\int_{C_1} xy\, dx + x^2\, dy = 0;$$

C_2: $x = t$, $y = 1$, $0 \le t \le 1$;

$$\int_{C_2} xy\, dx + x^2\, dy = \int_0^1 t\, dt = \frac{1}{2}.$$ Thus

$$\int_C xy\, dx + x^2\, dy = 0 + \frac{1}{2} = \frac{1}{2}.$$

39 Let C be the given curve. Then the area of the fence is $\int_C \dfrac{1}{1 + x^2}\, ds$. Since $y = \dfrac{x^2}{2}$, $dy = x\, dx$ and $ds = \sqrt{dx^2 + dy^2} = \sqrt{1 + x^2}\, dx$, so the area is $\int_{-\sqrt{3}}^{\sqrt{3}} \dfrac{1}{1 + x^2}\sqrt{1 + x^2}\, dx = \int_{-\sqrt{3}}^{\sqrt{3}} \dfrac{dx}{\sqrt{1 + x^2}} =$

$\ln|x + \sqrt{1 + x^2}|\Big|_{-\sqrt{3}}^{\sqrt{3}} = 2\ln(2 + \sqrt{3})$ square meters.

16.3 Four Applications of Line Integrals

1 We have $\int_C \mathbf{F}\cdot d\mathbf{r} = \int_C (2x\mathbf{i} + 0\mathbf{j})\cdot(dx\,\mathbf{i} + dy\,\mathbf{j})$

$= \int_C 2x\, dx$. We are given the parameterization of C: $x = 3\cos\theta$, $y = 3\sin\theta$, $0 \le \theta \le \pi$. So $dx = -3\sin\theta\, d\theta$ and $\int_C 2x\, dx =$

$$\int_0^\pi 2(3\cos\theta)(-3\sin\theta\, d\theta) = -9\int_0^\pi \sin 2\theta\, d\theta$$

$$= -9\left(-\frac{\cos 2\theta}{2}\right)\Big|_0^\pi = -9\left(-\frac{1}{2} - \left(-\frac{1}{2}\right)\right) = -9\cdot 0$$

$= 0$.

3 We know $\int_C \mathbf{F}\cdot d\mathbf{r} =$

$$\int_C (x\mathbf{i} + y\mathbf{j} + z\mathbf{k})\cdot(dx\,\mathbf{i} + dy\,\mathbf{j} + dz\,\mathbf{k})$$

$= \int_C x\, dx + y\, dy + z\, dz$. Also, the parameterization of C is $x = \cos t$, $y = \sin t$, $z = 3t$, $0 \le t \le 4\pi$. So $dx = -\sin t\, dt$, $dy = \cos t\, dt$, and $dz = 3\, dt$. Thus $\int_C x\, dx + y\, dy + z\, dz =$

$$\int_0^{4\pi} \cos t\,(-\sin t\, dt) + \sin t\,(\cos t\, dt) + 3t\,(3\, dt)$$

$$= \int_0^{4\pi} 9t\, dt = \frac{9t^2}{2}\Big|_0^{4\pi} = \frac{9\cdot 16\pi^2}{2} = 72\pi^2.$$

5 Note that $\mathbf{F}(\mathbf{r}) = \dfrac{\hat{\mathbf{r}}}{|\mathbf{r}|^2} = \dfrac{\mathbf{r}}{|\mathbf{r}|^3} =$

$\dfrac{x\mathbf{i} + y\mathbf{j} + z\mathbf{k}}{(x^2 + y^2 + z^2)^{3/2}}$. So $\int_C \mathbf{F}\cdot d\mathbf{r} =$

$$\int_C P\, dx + Q\, dy + R\, dz =$$

$\int_C \frac{x\,dx + y\,dy + z\,dz}{(x^2 + y^2 + z^2)^{3/2}}$. We are given $x = 2t$, $y = 3t$, and $z = 4t$ for $1 \le t \le 2$. Thus $dx = 2\,dt$, $dy = 3\,dt$, and $dz = 4\,dt$. So

$$\int_C \frac{x\,dx + y\,dy + z\,dz}{(x^2 + y^2 + z^2)^{3/2}}$$

$$= \int_1^2 \frac{2t\cdot 2 + 3t\cdot 3 + 4t\cdot 4}{[(2t)^2 + (3t)^2 + (4t)^2]^{3/2}}\,dt = \int_1^2 \frac{29t\,dt}{(29t^2)^{3/2}}$$

$$= \int_1^2 \frac{29t}{29^{3/2}t^3}\,dt = \frac{1}{\sqrt{29}}\left(-\frac{1}{t}\right)\Big|_1^2 = \frac{1}{\sqrt{29}}\left(-\frac{1}{2} + 1\right)$$

$$= \frac{1}{2\sqrt{29}}.$$

7 $\mathbf{F} = x^2y\mathbf{i} + y\mathbf{j}$ and the path of integration is from (0, 0) to (2, 4) on the curve $y = x^2$. The work done by $\mathbf{F}$ along the curve is $\int_C x^2y\,dx + y\,dy$

$$= \int_0^2 x^2(x^2)\,dx + \int_0^4 y\,dy = \frac{x^5}{5}\Big|_0^2 + \frac{y^2}{2}\Big|_0^4$$

$$= \frac{32}{5} + 8 = \frac{72}{5}.$$

9 We break the path (call it C) into two smaller paths C_1 and C_2, where C_1 is parameterized by $(x, y) = (t, 0)$ for t in [0, 2] and C_2 is parameterized by $(x, y) = (2, t)$ for t in [0, 4]. Then the work done by $\mathbf{F}$ along C is $\int_C x^2y\,dx + y\,dy =$

$\int_{C_1} x^2y\,dx + y\,dy + \int_{C_2} x^2y\,dx + y\,dy$. Since $y = 0$ along C_1, $x^2y\,dx + y\,dy = 0$ on that path and the work done along C_1 is 0. So $\int_C x^2y\,dx + y\,dy$

$$= \int_{C_2} x^2y\,dx + y\,dy = \int_0^4 4t(0\,dt) + t(1\,dt) =$$

$$\int_0^4 t\,dt = \frac{t^2}{2}\Big|_0^4 = 8.$$

11 (a) Work $= \int_C \mathbf{F}\cdot d\mathbf{r}$

$$= \int_C \frac{-x\,dx}{(x^2 + y^2)^{3/2}} + \frac{-y\,dy}{(x^2 + y^2)^{3/2}}$$

$$= \int_0^{\pi/2} \frac{4\cos t\sin t - \sin t\cos t}{(4\cos^2 t + \sin^2 t)^{3/2}}\,dt = \frac{1}{2}$$

(b) Work $= \int_0^1 \frac{4 - 5t}{(5t^2 - 8t + 4)^{3/2}}\,dt = \frac{1}{2}$

13 (a) $f(x, y) = (x + 1)^2\mathbf{i} + y\mathbf{j}$

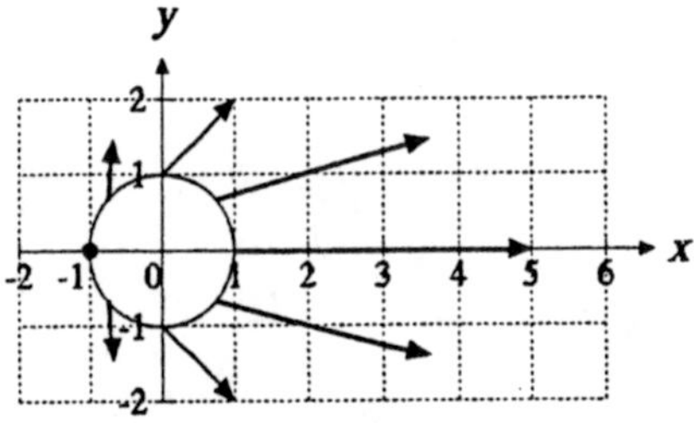

(b) The net flow certainly appears to be outward.

(c) C is parameterized by $x = \cos t$ and $y = \sin t$ as t goes from 0 to 2π, so $dx = -\sin t\,dt$ and $dy = \cos t\,dt$. To find the net outward flow, we wish to compute the integral of the normal component of $\mathbf{F}$ along C:

$$\oint_C \mathbf{F}\cdot\mathbf{n}\,ds = \oint_C -Q\,dx + P\,dy =$$

$$\oint_C -y\,dx + (x + 1)^2\,dy =$$

$$\int_0^{2\pi} (-\sin t)(-\sin t\,dt) + (\cos t + 1)^2(\cos t\,dt)$$

$$= \int_0^{2\pi} (\sin^2 t + \cos^3 t + 2\cos^2 t + \cos t)\,dt$$

$= \pi + 0 + 2\pi + 0 = 3\pi$. The net outward flow is positive, as expected.

17 See p. 961, "Circulation of a Fluid."

19 $\oint_C \mathbf{F}\cdot d\mathbf{r}$ represents the circulation of a fluid along C. Since $\oint_C \mathbf{F}\cdot d\mathbf{r} > 0$, the fluid will tend to circulate in the direction of C, counterclockwise.

21 By inspection of the figure, the angle subtended at the origin by the line segment is $\theta = \tan^{-1}(3/2) \approx$

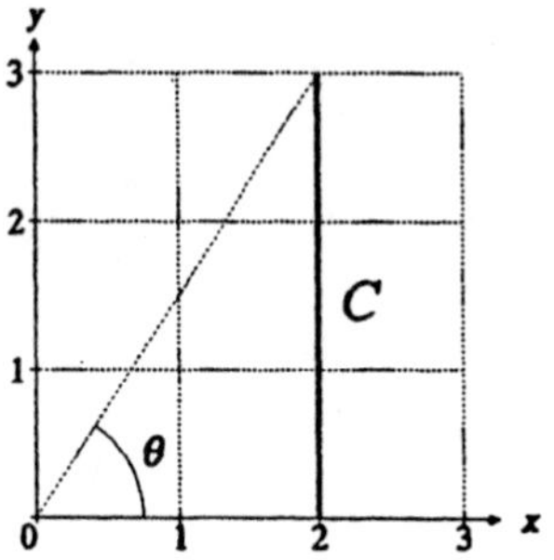

$56.31°$. According to (2), $\theta = \int_C \frac{\mathbf{n}\cdot\hat{\mathbf{r}}}{\|\mathbf{r}\|}\, ds =$

$$\int_0^3 \frac{\mathbf{i}\cdot(2\mathbf{i} + y\mathbf{j})}{\left(\sqrt{2^2 + y^2}\right)^2}\, dy = \int_0^3 \frac{2\, dy}{4 + y^2}$$

$$= \frac{1}{2}\int_0^3 \frac{4\, dy}{4 + y^2} = \frac{1}{2}\left(2\tan^{-1}\frac{y}{2}\right)\Big|_0^3$$

$= \tan^{-1}(3/2) - \tan^{-1} 0 = \tan^{-1}(3/2) \approx 56.31°$. Since this matches the direct calculation, (2) is verified.

23 (a) Let the portion of the circle from $(0, 3)$ to $(3, 0)$ (call it C) be parameterized by $(x, y) = (3 \sin t, 3 \cos t)$ for t in $[0, \pi/2]$. Then, since $\mathbf{F} = -3\mathbf{j}$, $\int_C \mathbf{F}\cdot d\mathbf{r} = \int_C (-3)\, dy =$

$$\int_0^{\pi/2} (-3)(-3 \sin t\, dt) = 9\int_0^{\pi/2} \sin t\, dt =$$

$$9(-\cos t)\big|_0^{\pi/2} = 9(0 - (-1)) = 9.$$

(b) Let the portion of the line from $(0, 3)$ to $(3, 0)$ (call it C) be parameterized by $(x, y) = (t, -t + 3)$ for t in $[0, 3]$. Then $\int_C \mathbf{F}\cdot d\mathbf{r}$

$$= \int_C (-3)\, dy = \int_0^3 (-3)(-1\, dt) = 3\int_0^3 dt$$

$= 3\cdot 3 = 9$.

(c) Any curve C from $(0, 3)$ to $(3, 0)$ can be parameterized by $(x, y) = (f(t), g(t))$ for t in $[0, 3]$ and some $f(t)$ and $g(t)$ (where $f(0) = 0$, $f(3) = 3$, $g(0) = 3$, and $g(3) = 0$). So

$$\int_C \mathbf{F}\cdot d\mathbf{r} = \int_C (-3)\, dy = -3\int_0^3 g'(t)\, dt$$

$$= -3g(t)\big|_0^3 = -3(g(3) - g(0))$$

$= -3(0 - 3) = 9$. Hence for any curve C from $(0, 3)$ to $(3, 0)$, $\int_C \mathbf{F}\cdot d\mathbf{r} = 9$.

25 See Figure 12 in Sec. 16.3. Let C be parameterized in polar coordinates by $(r, \theta) = (f(\theta), \theta)$ for $0 \le \theta$

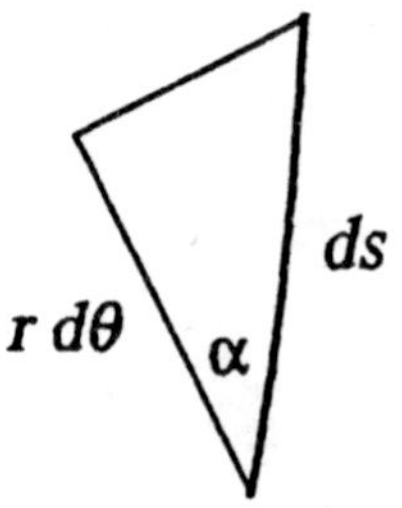

$\le 2\pi$ with the origin at O. (Note that f is continuous since C is a closed curve.) We note that $\frac{\mathbf{n}\cdot\hat{\mathbf{r}}}{\|\mathbf{r}\|}\, ds = \frac{\cos\alpha\, ds}{r}$. But $\cos\alpha\, ds = r\, d\theta$ by the definition of radian measure, where $r = \|\mathbf{r}\|$. So $d\theta = \frac{\cos\alpha\, ds}{r} = \frac{\mathbf{n}\cdot\hat{\mathbf{r}}}{\|\mathbf{r}\|}\, ds$ and the value of

$\oint_C \frac{\mathbf{n}\cdot\hat{\mathbf{r}}}{\|\mathbf{r}\|}\, ds$ is $\int_0^{2\pi} d\theta = 2\pi$.

27 Since $Q(t)$ is the total mass of fluid in R at time t, $\frac{dQ}{dt}$ is the rate of change of mass of fluid in R with respect to time. That is, $\frac{dQ}{dt}$ is the measure of the flow of fluid into R, or the net gain of fluid in R. So $\frac{dQ}{dt} = -\oint_C \mathbf{F}\cdot\mathbf{n}\,ds$.

29 (a) The field and the region both have radial symmetry. The vector field is perpendicular to C along the curved portions of C and in opposite directions to the orientation of C on the straight portions. Hence the circulation must be 0.

(b) The vector field is parallel to C along the straight portions of C, so there is no contribution to the flux there. The longer curved portion of C experiences a flux of lesser magnitude than the shorter curved portion (because the vector field is stronger toward the origin), so it appears that the flux across the curved portions of C cancels out. The flux must be close to 0.

16.4 Green's Theorem

1 $\mathbf{F} = 3x\mathbf{i} + 2y\mathbf{j}$, so $\nabla\cdot\mathbf{F} = 3 + 2 = 5$. Thus $\int_{\mathcal{A}} \nabla\cdot\mathbf{F}\,dA = 5\int_{\mathcal{A}} dA$, which equals 5 times the area of $\mathcal{A}$; $\mathcal{A}$ is the unit disk, so the result is 5π. Parameterize the circumference C of the disk by $x = \cos\theta$ and $y = \sin\theta$; then $\oint_C \mathbf{F}\cdot\mathbf{n}\,ds = \oint_C -Q\,dx + P\,dy = \oint_C -2y\,dx + 3x\,dy =$

$\int_0^{2\pi} (-2\sin\theta)(-\sin\theta\,d\theta) + (3\cos\theta)(\cos\theta\,d\theta)$

$= \int_0^{2\pi} (2\sin^2\theta + 3\cos^2\theta)\,d\theta = 2\pi + 3\pi$

$= 5\pi$, as expected.

3 $\mathbf{F} = xy\mathbf{i} + x^2y\mathbf{j}$, so $\nabla\cdot\mathbf{F} = y + x^2$. Thus

$\int_{\mathcal{A}} \nabla\cdot\mathbf{F}\,dA = \int_0^b \int_0^a (y + x^2)\,dx\,dy$

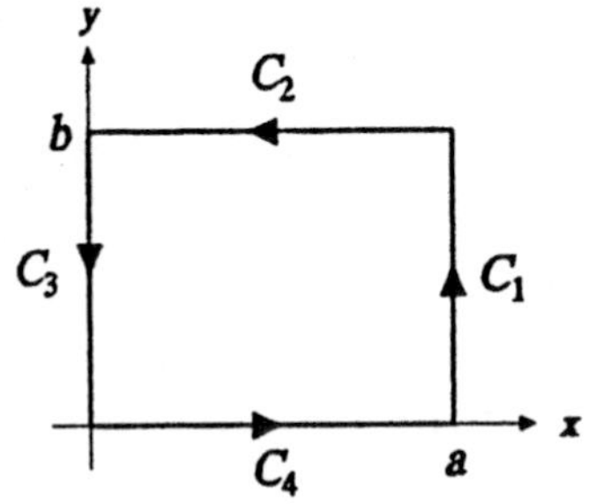

$= \int_0^b \left(yx + \frac{x^3}{3}\right)\Big|_0^a dy = \int_0^b \left(ay + \frac{a^3}{3}\right) dy$

$= \left(\frac{ay^2}{2} + \frac{a^3y}{3}\right)\Big|_0^b = \frac{ab^2}{2} + \frac{a^3b}{3}$. Parameterize the rectangle as shown in the figure. On C_1, $x = a$ as y goes from 0 to b. On C_2, $y = b$ as x goes from a to 0. On C_3, $x = 0$ as y goes from b to 0. On C_4, $y = 0$ as x goes from 0 to a. Thus

$\oint_C \mathbf{F}\cdot\mathbf{n}\,ds = \int_{C_1} \mathbf{F}\cdot\mathbf{n}\,ds + \int_{C_2} \mathbf{F}\cdot\mathbf{n}\,ds + \int_{C_3} \mathbf{F}\cdot\mathbf{n}\,ds + \int_{C_4} \mathbf{F}\cdot\mathbf{n}\,ds = \int_0^b xy\,dy + \int_a^0 (-x^2y)\,dx + \int_b^0 xy\,dy + \int_0^a -x^2y\,dx = \frac{ab^2}{2} + \frac{a^3b}{3}$, as expected.

5 $\mathbf{F} = e^x \sin y\,\mathbf{i} + e^{2x}\cos y\,\mathbf{j}$, so $\nabla\cdot\mathbf{F} = e^x \sin y - e^{2x}\sin y$. The region $\mathcal{A}$ is described by $0 \le x \le 1$,

$0 \le y \le \frac{\pi}{2}$. By Green's theorem, $\oint_C \mathbf{F}\cdot\mathbf{n}\, ds =$

$\int_{\mathcal{A}} \nabla\cdot\mathbf{F}\, dA = \int_0^{\pi/2}\int_0^1 (e^x - e^{2x}) \sin y\, dx\, dy$

$= \int_0^{\pi/2} \left(e^x - \frac{1}{2}e^{2x}\right) \sin y \Big|_0^1 dy$

$= \int_0^{\pi/2} \left(e - \frac{1}{2}e^2 - 1 + \frac{1}{2}\right) \sin y\, dy$

$= \left(e - \frac{e^2}{2} - \frac{1}{2}\right)(-\cos y)\Big|_0^{\pi/2} = e - \frac{e^2}{2} - \frac{1}{2}.$

7 $\mathbf{F} = 2x^3y\mathbf{i} - 3x^2y^2\mathbf{j}$, so $\nabla\cdot\mathbf{F} = 6x^2y - 6x^2y = 0$.

By Green's Theorem, $\oint_C \mathbf{F}\cdot\mathbf{n}\, ds = \int_{\mathcal{A}} \nabla\cdot\mathbf{F}\, dA$

$= \int_{\mathcal{A}} 0\, dA = 0.$

9 (a) As suggested by Figure 11 in the text, assume that **F** is pointing outward and is not tangent to C for all points on C. Then $\mathbf{F}\cdot\mathbf{n} > 0$ for all points on C. So $\int_{\mathcal{A}} \nabla\cdot\mathbf{F}\, dA = \oint_C \mathbf{F}\cdot\mathbf{n}\, ds > 0$.

(b) As indicated by Figure 12 in the text, **F** is always tangent to C. Then $\mathbf{F}\cdot\mathbf{n} = 0$ for all points on C. Hence $\int_{\mathcal{A}} \nabla\cdot\mathbf{F}\, dA = \oint_C \mathbf{F}\cdot\mathbf{n}\, ds = 0$.

11 The region $\mathcal{A}$ is a parabolic sector with description $0 \le x \le 2$, $0 \le y \le 2x - x^2$. Recall that $\oint_C \mathbf{F}\cdot\mathbf{n}\, ds = \oint_C -Q\, dx + P\, dy$, so

$\oint_C xy\, dx + e^x\, dy$ is the integral of the normal component of $\mathbf{F} = e^x\mathbf{i} - xy\mathbf{j}$. $\nabla\cdot\mathbf{F} = e^x - x$, so we apply Green's theorem: $\oint_C xy\, dx + e^x\, dy =$

$\oint_C \mathbf{F}\cdot\mathbf{n}\, ds = \int_{\mathcal{A}} \nabla\cdot\mathbf{F}\, dA$

$= \int_0^2 \int_0^{2x-x^2} (e^x - x)\, dy\, dx$

$= \int_0^2 (e^x - x)(2x - x^2)\, dx$

$= \int_0^2 (2xe^x - x^2e^x - 2x^2 + x^3)\, dx$

$= \left[2(x-1)e^x - (x^2 - 2x + 2)e^x - \frac{2x^3}{3} + \frac{x^4}{4}\right]\Big|_0^2$

$= \left(2e^2 - 2e^2 - \frac{16}{3} + 4\right) - (-2 - 2 - 0 + 0)$

$= \frac{8}{3}.$

13 We are given that $\mathbf{F} = \frac{\hat{\mathbf{r}}}{\|\mathbf{r}\|}$ is divergence-free in the xy plane; that is, $\nabla\cdot\mathbf{F} = 0$. (See Exercise 32(a) of Sec. 16.1.) Let $\mathcal{A}$ by the square region enclosed by C. Since **F** is defined throughout $\mathcal{A}$, we have $\oint_C \mathbf{F}\cdot\mathbf{n}\, ds = \int_{\mathcal{A}} \nabla\cdot\mathbf{F}\, dA = 0$ by Green's theorem.

17 Let $\mathcal{A}$ be the given disk and let C be the curve that contains $\mathcal{A}$ (taken counterclockwise). Recall that $\oint_C \mathbf{F}\cdot\mathbf{n}\, ds$ represents the rate of outflow. Let d be the diameter of $\mathcal{A}$. Since $\nabla\cdot\mathbf{F}(P) = \lim_{d\to 0} \frac{\oint_C \mathbf{F}\cdot\mathbf{n}\, ds}{\text{Area of } \mathcal{A}}$, and the diameter of $\mathcal{A}$ is small,

we may conclude that $\nabla\cdot\mathbf{F}(P) \approx \frac{\oint_C \mathbf{F}\cdot\mathbf{n}\, ds}{\text{Area of } \mathcal{A}}$, or

$\oint_C \mathbf{F}\cdot\mathbf{n}\, ds \approx \nabla\cdot\mathbf{F}(P)(\text{Area of } \mathcal{A}) = 4\pi(0.02)^2 =$

$0.0016\pi \approx 5.027 \times 10^{-3}$.

19 We are asked to estimate $\oint_C \mathbf{F}\cdot\mathbf{n}\, ds$. Since the diameter of $\mathcal{A}$ is small, we have from Exercise 17 that $\oint_C \mathbf{F}\cdot\mathbf{n}\, ds \approx \nabla\cdot\mathbf{F}(1, 1)(\text{Area of } \mathcal{A})$. Here $\nabla\cdot\mathbf{F}(1, 1) = 5$, and the area of $\mathcal{A}$ is $(0.02)^2 = 0.0004$ since $\mathcal{A}$ is a square with sides of length 0.02. Thus $\oint_C \mathbf{F}\cdot\mathbf{n}\, ds \approx 5(0.0004) = 2.000 \times 10^{-3}$.

21 Let $\mathbf{F}(x, y) = x\mathbf{i} + y\mathbf{j}$. Then $\frac{1}{2}\oint_C (-y\, dx + x\, dy)$

$$= \frac{1}{2}\oint_C \mathbf{F}\cdot\mathbf{n}\, ds = \frac{1}{2}\int_{\mathcal{A}} \nabla\cdot\mathbf{F}\, dA = \frac{1}{2}\int_{\mathcal{A}} 2\, dA$$

$$= \int_{\mathcal{A}} dA = \text{Area of } \mathcal{A}.$$

23 (a)

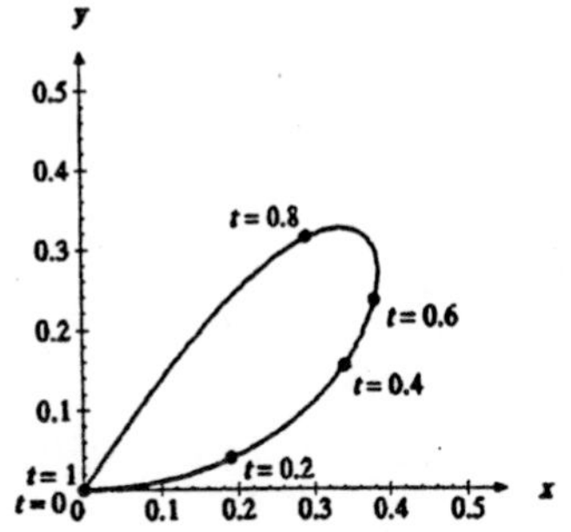

(b) y is not known as a function of x.

(c) We have $x = t(1 - t^2) = t - t^3$, so $dx = (1 - 3t^2)\, dt$; also, $y = t^2(1 - t^3) = t^2 - t^5$, so $dy = (2t - 5t^4)\, dt$. Then with C as the curve enclosing $\mathcal{A}$ (taken counterclockwise),

Area of $\mathcal{A} = \frac{1}{2}\oint_C (-y\, dx + x\, dy) =$

$$\frac{1}{2}\int_0^1 \left[-(t^2 - t^5)(1 - 3t^2) + (t - t^3)(2t - 5t^4)\right] dt$$

$$= \frac{1}{2}\int_0^1 (2t^7 - 4t^5 + t^4 + t^2)\, dt$$

$$= \frac{1}{2}\left(\frac{t^8}{4} - \frac{2t^6}{3} + \frac{t^5}{5} + \frac{t^3}{3}\right)\Bigg|_0^1$$

$$= \frac{1}{2}\left(\frac{1}{4} - \frac{2}{3} + \frac{1}{5} + \frac{1}{3}\right) = \frac{7}{120}.$$

25 Suppose a region is bounded by a curve of the form $r = f(\theta)$ in polar coordinates, where the parameter θ varies from α to β. Then $x = r\cos\theta$ and $y = r\sin\theta$, so $dx = -r\sin\theta\, d\theta + r'\cos\theta\, dr$ and $dy = r\cos\theta\, d\theta + r'\sin\theta\, dr$. We then have $-y\, dx + x\, dy = r^2\sin^2\theta\, d\theta - rr'\cos\theta\sin\theta\, dr + r^2\cos^2\theta\, d\theta + rr'\cos\theta\sin\theta\, dr$

$= r^2\cos^2\theta\, d\theta + r^2\sin^2\theta\, d\theta$

$= r^2(\cos^2\theta + \sin^2\theta)\, d\theta = r^2\, d\theta$. Thus

$\oint_C -y\, dx + x\, dy = \int_\alpha^\beta r^2\, d\theta$, and dividing through by 2 gives the formula for area in polar coordinates.

27 The proof is analogous to the second proof given in the text. Describe the region $\mathcal{A}$ suitably, switch P's and Q's, change x's to y's and y_i's to x_i's, and be alert for changes in the signs of various terms.

29 (a) On C_1, $\mathbf{n}\cdot\hat{\mathbf{r}} = -1$ and $\mathbf{n}\cdot\hat{\boldsymbol{\theta}} = 0$. Now since $d\theta$ is small, we may assume that $\mathbf{F}$ is constant at its value at $(r, \theta + d\theta/2)$ on C_1. Then $\mathbf{F}\cdot\mathbf{n} \approx \mathbf{F}(r, \theta + d\theta/2)\cdot\mathbf{n}$

$= \left[A(r, \theta + d\theta/2)\hat{\mathbf{r}} + B(r, \theta + d\theta/2)\hat{\boldsymbol{\theta}}\right]\cdot\mathbf{n}$

$= -A(r, \theta + d\theta/2)$ and $\int_{C_1} \mathbf{F}\cdot\mathbf{n}\, ds \approx$

$\int_{C_1} -A(r, \theta + d\theta/2)\, ds$

$= -A(r, \theta + d\theta/2)\int_{C_1} ds$

$= -A(r, \theta + d\theta/2)r\, d\theta.$

(b) On C_2, $\mathbf{n}\cdot\hat{\mathbf{r}} = 0$ and $\mathbf{n}\cdot\hat{\boldsymbol{\theta}} = -1$. Since dr is small, assume that $\mathbf{F} = \mathbf{F}(r + dr/2, \theta)$ on C_2. Then $\mathbf{F}\cdot\mathbf{n} \approx \mathbf{F}(r + dr/2, \theta)\cdot\mathbf{n} = -B(r + dr/2, \theta)$, and $\int_{C_2} \mathbf{F}\cdot\mathbf{n}\, ds \approx$

$$-B(r + dr/2, \theta) \int_{C_2} ds$$

$= -B(r + dr/2, \theta)\, dr$. On C_3, $\mathbf{n}\cdot\hat{\mathbf{r}} = 1$ and $\mathbf{n}\cdot\hat{\boldsymbol{\theta}} = 0$. Proceeding as in part (a), we see that $\int_{C_3} \mathbf{F}\cdot\mathbf{n}\, ds \approx$

$$\int_{C_3} A(r + dr, \theta + d\theta/2)\, ds$$

$$= A(r + dr, \theta + d\theta/2) \int_{C_3} ds$$

$= A(r + dr, \theta + d\theta/2)(r + dr)\, d\theta$. On C_4, $\mathbf{n}\cdot\hat{\mathbf{r}} = 0$ and $\mathbf{n}\cdot\hat{\boldsymbol{\theta}} = 1$. Calculations similar to those used in finding $\int_{C_2} \mathbf{F}\cdot\mathbf{n}\, ds$ give

$$\int_{C_4} \mathbf{F}\cdot\mathbf{n}\, ds \approx B(r + dr/2, \theta + d\theta)\, ds$$

$$= B(r + dr/2, \theta + d\theta) \int_{C_4} ds$$

$= B(r + dr/2, \theta + d\theta)\, dr$. Combining the results above and part (a), we have

$$\oint_C \mathbf{F}\cdot\mathbf{n}\, ds = \int_{C_1} \mathbf{F}\cdot\mathbf{n}\, ds + \int_{C_2} \mathbf{F}\cdot\mathbf{n}\, ds + \int_{C_3} \mathbf{F}\cdot\mathbf{n}\, ds + \int_{C_4} \mathbf{F}\cdot\mathbf{n}\, ds \approx$$

$$-A(r, \theta + d\theta/2)r\, d\theta + (-B(r + dr/2, \theta)\, dr) + A(r + dr, \theta + d\theta/2)(r + dr)\, d\theta + B(r + dr/2, \theta + d\theta)\, dr$$

$$= [A(r + dr, \theta + d\theta/2) - A(r, \theta + d\theta/2)]r\, d\theta + A(r + dr, \theta + d\theta/2)\, dr\, d\theta + [B(r + dr/2, \theta + d\theta) - B(r + dr/2, \theta)]\, dr.$$

(c) Assume that A and B are differentiable functions. By the mean-value theorem, there exists some r_0 in $[r, r + dr]$ such that $\left.\frac{\partial A}{\partial r}\right|_{r=r_0}$

$$= \frac{A(r + dr, \theta + d\theta/2) - A(r, \theta + d\theta/2)}{dr}.$$

Thus $[A(r + dr, \theta + d\theta/2) - A(r, \theta + d\theta/2)]r\, d\theta = \left(\left.\frac{\partial A}{\partial r}\right|_{r=r_0}\right) r\, dr\, d\theta$. Similarly,

$[B(r + dr/2, \theta + d\theta) - B(r + dr/2, \theta)]\, dr = \left(\left.\frac{\partial B}{\partial \theta}\right|_{\theta=\theta_0}\right) dr\, d\theta$ for some θ_0 in $[\theta, \theta + d\theta]$.

Thus the sum in (b) becomes $\left(\left.\frac{\partial A}{\partial r}\right|_{r=r_0}\right) r\, dr\, d\theta$

$+ A(r + dr, \theta + d\theta/2)\, dr\, d\theta +$

$\left(\left.\frac{\partial B}{\partial \theta}\right|_{\theta=\theta_0}\right) dr\, d\theta$. Since A is small, $\left(\left.\frac{\partial A}{\partial r}\right|_{r=r_0}\right)$

$\approx \frac{\partial A}{\partial r}$, $A(r + dr, \theta + d\theta/2) \approx A$, and

$\left(\left.\frac{\partial B}{\partial \theta}\right|_{\theta=\theta_0}\right) \approx \frac{\partial B}{\partial \theta}$. So the sum in (b) is approximately

$$\frac{\partial A}{\partial r} r\, dr\, d\theta + A\, dr\, d\theta + \frac{\partial B}{\partial \theta}\, dr\, d\theta,$$ as claimed.

(d) Let d be the diameter of $\mathcal{A}$. Using part (c), we have $\operatorname{div} \mathbf{F} = \lim_{d\to 0} \dfrac{\oint_C \mathbf{F}\cdot\mathbf{n}\, ds}{\text{Area of } \mathcal{A}}$

$$= \lim_{dr, d\theta \to 0} \frac{A_r r\, dr\, d\theta + A\, dr\, d\theta + B_\theta\, dr\, d\theta}{r\, dr\, d\theta}$$

$$= A_r + \frac{A}{r} + \frac{1}{r}B_\theta.$$

16.5 Applications of Green's Theorem

1 $\mathbf{F} = x\mathbf{i} + y^2\mathbf{j}$, so $\nabla\cdot\mathbf{F} = 1 + 2y$. The region $\mathcal{A}$ is conveniently described in polar coordinates as $0 \le \theta \le 2\pi$, $1 \le r \le 2$. We have $\int_{\mathcal{A}} \nabla\cdot\mathbf{F}\, dA$

$$= \int_{\mathcal{A}} (1 + 2y)\, dA$$

$$= \int_0^{2\pi} \int_1^2 (1 + 2r \sin\theta)\, r\, dr\, d\theta$$

$$= \int_0^{2\pi} \left(\frac{r^2}{2} + \frac{2r^3}{3}\sin\theta\right)\Bigg|_1^2 d\theta$$

$$= \int_0^{2\pi} \left(\frac{3}{2} + \frac{14}{3}\sin\theta\right) d\theta$$

$$= \left(\frac{3}{2}\theta - \frac{14}{3}\cos\theta\right)\Bigg|_0^{2\pi} = 3\pi - \frac{14}{3} - 0 + \frac{14}{3}$$

$= 3\pi$. Parameterize the inner boundary, C_1 by $x = \cos\theta$ and $y = \sin\theta$. Observe that the exterior normal $\mathbf{n}$ points away from $\mathcal{A}$ and toward the origin; in fact, $\mathbf{n} = -\mathbf{r} = -x\mathbf{i} - y\mathbf{j}$, so $\mathbf{F}\cdot\mathbf{n} = -x^2 - y^3$. It follows that $\oint_{C_1} \mathbf{F}\cdot\mathbf{n}\, ds =$

$$\oint_{C_1} (-x^2 - y^3)\, ds = \int_0^{2\pi} (-\cos^2\theta - \sin^3\theta)\, d\theta$$

$= -\pi + 0 = -\pi$. (We used the fact that $ds = d\theta$; recall that $s = r\theta$ and $r = 1$ on C_1.) Parameterize the outer boundary, C_2, by $x = 2\cos\theta$ and $y = 2\sin\theta$. The exterior normal $\mathbf{n}$ is equal to $\frac{1}{2}\mathbf{r} = \frac{1}{2}(x\mathbf{i} + y\mathbf{j})$, so $\mathbf{F}\cdot\mathbf{n} = \frac{1}{2}(x^2 + y^3)$.

On C_2, arc length s is equal to 2θ, so $ds = 2\, d\theta$.

Thus $\oint_{C_2} \mathbf{F}\cdot\mathbf{n}\, ds = \oint_{C_2} \frac{1}{2}(x^2 + y^3)\, ds$

$$= \int_0^{2\pi} \frac{1}{2}(4\cos^2\theta + 8\sin^3\theta)\, 2\, d\theta = 4\cdot\pi + 8\cdot 0$$

$= 4\pi$. Adding the line integrals gives $-\pi + 4\pi = 3\pi$, same as the value of the planar integral.

3 Since $\nabla\cdot\mathbf{F} = 0$ everywhere in the region $\mathcal{A}$ enclosed by C_2, $\oint_{C_2} \mathbf{F}\cdot\mathbf{n}\, ds = \int_{\mathcal{A}} \nabla\cdot\mathbf{F}\, dA = 0$ by Green's theorem. The curves C_1 and C_3 enclose P_1, where $\nabla\cdot\mathbf{F}$ is undefined. Since $\mathbf{F}$ is divergence-free, we have $\oint_{C_3} \mathbf{F}\cdot\mathbf{n}\, ds = \oint_{C_1} \mathbf{F}\cdot\mathbf{n}\, ds = 5$ by Corollary 2.

5 (a) $\mathbf{F} = \mathbf{r}/\|\mathbf{r}\|^2 = \mathbf{r}/r^2$; observe that $\mathbf{n} = \mathbf{r}/r$ is a unit vector that points out radially from the origin and is thus perpendicular to any circle centered at the origin. We have $\mathbf{F}\cdot\mathbf{n} = \dfrac{\mathbf{r}\cdot\mathbf{r}}{r^3}$

$= \dfrac{r^2}{r^3} = \dfrac{1}{r}$. On a circle of radius a we see that $\mathbf{F}\cdot\mathbf{n} = \dfrac{1}{a}$. Furthermore, on such a circle $s = a\theta$ and $ds = a\, d\theta$. Thus $\oint_C \mathbf{F}\cdot\mathbf{n}\, ds =$

$$\int_0^{2\pi} \frac{1}{a} a\, d\theta = 2\pi.$$

(b) This result does not contradict Green's theorem because the divergence of $\mathbf{F}$ does not exist everywhere within C. It is undefined at the origin.

(c) The two-curve version of Green's theorem shows that the integral must have the same value on every circle centered at (0, 0). This is because the divergence of **F** is 0 within every annular region, so the line integrals of the normal component of **F** over the inner and outer boundaries are equal.

7 $\mathbf{F}(x, y) = 2x\mathbf{i}$, so $\nabla \times \mathbf{F} = \begin{vmatrix} \mathbf{i} & \mathbf{j} & \mathbf{k} \\ \frac{\partial}{\partial x} & \frac{\partial}{\partial y} & \frac{\partial}{\partial z} \\ 2x & 0 & 0 \end{vmatrix} = \mathbf{0}$ and

$(\nabla \times \mathbf{F})\cdot\mathbf{k} = 0$. Hence $\int_{\mathcal{A}} (\nabla\times\mathbf{F})\cdot\mathbf{k}\, dA = 0$.

Also, with the given parameterization of C, $x = 3\cos\theta$ and $dx = -3\sin\theta\, d\theta$. Thus, $\oint_C \mathbf{F}\cdot d\mathbf{r} =$

$$\oint_C 2x\, dx = \int_0^{2\pi} 6\cos\theta\,(-3\sin\theta)\, d\theta$$

$$= -18\int_0^{2\pi} \sin\theta\cos\theta\, d\theta = -9\int_0^{2\pi} \sin 2\theta\, d\theta$$

$$= -9\left(-\frac{\cos 2\theta}{2}\right)\Bigg|_0^{2\pi} = 0, \text{ as expected.}$$

9 Since C_1 contains a point where **F** is undefined, nothing can be said about $\oint_{C_1} \mathbf{F}\cdot d\mathbf{r}$. This is not the case with C_2, and we have $\oint_{C_2} \mathbf{F}\cdot d\mathbf{r} =$

$\int_{\mathcal{A}} (\nabla\times\mathbf{F})\cdot\mathbf{k}\, dA = 0$ by Stokes' theorem in the plane, where $\mathcal{A}$ is the region that C_2 contains. Now let $\mathcal{A}'$ be the region between C_1 and C_3. Then, since $\nabla \times \mathbf{F} = \mathbf{0}$ throughout $\mathcal{A}'$, we know from Corollary 3 that $\oint_{C_3} \mathbf{F}\cdot d\mathbf{r} = \oint_{C_1} \mathbf{F}\cdot d\mathbf{r}$.

11 (a) $\mathbf{F}(x, y) = \dfrac{\hat{\mathbf{r}}}{|\mathbf{r}|^3} = \dfrac{\mathbf{r}}{|\mathbf{r}|^4} = \dfrac{x\mathbf{i} + y\mathbf{j}}{(x^2 + y^2)^2}$, so

$$\nabla \times \mathbf{F} = \begin{vmatrix} \mathbf{i} & \mathbf{j} & \mathbf{k} \\ \frac{\partial}{\partial x} & \frac{\partial}{\partial y} & \frac{\partial}{\partial z} \\ \frac{x}{(x^2+y^2)^2} & \frac{y}{(x^2+y^2)^2} & 0 \end{vmatrix} =$$

$$\left(-\frac{\partial}{\partial z}\left(\frac{y}{(x^2+y^2)^2}\right)\right)\mathbf{i} + \left(\frac{\partial}{\partial z}\left(\frac{x}{(x^2+y^2)^2}\right)\right)\mathbf{j} +$$

$$\left[\frac{\partial}{\partial x}\left(\frac{y}{(x^2+y^2)^2}\right) - \frac{\partial}{\partial y}\left(\frac{x}{(x^2+y^2)^2}\right)\right]\mathbf{k} =$$

$$\left[-\frac{4xy}{(x^2+y^2)^3} - \left(-\frac{4xy}{(x^2+y^2)^3}\right)\right]\mathbf{k} = \mathbf{0}.$$

(b) Parameterize C_1 by $x = \cos\theta$ and $y = \sin\theta$ for $0 \le \theta \le 2\pi$. Then $dx = -\sin\theta\, d\theta$ and $dy = \cos\theta\, d\theta$. Hence $\oint_{C_1} \mathbf{F}\cdot d\mathbf{r} =$

$$\oint_{C_1} P\, dx + Q\, dy = \oint_{C_1} \frac{x\, dx + y\, dy}{(x^2 + y^2)^2} =$$

$$\int_0^{2\pi} \frac{\cos\theta\,(-\sin\theta) + \sin\theta\,(\cos\theta)}{(\cos^2\theta + \sin^2\theta)^2}\, d\theta = 0.$$

(c) Note that C_2 encloses C_1, which contains the origin (where **F** is undefined). Also, from part (a), we have $\nabla \times \mathbf{F} = \mathbf{0}$ throughout the region between C_1 and C_2. So by Corollary 3,

$$\oint_{C_2} \mathbf{F}\cdot d\mathbf{r} = \oint_{C_1} \mathbf{F}\cdot d\mathbf{r} = 0.$$

13 Denote the curves enclosing regions $\mathcal{A}_3$ and $\mathcal{A}_4$ by C_3 and C_4, respectively. (See the figure.) By Green's theorem, $\oint_{C_3} \mathbf{F}\cdot\mathbf{n}\, ds = \int_{\mathcal{A}_3} \nabla\cdot\mathbf{F}\, dA$ and

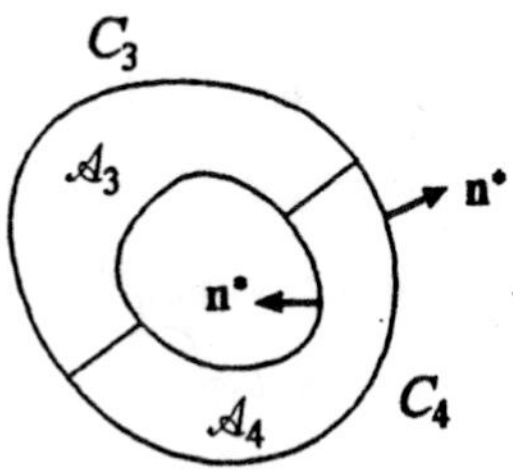

$\oint_{C_4} \mathbf{F}\cdot\mathbf{n}\, ds = \int_{\mathcal{A}_4} \nabla\cdot\mathbf{F}\, dA$. So by the cancellation principle, $\oint_{C_2} \mathbf{F}\cdot\mathbf{n}\, ds - \oint_{C_1} \mathbf{F}\cdot\mathbf{n}\, ds$

$= \oint_{C_1} \mathbf{F}\cdot\mathbf{n}^*\, ds + \oint_{C_2} \mathbf{F}\cdot\mathbf{n}^*\, ds$

$= \oint_{C_3} \mathbf{F}\cdot\mathbf{n}\, ds + \oint_{C_4} \mathbf{F}\cdot\mathbf{n}\, ds = \int_{\mathcal{A}_3} \nabla\cdot\mathbf{F}\, dA +$

$\int_{\mathcal{A}_4} \nabla\cdot\mathbf{F}\, dA = \int_{\mathcal{A}} \nabla\cdot\mathbf{F}\, dA$, where $\mathbf{n}^*$ is the exterior normal to the region $\mathcal{A}$ contained by the inner curve C_1 and the outer curve C_2 (both taken counterclockwise).

15 The integral over a curve of the normal component of the curl of a vector field equals the integral over the boundary of the curve of the tangential component of the vector field.

17 (a) Since $\mathbf{F} = \dfrac{\hat{\mathbf{r}}}{|\mathbf{r}|}$ is not defined at the origin, Green's theorem tells us nothing about

$\oint_C \mathbf{F}\cdot\mathbf{n}\, ds = \oint_C \left(\dfrac{\hat{\mathbf{r}}}{|\mathbf{r}|}\right)\cdot\mathbf{n}\, ds$. See Exercise 25 of Sec. 16.3 for the solution.

(b) If $\mathbf{F} = \dfrac{\hat{\mathbf{r}}}{|\mathbf{r}|}$, then $\mathbf{F}$ is defined throughout the region $\mathcal{A}$ contained in C and $\nabla\cdot\mathbf{F} = 0$. By Green's theorem, $\oint_C \mathbf{F}\cdot\mathbf{n}\, ds =$

$\oint_C \left(\dfrac{\hat{\mathbf{r}}}{|\mathbf{r}|}\right)\cdot\mathbf{n}\, ds = \int_{\mathcal{A}} \nabla\cdot\mathbf{F}\, dA = 0.$

19 Suppose that $f(p) \neq 0$ for some number p in (c, d). Let $f(p) = k$ and suppose $k > 0$. (The case $k < 0$ is similar.) Since f is continuous, there is some interval I containing p such that $f(x) > k/2$ for all x in I. (If I is not contained in (c, d), we replace I by its intersection with (c, d).) Then $\int_I f(x)\, dx \geq$

$\int_I \dfrac{k}{2}\, dx = \dfrac{k}{2}(\text{Length of } I) > 0$, contradicting the assumption that $\int_a^b f(x)\, dx = 0$ for all intervals $[a, b]$ contained in (c, d). Hence $f(x)$ must be 0 for all x in (c, d).

21 By Green's theorem, $\int_{\mathcal{A}} \nabla\cdot\mathbf{F}\, dA = \oint_C \mathbf{F}\cdot\mathbf{n}\, ds =$ 0 for any curve C, where $\mathcal{A}$ is the region contained in C. So $\int_{\mathcal{A}} \nabla\cdot\mathbf{F}\, dA = 0$ for any region $\mathcal{A}$ in the plane. By the "vanishing-integrals" principle, $\nabla\cdot\mathbf{F} = 0$ everywhere in the plane.

23 (a) Mass is the integral of density, $m(t) =$ $\int_{\mathcal{A}} \sigma\, dA$.

(b) $\dfrac{dm}{dt}$ is the rate at which the mass in $\mathcal{A}$ changes; when $\dfrac{dm}{dt} < 0$, $\dfrac{dm}{dt}$ is the rate of net loss of fluid in $\mathcal{A}$, so $\dfrac{dm}{dt} = -\oint_C \mathbf{F}\cdot\mathbf{n}\, ds$.

(c) From part (b) and Green's theorem, $\dfrac{dm}{dt} =$

$-\oint_C \mathbf{F}\cdot\mathbf{n}\, ds = -\int_{\mathcal{A}} \nabla\cdot\mathbf{F}\, dA.$

(d) When Δt is small, $\frac{m(t + \Delta t) - m(t)}{\Delta t} \approx \frac{dm}{dt}$

and $\frac{\sigma(x, y, t + \Delta t) - \sigma(x, y, t)}{\Delta t} \approx \frac{\partial\sigma}{\partial t}$. So the equation does indeed suggest that, in the limit as $\Delta t \to 0$, $\frac{dm}{dt} = \int_{\mathcal{A}} \frac{\partial\sigma}{\partial t}\, dA$.

(e) From parts (c) and (d), $\frac{dm}{dt} = \int_{\mathcal{A}} (-\nabla\cdot\mathbf{F})\, dA$ $= \int_{\mathcal{A}} \frac{\partial\sigma}{\partial t}\, dA$. So $\frac{\partial\sigma}{\partial t} = -\nabla\cdot\mathbf{F}$.

(f) The divergence of **F** at a point P in a small region $\mathcal{A}$ is a measure of the rate at which fluid tends to leave $\mathcal{A}$. The equation of continuity claims that the rate at which the density σ of fluid in a region $\mathcal{A}$ increases is equal to the rate at which fluid enters $\mathcal{A}$, a reasonable expectation.

25 (a) By Exercise 11(a), $\nabla \times \mathbf{F} = \mathbf{0}$.

(b) Replace C_1 with a circle of radius 1; call it C_3. Then since **F** is not defined at (0, 0), we have $\oint_{C_1} \mathbf{F}\cdot d\mathbf{r} = \oint_{C_3} \mathbf{F}\cdot d\mathbf{r} = 0$, where the result from Exercise 11(b) was used. By Stokes' theorem in the plane, we have $\oint_{C_2} \mathbf{F}\cdot d\mathbf{r} = \int_{\mathcal{A}} (\nabla\times\mathbf{F})\cdot\mathbf{k}\, dA = 0$, where $\mathcal{A}$ is the square that C_2 encloses.

27 Let $\mathcal{A}$ be a region in the plane bounded by two curves, C_1 and C_2. Let **F** be defined on $\mathcal{A}$. Then $\oint_{C_2} \mathbf{F}\cdot d\mathbf{r} - \oint_{C_1} \mathbf{F}\cdot d\mathbf{r} = \int_{\mathcal{A}} (\nabla\times\mathbf{F})\cdot\mathbf{k}\, dA$ (if C_1 is the inner boundary of $\mathcal{A}$).

16.6 Conservative Vector Fields

1 Let C be the curve formed by C_1 followed by C_2, and let $\mathcal{A}$ be the region C contains. So **F** is defined and $\nabla \times \mathbf{F} = \mathbf{0}$ for all points in $\mathcal{A}$, and $\mathcal{A}$ is a simply connected region. Hence Theorem 6 applies, and **F** is conservative. Therefore $\oint_C \mathbf{F}\cdot d\mathbf{r} =$ $\int_{C_1} \mathbf{F}\cdot d\mathbf{r} + \int_{C_2} \mathbf{F}\cdot d\mathbf{r} = 5 + \int_{C_2} \mathbf{F}\cdot d\mathbf{r} = 0$, and it follows that $\int_{C_2} \mathbf{F}\cdot d\mathbf{r} = -5$.

3 Let $\mathcal{A}$ be the region C_4 contains. Then **F** is defined on $\mathcal{A}$, $\nabla \times \mathbf{F} = \mathbf{0}$ throughout $\mathcal{A}$, and $\mathcal{A}$ is simply connected. Thus Theorem 6 applies, and **F** is conservative. Hence $\oint_{C_4} \mathbf{F}\cdot d\mathbf{r} = 0$.

5 $\mathbf{F} = 2xy\mathbf{i} + x^2\mathbf{j}$, so $P = 2xy$ and $Q = x^2$. Now $P_y = 2x$ and $Q_x = 2x$, so $P_y = Q_x$ and the curl of **F** is **0**. Since **F** is defined for the entire plane, **curl F** = **0** implies that **F** is conservative. Since **F** is conservative it is the gradient of some function $f(x, y)$. First, $f_x = P = 2xy$, so $f = \int f_x\, dx =$ $\int 2xy\, dx = x^2y + g(y)$; the arbitrary constant of integration is written as a function of y because it is constant merely with respect to x. We now have $Q = f_y = x^2 + g'(y)$, but $Q = x^2$, so $x^2 =$ $x^2 + g'(y)$. Therefore $g'(y) = 0$ and $g(y) = C$, where C is constant. **F** is the gradient of the function $f(x, y) = x^2y + C$, for any constant C.

7 $\mathbf{F} = (y + 1)\mathbf{i} + (x + 1)\mathbf{j}$, which is defined for the entire plane. $P = y + 1$ and $Q = x + 1$, so $P_y =$ 1 and $Q_x = 1$; hence $P_y = Q_x$, the curl of **F** is **0**, and **F** is a conservative vector field. Thus **F** is the gradient of some scalar field $f(x, y)$. Now $f_x = P =$

$y + 1$, so $f = \int f_x \, dx = \int (y + 1) \, dx =$ $(y + 1)x + g(y)$. Then $Q = f_y = x + g'(y)$; but $Q = x + 1$; hence $x + 1 = x + g'(y)$ and $g'(y) = 1$. So $g(y) = y + C$ and $f(x, y) = (y + 1)x + y + C = xy + x + y + C$.

9 (a) Let $f(x, y) = x^3y$. Then $df = 3x^2y \, dx + x^3 \, dy$, and $3x^2y \, dx + x^3 \, dy$ is an exact differential.

(b) $P \, dx + Q \, dy + R \, dz = 3x^2y \, dx + x^3 \, dy + xyz \, dz$, so $P = 3x^2y$, $Q = x^3$, and $R = xyz$. In order to be exact, the differential must correspond to a vector field $\mathbf{F} = P\mathbf{i} + Q\mathbf{j} + R\mathbf{k}$ whose curl is $\mathbf{0}$. But $P_z = 0$ while $R_x = yz$, so the $\mathbf{j}$ component of the curl would not be 0. The differential is not exact.

11 True, by Theorem 5.

13 True, since $\mathbf{curl}(\nabla f) = \mathbf{0}$ for all scalar functions f and hence for all gradient fields.

15 Since $\nabla \times \mathbf{F} = \mathbf{0}$, and $\mathbf{F}$ is defined everywhere except at the origin, we must ensure that the closed curve does not contain the origin. Then the region C encloses would be simply connected, and $\mathbf{F}$ would be conservative by Theorem 6. We could then safely claim that $\oint_C \mathbf{F} \cdot d\mathbf{r} = 0$.

17 See Exercise 27, Sec 16.5, for the statement of the theorem, and mimic the proof of the two curve version of Green's theorem given in Exercise 13, Sec. 16.5. Be sure to replace all "$\mathbf{F} \cdot \mathbf{n}$" and "$\nabla \cdot \mathbf{F}$" with "$\mathbf{F} \cdot d\mathbf{r}$" and "$(\nabla \times \mathbf{F}) \cdot \mathbf{k}$", respectively.

19 Let C be a closed curve and $\mathcal{A}$ the region it contains. Since $\mathbf{F}$ is conservative, $\mathbf{F} = \nabla f$ for some scalar function f by Theorem 4. Thus, by Stokes' theorem in the plane, $\int_{\mathcal{A}} (\nabla \times \mathbf{F}) \cdot \mathbf{k} \, dA =$ $\oint_C \mathbf{F} \cdot d\mathbf{r} = \oint_C \nabla f \cdot d\mathbf{r} = 0$ by Theorem 3, so

$\int_{\mathcal{A}} (\nabla \times \mathbf{F}) \cdot \mathbf{k} \, dA = 0$ for any region $\mathcal{A}$ by choosing an appropriate closed curve C. By the "vanishing-integrals" principle, $(\nabla \times \mathbf{F}) \cdot \mathbf{k} = 0$. Hence $\nabla \times \mathbf{F} = \mathbf{0}$.

21 (a) If $P \, dx + Q \, dy$ is exact, then $P = f_x$ and $Q = f_y$ for some scalar function f. So $P_y = f_{xy} = f_{yx} = Q_x$ follows.

(b) We need only find P and Q such $P_y \neq Q_x$. For example, let $P = y^3$ and $Q = x^3$. Then $P_y = 3y^2 \neq 3x^2 = Q_x$. Thus $P \, dx + Q \, dy = y^3 \, dx + x^3 \, dy$ is not exact.

23 By the first fundamental theorem of calculus, $g(r)$

$$= \int_a^r \frac{dg}{du} \, du = \int_a^r f(u) \, du,$$ where a is a constant. So let $h(x, y) = g(r)$, and let $a = 0$.

Then $$\nabla h = \nabla\left(\int_0^{\sqrt{x^2+y^2}} f(u) \, du\right) =$$

$$\frac{\partial}{\partial x}\left(\int_0^{\sqrt{x^2+y^2}} f(u) \, du\right)\mathbf{i} + \frac{\partial}{\partial y}\left(\int_0^{\sqrt{x^2+y^2}} f(u) \, du\right)\mathbf{j} =$$

$$f\left(\sqrt{x^2 + y^2}\right) \cdot \frac{2x}{2\sqrt{x^2 + y^2}}\mathbf{i} +$$

$$f\left(\sqrt{x^2 + y^2}\right) \cdot \frac{2y}{2\sqrt{x^2 + y^2}}\mathbf{j} = f(r)\hat{\mathbf{r}} = \mathbf{F}.$$ So $h(x, y)$

$$= \int_0^{\sqrt{x^2+y^2}} f(u) \, du$$ is one such function.

16.S Guide Quiz

1 (a) See "Work Along a Curve," p. 959 and "Circulation of a Fluid," p. 961.

(b) See "Circulation of a Fluid," p. 961.

2 (a) See "Loss or Gain of a Fluid (Flux)," p. 961.

(b) See "Loss or Gain of a Fluid (Flux)," p. 961.

3 See "Why $\nabla\cdot\mathbf{F}$ is Called Divergence," p. 972.

4 See "Why Curl is Called Curl," p. 984.

5 The divergence of $\dfrac{\hat{\mathbf{r}}}{|\mathbf{r}|^2}$ is 0, by Example 3 of Sec. 16.1. To compute the curl of $\dfrac{\hat{\mathbf{r}}}{|\mathbf{r}|^2}$, let $\mathbf{F} = \dfrac{\hat{\mathbf{r}}}{|\mathbf{r}|^2} = \dfrac{\mathbf{r}}{|\mathbf{r}|^3}$, and define $r = |\mathbf{r}| = \sqrt{x^2+y^2+z^2}$.

Then $\mathbf{F} = \dfrac{x\mathbf{i}+y\mathbf{j}+z\mathbf{k}}{(x^2+y^2+z^2)^{3/2}} = \dfrac{x}{r^3}\mathbf{i} + \dfrac{y}{r^3}\mathbf{j} + \dfrac{z}{r^3}\mathbf{k}$

and $\nabla\times\mathbf{F} = \left[\dfrac{\partial}{\partial y}\left(\dfrac{z}{r^3}\right) - \dfrac{\partial}{\partial z}\left(\dfrac{y}{r^3}\right)\right]\mathbf{i} + \left[\dfrac{\partial}{\partial z}\left(\dfrac{x}{r^3}\right) - \dfrac{\partial}{\partial x}\left(\dfrac{z}{r^3}\right)\right]\mathbf{j} + \left[\dfrac{\partial}{\partial x}\left(\dfrac{y}{r^3}\right) - \dfrac{\partial}{\partial y}\left(\dfrac{x}{r^3}\right)\right]\mathbf{k}$.

Now $\dfrac{\partial}{\partial x}\left(\dfrac{1}{r^3}\right) = \dfrac{\partial}{\partial x}[(x^2+y^2+z^2)^{-3/2}] = -\dfrac{3}{2}(x^2+y^2+z^2)^{-5/2}(2x) = -\dfrac{3x}{(x^2+y^2+z^2)^{5/2}} = -\dfrac{3x}{r^5}$. Similarly, $\dfrac{\partial}{\partial y}\left(\dfrac{1}{r^3}\right) = -\dfrac{3y}{r^5}$ and $\dfrac{\partial}{\partial z}\left(\dfrac{1}{r^3}\right) = -\dfrac{3z}{r^5}$. Thus $\nabla\times\mathbf{F} = \left[z\left(-\dfrac{3y}{r^5}\right) - y\left(-\dfrac{3z}{r^5}\right)\right]\mathbf{i} + \left[x\left(-\dfrac{3z}{r^5}\right) - z\left(-\dfrac{3x}{r^5}\right)\right]\mathbf{j} + \left[y\left(-\dfrac{3x}{r^5}\right) - x\left(-\dfrac{3y}{r^5}\right)\right]\mathbf{k} = \mathbf{0}$.

6 The gradient of a scalar field is conservative, so taking gradients is a simple way to generate conservative fields.

7 If the domain of **F** is simply connected, the quickest way to show that **F** is conservative is to show that $\nabla\times\mathbf{F} = \mathbf{0}$. Another way to show that **F** is conservative is to construct a scalar function f such that $\mathbf{F} = \nabla f$. If $\nabla\times\mathbf{F} \neq \mathbf{0}$, then **F** cannot be conservative, regardless of the domain.

8 $\text{div}\,\mathbf{F} = \lim_{d\to 0} \dfrac{\oint_C \mathbf{F}\cdot\mathbf{n}\,ds}{\text{Area of }\mathcal{A}}$, where $\mathcal{A}$ is the region enclosed by the curve C and d is the diameter of $\mathcal{A}$.

9 The average value of div **F** on $\mathcal{A}$ is given by $\dfrac{\int_{\mathcal{A}} \nabla\cdot\mathbf{F}\,dA}{\text{Area of }\mathcal{A}}$. But $\int_{\mathcal{A}} \nabla\cdot\mathbf{F}\,dA = \oint_C \mathbf{F}\cdot\mathbf{n}\,ds = 20$ by Green's theorem and Area of $\mathcal{A} = \pi\cdot 5^2 = 25\pi$. The average value of div **F** on $\mathcal{A}$ is $\dfrac{20}{25\pi} = \dfrac{4}{5\pi}$.

10 (a) Let C be the circle of radius a, parameterized by $x = a\cos\theta$, $y = a\sin\theta$ for $0 \le \theta \le 2\pi$. On C, $\mathbf{n} = \hat{\mathbf{r}}$ and hence $\mathbf{F}\cdot\mathbf{n} = \dfrac{\hat{\mathbf{r}}}{|\mathbf{r}|}\cdot\hat{\mathbf{r}} = \dfrac{|\hat{\mathbf{r}}|^2}{|\mathbf{r}|} = \dfrac{1}{a}$. Also $ds = r\,d\theta = a\,d\theta$. Thus

$\oint_C \mathbf{F}\cdot\mathbf{n}\,ds = \int_0^{2\pi}\left(\dfrac{1}{a}\right)(a\,d\theta) = \int_0^{2\pi} d\theta = 2\pi$. (Note that, though $\nabla\cdot\mathbf{F} = 0$, Green's theorem cannot be applied because it contains the origin, where **F** is undefined.)

(b) Let C_1 be the boundary of the square. Since C_1 encloses the origin, where **F** is undefined, and $\nabla\cdot\mathbf{F} = 0$, we have $\oint_{C_1} \mathbf{F}\cdot\mathbf{n}\,ds =$

$\oint_C \mathbf{F}\cdot\mathbf{n}\, ds = 2\pi$ by Corollary 2 of Sec. 16.5, where C is the circle given in part (a).

(c) Let C be the given boundary and $\mathcal{A}$ be the region C contains. Since $\nabla\cdot\mathbf{F} = 0$ and $\mathbf{F}$ is defined throughout $\mathcal{A}$, we have $\oint_C \mathbf{F}\cdot\mathbf{n}\, ds = \int_{\mathcal{A}} \nabla\cdot\mathbf{F}\, dA = 0$ by Green's theorem.

11 We first show that $\mathbf{F} = \sin x\,\mathbf{i} + \cos y\,\mathbf{j} = \nabla f$ for some scalar function f (thereby showing that $\mathbf{F}$ is conservative). Here $P = f_x = \sin x$, so $f = \int f_x\, dx = -\cos x + g(y)$, where g is a function of y alone. Then $f_y = g'(y) = Q = \cos y$, and it follows that $g(y) = \int \cos y\, dy = \sin y + C$, where C is a constant. Therefore $f(x, y) = -\cos x + \sin y$ is a scalar function such that $\mathbf{F} = \nabla f$. Hence, by Theorem 2 of Sec. 16.6, $\int_C \mathbf{F}\cdot d\mathbf{r} = \int_C \nabla f\cdot d\mathbf{r} = f(2, 4) - f(1, 1)$ for any curve C connecting (1, 1) to (2, 4). So for both the curves given in (a) and (b), $\int_C \mathbf{F}\cdot d\mathbf{r} = f(2, 4) - f(1, 1)$

$= -\cos 2 + \sin 4 - (-\cos 1 + \sin 1)$

$= \sin 4 - \sin 1 + \cos 1 - \cos 2.$

16.S Review Exercises

1 $\mathbf{F} = (x - 5y)\mathbf{i} + xy\mathbf{j}$, $\nabla\cdot\mathbf{F} = 1 + x$, and $\mathcal{A}$ is described by $0 \le x \le 1$, $x^2 \le y \le \sqrt{x}$. Hence

$$\int_{\mathcal{A}} \nabla\cdot\mathbf{F}\, dA = \int_0^1 \int_{x^2}^{\sqrt{x}} (1 + x)\, dy\, dx$$

$$= \int_0^1 (1 + x)y\Big|_{x^2}^{\sqrt{x}}\, dx$$

$$= \int_0^1 (\sqrt{x} + x^{3/2} - x^2 - x^3)\, dx$$

$$= \left(\frac{x^{3/2}}{3/2} + \frac{x^{5/2}}{5/2} - \frac{x^3}{3} - \frac{x^4}{4}\right)\Bigg|_0^1$$

$= \frac{2}{3} + \frac{2}{5} - \frac{1}{3} - \frac{1}{4} = \frac{29}{60}$. C is the boundary curve of $\mathcal{A}$. Let C_1 be the portion of C that lies on the curve $y = x^2$. Then $dy = 2x\, dx$ as x goes from 0 to 1, so $\int_{C_1} \mathbf{F}\cdot\mathbf{n}\, ds = \int_{C_1} -Q\, dx + P\, dy$

$$= \int_{C_1} -xy\, dx + (x - 5y)\, dy$$

$$= \int_0^1 -x^3\, dx + (x - 5x^2)\, 2x\, dx$$

$$= \int_0^1 (2x^2 - 11x^3)\, dx = \left(\frac{2x^3}{3} - \frac{11x^4}{4}\right)\Bigg|_0^1$$

$= \frac{2}{3} - \frac{11}{4} = -\frac{25}{12}$. Now let C_2 be the portion of C on the curve $y = \sqrt{x}$. Then $x = y^2$, so $dx = 2y\, dy$ as y goes from 1 to 0. Hence $\int_{C_2} \mathbf{F}\cdot\mathbf{n}\, ds =$

$$\int_{C_2} -Q\, dx + P\, dy = \int_{C_2} -xy\, dx + (x - 5y)\, dy$$

$$= \int_1^0 -y^3\, 2y\, dy + (y^2 - 5y)\, dy$$

$$= \int_1^0 (-2y^4 + y^2 - 5y)\, dy$$

$$= \left(-\frac{2y^5}{5} + \frac{y^3}{3} - \frac{5y^2}{2}\right)\Bigg|_1^0 = \frac{2}{5} - \frac{1}{3} + \frac{5}{2} = \frac{77}{30}.$$

Therefore $\oint_C \mathbf{F}\cdot\mathbf{n}\, ds = -\frac{25}{12} + \frac{77}{30} = \frac{29}{60} = \int_{\mathcal{A}} \nabla\cdot\mathbf{F}\, dA$, as claimed.

3 (a) If $\mathbf{F} = \nabla f$, $P = f_x$ and $Q = f_y$, so $P_y = f_{xy} = f_{yx} = Q_x$.

(b) Since $\frac{\partial}{\partial x}(x^2y) \neq \frac{\partial}{\partial y}(-x^2y)$, $f(x, y) = x^2y\mathbf{i} - xy^2\mathbf{j}$ is not a gradient field.

5 (a) No. See Example 2 of Sec 16.6, where since $\mathbf{F}$ is not conservative, $\mathbf{F}$ cannot be expressed as the gradient of a scalar field though $\nabla \times \mathbf{F} = \mathbf{0}$.

(b) Yes. Since $\mathbf{F}$ is the gradient of some scalar field f, $\mathbf{F}$ is conservative by Theorem 3 of Sec. 16.6. So **curl** $\mathbf{F} = \mathbf{0}$ by Theorem 5 of that section.

7 (a) $\nabla \times \mathbf{F} = \begin{vmatrix} \mathbf{i} & \mathbf{j} & \mathbf{k} \\ \frac{\partial}{\partial x} & \frac{\partial}{\partial y} & \frac{\partial}{\partial z} \\ P(x, y) & Q(x, y) & 0 \end{vmatrix}$

$= \left(\frac{\partial}{\partial y}(0) - \frac{\partial}{\partial z}Q(x, y)\right)\mathbf{i} + \left(\frac{\partial}{\partial z}P(x, y) - \frac{\partial}{\partial x}(0)\right)\mathbf{j} + \left(\frac{\partial}{\partial x}Q(x, y) - \frac{\partial}{\partial y}P(x, y)\right)\mathbf{k}$

$= \left(\frac{\partial}{\partial x}Q(x, y) - \frac{\partial}{\partial y}P(x, y)\right)\mathbf{k}$, which is parallel to $\mathbf{k}$.

(b) Let $\mathcal{A}$ be the region contained in C, the circle of radius 0.01, and let A be its area. The average value of $\|\nabla\times\mathbf{F}\| = (\nabla \times \mathbf{F})\cdot\mathbf{k}$ over $\mathcal{A}$ is $\frac{1}{A}\int_{\mathcal{A}} (\nabla\times\mathbf{F})\cdot\mathbf{k}\, dA = \frac{1}{A}\oint_C \mathbf{F}\cdot d\mathbf{r} = \frac{-0.002}{\pi(0.01)^2} = -\frac{20}{\pi}$, where Stokes' theorem in the plane was used. So $\nabla \times \mathbf{F} \approx -\frac{20}{\pi}\mathbf{k}$ at points within the circle.

9 See Figure 3 of Sec. 16.4 and the discussion on p. 969.

11 Let C be the straight line from (1, 0) to (2, 0), parameterized as $y = 0$ while x goes from 1 to 2. Then if $\mathbf{E} = P\mathbf{i} + Q\mathbf{j}$, we have $\int_C \mathbf{E}\cdot\mathbf{T}\, ds = \int_C P\, dx + Q\, dy = \int_C P\, dx$. Now $\mathbf{E} = \frac{q\mathbf{u}}{4\pi\epsilon_0 r^2} = \frac{q\mathbf{r}}{4\pi\epsilon_0 r^3} = \frac{qx\mathbf{i} + qy\mathbf{j}}{4\pi\epsilon_0 r^3}$, so $P = \frac{qx}{4\pi\epsilon_0(x^2 + y^2)^{3/2}} = \frac{qx}{4\pi\epsilon_0 x^3} = \frac{q}{4\pi\epsilon_0 x^2}$ on C.

Then $\int_C \mathbf{E}\cdot\mathbf{T}\, ds = \int_C P\, dx = \int_1^2 \frac{q}{4\pi\epsilon_0 x^2}\, dx = \left.\frac{-q}{4\pi\epsilon_0 x}\right|_1^2 = -\frac{q}{4\pi\epsilon_0}\left(\frac{1}{2} - 1\right) = \frac{q}{8\pi\epsilon_0}$. It was shown above that $\mathbf{E}$ is a scalar multiple of the vector field $\mathbf{r}/r^3$. Recall that this is a conservative field, since $\mathbf{r}/r^3 = \nabla f$, where $f = -\frac{1}{r} = -(x^2 + y^2)^{-1/2}$. Hence the line integral depends only on the endpoints of the path and not on the path itself. Thus $\frac{q}{8\pi\epsilon_0}$ is the answer to both parts (a) and (b). There is no difference. (If you worked this out the long way, you may find it useful to know that the line integral from (1, 0) to $\left(1, \frac{1}{2}\right)$ is

$\frac{q}{4\pi\epsilon_0}\left(1 - \frac{2}{\sqrt{5}}\right)$; from $\left(1, \frac{1}{2}\right)$ to $\left(2, \frac{1}{2}\right)$ it is

$\frac{q}{4\pi\epsilon_0}\left(\frac{2}{\sqrt{5}} - \frac{2}{\sqrt{17}}\right)$; and from $\left(2, \frac{1}{2}\right)$ to (2, 0) it is

$\frac{q}{4\pi\epsilon_0}\left(\frac{2}{\sqrt{17}} - \frac{1}{2}\right)$. These add up to $\frac{q}{8\pi\epsilon_0}$.)

13 (a) $f(x, y) = \frac{\hat{\mathbf{r}}}{r^k} = \frac{\mathbf{r}}{r^{k+1}} = \frac{x\mathbf{i} + y\mathbf{j}}{r^{k+1}}$, and curl **F**

$= \left(\frac{\partial}{\partial x}\left(\frac{y}{r^{k+1}}\right) - \frac{\partial}{\partial y}\left(\frac{x}{r^{k+1}}\right)\right)\mathbf{k}$. Now $\frac{\partial}{\partial x}\left(\frac{1}{r^{k+1}}\right)$

$= -(k+1)r^{-(k+2)}\frac{x}{r} = \frac{-x(k+1)}{r^{k+3}}$. Similarly,

$\frac{\partial}{\partial y}\left(\frac{1}{r^{k+1}}\right) = -\frac{y(k+1)}{r^{k+3}}$. So curl **F** =

$\left[y\left(-\frac{x(k+1)}{r^{k+3}}\right) - x\left(-\frac{y(k+1)}{r^{k+3}}\right)\right]\mathbf{k} = \mathbf{0}$ for all

integers k.

(b) By Exercise 32 of Sec. 16.1, **F** is divergence-free if and only if $k = 1$.

15 See "Loss or Gain of a Fluid (Flux)," p. 961, and remove all references to Figure 7 and Figure 8.

17 Label the three segments of C as shown in the figure. On C_1, $y = 0$ as x goes from 0 to 1, so dy

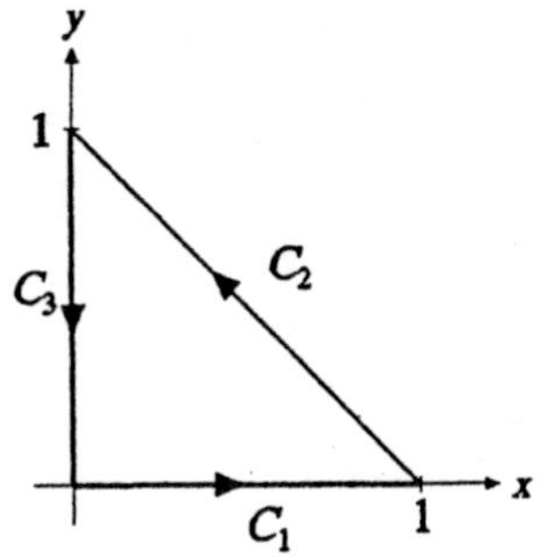

$= 0$ and $\int_{C_1} (3x^2 + y)\,dx + 4y^2\,dy = \int_0^1 3x^2\,dx$

$= x^3\big|_0^1 = 1$. On C_2, $x = 1 - y$ as y goes from 0 to 1, so $dx = -dy$ and $\int_{C_2} (3x^2 + y)\,dx + 4y^2\,dy$

$= \int_0^1 [-(3(1-y)^2 + y) + 4y^2]\,dy$

$= \int_0^1 (y^2 + 5y - 3)\,dy = \left(\frac{y^3}{3} + \frac{5y^2}{2} - 3y\right)\Big|_0^1$

$= \frac{1}{3} + \frac{5}{2} - 3 = -\frac{1}{6}$. On C_3, $x = 0$ as y goes from 1 to 0, so $dx = 0$ and

$\int_{C_3} (3x^2 + y)\,dx + 4y^2\,dy = \int_1^0 4y^2\,dy =$

$\frac{4y^3}{3}\Big|_1^0 = -\frac{4}{3}$. Therefore

$\oint_C (3x^2 + y)\,dx + 4y^2\,dy = 1 - \frac{1}{6} - \frac{4}{3}$

$= -\frac{1}{2}$. By Green's theorem this should equal

$\int_{\mathcal{A}} (P_x + Q_y)\,dA$, where $Q = -(3x^2 + y)$ and $P = 4y^2$ are found by comparing

$\oint_C (3x^2 + y)\,dy + 4y^2\,dy$ and

$\oint_C -Q\,dx + P\,dy$. Thus $P_x = 0$ and $Q_y = -1$, so

$\int_{\mathcal{A}} (P_x + Q_y)\,dA = \int_{\mathcal{A}} (-1)\,dA = -(\text{Area of } \mathcal{A})$

$= -\frac{1}{2}$.

19 Since **F** is tangent to C_1 at all points on C_1, $\mathbf{F}\cdot\mathbf{n} = 0$ on C_1 and $\oint_{C_1} \mathbf{F}\cdot\mathbf{n}\,ds = 0$. So, since **F** is

undefined at (0, 0), $\nabla\cdot\mathbf{F} = 0$, and C_1 and C_2 enclose (0, 0), we have $\oint_{C_2} \mathbf{F}\cdot\mathbf{n}\, ds = \oint_{C_1} \mathbf{F}\cdot\mathbf{n}\, ds$ $= 0$. Green's theorem applies to C_3 and the region $\mathcal{A}$ it bounds because $\mathbf{F}$ is defined throughout $\mathcal{A}$. Hence $\oint_{C_3} \mathbf{F}\cdot\mathbf{n}\, ds = \int_{\mathcal{A}} \nabla\cdot\mathbf{F}\, dA = 0$ since $\nabla\cdot\mathbf{F}$ $= 0$.

21 (a) $x = 2\cos t$ and $y = 3\sin t$, so $\left(\frac{x}{2}\right)^2 + \left(\frac{y}{3}\right)^2$ $= \cos^2 t + \sin^2 t = 1$; C is the ellipse $\frac{x^2}{4} + \frac{y^2}{9} = 1$. (See the accompanying figure.)

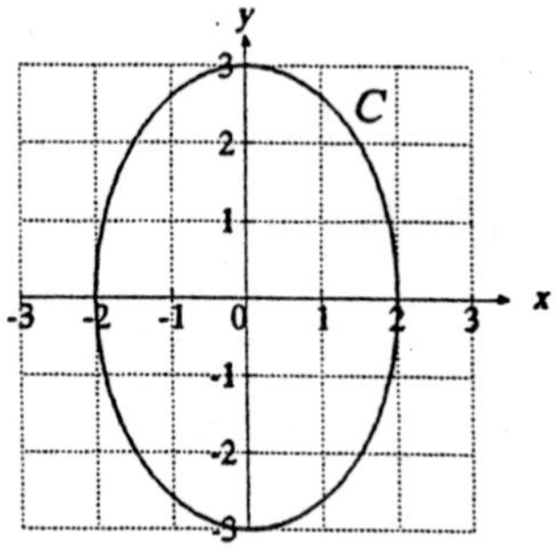

(b) $\oint_C x^2\, dx + (y + 1)\, dy$

$= \int_0^{2\pi} (4\cos^2 t)(-2\sin t\, dt) +$ $(3\sin t + 1)(3\cos t\, dt)$

$= \left(\frac{8\cos^3 t}{3} + \frac{1}{2}(3\sin t + 1)^2\right)\Big|_0^{2\pi}$

$= \frac{8}{3} + \frac{1}{2} - \frac{8}{3} - \frac{1}{2} = 0.$

(c) Find the work done by the force $\mathbf{F} = x^2\mathbf{i} + (y + 1)\mathbf{j}$ in moving a unit mass once around the ellipse $\frac{x^2}{4} + \frac{y^2}{9} = 1$.

(d) The vector field $\mathbf{F} = (y + 1)\mathbf{i} - x^2\mathbf{j}$ describes a fluid flow. Compute the net flow across C, the ellipse $\frac{x^2}{4} + \frac{y^2}{9} = 1$.

23 (a) $\nabla \times \mathbf{F} = \left(\frac{\partial}{\partial x}(x^2) - \frac{\partial}{\partial y}(2xy)\right)\mathbf{k}$ $= (2x - 2x)\mathbf{k} = \mathbf{0}$. So since the plane is simply connected and **curl F** $= \mathbf{0}$, **F** is conservative.

(b) Let (a, b) be any point in the plane where $a \neq 0$ and $b \neq 0$. Let C_1 be the line from (0, 0) to $(a, 0)$. On C_1, $y = 0$, $dy = 0$, and x varies from 0 to a. Then $\int_{C_1} \mathbf{F}\cdot d\mathbf{r} =$ $\int_{C_1} 2xy\, dx + x^2\, dy = 0$. Now let C_2 be the line from $(a, 0)$ to (a, b). On C_2, $x = a$, dx $= 0$, and y varies from 0 to b. Now $\int_{C_2} \mathbf{F}\cdot d\mathbf{r}$ $= \int_{C_2} 2xy\, dx + x^2\, dy = \int_0^b a^2\, dy = a^2 b$.

Thus if C is the curve consisting of C_1 followed by C_2, we have $f(a, b) = a^2 b$, and it follows that $f(x, y) = x^2 y$ is one scalar function such that $\mathbf{F} = \nabla f$.

25 (b) From Figure 2 in the text, $\hat{\boldsymbol{\theta}} = \frac{-y\mathbf{i} + x\mathbf{j}}{\sqrt{x^2 + y^2}}$ when $\hat{\mathbf{r}} = \frac{x\mathbf{i} + y\mathbf{j}}{\sqrt{x^2 + y^2}}$. So $\mathbf{F} = |\mathbf{r}|^2\hat{\boldsymbol{\theta}}$

$= (x^2 + y^2)\left(\frac{-y\mathbf{i} + x\mathbf{j}}{\sqrt{x^2 + y^2}}\right)$

$= \sqrt{x^2 + y^2}(-y\mathbf{i} + x\mathbf{j})$, and $(\nabla \times \mathbf{F})\cdot\mathbf{k}$

$= \frac{\partial}{\partial x}\left(x\sqrt{x^2+y^2}\right) - \frac{\partial}{\partial y}\left(-y\sqrt{x^2+y^2}\right)$

$= \left(\sqrt{x^2+y^2} + \frac{x^2}{\sqrt{x^2+y^2}}\right) -$

$\left(-\sqrt{x^2+y^2} - \frac{y^2}{\sqrt{x^2+y^2}}\right) = 3\sqrt{x^2+y^2}$

$= 3r$, where $r = \|\mathbf{r}\|$. So by Stokes' theorem in the plane with $\mathcal{A}$ described by $r_1 \le r \le r_2$, $\theta_1 \le \theta \le \theta_2$ we have $\oint_C \mathbf{F}\cdot d\mathbf{r} =$

$\int_{\mathcal{A}} (\nabla\times\mathbf{F})\cdot\mathbf{k}\, dA = \int_{\theta_1}^{\theta_2}\int_{r_1}^{r_2} 3r^2\, dr\, d\theta$

$= (\theta_2 - \theta_1)\left(r_2^3 - r_1^3\right)$.

(c) Since $\oint_C \mathbf{F}\cdot d\mathbf{r} = (\theta_2 - \theta_1)\left(r_2^3 - r_1^3\right) > 0$, the circulation around C is counterclockwise. Therefore a paddle wheel will turn clockwise as viewed from the origin, or counterclockwise as viewed from above.

27 Let C_1^* be the path from B to A along C_1. Let C be the path from A to A that moves along C_2 and then C_1^*, and let $\mathcal{A}$ be the region bounded by C. Then

$\int_{C_2} G\,dx + H\,dy - \int_{C_1} G\,dx + H\,dy =$

$\oint_C G\,dx + H\,dy = \int_{\mathcal{A}} (H_x - G_y)\,dA \ge 0$, since $H_x - G_y \ge 0$. (We applied Green's theorem with $\mathbf{F} = H\mathbf{i} - G\mathbf{j}$.)

29 Let $\mathbf{v}$ be the air velocity of the plane. We are given that $v = \|\mathbf{v}\|$ is constant. The ground velocity of the plane is given by $\mathbf{G} = \mathbf{v} + \mathbf{W}$. Let $G = \|\mathbf{G}\|$ and $W = \|\mathbf{W}\|$. We are also given that $W < v$. Introduce a coordinate system so that $\mathbf{W}$ is parallel

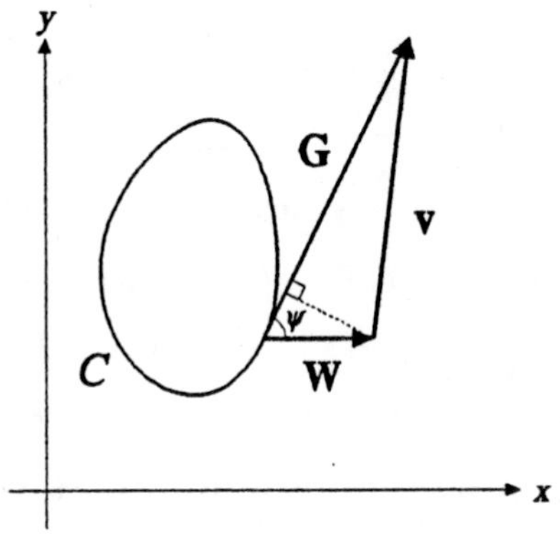

to the x axis and let C be the ground track of the plane. The ground speed is $\frac{ds}{dt} = G$, so the travel time for one trip is $\oint_C dt = \oint_C \frac{ds}{G}$. By the figure, $G = W\cos\psi + \sqrt{v^2 - W^2\sin^2\psi}$. Recall that $\frac{dx}{ds} = \cos\psi$, so $\oint_C \cos\psi\, ds = \oint_C dx = 0$. Hence

$$\oint_C \frac{ds}{G} = \oint_C \frac{ds}{W\cos\psi + \sqrt{v^2 - W^2\sin^2\psi}}$$

$$= \int_C \frac{W\cos\psi - \sqrt{v^2 - W^2\sin^2\psi}}{W^2\cos^2\psi - v^2 + W^2\sin^2\psi}\, ds$$

$$= \oint_C \frac{\sqrt{v^2 - W^2\sin^2\psi}}{v^2 - W^2}\, ds \ge \oint_C \frac{\sqrt{v^2 - W^2}}{v^2 - W^2}\, ds$$

$= \oint_C \frac{ds}{\sqrt{v^2 - W^2}} \ge \oint_C \frac{ds}{v}$, which is the time for a trip with $W = 0$.

31 $\frac{\partial f}{\partial r}$ is the directional derivative of f in the $\hat{\mathbf{r}} = \mathbf{n}$ direction, $D_{\mathbf{n}}f$. But $D_{\mathbf{n}}f = \nabla f\cdot\mathbf{n}$, so $\frac{\partial f}{\partial r} = \nabla f\cdot\mathbf{n}$.

33 From Exercise 32, $I(r) = k$ for all $r > 0$. So $\lim_{r\to 0^+} I(r) = k$. But when $r = 0$, the average value of points on the "circle" is merely $f(Q)$ since the "circle" consists of Q alone. Thus $k = f(Q)$.

35 (a) Consider any circle of nonzero radius contained in C with Q as its center. Since f is harmonic, the average value of f on this circle is $f(Q) = M$. However, $f(x, y) \le M$ on the circle, so $f(x, y) = M$ for all (x, y) on the circle. (Otherwise, the average value of f would be less than M, a contradiction.) So $f(x, y) = M$ throughout the disk formed by all concentric circles in C with center Q.

(b) Following the procedure in (a), we may expand the disk in (a) by including every disk in C with a center in the original disk, and $f(x, y) = M$ throughout this region. Continuing in this fashion, we eventually cover $\mathcal{A}$. So $f(x, y) = M$ throughout $\mathcal{A}$.

(c) Arguing similarly, $f(x, y) = m$ throughout $\mathcal{A}$. So $M = m = 0$ and $f(x, y) = 0$ for all (x, y) in $\mathcal{A}$.

37 (a) Since $\nabla\cdot\mathbf{F} = 0$ and $\mathbf{F}$ is defined throughout the region $\mathcal{A}$ contains, $\oint_C \mathbf{F}\cdot\mathbf{n}\,ds = \int_{\mathcal{A}} \nabla\cdot\mathbf{F}\,dA = 0$ by Green's theorem.

(b) Computing $\oint_C \mathbf{F}\cdot\mathbf{n}\,ds$ directly, we break C up into four curves: C_1, where $\theta = \theta_1$, $1 \le r \le r_2$; C_2, where $r = r_2$, $\theta_1 \le \theta \le \theta_2$; C_3, where $\theta = \theta_2$, $1 \le r \le r_2$; C_4, where $r = 1$, $\theta_1 \le \theta \le \theta_2$. On C_1 and C_3, $\mathbf{F}\cdot\mathbf{n} = f(r)\hat{\mathbf{r}}\cdot(\pm\hat{\boldsymbol{\theta}}) = 0$, so $\oint_{C_1} \mathbf{F}\cdot\mathbf{n}\,ds = \oint_{C_3} \mathbf{F}\cdot\mathbf{n}\,ds = 0$. On C_2, $\mathbf{n} = \hat{\mathbf{r}}$ and $r = r_2$ as θ goes from θ_1 to θ_2; hence $\oint_{C_2} \mathbf{F}\cdot\mathbf{n}\,ds = \int_{C_2} f(r)\,ds = \int_{\theta_1}^{\theta_2} f(r_2)r_2\,d\theta = r_2 f(r_2)(\theta_2 - \theta_1)$. On C_4, $\mathbf{n} = -\hat{\mathbf{r}}$ and $r = 1$ as θ goes from θ_2 to θ_1; thus $\oint_{C_4} \mathbf{F}\cdot\mathbf{n}\,ds = \int_{C_2} -f(r)\,ds = \int_{\theta_2}^{\theta_1} -f(1)\,d\theta = -f(1)(\theta_1 - \theta_2) = f(1)(\theta_2 - \theta_1)$. So $\oint_C \mathbf{F}\cdot\mathbf{n}\,ds = \oint_{C_1} \mathbf{F}\cdot\mathbf{n}\,ds + \oint_{C_2} \mathbf{F}\cdot\mathbf{n}\,ds + \oint_{C_3} \mathbf{F}\cdot\mathbf{n}\,ds + \oint_{C_4} \mathbf{F}\cdot\mathbf{n}\,ds$ $= [r_2 f(r_2) + f(1)](\theta_2 - \theta_1) = 0$ by part (a). Since $\theta_2 - \theta_1 > 0$, $r_2 f(r_2) + f(1) = 0$. So $f(r_2) = -\dfrac{f(1)}{r_2}$ for all $r_2 > 0$. Thus $f(r) = \dfrac{k}{r}$, where $k = -f(1)$.

17 The Divergence Theorem and Stokes' Theorem

17.1 Surface Integrals

1 We have $dA \approx \cos\gamma\, dS$ where $dA = 0.05$ and $\gamma = \pi/4$. Then $0.05 \approx \cos \pi/4\, dS$, so $dS \approx (0.05)\sqrt{2} \approx 0.07$.

3 (a)

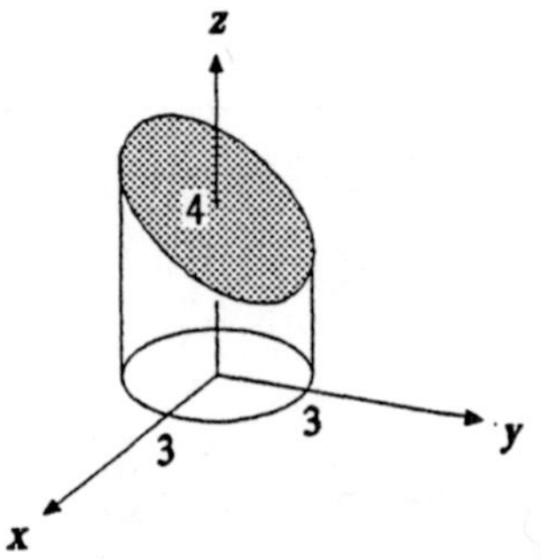

(b) The elliptical region $\mathcal{S}$ has as its projection on the xy plane the circle $\mathcal{A}$ of radius 3. Hence Area of $\mathcal{S} = \dfrac{1}{\cos\gamma}$(Area of $\mathcal{A}$), where γ is the angle between $\mathbf{N} = \mathbf{i} + 2\mathbf{j} + 3\mathbf{k}$ (a normal to $\mathcal{S}$) and $\mathbf{k}$ (a normal to $\mathcal{A}$). We have $\cos\gamma = \mathbf{N}\cdot\mathbf{k}/\|\mathbf{N}\| = 3/\sqrt{14}$, so Area of $\mathcal{S} = \dfrac{\sqrt{14}}{3}\cdot 9\pi = 3\pi\sqrt{14}$.

5 Let S be the area of $\mathcal{S}$. Then the average of $f(P)$ over $\mathcal{S}$ would be $\dfrac{1}{S}\int_{\mathcal{S}} f(P)\, dS$. Let $f(P) = x^2 + y^2 = r^2 = a^2\sin^2\phi$, the square of the distance to the z axis for points on the sphere $\rho = a$. Then

$$\frac{1}{4\pi a^2}\int_0^{2\pi}\int_0^{\pi} a^2\sin^2\phi\; a^2\sin\phi\, d\phi\, d\theta$$

$= \dfrac{a^2}{2}\displaystyle\int_0^{\pi}\sin^3\phi\, d\phi = \dfrac{a^2}{2}\cdot\dfrac{4}{3} = \dfrac{2}{3}a^2$. (Formula 73 from the table of integrals was used.)

7 $\mathbf{F} = x^2\mathbf{i} + y^2\mathbf{j} + z^2\mathbf{k}$ and the surface $\mathcal{S}$ is the sphere $x^2 + y^2 + z^2 = 9$; the unit exterior normal to $\mathcal{S}$ is $\mathbf{n} = \dfrac{1}{3}(x\mathbf{i} + y\mathbf{j} + z\mathbf{k})$, so $\mathbf{F}\cdot\mathbf{n} = \dfrac{1}{3}(x^3 + y^3 + z^3)$. By symmetry, points on opposite sides of $\mathcal{S}$ will have opposite values of $\mathbf{F}\cdot\mathbf{n}$. The integral must be 0.

9 Using the notation of Example 2, with $R(x, y, z) = x$, we have $\int_{\mathcal{S}} x\cos\gamma\, dS = \int_{\mathcal{A}} (x - x)\, dA = 0$.

11 $\int_{\mathcal{S}} x\cos\alpha\, dS = \int_{\mathcal{A}} (x_2 - x_1)\, dA =$ Volume of $\mathcal{S}$, where $\mathcal{A}$ is the projection of $\mathcal{S}$ onto the yz plane.

13 By Example 4 and symmetry, the area of $\mathcal{S}$ is $S = \dfrac{\pi}{6}[(1 + a^2)^{3/2} - 1]$. Also by Example 4, we know that $\dfrac{1}{|\cos\gamma|} = \sqrt{1 + r^2}$, so the moment with respect to the xy plane is $M_{xy} = \int_{\mathcal{S}} z\, dS =$

$$\int_{\mathcal{A}} \frac{z}{|\cos\gamma|}\, dA = \int_0^{\pi/2}\int_0^{a} z\sqrt{1 + r^2}\; r\, dr\, d\theta =$$

$\int_0^{\pi/2}\int_0^a (r^2\cos\theta\sin\theta)\sqrt{1+r^2}\,r\,dr\,d\theta =$

$\int_0^{\pi/2}\int_0^a (\cos\theta\sin\theta)\sqrt{1+r^2}\,r^3\,dr\,d\theta$, since $z = xy = (r\cos\theta)(r\sin\theta)$. If we rearrange the result to integrate with respect to θ first, we have

$\int_0^{\pi/2}\sin\theta\cos\theta\,d\theta = \left.\frac{\sin^2\theta}{2}\right|_0^{\pi/2} = \frac{1}{2}$, so $M_{xy} =$

$\frac{1}{2}\int_0^a \sqrt{1+r^2}\,r^3\,dr$. By parts, with $u = r^2$ and

$dv = \sqrt{1+r^2}\,r\,dr$, we have $du = 2r\,dr$ and $v =$

$\frac{1}{3}(1+r^2)^{3/2}$, so the integral becomes

$$\frac{1}{2}\left[\left.\frac{r^2}{3}(1+r^2)^{3/2}\right|_0^a - \int_0^a \frac{1}{3}(1+r^2)^{3/2}\,2r\,dr\right]$$

$$= \frac{1}{2}\left[\frac{a^2}{3}(1+a^2)^{3/2} - \left.\frac{1}{3}\frac{2}{5}(1+r^2)^{5/2}\right|_0^a\right]$$

$$= \frac{a^2}{6}(1+a^2)^{3/2} - \frac{1}{15}(1+a^2)^{5/2} + \frac{1}{15}.$$ Hence $\bar{z}$

$$= \frac{M_{xy}}{S} = \frac{\frac{a^2}{6}(1+a^2)^{3/2} - \frac{1}{15}(1+a^2)^{5/2} + \frac{1}{15}}{\frac{\pi}{6}[(1+a^2)^{3/2} - 1]}$$

$$= \frac{5a^2(1+a^2)^{3/2} - 2(1+a^2)^{5/2} + 2}{5\pi[(1+a^2)^{3/2} - 1]}.$$

15 $\mathcal{S}$ is part of the parabolic cylinder $z = \frac{1}{2}x^2$, which is a level surface of the function $f(x, y, z) = z - \frac{1}{2}x^2$, so $\mathbf{N} = \nabla f = -x\mathbf{i} + \mathbf{k}$ is a vector normal to $\mathcal{S}$. It follows that $\cos\gamma = \mathbf{k}\cdot\mathbf{N}/\|\mathbf{N}\| = \frac{1}{\sqrt{1+x^2}}$. The planes $y = 0$, $y = x$, and $x = 2$ bound $\mathcal{S}$, and they also bound the triangle $\mathcal{A}$ in the xy plane described by $0 \le x \le 2$, $0 \le y \le x$, which is the projection of $\mathcal{S}$ in the xy plane. The area of $\mathcal{S}$ is $\int_{\mathcal{S}} dS = \int_{\mathcal{A}} \frac{1}{|\cos\gamma|}\,dA$

$$= \int_0^2\int_0^x \sqrt{1+x^2}\,dy\,dx = \int_0^2 \sqrt{1+x^2}\,x\,dx$$

$$= \left.\frac{1}{3}(1+x^2)^{3/2}\right|_0^2 = \frac{1}{3}(5^{3/2} - 1).$$

17 (a) The vertices of the triangle in the xy plane are (1, 2), (3, 4), and (2, 5). The triangle is therefore half of the parallelogram spanned by $(2-1)\mathbf{i} + (5-2)\mathbf{j} = \mathbf{i} + 3\mathbf{j}$ and $(3-1)\mathbf{i} + (4-2)\mathbf{j} = 2\mathbf{i} + 2\mathbf{j}$. The area of the parallelogram is abs$\begin{vmatrix}1 & 3\\ 2 & 2\end{vmatrix} = |1\cdot 2 - 3\cdot 2| = 4$, so the area of the triangle is 2.

(b) Rewrite the equation of the plane $z = x + y$ as $x + y - z = 0$; the coefficients provide the components of a vector that is normal to the plane, $\mathbf{N} = \mathbf{i} + \mathbf{j} - \mathbf{k}$. A unit normal with positive z component is $\mathbf{n} = -\mathbf{N}/\|\mathbf{N}\|$ $= \frac{1}{\sqrt{3}}(-\mathbf{i} - \mathbf{j} + \mathbf{k})$. We now have $\cos\gamma = \mathbf{n}\cdot\mathbf{k} = \frac{1}{\sqrt{3}}$, where γ is the angle by which the plane $z = x + y$ is tilted with respect to the xy plane. Since $\cos\gamma$ is constant, the area of the triangle in the tilted plane is equal to the area of its projection in the xy plane multiplied by $\frac{1}{\cos\gamma}$. From (a) we know that the projection has area 2, so the triangle in the plane $z = x + y$ has area $\frac{2}{\cos\gamma} = 2\sqrt{3}$.

19 The cone $z^2 = x^2 + y^2$ is a level surface of the function $f(x, y, z) = z^2 - x^2 - y^2$, so a normal vector is provided by $\mathbf{N} = \nabla f = -2x\mathbf{i} - 2y\mathbf{j} + 2z\mathbf{k}$. The angle γ between the normal vector and $\mathbf{k}$ can be found from $\cos\gamma = \mathbf{k}\cdot\mathbf{N}/\|\mathbf{N}\| = \frac{2z}{2\sqrt{x^2+y^2+z^2}} = \frac{z}{\sqrt{2z^2}} = \frac{1}{\sqrt{2}}$. (We used $z^2 = x^2 + y^2$ for points on the cone.) Therefore, Area of $S = \int_{\mathcal{A}} \frac{1}{\cos\gamma}\,dA = \sqrt{2}(\text{Area of } \mathcal{A})$. As θ goes from $-\frac{\pi}{4}$ to $\frac{\pi}{4}$, one loop of the curve $r = \sqrt{\cos 2\theta}$ is traced out. The loop is symmetric, so we may find the area of $\mathcal{A}$ by integrating from 0 to $\frac{\pi}{4}$ and doubling the result: Area of $\mathcal{A}$ =

$$2\int_0^{\pi/4} \frac{1}{2}r^2\,d\theta = \int_0^{\pi/4} \cos 2\theta\,d\theta = \frac{1}{2}\sin 2\theta\Big|_0^{\pi/4}$$

$= \frac{1}{2}$. The area of S is equal to $\sqrt{2}\cdot\frac{1}{2} = \frac{1}{\sqrt{2}}$.

21 S is part of the surface $2z = x^2 + y^2$, which is a level surface for the function $f(x, y, z) = z - \frac{1}{2}(x^2 + y^2)$. Hence $\mathbf{N} = \nabla f = -x\mathbf{i} - y\mathbf{j} + \mathbf{k}$ is normal to S. Therefore $\cos\gamma = \mathbf{k}\cdot\mathbf{N}/\|\mathbf{N}\| = \frac{1}{\sqrt{x^2+y^2+1}} = \frac{1}{\sqrt{r^2+1}}$, so $\frac{1}{\cos\gamma} = \sqrt{r^2+1}$. When $z = 9$, $x^2 + y^2 = 2z = 18$, so the projection of S onto the xy plane is the disk $\mathcal{A}$ enclosed by the circle $x^2 + y^2 = 18$ of radius $3\sqrt{2}$. $\mathcal{A}$ is described in polar coordinates by $0 \le r \le 3\sqrt{2}$, $0 \le \theta \le 2\pi$. Therefore Area of $S = S = \int_S dS = \int_{\mathcal{A}} \frac{1}{\cos\gamma}\,dA$

$$= \int_0^{2\pi}\int_0^{3\sqrt{2}} \sqrt{r^2+1}\; r\,dr\,d\theta =$$

$$2\pi\left[\frac{1}{2}\cdot\frac{(r^2+1)^{3/2}}{3/2}\right]\Bigg|_0^{3\sqrt{2}} = \frac{2\pi}{3}(19^{3/2} - 1).$$ We now need to compute $\int_S z\,dS$, where $z = \frac{1}{2}(x^2 + y^2)$

$= \frac{1}{2}r^2$: $\int_S z\,dS = \int_{\mathcal{A}} \frac{z}{\cos\gamma}\,dA$

$$= \int_0^{2\pi}\int_0^{3\sqrt{2}} \frac{1}{2}r^2\sqrt{r^2+1}\; r\,dr\,d\theta$$

$$= 2\pi\cdot\frac{1}{2}\int_0^{3\sqrt{2}} r^3\sqrt{r^2+1}\,dr$$

$$= \pi\left[\frac{1}{5}(r^2+1)^{5/2} - \frac{1}{3}(r^2+1)^{3/2}\right]\Bigg|_0^{3\sqrt{2}}$$

$$= \pi\left[\frac{1}{5}\cdot 19^{5/2} - \frac{1}{3}\cdot 19^{3/2} - \frac{1}{5} + \frac{1}{3}\right]$$

$= \frac{\pi}{15}[3\cdot 19^{5/2} - 5\cdot 19^{3/2} + 2]$. Thus $\bar{z}$ =

$$\frac{1}{S}\int_S z\,dS = \frac{\frac{\pi}{15}[3\cdot 19^{5/2} - 5\cdot 19^{3/2} + 2]}{\frac{2\pi}{3}(19^{3/2} - 1)}$$

$$= \frac{3\cdot 19^{5/2} - 5\cdot 19^{3/2} + 2}{10(19^{3/2} - 1)}.$$

23 (a) By symmetry, $\int_S x\, dS = 0$.

(b) By symmetry, $\int_S x^3\, dS = 0$.

(c) $\int_S \frac{2x + 4y^5}{\sqrt{2 + x^2 + 3y^2}}\, dS =$

$\int_S \frac{2x}{\sqrt{2 + x^2 + 3y^2}}\, dS +$

$\int_S \frac{4y^5}{\sqrt{2 + x^2 + 3y^2}}\, dS$; the expression in the radical in the denominator is even with respect to x and y—it is always positive. Since the numerators are odd functions, symmetry says that the integrals are 0.

25 Let S be the sphere $\rho = a$; then $dS = a^2 \sin\phi\, d\phi\, d\theta$. For points on S the radiated power is $\delta(P) = k(\sin^2\phi)/\rho^2 = k(\sin^2\phi)/a^2$. The total power is

$\int_S \delta(P)\, dS = \int_0^{2\pi}\int_0^{\pi} \frac{k}{a^2}\sin^2\phi\; a^2\sin\phi\, d\phi\, d\theta$

$= 2\pi k \int_0^{\pi} \sin^3\phi\, d\phi = 2\pi k\cdot\frac{4}{3} = \frac{8\pi k}{3}$.

27 The homogeneous sphere S has mass M and surface area $4\pi a^2$, so the constant density is equal to $\sigma = \frac{M}{4\pi a^2}$. As in Example 1, it is convenient to regard the sphere in spherical coordinates as the surface $\rho = a$ with surface element $dS = a^2 \sin\phi\, d\phi\, d\theta$. The integrand used to compute moment of inertia about the z axis is r^2, where $r = \rho\sin\phi = a\sin\phi$. Hence $I_z = \int_S r^2\sigma\, dS$

$= \sigma\int_0^{2\pi}\int_0^{\pi} a^2\sin^2\phi\; a^2\sin\phi\, d\phi\, d\theta$

$= 2\pi\sigma a^4 \int_0^{\pi}\sin^3\phi\, d\phi = 2\pi\sigma a^4\cdot\frac{4}{3}$

$= \frac{2\pi M a^4}{4\pi a^2}\cdot\frac{4}{3} = \frac{2Ma^2}{3}$.

29 Note that $q^2 = \overline{PA}^2 = a^2 + b^2 - 2ab\cos\phi$. (See Example 5 in Sec. 15.7.) Then $\int_S \delta(P)\, dS$

$= \int_S \frac{1}{q}\, dS$

$= \int_0^{2\pi}\int_0^{\pi} \frac{a^2\sin\phi}{\sqrt{(a^2 + b^2) - 2ab\cos\phi}}\, d\phi\, d\theta$

$= \int_0^{2\pi} \frac{a}{b}\left(\sqrt{a^2 + b^2 + 2ab} - \sqrt{a^2 + b^2 - 2ab}\right) d\theta$

$= \int_0^{2\pi} \frac{2a^2}{b}\, d\theta = \frac{4\pi a^2}{b}$. So the average of δ is

$\frac{4\pi a^2}{b}\cdot\frac{1}{4\pi a^2} = \frac{1}{b}$.

31 The cone of radius a and height h has curved surface area $A = \pi a s$, where $s = \sqrt{a^2 + h^2}$, and vertex angle $\alpha = \tan^{-1}\frac{a}{h}$, so $\sin\alpha = \frac{a}{\sqrt{a^2 + h^2}}$.

The distance to the z axis is $r = \rho\sin\alpha$ for points on the surface S, so the average distance is

$\frac{1}{A}\int_S r\, dS$

$= \frac{1}{\pi a s}\int_0^{2\pi}\int_0^{s} \rho\sin\alpha\;(\rho\sin\alpha\, d\rho\, d\theta)$

$= \frac{1}{\pi a s}\int_0^{2\pi} \frac{s^3}{3}\sin^2\alpha\, d\theta = \frac{s^2}{3\pi a}\sin^2\alpha\; 2\pi$

$= \frac{a^2 + h^2}{3\pi a}\,\frac{a^2}{a^2 + h^2}\,2\pi = \frac{2}{3}a$.

33 Because the cone has radius and height both equal to 1, the conical surface $\mathcal{S}$ is described by $0 \le \theta \le 2\pi$, $0 \le \rho \le \sqrt{2}$, $\phi = \frac{\pi}{4}$. Now $x = \rho \sin\phi \cos\theta$, so $\int_{\mathcal{S}} x^2\, dS$

$$= \int_0^{2\pi} \int_0^{\sqrt{2}} (\rho \sin\alpha \cos\theta)^2\, \rho \sin\alpha\, d\rho\, d\theta$$

$$= \int_0^{2\pi} \int_0^{\sqrt{2}} \rho^3 \sin^3\alpha \cos^2\theta\, d\rho\, d\theta$$

$$= \frac{(\sqrt{2})^4}{4} \sin^3\frac{\pi}{4} \int_0^{2\pi} \cos^2\theta\, d\theta = \frac{\pi}{2\sqrt{2}}.$$

35 The cylindrical surface $\mathcal{S}$ is described by $0 \le \theta \le 2\pi$, $0 \le z \le h$, $r = a$. Choose the diameter that lies on the y axis so that the square of the distance to the diameter is $x^2 + z^2$. The area of $\mathcal{S}$ is $2\pi ah$, so the average of the squared distance is

$$\frac{1}{2\pi ah} \int_{\mathcal{S}} (x^2 + z^2)\, dS$$

$$= \frac{1}{2\pi ah} \int_0^{2\pi} \int_0^{h} (a^2 \cos^2\theta + z^2)\, a\, dz\, d\theta$$

$$= \frac{1}{2\pi h} \int_0^{2\pi} \left(a^2 h \cos^2\theta + \frac{h^3}{3}\right) d\theta$$

$$= \frac{1}{2\pi h}\left(a^2 h\pi + \frac{h^3}{3} 2\pi\right) = \frac{a^2}{2} + \frac{h^2}{3}.$$

17.2 Surface Integrals

1 See page 967.

3 $\mathbf{F} = x\mathbf{i} + y\mathbf{j} + z\mathbf{k}$, so $\nabla\cdot\mathbf{F} = 1 + 1 + 1 = 3$. $\mathcal{V}$ is the ball of radius a, so $\int_{\mathcal{V}} \nabla\cdot\mathbf{F}\, dV = 3(\text{Volume of } \mathcal{V}) = 3\cdot\frac{4\pi a^3}{3} = 4\pi a^3$. Now observe that the exterior unit normal for the sphere is $\mathbf{n} = \frac{1}{a}(x\mathbf{i} + y\mathbf{j} + z\mathbf{k})$, so $\mathbf{F}\cdot\mathbf{n} = \frac{1}{a}(x^2 + y^2 + z^2) = \frac{1}{a}(a^2) = a$; thus $\int_{\mathcal{S}} \mathbf{F}\cdot\mathbf{n}\, dS = \int_{\mathcal{S}} a\, dS = a(\text{Area of } \mathcal{S}) = a(4\pi a^2) = 4\pi a^3$, in agreement with the previous result and confirming the divergence theorem in this instance.

5 $\mathbf{F} = x^2\mathbf{i}$, so $\nabla\cdot\mathbf{F} = 2x$. We then have $\int_{\mathcal{S}} \mathbf{F}\cdot\mathbf{n}\, dS$

$$= \int_{\mathcal{V}} \nabla\cdot\mathbf{F}\, dV = \int_0^2 \int_0^3 \int_0^4 2x\, dz\, dy\, dx =$$

$$\int_0^2 24x\, dx = 12x^2\Big|_0^2 = 48.$$

7 $\mathbf{F} = 3x\mathbf{i} + 2y\mathbf{j} + 6z\mathbf{k}$, so $\nabla\cdot\mathbf{F} = 3 + 2 + 6 = 11$. Recall that the volume of a tetrahedron is $\frac{1}{3}Ah$, where A is the area of the base and h is its height. The vertices of $\mathcal{V}$ are $(0, 0, 0)$, $(1, 0, 0)$, $(0, 2, 0)$, and $(0, 0, 3)$, so the base has area $\frac{1}{2}\cdot 1\cdot 2 = 1$ and the height is 3. Hence the volume of $\mathcal{V}$ is $\frac{1}{3}\cdot 1\cdot 3 = 1$, and $\int_{\mathcal{S}} \mathbf{F}\cdot\mathbf{n}\, dS = \int_{\mathcal{S}} \nabla\cdot\mathbf{F}\, dV$

$$= \int_{\mathcal{V}} 11\, dV = 11(\text{Volume of } \mathcal{V}) = 11\cdot 1 = 11.$$

9 $\mathbf{F} = x^3\mathbf{i} + y^3\mathbf{j} + 3z\mathbf{k}$, so $\nabla\cdot\mathbf{F} = 3x^2 + 3y^2 + 3 = 3(x^2 + y^2 + 1) = 3(r^2 + 1)$. $\mathcal{V}$ lies above the xy plane, below the surface $z = x^2 = r^2\cos^2\theta$, and within the cylinder $r = \sin\theta$; $\mathcal{V}$ can therefore be described by $0 \le \theta \le \pi$, $0 \le r \le \sin\theta$, $0 \le z \le r^2\cos^2\theta$. Hence $\int_{\mathcal{S}} \mathbf{F}\cdot\mathbf{n}\, dS = \int_{\mathcal{V}} \nabla\cdot\mathbf{F}\, dV$

$$= \int_0^{\pi} \int_0^{\sin\theta} \int_0^{r^2\cos^2\theta} 3(r^2 + 1)\, r\, dz\, dr\, d\theta$$

$$= \int_0^\pi \int_0^{\sin\theta} 3r^3 \cos^2\theta\,(r^2 + 1)\,dr\,d\theta$$

$$= \int_0^\pi \int_0^{\sin\theta} (\cos^2\theta)(3r^5 + 3r^3)\,dr\,d\theta$$

$$= \int_0^\pi (\cos^2\theta)\left(\frac{r^6}{2} + \frac{3r^4}{4}\right)\Bigg|_0^{\sin\theta} d\theta$$

$$= \frac{1}{4}\int_0^\pi (\cos^2\theta)(2\sin^6\theta + 3\sin^4\theta)\,d\theta =$$

$$\frac{1}{4}\int_0^\pi (2\sin^6\theta - 2\sin^8\theta + 3\sin^4\theta - 3\sin^6\theta)\,d$$

$$= \frac{2}{4}\int_0^{\pi/2} (3\sin^4\theta - \sin^6\theta - 2\sin^8\theta)\,d\theta$$

$$= \frac{1}{2}\left[3\cdot\frac{1\cdot 3}{2\cdot 4}\cdot\frac{\pi}{2} - \frac{1\cdot 3\cdot 5}{2\cdot 4\cdot 6}\cdot\frac{\pi}{2} - 2\cdot\frac{1\cdot 3\cdot 5\cdot 7}{2\cdot 4\cdot 6\cdot 8}\cdot\frac{\pi}{2}\right]$$

$= \frac{\pi}{4}\left[\frac{9}{8} - \frac{15}{48} - \frac{105}{192}\right] = \frac{17\pi}{256}$. (We used Formula 73 in the text's table of antiderivatives.)

11 If $\mathbf{F}$ is a constant vector field, then $\nabla\cdot\mathbf{F} = 0$. Suppose $\mathcal{S}$ is a surface bounding a region $\mathcal{V}$. Then $\int_{\mathcal{S}} \mathbf{F}\cdot\mathbf{n}\,dS = \int_{\mathcal{V}} \nabla\cdot\mathbf{F}\,dV = \int_{\mathcal{V}} 0\,dV = 0$, as claimed.

13 Since $\nabla\cdot\mathbf{F} = \frac{\partial}{\partial x}(y^3) + \frac{\partial}{\partial y}(z^3) + \frac{\partial}{\partial z}(x^3) = 0$, the integral is $\int_{\mathcal{S}} \mathbf{F}\cdot\mathbf{n}\,dS = \int_{\mathcal{V}} \nabla\cdot\mathbf{F}\,dV = 0$.

15 Since $\nabla\cdot\mathbf{F} = \frac{\partial}{\partial x}\left(z\sqrt{x^2+z^2}\right) + \frac{\partial}{\partial y}(y + 3) + \frac{\partial}{\partial z}\left(-x\sqrt{x^2+z^2}\right) = \frac{xz}{\sqrt{x^2+z^2}} + 1 - \frac{xz}{\sqrt{x^2+z^2}} = 1$ and $\mathcal{V}$ can be described in cylindrical coordinates by $0 \le \theta \le 2\pi$, $0 \le r \le 4\cos\theta$, $r^2 \le z \le 4r\cos\theta$, we have $\int_{\mathcal{S}} \mathbf{F}\cdot\mathbf{n}\,dS = \int_{\mathcal{V}} \nabla\cdot\mathbf{F}\,dV = \int_{\mathcal{V}} 1\,dV = \int_0^\pi \int_0^{4\cos\theta} \int_{r^2}^{4r\cos\theta} r\,dz\,dr\,d\theta = 8\pi$.

17 Let $\mathcal{T}$ be the missing face; $\mathcal{S}$ and $\mathcal{T}$ together make up the cube's surface, so $\int_{\mathcal{S}} \mathbf{F}\cdot\mathbf{n}\,dS + \int_{\mathcal{T}} \mathbf{F}\cdot\mathbf{n}\,dS = \int_{\mathcal{V}} \nabla\cdot\mathbf{F}\,dV$. Solving for $\int_{\mathcal{S}} \mathbf{F}\cdot\mathbf{n}\,dS$, we have $\int_{\mathcal{S}} \mathbf{F}\cdot\mathbf{n}\,dS = \int_{\mathcal{V}} \nabla\cdot\mathbf{F}\,dV - \int_{\mathcal{T}} \mathbf{F}\cdot\mathbf{n}\,dS$. Now $\nabla\cdot\mathbf{F} = \frac{\partial}{\partial x}(4xz) + \frac{\partial}{\partial y}(-y^2) + \frac{\partial}{\partial z}(yz) = 4z - 2y + y = 4z - y$. Hence $\int_{\mathcal{V}} \nabla\cdot\mathbf{F}\,dV$

$$= \int_0^1 \int_0^1 \int_0^1 (4z - y)\,dz\,dy\,dx$$

$$= \int_0^1 \int_0^1 (2z^2 - yz)\Big|_0^1\,dy\,dx$$

$$= \int_0^1 \int_0^1 (2 - y)\,dy\,dx =$$

$\int_0^1 \left(2y - \frac{1}{2}y^2\right)\Big|_0^1\,dx = \int_0^1 \frac{3}{2}\,dx = \frac{3}{2}$. On $\mathcal{T}$, $\mathbf{n} = \mathbf{i}$, so $\mathbf{F}\cdot\mathbf{n} = 4xz = 4z$. Hence $\int_{\mathcal{T}} \mathbf{F}\cdot\mathbf{n}\,dS =$

$$\int_0^1 \int_0^1 4z\,dz\,dy = \int_0^1 2z^2\Big|_0^1\,dy = \int_0^1 2\,dy = 2.$$

Hence $\int_{\mathcal{S}} \mathbf{F}\cdot\mathbf{n}\,dS = \frac{3}{2} - 2 = -\frac{1}{2}$.

19 If $\|\mathbf{F}\| \le 5$, then $\mathbf{F}\cdot\mathbf{n} = \|\mathbf{F}\|\,\|\mathbf{n}\|\cos\theta = \|\mathbf{F}\|\cos\theta \le 5$. By the divergence theorem, abs $\int_{\mathcal{V}} \nabla\cdot\mathbf{F}\,dV =$ abs $\int_{\mathcal{S}} \mathbf{F}\cdot\mathbf{n}\,dS \le \int_{\mathcal{S}} 5\,dS =$ 5(Area of $\mathcal{S}$).

21 By Example 3 of Sec. 16.1, the divergence of $\mathbf{F} = \frac{\hat{\mathbf{r}}}{\|\mathbf{r}\|^2}$ is equal to 0 everywhere except at the

origin, where it is undefined. If $\mathcal{S}$ is the sphere of radius 2 and center (5, 3, 1), then the origin is not within $\mathcal{S}$ and $\nabla\cdot\mathbf{F} = 0$ throughout the region $\mathcal{V}$ enclosed by $\mathcal{S}$. Therefore $\int_{\mathcal{S}} \mathbf{F}\cdot\mathbf{n}\, dS = \int_{\mathcal{V}} \nabla\cdot\mathbf{F}\, dV$ $= \int_{\mathcal{V}} 0\, dV = 0$.

23 $\mathcal{S}$ does not enclose the origin. As in Exercise 21, $\int_{\mathcal{S}} \mathbf{F}\cdot\mathbf{n}\, dS = 0$.

25 (a) $\mathbf{E}$ should be radially outward from the line (no component parallel to L).

(b) By symmetry, $\mathbf{E}\cdot\mathbf{n} = 0$ on the two circular ends of the cylinder. On the curved surface $\mathcal{S}$ of the cylinder, $\mathbf{E}\cdot\mathbf{n} = \|\mathbf{E}\|$ because $\mathbf{E}$ and $\mathbf{n}$ are parallel. By symmetry, $\|\mathbf{E}\|$ is constant on $\mathcal{S}$; hence $\int_{\mathcal{S}} \mathbf{E}\cdot\mathbf{n}\, dS = \|\mathbf{E}\|(\text{Area of } \mathcal{S}) =$ $\|\mathbf{E}\| 2\pi r x$, which, by Gauss's law, equals $\frac{qx}{\epsilon_0}$. Therefore $\|\mathbf{E}\| = \frac{q}{2\pi r \epsilon_0}$.

29 (a) $\mathbf{F}(x, y, z) = z\mathbf{k}$, so $\nabla\cdot\mathbf{F} = 1$. Hence $\int_{\mathcal{V}} \nabla\cdot\mathbf{F}\, dV = \text{Volume of } \mathcal{V}$. By the divergence theorem, $\int_{\mathcal{S}} \mathbf{F}\cdot\mathbf{n}\, dS = \text{Volume of}$ $\mathcal{V}$.

(b) $\int_{\mathcal{S}} \mathbf{F}\cdot\mathbf{n}\, dS = \int_{\mathcal{S}} z\mathbf{k}\cdot\mathbf{n}\, dS = \int_{\mathcal{S}} z\cos\gamma\, dS$ $= \text{Volume of } \mathcal{V}$, by Exercise 10 of Sec. 17.1.

31 Let $\mathcal{S}$ be a sphere of radius r, centered at the origin. Since div $\mathbf{F} = 0$, it follows from Corollary 1 of this section that $\int_{\mathcal{S}_1} \mathbf{F}\cdot\mathbf{n}\, dS = \int_{\mathcal{S}} \mathbf{F}\cdot\mathbf{n}\, dS \le$ $\int_{\mathcal{S}} \|\mathbf{F}\|\, dS = \|\mathbf{F}\| 4\pi r^2$, which approaches 0 as $r \to$ ∞ since $\lim_{|\mathbf{r}|\to\infty} \mathbf{F}(\mathbf{r})|\mathbf{r}|^2 = \mathbf{0}$. Thus $\int_{\mathcal{S}_1} \mathbf{F}\cdot\mathbf{n}\, dS$ must be 0.

33 The integral gives the steradian measure of the sphere as seen from a point on its surface. Hence its value is 2π (just as the horizon line of the earth separates the equal measures of sky and ground).

35 (a) The three faces that do not contain the chosen vertex together subtend $\frac{1}{8}(4\pi) = \frac{\pi}{2}$

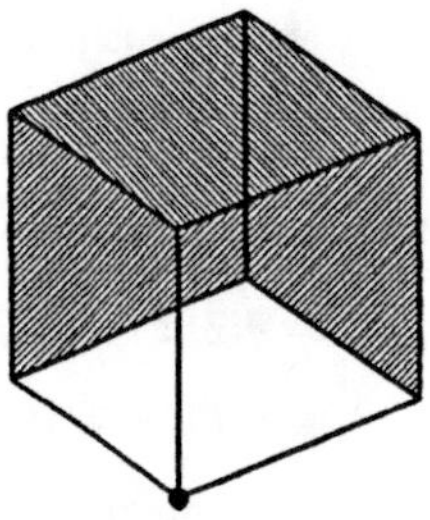

steradians. Since each face has an equal effect, each subtends $\frac{1}{3}\cdot\frac{\pi}{2} = \frac{\pi}{6}$ steradians.

(b) Each of the six faces subtends an equal share: $\frac{1}{6}(4\pi) = \frac{2\pi}{3}$ steradians.

37 By Exercise 36, if $\mathcal{V}$ is the volume of the region enclosed by $\mathcal{S}$, then $\nabla\cdot\mathbf{F} = \lim_{V\to 0} \frac{1}{V} \int_{\mathcal{V}} \mathbf{F}\cdot\mathbf{n}\, dS$. By

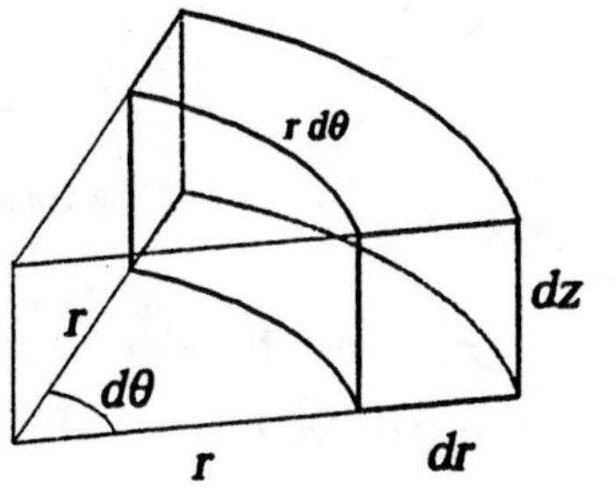

analogy with Exercise 29 of Sec 16.4, we approximate the flux over each face of the solid shown in the figure, which is described by cylindrical coordinates ranging from r to $r + dr$, θ to $\theta + d\theta$, and z to $z + dz$. If we assume that $\mathbf{F} = A\hat{\mathbf{r}} + B\hat{\boldsymbol{\theta}} + C\mathbf{k}$, then $\mathbf{F}\cdot(-\hat{\mathbf{r}}) = -A$ on the inner curved surface and $\mathbf{F}\cdot\hat{\mathbf{r}} = A$ on the outer curved surface (since $-\hat{\mathbf{r}}$ and $\hat{\mathbf{r}}$ are the outward pointing unit normals for those faces). The inner surface has approximate area $r\,d\theta\,dz$ and the outer surface has approximate area $(r + dr)\,d\theta\,dz$. Evaluating the scalar function A at the midpoints of the two surfaces, we can approximate the net flux through the surfaces as

$-A(r, \theta + d\theta/2, z + dz/2)\,r\,d\theta\,dz$

$+ A(r + dr, \theta + d\theta/2, z + dz/2)(r + dr)\,d\theta\,dz$

$= [A(r + dr, \theta + d\theta/2, z + dz/2)$

$- A(r, \theta + d\theta/2, z + dz/2)]\,r\,d\theta\,dz$

$+ A(r + dr, \theta + d\theta/2, z + dz/2)\,dr\,d\theta\,dz$

$\approx \dfrac{\partial A}{\partial r}\,r\,dr\,d\theta\,dz + A\,dr\,d\theta\,dz$. Similarly, on the two rectangular faces,

$[B(r + dr/2, \theta + d\theta, z + dz/2)$

$- B(r + dr/2, \theta, z + dz/2)]\,dr\,dz$

$\approx \dfrac{\partial B}{\partial \theta}\,dr\,d\theta\,dz = \dfrac{\partial B}{d\theta}\cdot\dfrac{r\,dr\,d\theta\,dz}{r}$, and on the two remaining faces,

$[C(r + dr/2, \theta + d\theta/2, z + dz)$

$- C(r + dr/2, \theta + d\theta/2, z)](r + dr/2)\,d\theta\,dr$

$\approx \dfrac{\partial C}{dz}\,r\,dr\,d\theta\,dz$. Since the volume of the little solid is approximately $r\,dr\,d\theta\,dz$, adding our results and dividing by the volume yields $\nabla\cdot\mathbf{F} = \dfrac{\partial A}{\partial r} + \dfrac{A}{r} + \dfrac{1}{r}\cdot\dfrac{\partial B}{\partial \theta} + \dfrac{\partial C}{\partial z}$.

17.3 Stokes' Theorem

1 The integral over a surface of the normal component of the curl of a vector field equals the integral over the surface's boundary curve of the tangential component of the vector field.

3 $\nabla \times \mathbf{F} = \begin{vmatrix} \mathbf{i} & \mathbf{j} & \mathbf{k} \\ \dfrac{\partial}{\partial x} & \dfrac{\partial}{\partial y} & \dfrac{\partial}{\partial z} \\ xy^2 & y^3 & y^2z \end{vmatrix} = 2yz\mathbf{i} - 2xy\mathbf{k}$. On $\mathcal{S}$, $\mathbf{n} = \mathbf{r} = x\mathbf{i} + y\mathbf{j} + z\mathbf{k}$, so $(\nabla \times \mathbf{F})\cdot\mathbf{n} = 2xyz - 2xyz = 0$. Hence $\int_{\mathcal{S}} (\nabla\times\mathbf{F})\cdot\mathbf{n}\,dS = 0$. C is the unit circle, $x^2 + y^2 = 1$. On C, $z = 0$, so $\mathbf{F} = xy^2\mathbf{i} + y^3\mathbf{j}$. Also, $\mathbf{T} = -y\mathbf{i} + x\mathbf{j}$ (since $\mathbf{T}$ is obtained by rotating $\mathbf{n} = x\mathbf{i} + y\mathbf{j}$ counterclockwise through a right angle). Hence $\mathbf{F}\cdot\mathbf{T} = -xy^3 + xy^3 = 0$, so $\oint_C \mathbf{F}\cdot\mathbf{T}\,ds = 0$.

5 $\nabla \times \mathbf{F} = \begin{vmatrix} \mathbf{i} & \mathbf{j} & \mathbf{k} \\ \dfrac{\partial}{\partial x} & \dfrac{\partial}{\partial y} & \dfrac{\partial}{\partial z} \\ y^5 & x^3 & z^4 \end{vmatrix} = (3x^2 - 5y^4)\mathbf{k}$. Since $\mathcal{S}$ is part of a level surface of $g(x, y, z) = z - x^2 - y^2$, a normal vector is given by $\nabla g = -2x\mathbf{i} - 2y\mathbf{j} + \mathbf{k}$; a unit normal is $\mathbf{n} = \dfrac{-2x\mathbf{i} - 2y\mathbf{j} + \mathbf{k}}{\sqrt{(2x)^2 + (2y)^2 + 1^2}} = \dfrac{-2x\mathbf{i} - 2y\mathbf{j} + \mathbf{k}}{\sqrt{4z + 1}}$. Hence $(\nabla \times \mathbf{F})\cdot\mathbf{n} = \dfrac{3x^2 - 5y^4}{\sqrt{4z + 1}}$, so we must find $\int_{\mathcal{S}} \dfrac{3x^2 - 5y^4}{\sqrt{4z + 1}}\,dS$.

Let $\mathcal{A}$ be the projection of $\mathcal{S}$ on the xy plane; thus $\mathcal{A}$

is the unit disk, $x^2 + y^2 \le 1$. By Theorem 2(b) of Sec. 17.1, since $\mathcal{S}$ is described by $z = f(x, y)$, where $f(x, y) = x^2 + y^2$, we have $1/|\cos\gamma| = \sqrt{f_x^2 + f_y^2 + 1} = \sqrt{(2x)^2 + (2y)^2 + 1} = \sqrt{4z + 1}$,

so $\int_{\mathcal{S}} \frac{3x^2 - 5y^4}{\sqrt{4z+1}}\, dS =$

$$\int_{\mathcal{A}} \frac{3x^2 - 5y^4}{\sqrt{4z+1}} \cdot \sqrt{4z+1}\, dA = \int_{\mathcal{A}} (3x^2 - 5y^4)\, dA.$$

$\mathcal{A}$ is described in polar coordinates by $0 \le \theta \le 2\pi$, $0 \le r \le 1$, so $\int_{\mathcal{A}} (3x^2 - 5y^4)\, dA$

$$= \int_0^{2\pi} \int_0^1 (3(r\cos\theta)^2 - 5(r\sin\theta)^4)\, r\, dr\, d\theta$$

$$= \int_0^{2\pi} \int_0^1 [(3\cos^2\theta) r^3 - (5\sin^4\theta) r^5]\, dr\, d\theta$$

$$= \int_0^{2\pi} \left(\frac{3}{4}\cos^2\theta - \frac{5}{6}\sin^4\theta\right) d\theta$$

$= \frac{3}{4}\pi - \frac{5}{6}\cdot 4 \cdot \frac{1\cdot 3}{2\cdot 4}\cdot\frac{\pi}{2} = \frac{\pi}{8}$. The boundary C of $\mathcal{S}$ is the circle $x^2 + y^2 = 1$, $z = 1$. On C, $\mathbf{F} = y^5\mathbf{i} + x^3\mathbf{j} + \mathbf{k}$ and $\mathbf{T} = -y\mathbf{i} + x\mathbf{j}$, so $\mathbf{F}\cdot\mathbf{T} = x^4 - y^6$. C can be parameterized by $x = \cos\theta$, $y = \sin\theta$, $z = 1$ for $0 \le \theta \le 2\pi$; thus $ds = d\theta$, so

$$\oint_C \mathbf{F}\cdot\mathbf{T}\, ds = \int_0^{2\pi} (\cos^4\theta - \sin^6\theta)\, d\theta$$

$$= 4\cdot\frac{1\cdot 3}{2\cdot 4}\cdot\frac{\pi}{2} - 4\cdot\frac{1\cdot 3\cdot 5}{2\cdot 4\cdot 6}\cdot\frac{\pi}{2} = \frac{\pi}{8}.$$

7 $\mathbf{F} = x\mathbf{i} - y\mathbf{j}$, so $\nabla\cdot\mathbf{F} = 0$. Let $\mathcal{S}^*$ be the entire surface of the unit cube; denote by $\mathcal{S}_1$ the face in the plane $x = 1$, and let $\mathcal{S}$ be the rest of the surface. Then $\int_{\mathcal{S}} \mathbf{F}\cdot\mathbf{n}\, dS + \int_{\mathcal{S}_1} \mathbf{F}\cdot\mathbf{n}\, dS =$

$$\int_{\mathcal{S}^*} \mathbf{F}\cdot\mathbf{n}\, dS = \int_{\mathcal{V}} \nabla\cdot\mathbf{F}\, dV = \int_{\mathcal{V}} 0\, dV = 0, \text{ so}$$

$\int_{\mathcal{S}} \mathbf{F}\cdot\mathbf{n}\, dS = -\int_{\mathcal{S}_1} \mathbf{F}\cdot\mathbf{n}\, dS$. On $\mathcal{S}_1$, $\mathbf{n} = \mathbf{i}$ so $\mathbf{F}\cdot\mathbf{n} = x$, which is equal to 1; therefore $\int_{\mathcal{S}} \mathbf{F}\cdot\mathbf{n}\, dS =$

$$-\int_{\mathcal{S}_1} \mathbf{F}\cdot\mathbf{n}\, dS = -\int_{\mathcal{S}_1} dS = -1.$$

9 $\mathbf{F} = \sin xy\, \mathbf{i}$, so $\nabla \times \mathbf{F} = P_z\mathbf{j} - P_y\mathbf{k} = -x\cos xy\, \mathbf{k}$. C is the intersection of the plane $x + y + z = 1$ and the cylinder $x^2 + y^2 = 1$, so it bounds an ellipse $\mathcal{S}$. The projection of $\mathcal{S}$ onto the xy plane is the unit disk $\mathcal{A}$. From the equation of the plane $x + y + z = 1$, we have $\mathbf{n} = \frac{1}{\sqrt{3}}(\mathbf{i} + \mathbf{j} + \mathbf{k})$ as the unit normal to $\mathcal{S}$, so

$(\nabla \times \mathbf{F})\cdot\mathbf{n} = (-x\cos xy)/\sqrt{3}$ and $\cos\gamma = \mathbf{n}\cdot\mathbf{k} = \frac{1}{\sqrt{3}}$. Then $\oint_C \mathbf{F}\cdot d\mathbf{r} = \int_{\mathcal{S}} (\nabla\times\mathbf{F})\cdot\mathbf{n}\, dS$

$$= \int_{\mathcal{S}} (-x\cos xy)/\sqrt{3}\, dS = \int_{\mathcal{A}} \frac{-x\cos xy}{\sqrt{3}\cos\gamma}\, dA$$

$= \int_{\mathcal{A}} (-x\cos xy)\, dA$. But $-x\cos xy$ is an odd function with respect to x while $\mathcal{A}$ is a region symmetric with respect to x, so the negative values cancel the positive values and the integral is 0.

11 $\mathbf{F} = xy\mathbf{k}$, so $\nabla \times \mathbf{F} = R_y\mathbf{i} - R_x\mathbf{j} = x\mathbf{i} - y\mathbf{j}$. C is the intersection of the plane $z = y$ and the cylinder $x^2 - 2x + y^2 = 0$, so it bounds an ellipse $\mathcal{S}$ whose projection into the xy plane is the unit disk $\mathcal{A}$ with center $(1, 0)$. (Note that $x^2 - 2x + y^2 = 0$ is equivalent to $(x - 1)^2 + y^2 = 1$.) The unit normal $\mathbf{n}$ to the surface $\mathcal{S}$ is the unit normal to the plane $z - y = 0$, so $\mathbf{n} = \frac{1}{\sqrt{2}}(-\mathbf{j} + \mathbf{k})$ and $(\nabla \times \mathbf{F})\cdot\mathbf{n} =$

$\frac{y}{\sqrt{2}}$. Also, $\cos\gamma = \mathbf{n}\cdot\mathbf{k} = \frac{1}{\sqrt{2}}$ and $\oint_C \mathbf{F}\cdot d\mathbf{r} =$

$\int_S (\nabla\times\mathbf{F})\cdot\mathbf{n}\, dS = \int_S \frac{y}{\sqrt{2}}\, dS = \int_{\mathcal{A}} \frac{y/\sqrt{2}}{\cos\gamma}\, dA =$

$\int_{\mathcal{A}} y\, dA$. But y is an odd function with respect to y and $\mathcal{A}$ is a region symmetric with respect to y, so the integral is 0.

13 Let C denote the "equator" of the sphere, which is then a common boundary to S_1 and S_2. If C is oriented counterclockwise as viewed from above, then $\int_{S_1} (\nabla\times\mathbf{F})\cdot\mathbf{n}\, dS = \oint_C \mathbf{F}\cdot d\mathbf{r}$, where $\mathbf{n}$ is the exterior normal to the sphere. On S_2, however, $\mathbf{n}$ has a negative $\mathbf{k}$ component; the right-hand rule applied to S_2 and $\mathbf{n}$ would reverse the orientation of C. Therefore, if we use $-\mathbf{n}$, we have

$\int_{S_2} (\nabla\times\mathbf{F})\cdot(-\mathbf{n})\, dS = \oint_C \mathbf{F}\cdot d\mathbf{r}$. But

$\int_{S_2} (\nabla\times\mathbf{F})\cdot\mathbf{n}\, dS = -\int_{S_2} (\nabla\times\mathbf{F})\cdot\mathbf{n}\, dS$, so

$-\int_{S_2} (\nabla\times\mathbf{F})\cdot\mathbf{n}\, dS = \int_{S_1} (\nabla\times\mathbf{F})\cdot\mathbf{n}\, dS$.

15 (a) No. For example, let $\mathbf{F} =$

$-\frac{y}{x^2+y^2}\mathbf{i} + \frac{x}{x^2+y^2}\mathbf{j}$, and let C_1 be the unit circle oriented counterclockwise. By Example 2 of Sec. 16.6, $\nabla\times\mathbf{F} = \mathbf{0}$, but

$\oint_{C_1} \mathbf{F}\cdot\mathbf{T}\, ds = 2\pi \neq 0$.

(b) Yes. They are equal by Corollary 3.

17 (a), (b), (d), (e), (f), (h), (i), (j), and (k) are connected; (b), (f), and (h) are simply connected.

19 On S_1, $\nabla\times\mathbf{F} = \begin{vmatrix} \mathbf{i} & \mathbf{j} & \mathbf{k} \\ \frac{\partial}{\partial x} & \frac{\partial}{\partial y} & \frac{\partial}{\partial z} \\ y & xz & x+2y \end{vmatrix} = (2-x)\mathbf{i} - \mathbf{j}$

$+ (z-1)\mathbf{k} = (2-x)\mathbf{i} - \mathbf{j} + (x+1)\mathbf{k}$. Since S_1 is part of a level surface of $g(x, y, z) = z - x$, a normal is given by $\nabla g = -\mathbf{i} + \mathbf{k}$; the outward unit normal is $\mathbf{n} = \frac{1}{\sqrt{2}}(-\mathbf{i}+\mathbf{k})$. Hence $(\nabla\times\mathbf{F})\cdot\mathbf{n} =$

$\frac{1}{\sqrt{2}}(2x-1)$. From Theorem 2(a) of Sec. 17.1, we have $1/|\cos\gamma| = \frac{\sqrt{g_x^2 + g_y^2 + g_z^2}}{|g_z|} = \sqrt{2}$. Hence, if $\mathcal{A}$ is the unit circle in the xy plane, $\oint_C \mathbf{F}\cdot d\mathbf{r} =$

$\int_{S_1} (\nabla\times\mathbf{F})\cdot\mathbf{n}\, dS = \int_{\mathcal{A}} \frac{(2x-1)/\sqrt{2}}{1/\sqrt{2}}\, dA =$

$\int_{\mathcal{A}} (2x-1)\, dA$. By symmetry, $\int_{\mathcal{A}} x\, dA = 0$.

Hence $\oint_C \mathbf{F}\cdot d\mathbf{r} = -\int_{\mathcal{A}} dA = -(\text{Area of } \mathcal{A})$

$= -\pi$.

21 Since a gradient field $\mathbf{F} = \nabla f$ is conservative, $\nabla\times\mathbf{F} = \nabla\times\nabla f = \mathbf{0}$ and $\oint_C \mathbf{F}\cdot d\mathbf{r} = \oint_C \nabla f\cdot d\mathbf{r}$

$= 0$. Hence both $\int_S (\nabla\times\mathbf{F})\cdot\mathbf{n}\, dS$ and $\oint_C \mathbf{F}\cdot d\mathbf{r}$ are zero.

23 Let C^* be the unit circle on the xy plane with center at the origin. By Example 2 of Sec. 16.6, $\nabla\times\mathbf{F}$ $= \mathbf{0}$ and $\oint_C \mathbf{F}\cdot d\mathbf{r} = \oint_{C^*} \mathbf{F}\cdot d\mathbf{r} = 2\pi$ by Corollary 3.

25 (a) $\mathbf{F} = \mathbf{r}/\|\mathbf{r}\|^a$, where $\mathbf{r} = x\mathbf{i} + y\mathbf{j} + z\mathbf{k}$ and $r = \|\mathbf{r}\| = \sqrt{x^2 + y^2 + z^2}$. Observe that $\mathbf{F} = P\mathbf{i} + Q\mathbf{j} + R\mathbf{k} = (x/r^a)\mathbf{i} + (y/r^a)\mathbf{j} + (z/r^a)\mathbf{k}$, so $P = x/r^a$, $Q = y/r^a$, and $R = z/r^a$. We will show that the x component of $\nabla \times \mathbf{F}$ is 0 by showing that $R_y - Q_z = 0$. Recall that $r_y = y/r$ and $r_z = z/r$; then $R_y = (zr^{-a})_y = z(-ar^{-a-1})r_y = -azr^{-a-1}(y/r) = -ayzr^{-a-2}$ and $Q_z = (yr^{-a})_z = y(-ar^{-a-1})r_z = -ayr^{-a-1}(z/r) = -ayzr^{-a-2} = R_y$. Thus $R_y - Q_z = 0$ and $\nabla \times \mathbf{F}$ has 0 for its x component. Similarly, $R_x - P_z = 0$ and $Q_x - P_y = 0$, so $\nabla \times \mathbf{F} = \mathbf{0}$.

(b) The domain of $\mathbf{F}$ is simply connected, so $\nabla \times \mathbf{F} = \mathbf{0}$ implies that $\mathbf{F}$ is conservative.

(c) We wish to find a function f such that $\mathbf{F} = \nabla f$. In particular, then, we must have $f_x = xr^{-a}$ where $r = \|\mathbf{r}\| = \sqrt{x^2 + y^2 + z^2}$. But

$$xr^{-a} = \frac{x}{r}r^{-a+1} = r_x r^{-a+1} = r^{-a+1}r_x = \left(\frac{r^{-a+2}}{-a+2}\right)_x,$$

so $f = \dfrac{r^{2-a}}{2-a}$ is a possible solution. Since $f_y = y/r^a$ and $f_z = z/r^a$, we find that this is in fact a solution. Note that f can be written as $\dfrac{1}{2-a}(x^2 + y^2 + z^2)^{(2-a)/2}$.

27 The intersection of $z = x$ and $z = x^2 + y^2$ yields $x^2 + y^2 = x$, or $r = \cos\theta$, as the projection $\mathcal{A}$ in the xy plane of the surface $\mathcal{S}$ which lies in the plane $z = x$ and is bounded by the paraboloid $z = x^2 + y^2$. Note that

$$\nabla \times \mathbf{F} = \begin{vmatrix} \mathbf{i} & \mathbf{j} & \mathbf{k} \\ \dfrac{\partial}{\partial x} & \dfrac{\partial}{\partial y} & \dfrac{\partial}{\partial z} \\ xyz & x^2 & xz \end{vmatrix} = (xy - z)\mathbf{j} + (2x - xz)\mathbf{k}$$

while a unit normal to the plane $-x + z = 0$ is $\mathbf{n} = -\dfrac{1}{\sqrt{2}}\mathbf{i} + \dfrac{1}{\sqrt{2}}\mathbf{k}$ (where the sign of $\mathbf{n}$ is chosen to be consistent with the orientation of C). Hence $(\nabla \times \mathbf{F})\cdot\mathbf{n} =$

$$\langle 0, xy - z, 2x - xz\rangle \cdot \left\langle -\frac{1}{\sqrt{2}}, 0, \frac{1}{\sqrt{2}}\right\rangle = \frac{2x - xz}{\sqrt{2}},$$

which equals $\dfrac{2x - x^2}{\sqrt{2}}$ on the plane $z = x$. Since $\cos\gamma = \mathbf{n}\cdot\mathbf{k} = \dfrac{1}{\sqrt{2}}$, we have $\oint_C \mathbf{F}\cdot d\mathbf{r} =$

$$\int_{\mathcal{S}} (\nabla\times\mathbf{F})\cdot\mathbf{n}\, dS = \int_{\mathcal{A}} \frac{2x - x^2}{\sqrt{2}}\sqrt{2}\, dA$$

$$= \int_0^\pi \int_0^{\cos\theta} (2r\cos\theta - r^2\cos^2\theta)\, r\, dr\, d\theta$$

$$= \int_0^\pi \int_0^{\cos\theta} (2r^2\cos\theta - r^3\cos^2\theta)\, dr\, d\theta$$

$$= \int_0^\pi \left(\frac{2}{3}r^3\cos\theta - \frac{r^4}{4}\cos^2\theta\right)\Bigg|_0^{\cos\theta} d\theta$$

$$= \int_0^\pi \left(\frac{2}{3}\cos^4\theta - \frac{1}{4}\cos^6\theta\right) d\theta$$

$$= 2\left(\frac{2}{3}\cdot\frac{1\cdot3}{2\cdot4}\frac{\pi}{2} - \frac{1}{4}\cdot\frac{1\cdot3\cdot5}{2\cdot4\cdot6}\frac{\pi}{2}\right) = \frac{11\pi}{64}.$$

29 Try it and see for yourself.

31 (a) Since curl $\mathbf{F} = \mathbf{0}$ and the domain of $\mathbf{F}$ is simply connected, $\mathbf{F}$ is conservative by Theorem 6 of Sec. 16.6.

(b) If you begin to trace out C from within the two spheres, you see that C must cross the gap between the spheres an even number of times (since C is closed and you have to end up inside again). Each outward-bound segment can therefore be paired with an inward-bound segment, suggesting that they tend to cancel out between the two spheres. Since $\mathbf{F}$ is a central field, this argument works for any sufficiently close pair of spheres, suggesting that, overall, $\oint_C \mathbf{F}\cdot d\mathbf{r} = 0$.

17.4 Applications of Stokes' Theorem

1 $(\text{curl } \mathbf{F})\cdot\mathbf{n} = \langle 1, -1, 1\rangle \cdot \dfrac{\langle 3, 1, 1\rangle}{\sqrt{11}} = \dfrac{3}{\sqrt{11}}$. Since $\mathcal{S}$ is small, curl $\mathbf{F}\cdot\mathbf{n}$ is approximately constant over its flat surface. Hence $\int_{\mathcal{S}} \text{curl}\,\mathbf{F}\cdot\mathbf{n}\; dS \approx \dfrac{3}{\sqrt{11}}(\text{Area of } \mathcal{S}) = \dfrac{3}{\sqrt{11}}\pi(0.03)^2 \approx 0.0026$. By Stokes' theorem, this is also approximately $\oint_C \mathbf{F}\cdot d\mathbf{r}$.

5 (a) Law (1′) states that $\nabla\cdot\mathbf{E} = q/\epsilon_0$. By the divergence theorem, $\int_{\mathcal{S}} \mathbf{E}\cdot\mathbf{n}\; dS = \int_{\mathcal{V}} \nabla\cdot\mathbf{E}\; dV = \int_{\mathcal{V}} (q/\epsilon_0)\; dV$.

(b) The total charge Q is the integral of the charge density q; that is, $Q = \int_{\mathcal{V}} q\; dV$. Therefore, by the result in (a), $\int_{\mathcal{S}} \mathbf{E}\cdot\mathbf{n}\; dS = \dfrac{1}{\epsilon_0}\int_{\mathcal{V}} q\; dV = Q/\epsilon_0$, which is Law (1).

7 $\int_{\mathcal{S}} \mathbf{B}\cdot\mathbf{n}\; dS = 0$, so the divergence theorem says $\int_{\mathcal{V}} \nabla\cdot\mathbf{B}\; dV = 0$, where $\mathcal{V}$ is the volume enclosed by $\mathcal{S}$. For a small volume, $\nabla\cdot\mathbf{B}$ is approximately constant, so $\int_{\mathcal{V}} \nabla\cdot\mathbf{B}\; dV \approx \nabla\cdot\mathbf{B}\,(\text{Volume of } \mathcal{V})$. Thus $\nabla\cdot\mathbf{B} \approx 0$. In the limit, $\nabla\cdot\mathbf{B} = 0$.

9 $\oint_C \mathbf{E}\cdot d\mathbf{r} = 0$, so Stokes' theorem says $\int_{\mathcal{S}} (\nabla\times\mathbf{E})\cdot\mathbf{n}\; dS = 0$, where $\mathcal{S}$ is any surface bounded by C. Whenever $\mathcal{S}$ is a small, flat surface, $(\nabla\times\mathbf{E})\cdot\mathbf{n}$ is approximately constant, so $0 \approx (\nabla\times\mathbf{E})\cdot\mathbf{n}(\text{Area of } \mathcal{S})$. Since this is true whatever direction $\mathbf{n}$ may have, it follows that in the limit $\nabla\times\mathbf{E} = \mathbf{0}$.

11 $c^2 \oint_C \mathbf{B}\cdot d\mathbf{r} = \dfrac{1}{\epsilon_0}\int_{\mathcal{S}} \mathbf{j}\cdot\mathbf{n}\; dS$, so by applying Stokes' theorem to the left side we have $c^2\int_{\mathcal{S}} (\nabla\times\mathbf{B})\cdot\mathbf{n}\; dS = \dfrac{1}{\epsilon_0}\int_{\mathcal{S}} \mathbf{j}\cdot\mathbf{n}\; dS$. Therefore, $\int_{\mathcal{S}} \left(c^2(\nabla\times\mathbf{B}) - \dfrac{1}{\epsilon_0}\mathbf{j}\right)\cdot\mathbf{n}\; dS = 0$. By the same argument as in Exercise 9, $c^2(\nabla\times\mathbf{B}) - \dfrac{1}{\epsilon_0}\mathbf{j} = \mathbf{0}$.

13 (a)

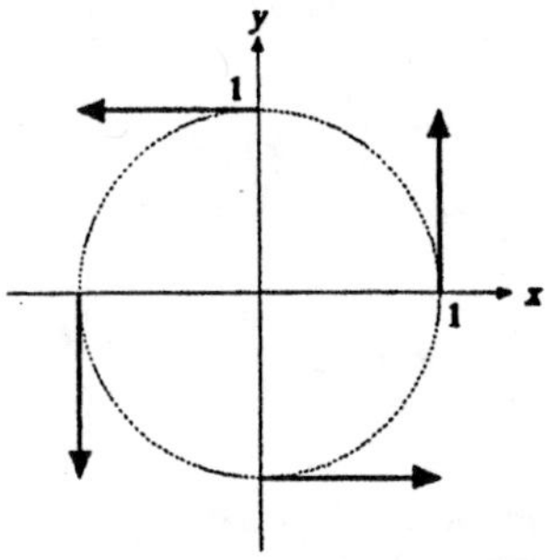

(b) $\mathbf{F} = \mathbf{k} \times (x\mathbf{i} + y\mathbf{j} + z\mathbf{k}) = -y\mathbf{i} + x\mathbf{j}$, so

$$\nabla \times \mathbf{F} = \begin{vmatrix} \mathbf{i} & \mathbf{j} & \mathbf{k} \\ \frac{\partial}{\partial x} & \frac{\partial}{\partial y} & \frac{\partial}{\partial z} \\ -y & x & 0 \end{vmatrix} = 2\mathbf{k}.$$

(d) If the axis is parallel to the z axis, the wheel will spin counterclockwise. If the axis is perpendicular to the z axis, the wheel won't spin.

17.S Guide Quiz

1 (a) See the discussion on pp. 982–983.

(b) See the discussion on p. 1025.

2 (a) See Examples 2 and 3 of Sec. 17.1. If $\mathcal{A}$ is the projection of $\mathcal{S}$ onto the yz plane, then $\int_{\mathcal{S}} x \cos\alpha \, dS = \int_{\mathcal{A}} (x_2(P) - x_1(P)) \, dA =$ Volume of $\mathcal{S}$.

(b) If $\mathcal{A}$ is the projection of $\mathcal{S}$ onto the yz plane, then $\int_{\mathcal{S}} y \cos\alpha \, dS = \int_{\mathcal{A}} (y_2(P) - y_1(P)) \, dA$.

But for any projection parallel to the x axis, $y_2 = y_1$, so $y_2 - y_1 = 0$ and the integral is also 0.

3 Label the six sides of the box as shown in the figure. Observe that $\int_{S_n} \mathbf{F} \cdot \mathbf{n} \, dS = 0$ whenever n is

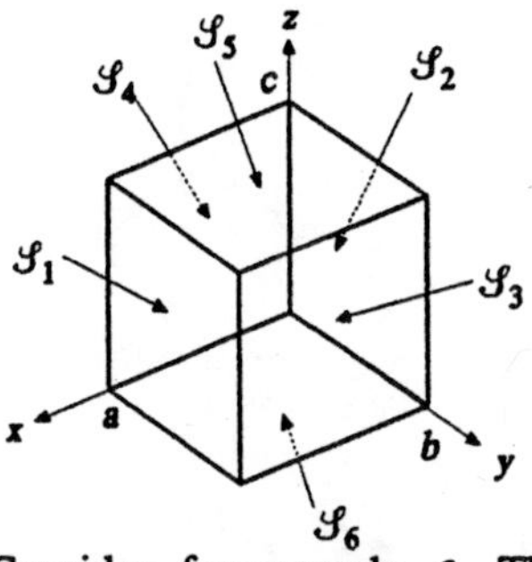

even. Consider, for example, S_2. The exterior unit normal on S_2 is $-\mathbf{i}$; $\mathbf{F} = x^2\mathbf{i} + y^2\mathbf{j} + z^2\mathbf{k}$, so $\mathbf{F}\cdot\mathbf{n} = -x^2 = 0$, since $x = 0$ on S_2. On S_1, $\mathbf{n} = \mathbf{i}$, so $\mathbf{F}\cdot\mathbf{n} = x^2 = a^2$; the area of S_1 is bc and thus $\int_{S_1} \mathbf{F}\cdot\mathbf{n} \, dS = \int_{S_1} a^2 \, dS = a^2bc$. On S_3, $\mathbf{n} = \mathbf{j}$, $\mathbf{F}\cdot\mathbf{n} = y^2 = b^2$, and the area of S_3 is ac, so $\int_{S_3} \mathbf{F}\cdot\mathbf{n} \, dS = \int_{S_3} b^2 \, dS = b^2ac$. On S_5, $\mathbf{n} = \mathbf{k}$, $\mathbf{F}\cdot\mathbf{n} = z^2 = c^2$, and the area of S_5 is ab, so $\int_{S_5} \mathbf{F}\cdot\mathbf{n} \, dS = \int_{S_5} c^2 \, dS = c^2ab$. Thus $\int_S \mathbf{F}\cdot\mathbf{n} \, dS = a^2bc + ab^2c + abc^2 = abc(a + b + c)$.

Observe that $\nabla\cdot\mathbf{F} = 2x + 2y + 2z$, so $\int_V \nabla\cdot\mathbf{F} \, dV$

$$= \int_0^a \int_0^b \int_0^c (2x + 2y + 2z) \, dz \, dy \, dx$$

$$= \int_0^a \int_0^b (2xc + 2yc + c^2) \, dy \, dx$$

$$= \int_0^a (2xbc + b^2c + bc^2) \, dx = a^2bc + ab^2c + abc^2 = abc(a + b + c) = \int_S \mathbf{F}\cdot\mathbf{n} \, dS,$$ as claimed.

4 Label the three segments of C as shown in the figure. Parameterize C_1 by $x = y = z = t$ as t goes from 0 to 1. Then $ds^2 = dx^2 + dy^2 + dz^2 = 3\,dt^2$

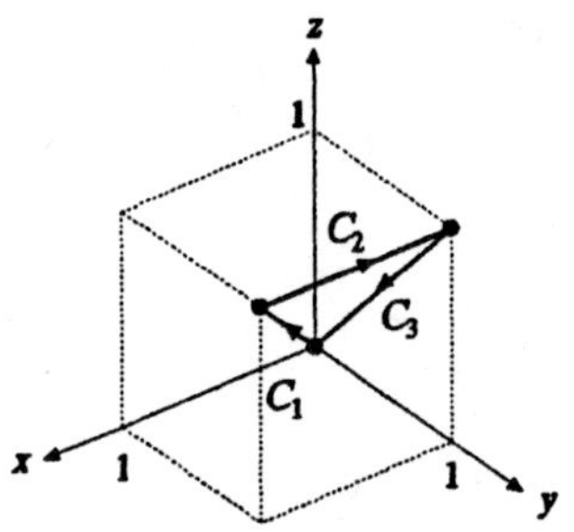

and $\mathbf{T} = \frac{dx}{ds}\mathbf{i} + \frac{dy}{ds}\mathbf{j} + \frac{dz}{ds}\mathbf{k} =$

$\left(\frac{dx}{dt}\mathbf{i} + \frac{dy}{dt}\mathbf{j} + \frac{dz}{dt}\mathbf{k}\right)\frac{dt}{ds} = \frac{1}{\sqrt{3}}(\mathbf{i} + \mathbf{j} + \mathbf{k})$. Now $\mathbf{F}$ $= xy\mathbf{i} + 3z\mathbf{j} + y\mathbf{k} = t^2\mathbf{i} + 3t\mathbf{j} + t\mathbf{k}$, so $\mathbf{F}\cdot\mathbf{T} =$ $\frac{1}{\sqrt{3}}(t^2 + 4t)$ and $\int_{C_1} \mathbf{F}\cdot\mathbf{T}\, ds = \int_0^1 \frac{t^2 + 4t}{\sqrt{3}}\sqrt{3}\, dt$

$= \int_0^1 (t^2 + 4t)\, dt = \left(\frac{t^3}{3} + 2t^2\right)\Big|_0^1 = \frac{1}{3} + 2 =$

$\frac{7}{3}$. On C_2 x varies, so we use it as a parameter, letting it go from 0 to 1 while $y = z = 1$. Furthermore, $ds = dx$, $\mathbf{T} = -\mathbf{i}$, $\mathbf{F} = x\mathbf{i} + 3\mathbf{j} + \mathbf{k}$, and $\mathbf{F}\cdot\mathbf{T} = -x$, so $\int_{C_2} \mathbf{F}\cdot\mathbf{T}\, ds = \int_0^1 -x\, dx$

$= -\frac{x^2}{2}\Big|_0^1 = -\frac{1}{2}$. (Notice that we did not integrate from $x = 1$ to $x = 0$, but rather from $x = 0$ to $x = 1$. This does not contradict the orientation of C_2, which goes from (1, 1, 1) to (0, 0, 0), because we used $\mathbf{T} = -\mathbf{i}$, which takes the orientation into account. Had we instead used the form $\int_{C_2} P\, dx$ $= \int_{C_2} x\, dx$ (in this case), it would have been necessary to integrate from 1 to 0 to get the sign to agree with the above calculation.) On C_3, $x = 0$ while $y = z = t$ as t goes from 0 to 1, so $ds^2 =$ $2\, dt^2$, $\mathbf{T} = \frac{1}{\sqrt{2}}(-\mathbf{j} - \mathbf{k})$, $\mathbf{F} = 3t\mathbf{j} + t\mathbf{k}$, and $\mathbf{F}\cdot\mathbf{T}$ $= -4t/\sqrt{2}$. Therefore $\int_{C_3} \mathbf{F}\cdot\mathbf{T}\, ds =$

$\int_0^1 \frac{-4t}{\sqrt{2}}\sqrt{2}\, dt = \int_0^1 (-4t)\, dt = -2t^2\Big|_0^1 = -2.$

Finally $\oint_C \mathbf{F}\cdot\mathbf{T}\, ds$ is the sum of the line integrals on C_1, C_2 and C_3, which equals $\frac{7}{3} - \frac{1}{2} - 2 =$ $-\frac{1}{6}$.

5 (a) By symmetry, $\mathbf{G}$ must be of the form $f(r)\hat{\mathbf{r}}$, a central field.

(b) Let S be a sphere of radius r centered at (0, 0, 0): Then $km = \int_S \mathbf{G}\cdot\mathbf{n}\, dS =$

$\int_S f(r)\hat{\mathbf{r}}\cdot\hat{\mathbf{r}}\, dS = \int_S f(r)\, dS = f(r)(\text{Area of } S)$

$= 4\pi r^2 f(r)$, so $f(r) = \frac{km}{4\pi r^2}$. Thus $\mathbf{G} =$ $\frac{km}{4\pi r^2}\hat{\mathbf{r}}$.

6 S is the upper half of the ellipsoid $\frac{x^2}{4} + \frac{y^2}{4} + \frac{z^2}{25}$ $= 1$. Observe that the boundary of S is simply the circle C of radius 2 and center (0, 0). Let T be the disk in the xy plane bounded by C. By Stokes' theorem (twice), $\int_S (\nabla\times\mathbf{F})\cdot\mathbf{n}\, dS = \oint_C \mathbf{F}\cdot\mathbf{T}\, ds$ $= \int_T (\nabla\times\mathbf{F})\cdot\mathbf{n}\, dS$. But, on T, $\mathbf{n} = \mathbf{k}$, so

$(\nabla \times \mathbf{F})\cdot\mathbf{n}$ is the $\mathbf{k}$ component of $\nabla \times \mathbf{F}$, namely $\frac{\partial}{\partial x}(y^2 + x) - \frac{\partial}{\partial y}(x^2 + y) = 0$. Hence the integral is 0.

7 (a) By Exercise 36(a) of Sec. 16.1, all central fields in the plane have curl $\mathbf{0}$.

(b) By Exercise 36(b) of Sec 16.1, all central fields in space have curl $\mathbf{0}$.

(c) If $\nabla\cdot\mathbf{F} = 0$ for $f(x, y) = f(r)\hat{\mathbf{r}} = f(r)\frac{x\mathbf{i} + y\mathbf{j}}{r}$ $= f(r)\frac{x}{r}\mathbf{i} + f(r)\frac{y}{r}\mathbf{j}$, then $\frac{\partial}{\partial x}(f(r)\frac{x}{r}) + \frac{\partial}{\partial y}(f(r)\frac{y}{r}) = 0$. Since $\frac{\partial}{\partial x}(r) = \frac{x}{r}$, we have $\frac{\partial}{\partial x}(f(r)\frac{x}{r}) =$

$$f(r)\frac{r\cdot 1 - x\cdot\frac{x}{r}}{r^2} + f'(r)\frac{x}{r}\cdot\frac{x}{r} =$$

$f(r)\frac{r^2 - x^2}{r^3} + f'(r)\frac{x^2}{r^2}$. Similarly, $\frac{\partial}{\partial y}(f(r)\frac{y}{r})$ $= f(r)\frac{r^2 - y^2}{r^3} + f'(r)\frac{y^2}{r^2}$, so $\nabla\cdot\mathbf{F} =$

$$f(r)\frac{2r^2 - x^2 - y^2}{r^3} + f'(r)\left(\frac{x^2}{r^2} + \frac{y^2}{r^2}\right) = 0;$$

that is, $f(r)\frac{1}{r} + f'(r) = 0$. Hence $\frac{f'(r)}{f(r)} = -\frac{1}{r}$, so $f(r) = \frac{k}{r}$ for some constant k. We must have $\mathbf{F} = \frac{k\hat{\mathbf{r}}}{r}$.

(d) By Exercise 34(b) of Sec. 16.1, $\mathbf{F} = \frac{k\hat{\mathbf{r}}}{r^2}$ for some constant k.

8 (a) $\int_C \frac{\mathbf{n}\cdot\hat{\mathbf{r}}}{\|\mathbf{r}\|}\,ds$ gives the radian measure of the angle subtended at the origin by the curve C.

(b) $\int_{\mathcal{S}} \frac{\mathbf{n}\cdot\hat{\mathbf{r}}}{\|\mathbf{r}\|^2}\,dS$ gives the steradian measure of the solid angle subtended at the origin by the surface $\mathcal{S}$.

9 (a) If $\nabla \times \mathbf{F} = \mathbf{0}$ everywhere on some surface $\mathcal{S}$ with boundary C, then $\oint_C \mathbf{F}\cdot\mathbf{T}\,ds =$ $\int_{\mathcal{S}} (\nabla\times\mathbf{F})\cdot\mathbf{n}\,dS = \int_{\mathcal{S}} 0\,dS = 0$, by Stokes' theorem.

(b) If C is a plane curve and $\nabla\cdot\mathbf{F} = 0$ everywhere in the region $\mathcal{A}$ bounded by C, then, by Green's theorem, $\oint_C \mathbf{F}\cdot\mathbf{n}\,ds = \int_{\mathcal{A}} \nabla\cdot\mathbf{F}\,dA$ $= 0$.

(c) If $\nabla\cdot\mathbf{F} = 0$ everywhere in the region $\mathcal{V}$ bounded by $\mathcal{S}$, then, by the divergence theorem, $\int_{\mathcal{S}} \mathbf{F}\cdot\mathbf{n}\,dS = \int_{\mathcal{V}} \nabla\cdot\mathbf{F}\,dV = 0$.

(d) If $\nabla\cdot\mathbf{F} = 0$ everywhere in $\mathcal{V}$, then $\int_{\mathcal{S}_1} \mathbf{F}\cdot\mathbf{n}\,dS$ $= \int_{\mathcal{S}_2} \mathbf{F}\cdot\mathbf{n}\,dS$ by Corollary 1 in Sec. 17.2.

(e) If $\nabla \times \mathbf{F} = \mathbf{0}$, C_1 and C_2 bound an orientable surface $\mathcal{S}$ where $\mathbf{F}$ is defined, and C_1 and C_2 are similarly oriented, then $\oint_{C_1} \mathbf{F}\cdot d\mathbf{r} =$ $\oint_{C_2} \mathbf{F}\cdot d\mathbf{r}$.

10 (a) $\int_{\mathcal{V}(t)} f(P)\, dV$ is the integral of the density function, so it gives the mass of the fluid at time t.

(b) The derivative of the mass integral with respect to t gives the rate of change of mass with respect to time.

(c) $\dfrac{\partial f}{\partial t}$ gives the rate of change of fluid density with respect to time.

(d) $\int_{\mathcal{V}(t)} \dfrac{\partial f}{\partial t}\, dV$ gives the rate of change of mass with respect to time within the region $\mathcal{V}(t)$.

(e) $\int_{\mathcal{S}(t)} f(P)(\mathbf{W} \cdot \mathbf{n})\, dS$ is the flux of mass through the surface $\mathcal{S}(t)$.

(f) The general transport theorem says that the rate of change of the fluid mass within $\mathcal{V}(t)$ equals the rate at which the density is changing with the region minus the fluid mass that passes through the surface $\mathcal{S}(t)$.

11 Yes. A torus can easily be split into two pieces, each of which has a surface $\mathcal{S}$ that can be set in a coordinate system so that each line parallel to an axis intersects $\mathcal{S}$ in at most two points. The proof of Sec. 17.2 can then be applied to each half, showing that the divergence theorem is true for both halves together.

12 Suppose the circle C has radius a, where $a \le 0.01$. By assumption, $\oint_C \mathbf{F} \cdot d\mathbf{r} = 0$. By Stokes' theorem, $\int_{\mathcal{S}} (\nabla \times \mathbf{F}) \cdot \mathbf{n}\, dS = 0$, where $\mathcal{S}$ is the disk bounded by C. $\mathcal{S}$ is small, so $\nabla \times \mathbf{F}$ is approximately constant on $\mathcal{S}$. Thus $(\nabla \times \mathbf{F})\cdot\mathbf{n}$ (Area of $\mathcal{S}$) $= (\nabla \times \mathbf{F})\cdot\mathbf{n}(\pi a^2) \approx 0$, where we evaluate $\nabla \times \mathbf{F}$ at the center of $\mathcal{S}$. This shows that $(\nabla \times \mathbf{F})\cdot\mathbf{n}$ is approximately 0 no matter what orientation we choose for our small circles. Hence, in the limit, we must have $\nabla \times \mathbf{F} = \mathbf{0}$. Since the domain of $\mathbf{F}$ is simply connected, $\mathbf{F}$ is conservative and it follows that $\oint_C \mathbf{F} \cdot d\mathbf{r} = 0$ for all large circles (or any other closed path, for that matter).

17.S Review Exercises

1 See Sec. 17.2.

3 (a) $\mathcal{S}$ is that part of the surface $z = x^2 + 3y^2$ that lies below the paraboloid $z = 1 - x^2 - y^2$. Points on the intersection of the two surfaces must satisfy $x^2 + 3y^2 = 1 - x^2 - y^2$, so $2x^2 + 4y^2 = 1$, which is the equation of an ellipse in the xy plane, the interior of which is the projection of $\mathcal{S}$, denoted $\mathcal{A}$. Since $\mathcal{S}$ is part of the level surface $z - x^2 - 3y^2 = 0$, we see that $\mathbf{N} = -2x\mathbf{i} - 6y\mathbf{j} + \mathbf{k}$ is a vector normal to $\mathcal{S}$. Therefore, $1/(\cos\gamma) = \|\mathbf{N}\|/\mathbf{k}\cdot\mathbf{N} = \sqrt{4x^2 + 36y^2 + 1}$. So the area of $\mathcal{S}$ equals $\int_{\mathcal{A}} \sqrt{4x^2 + 36y^2 + 1}\, dA$.

(b) The boundary of $\mathcal{A}$ can be expressed as $y^2 = \frac{1}{4}(1 - 2x^2)$; thus $\mathcal{A}$ can be described by $-\frac{1}{\sqrt{2}} \le x \le \frac{1}{\sqrt{2}}$, $-\frac{1}{2}\sqrt{1 - 2x^2} \le y \le \frac{1}{2}\sqrt{1 - 2x^2}$. The area of $\mathcal{S}$ is thus

$$\int_{\mathcal{A}} \frac{1}{\cos \gamma}\, dA$$

$$= \int_{-1/\sqrt{2}}^{1/\sqrt{2}} \int_{-\frac{1}{2}\sqrt{1-2x^2}}^{\frac{1}{2}\sqrt{1-2x^2}} \sqrt{4x^2 + 36y^2 + 1}\, dy\, dx$$

$$= 4 \int_0^{1/\sqrt{2}} \int_0^{\frac{1}{2}\sqrt{1-2x^2}} \sqrt{4x^2 + 36y^2 + 1}\, dy\, dx.$$

To describe $\mathcal{A}$ in polar coordinates, first substitute $x = r\cos\theta$ and $y = r\sin\theta$ into the equation of the boundary: $1 = 2x^2 + 4y^2 = 2r^2\cos^2\theta + 4r^2\sin^2\theta = 2r^2(\cos^2\theta + 2\sin^2\theta) = 2r^2(1 + \sin^2\theta)$, so $r^2 = \dfrac{1}{2(1 + \sin^2\theta)}$ or $r = [2(1 + \sin^2\theta)]^{-1/2}$. Now consider the expression inside the radical for $\cos\gamma$: $4x^2 + 36y^2 + 1$

$= 4r^2\cos^2\theta + 36r^2\sin^2\theta + 1$

$= 4r^2(\cos^2\theta + 9\sin^2\theta) + 1$

$= 4r^2(1 + 8\sin^2\theta) + 1$. So, in polar coordinates, the integral for the area of $\mathcal{S}$ is

$$\int_0^{2\pi} \int_0^{1/\sqrt{2(1+\sin^2\theta)}} \sqrt{1 + 4r^2(1 + 8\sin^2\theta)}\, r\, dr\, d\theta,$$

which, by symmetry, we may write as

$$4\int_0^{\pi/2} \int_0^{1/\sqrt{2(1+\sin^2\theta)}} \sqrt{1 + 4r^2(1 + 8\sin^2\theta)}\, r\, dr\, d\theta.$$

5 (a) $\mathbf{F}$ is conservative and $\oint_{C_1} \mathbf{F}\cdot\mathbf{T}\, ds = \oint_{C_2} \mathbf{F}\cdot\mathbf{T}\, ds = 0.$

(b) $\oint_{C_1} \mathbf{F}\cdot\mathbf{T}\, ds = \oint_{C_2} \mathbf{F}\cdot\mathbf{T}\, ds$

7 (a) It was shown in Exercise 14 of Chapter 16 Review Exercises that the curl of $\mathbf{r}/\|\mathbf{r}\|^n$ is $\mathbf{0}$ for all real numbers n (not just integers).

(b) It was shown in Exercise 14 of Chapter 16 Review Exercises that the divergence of $\mathbf{r}/\|\mathbf{r}\|^n$ is 0 only for $n = 3$.

9 $\mathbf{r} = t\mathbf{i} + t^2\mathbf{j} + t^3\mathbf{k}$ for $1 \le t \le 2$, so $d\mathbf{r} = (\mathbf{i} + 2t\mathbf{j} + 3t^2\mathbf{k})\, dt$ and

$$\int_C \ln z\, dx + \frac{1}{xy}\, dy + \sqrt{xy + 1}\, dz$$

$$= \int_1^2 \ln t^3\, dt + \frac{1}{t^3}\, 2t\, dt + \sqrt{t^3 + 1}\, 3t^2\, dt$$

$$= \int_1^2 \left[3\ln t + 2t^{-2} + (t^3 + 1)^{1/2}\, 3t^2\right] dt$$

$$= \left[3(t\ln t - t) - \frac{2}{t} + \frac{2}{3}(t^3 + 1)^{3/2}\right]\Bigg|_1^2$$

$= 3(2\ln 2 - 2) - 1 + \frac{2}{3}\cdot 27 - \left[3(-1) - 2 + \frac{2}{3}\cdot 2^{3/2}\right] = 6\ln 2 + 16 - \frac{4}{3}\sqrt{2}.$

11 $\mathbf{r} = \cos t\,\mathbf{i} + \sin t\,\mathbf{j} + \tan t\,\mathbf{k}$ for $0 \le t \le \pi/4$, so $d\mathbf{r} = (-\sin t\,\mathbf{i} + \cos t\,\mathbf{j} + \sec^2 t\,\mathbf{k})\, dt$ and

$\int_C (-xy^2)\, dx + x^2y\, dy + z^2\, dz =$

$\int_0^{\pi/4} (-\cos t\sin^2 t)(-\sin t\, dt) + \cos^2 t\sin t\,(\cos t\, dt) + \tan^2 t\,(\sec^2 t\, dt) =$

$\int_0^{\pi/4} (\cos t\sin^3 t + \cos^3 t\sin t + \tan^2 t\sec^2 t)\, dt$

$= \int_0^{\pi/4} [\cos t\sin t\,(\sin^2 t + \cos^2 t) + \tan^2 t\sec^2 t]\, dt$

$= \int_0^{\pi/4} (\cos t\sin t + \tan^2 t\sec^2 t)\, dt$

$$= \frac{1}{2}\sin^2 t + \frac{1}{3}\tan^3 t\Big|_0^{\pi/4} = \frac{1}{2}\cdot\frac{1}{2} + \frac{1}{3} = \frac{7}{12}.$$

13 Let C denote the "equator" of the surface S, which is then a common boundary to S_1 and S_2. If C is oriented counterclockwise as viewed from above, then $\int_{S_1} (\nabla\times\mathbf{F})\cdot\mathbf{n}\,dS = \oint_C \mathbf{F}\cdot d\mathbf{r}$, where $\mathbf{n}$ is the exterior normal to the sphere. On S_2, however, $\mathbf{n}$ has a negative $\mathbf{k}$ component; the right-hand rule applied to S_2 and $\mathbf{n}$ would reverse the orientation of C. Therefore, if we use $-\mathbf{n}$, we have

$\int_{S_2} (\nabla\times\mathbf{F})\cdot(-\mathbf{n})\,dS = \oint_C \mathbf{F}\cdot d\mathbf{r}$. But

$\int_{S_2} (\nabla\times\mathbf{F})\cdot(-\mathbf{n})\,dS = -\int_{S_2} (\nabla\times\mathbf{F})\cdot\mathbf{n}\,dS$, so

$-\int_{S_2} (\nabla\times\mathbf{F})\cdot\mathbf{n}\,dS = \int_{S_1} (\nabla\times\mathbf{F})\cdot\mathbf{n}\,dS$.

15 (a) See the solution to Exercise 33 of Sec. 17.1 for a discussion of conical surfaces. The surface S can be described by $0 \le \theta \le 2\pi$, $1 \le \rho \le 2$, and $\phi = \pi/4$. The area of the cone is $\int_S dS = \int_0^{2\pi}\int_1^2 \rho\cdot\frac{1}{\sqrt{2}}\,d\rho\,d\theta$

$$= \frac{3\pi}{\sqrt{2}}.$$

(b) Let $\mathcal{A}$ be the region in the xy plane bounded outside by the circle of radius $\sqrt{2}$ and inside by the circle of radius $\frac{1}{\sqrt{2}}$. Since $z =$

$\sqrt{x^2 + y^2}$, $\cos\gamma = \frac{1}{\sqrt{2}}$ and $\int_S dS =$

$$\int_{\mathcal{A}} \sqrt{2}\,dA = \sqrt{2}\cdot\frac{3}{2}\pi = \frac{3\pi}{\sqrt{2}}.$$

17 (a) See the graph in the answer section of the text.

(b) Area of $S_1 = \int_{S_1} dS$

$$= \int_0^1\int_0^x \sqrt{(2x)^2 + 1}\,dy\,dx$$

$$= \frac{1}{12}(4x^2 + 1)^{3/2}\Big|_0^1 = \frac{1}{12}(5^{3/2} - 1).$$

Area of S_2 = Area of $S_1 = \frac{1}{12}(5^{3/2} - 1)$.

Area of S_3 = Area of $S_4 = \frac{1}{3}$. Area of $S_5 =$

1. So the total surface area is $\frac{5\sqrt{5}}{6} + \frac{3}{2}$.

19 $\mathbf{F} = xz\mathbf{i} + x^2\mathbf{j} + xy\mathbf{k}$; let the segments of the given path C be labeled as in the figure. On C_1, $y = 0$

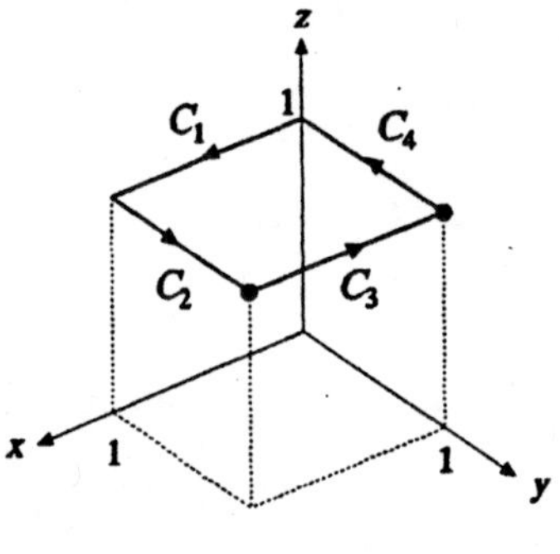

and $z = 1$ as x goes from 0 to 1, so $dy = dz = 0$. On C_2, $x = z = 1$ as y goes from 0 to 1, so $dx = dz = 0$. On C_3, $y = z = 1$ as x goes from 1 to 0, so $dy = dz = 0$. On C_4, $x = 0$ and $z = 1$ as y goes from 1 to 0, so $dx = dz = 0$. We have

$\oint_C \mathbf{F}\cdot d\mathbf{r} = \oint_C xz\,dx + x^2\,dy + xy\,dz$. Now

$$\int_{C_1} xz\,dx + x^2\,dy + xy\,dz = \int_0^1 x\,dx = \frac{x^2}{2}\Big|_0^1$$

$= \frac{1}{2}$; $\int_{C_2} xz\,dx + x^2\,dy + xy\,dz = \int_0^1 1^2\,dy =$

1; $\int_{C_3} xz\,dx + x^2\,dy + xy\,dz = \int_1^0 x\,dx =$

$\left.\frac{x^2}{2}\right|_1^0 = -\frac{1}{2}$; and $\int_{C_4} xz\,dx + x^2\,dy + xy\,dz =$

$\int_1^0 0\,dy = 0$, so $\oint_C \mathbf{F}\cdot d\mathbf{r} = \frac{1}{2} + 1 - \frac{1}{2} + 0 = 1$.

21 (a) Q is distributed uniformly on a sphere; by symmetry, **E** must look the same in all directions, so it is a central field.

(b) By (a), $\mathbf{E} = f(r)\hat{\mathbf{r}}$. By Gauss's law, if S is a sphere of radius $b > a$, then $\int_S \mathbf{E}\cdot\mathbf{n}\,dS =$ $\frac{Q}{\epsilon_0}$. Therefore $\int_S f(r)\hat{\mathbf{r}}\cdot\hat{\mathbf{r}}\,dS = \int_S f(r)\,dS =$ $f(b)(\text{Area of } S) = f(b)(4\pi b^2) = \frac{Q}{\epsilon_0}$, so $f(b) =$ $\frac{Q}{4\pi b^2\epsilon_0}$. Hence $\mathbf{E} = \frac{Q}{4\pi r^2\epsilon_0}\hat{\mathbf{r}}$.

(c) Gauss's law says that $\int_S \mathbf{E}\cdot\mathbf{n}\,dS = 0$ in this case, leading to $\mathbf{E} = \mathbf{0}$.

23 Note that $\mathbf{F} = e^x\mathbf{i} + e^y\mathbf{j} + \sin z\,\mathbf{k} = \nabla f$, where $f(x, y, z) = e^x + e^y - \cos z$. The helix C goes from (3, 0, 2) to (3, 0, 0). By Theorem 2 of Sec. 16.6, $\int_C \mathbf{F}\cdot d\mathbf{r} = f(3, 0, 0) - f(3, 0, 2) =$ $(e^3 + 1 - 1) - (e^3 + 1 - \cos 2) = -1 + \cos 2$.

25 (a) For points outside the ball of radius a, $\mathbf{E} =$ $\frac{Q}{4\pi\epsilon_0 r^2}\hat{\mathbf{r}}$, exactly as in Exercise 21(b), since all of the charge would be contained in our sphere S with radius $b > a$.

(b) The charge density is $q = \frac{Q}{4\pi a^3/3} = \frac{3Q}{4\pi a^3}$. If S is a sphere of radius r, $r < a$, then S contains a charge of $q\cdot\frac{4}{3}\pi r^3 = \frac{3Q}{4\pi a^3}\cdot\frac{4\pi r^3}{3}$ $= \frac{r^3}{a^3}Q$. Using this in place of Q in our result from Exercise 21(b), we have $\mathbf{E} =$ $\frac{r^3Q/a^3}{4\pi\epsilon_0 r^2}\hat{\mathbf{r}} = \frac{Qr}{4\pi\epsilon_0 a^3}\hat{\mathbf{r}}$.

27 (a) A gradient is always a conservative vector field.

(b) $\mathbf{F} = P\mathbf{i} + Q\mathbf{j} = \frac{-y\mathbf{i} + x\mathbf{j}}{(x^2 + y^2)^2}$; **F** cannot be conservative unless $P_y = Q_x$. To simplify the notation, let $r^2 = x^2 + y^2$, as usual, and recall that $r_x = \frac{x}{r}$ and $r_y = \frac{y}{r}$. Then $P = -y/r^4$ and $Q = x/r^4$, so $P_y = \frac{\partial}{\partial y}(-y/r^4) =$

$\frac{r^4\cdot(-1) - (-y)4r^3r_y}{r^8} = \frac{-r^4 + 4y^2r^2}{r^8} =$

$\frac{4y^2 - r^2}{r^6}$; $Q_x = \frac{\partial}{\partial x}(x/r^4) =$

$\frac{r^4\cdot 1 - x\cdot 4r^3r_x}{r^8} = \frac{r^4 - 4x^2r^2}{r^8} =$

$\frac{r^2 - 4x^2}{r^6}$. Now $4y^2 - r^2 = 4y^2 - x^2 - y^2 =$

$3y^2 - x^2$ while $r^2 - 4x^2 = x^2 + y^2 - 4x^2 = y^2 - 3x^2$. Therefore $P_y \neq Q_x$ and **F** is not conservative.

29 (a) Yes. Gradient fields are conservative.

(b) No. The scalar component of $\nabla \times \mathbf{F}$ along the vector **i** is

$$\frac{\partial}{\partial y}\left(-\frac{z}{x^2+y^2+z^2}\right) - \frac{\partial}{\partial z}\left(\frac{y}{x^2+y^2+z^2}\right) =$$

$$\frac{2yz}{(x^2+y^2+z^2)^2} + \frac{2yz}{(x^2+y^2+z^2)^2} =$$

$$\frac{4yz}{(x^2+y^2+z^2)^2} \neq 0 \text{ for } y \neq 0,\ z \neq 0.$$

Hence $\nabla \times \mathbf{F}$ is not identically zero, so **F** is not conservative.

31 No. The scalar component of $\nabla \times \mathbf{F}$ along the vector **i** is $\frac{\partial}{\partial y}(x^2y + yz^2) - \frac{\partial}{\partial z}(2x^2yz + \frac{3}{2}x^2y^2) = x^2 + z^2 - 2x^2y$. Since $\nabla \times \mathbf{F}$ is not zero everywhere, **F** is not conservative.

33 $\mathbf{F} = (\cos^2 r)\hat{\mathbf{r}} =$

$\frac{x}{r}\cos^2 r\,\mathbf{i} + \frac{y}{r}\cos^2 r\,\mathbf{j} + \frac{z}{r}\cos^2 r\,\mathbf{k}$. Recalling that $\frac{\partial r}{\partial x} = \frac{x}{r}$, we check the **k** component of $\nabla \times \mathbf{F}$: $\frac{\partial}{\partial x}\left(\frac{y}{r}\cos^2 r\right) - \frac{\partial}{\partial y}\left(\frac{x}{r}\cos^2 r\right)$. The first term is $y\frac{\partial}{\partial x}\left(\frac{\cos^2 r}{r}\right) =$

$$y\frac{r(2\cos r)(-\sin r)\frac{x}{r} - (\cos^2 r)\frac{x}{r}}{r^2} =$$

$-\frac{xy}{r^3}(2r\cos\theta\sin\theta + \cos^2 r)$. By symmetry, we see that this equals $\frac{\partial}{\partial y}\left(\frac{x}{r}\cos^2 r\right)$, so the **k** component of $\nabla \times \mathbf{F}$ is 0. Similarly, so are the other two components. Hence $\nabla \times \mathbf{F} = \mathbf{0}$; since the domain of **F** is simply connected, **F** is conservative. Alternatively, let $f(r)$ be an antiderivative for $\cos^2 r$. Then $\mathbf{F} = \nabla f$, so **F** is conservative.

35 (a) $\mathbf{F} = \hat{\mathbf{r}}|\mathbf{r}|^2 = r^2\hat{\mathbf{r}} = r\mathbf{r} = xr\mathbf{i} + yr\mathbf{j} + zr\mathbf{k}$,

so $\nabla \times \mathbf{F} = \left(z\cdot\frac{y}{r} - y\cdot\frac{z}{r}\right)\mathbf{i} - \left(z\cdot\frac{x}{r} - x\cdot\frac{z}{r}\right)\mathbf{j} + \left(y\cdot\frac{x}{r} - x\cdot\frac{y}{r}\right)\mathbf{k} = \mathbf{0}$; the domain of **F** is simply connected, so **F** is conservative.

(b) $\mathbf{F} = xr\mathbf{i} + yr\mathbf{j} + zr\mathbf{k} =$

$r^2\cdot\frac{x}{r}\mathbf{i} + r^2\cdot\frac{y}{r}\mathbf{j} + r^2\cdot\frac{z}{r}\mathbf{k} = \nabla\left(\frac{r^3}{3}\right)$, so

$$f(x, y, z) = \frac{1}{3}(x^2 + y^2 + z^2)^{3/2}.$$

37 Recall that the volume of a cone in one-third the product of the cone's altitude and the area of its base. Consider the cone with apex at the origin and base $d\mathcal{S}$. If **F** is the vector from the apex to the base and **n** is the unit normal to the surface $d\mathcal{S}$, then $\mathbf{F}\cdot\mathbf{n}$ is the cone's altitude, so $\frac{1}{3}\ \mathbf{F}\cdot\mathbf{n}\ dS$ is the volume of the cone, where dS is the area of $d\mathcal{S}$. Using this local approximation for volume, dV, we have Volume of $\mathcal{V} = \int_{\mathcal{V}} dV = \frac{1}{3}\int_{\mathcal{S}} \mathbf{F}\cdot\mathbf{n}\ dS$, from which the desired result immediately follows.

39 Let $\mathbf{F} = P\mathbf{i} + Q\mathbf{j} + R\mathbf{k}$.

(a) Show $\nabla \times \mathbf{F} = \mathbf{0}$.

Show $\mathbf{F} = \nabla f$.

Show $\oint_C \mathbf{F}\cdot d\mathbf{r} = 0$ for all closed curves C.

(b) Show that any statement in (a) is false.

41 (a) Since $\frac{\partial f}{\partial x} = ze^{xz}$, $f(x, y, z) = e^{xz} + g(y, z)$ for some function g. Since $0 = \frac{\partial f}{\partial y} = 0 + \frac{\partial}{\partial x}(g(y, z))$, $g(y, z) = h(z)$ for some function h. Finally, $xe^{xz} = \frac{\partial f}{\partial z} = xe^{xz} + h'(z)$, so h is a constant, which we may take to be 0. Thus $f(x, y, z) = e^{xz}$ works.

(b) The curve C begins at $(0, 1, 1)$ and ends at $(4\pi^2, 1, e^{2\pi})$. By (a) and Theorem 2 of Sec. 16.6, $\oint_C \mathbf{F}\cdot\mathbf{T}\, ds = f(4\pi^2, 1, e^{2\pi}) - f(0, 1, 1)$

$= e^{4\pi^2 e^{2\pi}} - 1.$

43 $\mathbf{F} = 4xz\mathbf{i} - y^2\mathbf{j} + yz\mathbf{k}$, so div $\mathbf{F} = \nabla\cdot\mathbf{F} = 4z - 2y + y = 4z - y$. $\mathcal{V}$ is the unit cube, so

$$\int_{\mathcal{V}} \nabla\cdot\mathbf{F}\, dV = \int_0^1 \int_0^1 \int_0^1 (4z - y)\, dx\, dz\, dy$$

$$= \int_0^1 \int_0^1 (4z - y)\, dz\, dy = \int_0^1 (2z^2 - yz)\Big|_0^1 dy$$

$$= \int_0^1 (2 - y)\, dy = \left(2y - \frac{y^2}{2}\right)\Bigg|_0^1 = 2 - \frac{1}{2}$$

$= \frac{3}{2}$. Label the surface as shown in the figure.

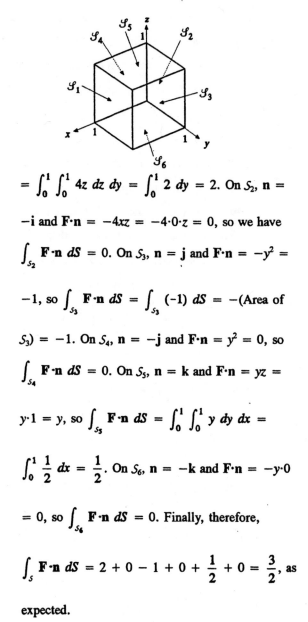

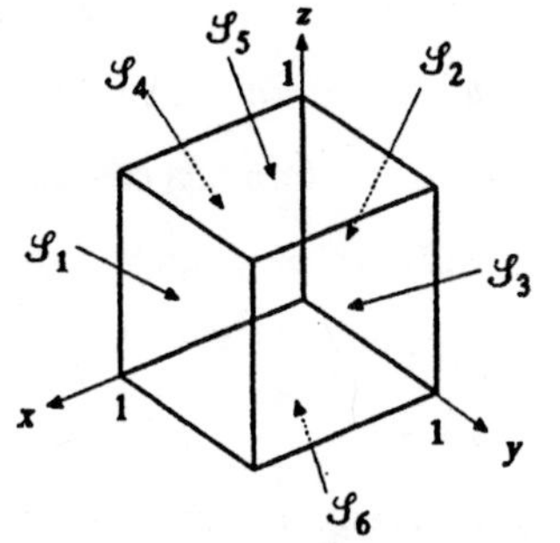

On $\mathcal{S}_1$, $\mathbf{n} = \mathbf{i}$ and $\mathbf{F}\cdot\mathbf{n} = 4xz = 4z$, so $\int_{\mathcal{S}_1} \mathbf{F}\cdot\mathbf{n}\, dS$ $= \int_0^1 \int_0^1 4z\, dz\, dy = \int_0^1 2\, dy = 2$. On $\mathcal{S}_2$, $\mathbf{n} = -\mathbf{i}$ and $\mathbf{F}\cdot\mathbf{n} = -4xz = -4\cdot 0\cdot z = 0$, so we have $\int_{\mathcal{S}_2} \mathbf{F}\cdot\mathbf{n}\, dS = 0$. On $\mathcal{S}_3$, $\mathbf{n} = \mathbf{j}$ and $\mathbf{F}\cdot\mathbf{n} = -y^2 = -1$, so $\int_{\mathcal{S}_3} \mathbf{F}\cdot\mathbf{n}\, dS = \int_{\mathcal{S}_3} (-1)\, dS = -(\text{Area of } \mathcal{S}_3) = -1$. On $\mathcal{S}_4$, $\mathbf{n} = -\mathbf{j}$ and $\mathbf{F}\cdot\mathbf{n} = y^2 = 0$, so $\int_{\mathcal{S}_4} \mathbf{F}\cdot\mathbf{n}\, dS = 0$. On $\mathcal{S}_5$, $\mathbf{n} = \mathbf{k}$ and $\mathbf{F}\cdot\mathbf{n} = yz = y\cdot 1 = y$, so $\int_{\mathcal{S}_5} \mathbf{F}\cdot\mathbf{n}\, dS = \int_0^1 \int_0^1 y\, dy\, dx = \int_0^1 \frac{1}{2}\, dx = \frac{1}{2}$. On $\mathcal{S}_6$, $\mathbf{n} = -\mathbf{k}$ and $\mathbf{F}\cdot\mathbf{n} = -y\cdot 0 = 0$, so $\int_{\mathcal{S}_6} \mathbf{F}\cdot\mathbf{n}\, dS = 0$. Finally, therefore,

$\int_{\mathcal{S}} \mathbf{F}\cdot\mathbf{n}\, dS = 2 + 0 - 1 + 0 + \frac{1}{2} + 0 = \frac{3}{2}$, as expected.

45 (a) From Figure 8, we see that $(r - b)^2 + z^2 = a^2$ is an equation in the "rz" plane of the circular cross section of the torus. Using v as the angle parameter for this circle, we have $r = b + a\cos v$ and $z = a\sin v$. Using this result for r, with u as the angle parameter in place of the usual θ, we have $x = r\cos u = (b + a\cos v)\cos u$ and $y = r\sin u = (b + a\cos v)\sin u$. Combining this with the previous result for z gives us a parameterization for the surface of the torus. (In terms of vectors this could be written as a vector with *two* parameters: $\mathbf{r}(u, v) = (b + a\cos v)\cos u\,\mathbf{i} + (b + a\cos v)\sin u\,\mathbf{j} + a\sin v\,\mathbf{k}$.)

(b) The region with u coordinate between u and $u + du$ and v coordinate between v and $v + dv$ is at distance $b + a\cos v$ from L. It is approximately a rectangle with sides $a\,dv$ and $(b + a\cos v)\,du$, so the element of area is $a(b + a\cos v)\,du\,dv$. The total area is $A = \int_0^{2\pi}\int_0^{2\pi} a(b + a\cos v)\,du\,dv = 4\pi^2 ab$.

The average distance from L is

$$\frac{1}{A}\int_0^{2\pi}\int_0^{2\pi}(b + a\cos v)\cdot a(b + a\cos v)\,du\,dv$$

$$= b + \frac{a^2}{2b}.$$

47 (a) As shown, the z component of force is $cz\mathbf{k}\cdot\mathbf{n}\,\Delta S$; using the informal approach, we see that the total vertical component of force should be $\int_S cz\mathbf{k}\cdot\mathbf{n}\,dS$.

(b) By the divergence theorem, $\int_S cz\mathbf{k}\cdot\mathbf{n}\,dS = \int_V \nabla\cdot(cz\mathbf{k})\,dV = \int_V c\,dV$, which equals the weight of the displaced water.

(c) The x component of the force on dS is $cz\mathbf{i}\cdot\mathbf{n}\,dS$, so the total force in the x direction is $\int_S cz\mathbf{i}\cdot\mathbf{n}\,dS = \int_V \nabla\cdot(cz\mathbf{i})\,dV = \int_V 0\,dV = 0$. Similarly, the y component is 0.

49 $\nabla f\cdot\nabla f = \|\nabla f\|^2 \geq 0$, so $\int_V (\nabla f\cdot\nabla f)\,dV = 0$ implies that $\|\nabla f\| = 0$. This can occur only if the three components of ∇f are 0; that is, $f_x = f_y = f_z = 0$. Thus f is constant with respect to each of the variables x, y, and z, so it is a constant function.

51 (a) $\int_S (h\nabla h)\cdot\mathbf{n}\,dS = \int_V (h\nabla^2 h + \nabla h\cdot\nabla h)\,dV$

$= \int_V h\nabla^2 h\,dV + \int_V \nabla h\cdot\nabla h\,dV$

(b) Since $\nabla^2 h = 0$ on $\mathcal{V}$, $\int_V h\nabla^2 h\,dV = 0$. Since $h = 0$ on $\mathcal{S}$, $\int_S (h\nabla h)\cdot\mathbf{n}\,dS = 0$. Therefore

$\int_V \nabla h\cdot\nabla h\,dV$

$= \int_S (h\nabla h)\cdot\mathbf{n}\,dS - \int_V h\nabla^2 h\,dV = 0$, and

$\nabla h = \mathbf{0}$.

(c) $h_x = h_y = h_z = 0$, and h is a constant on $\mathcal{V}$.

(d) Since $h = 0$ on $\mathcal{S}$ and is constant on $\mathcal{V}$, $h = 0$ on $\mathcal{V}$.

53 $\mathbf{F} = P\mathbf{i} + Q\mathbf{j} + R\mathbf{k}$, so $\int_V \mathbf{F}\,dV = \int_V P\,dV\,\mathbf{i} + \int_V Q\,dV\,\mathbf{j} + \int_V R\,dV\,\mathbf{k}$. Let $\mathbf{c} = a\mathbf{i} + b\mathbf{j} + c\mathbf{k}$; then $\int_V \mathbf{c}\cdot\mathbf{F}\,dV = \int_V (aP + bQ + cR)\,dV = a\int_V P\,dV + b\int_V Q\,dV + c\int_V R\,dV$

$= \mathbf{c}\cdot\int_V \mathbf{F}\,dV$. The corresponding result for surface integrals is proved similarly.

55 (a) By the divergence theorem, with $\mathbf{F} = f\mathbf{c}$,

$\int_V \nabla\cdot f\mathbf{c}\,dV = \int_S f\mathbf{c}\cdot\mathbf{n}\,dS$.

(b) Let $\mathbf{c} = a\mathbf{i} + b\mathbf{j} + c\mathbf{k}$; then $\nabla\cdot f\mathbf{c} = (fa)_x + (fb)_y + (fc)_z = f_x a + f_y b + f_z c = \nabla f\cdot\mathbf{c} = \mathbf{c}\cdot\nabla f$, so $\int_V \nabla\cdot f\mathbf{c}\,dV = \int_V \mathbf{c}\cdot\nabla f\,dV$.

(c) By Exercise 53, $\mathbf{c}\cdot\int_V \nabla f\, dV = \int_V \mathbf{c}\cdot\nabla f\, dV$; by (b), this latter integral is equal to $\int_V \nabla\cdot f\mathbf{c}\, dV$, which, by (a), is $\int_S f\mathbf{c}\cdot\mathbf{n}\, dS$. Since $f\mathbf{c}\cdot\mathbf{n} = \mathbf{c}\cdot(f\mathbf{n})$, another application of Exercise 53 produces $\mathbf{c}\cdot\int_S f\mathbf{n}\, dS$. Thus $\mathbf{c}\cdot\int_V \nabla f\, dV = \mathbf{c}\cdot\int_S f\mathbf{n}\, dS$, as claimed.

(d) Apply the result of (c) for $\mathbf{c} = \mathbf{i}, \mathbf{j}$, and $\mathbf{k}$. Thus $\int_V \nabla f\, dV$ and $\int_S f\mathbf{n}\, dS$ have identical x, y, and z components, so they must be equal.

57 Let S_i be a face of the tetrahedron and define $\mathbf{A}_i$ to be the exterior normal to S_i such that $\|\mathbf{A}_i\| = A_i =$ Area of S_i. Let S be the entire surface of the tetrahedron; then, by Exercise 56(b), $\mathbf{0} = \int_S \mathbf{n}\, dS$ $= \int_{S_1} \mathbf{n}\, dS + \int_{S_2} \mathbf{n}\, dS + \int_{S_3} \mathbf{n}\, dS + \int_{S_4} \mathbf{n}\, dS$ $= \mathbf{A}_1 + \mathbf{A}_2 + \mathbf{A}_3 + \mathbf{A}_4$, which gives the desired result.

59 Note that the boundary of S is the ellipse C in the xy plane with equation $\frac{x^2}{16} + \frac{y^2}{9} = 1$. Let S' be the region in the xy plane bounded by C; on S' we have $\mathbf{n} = \mathbf{k}$, so $\int_S (\nabla\times\mathbf{F})\cdot\mathbf{n}\, dS = \oint_C \mathbf{F}\cdot d\mathbf{r} =$ $\int_{S'} (\nabla\times\mathbf{F})\cdot\mathbf{k}\, dS$. To evaluate the last integral, all we need is the z component of $\nabla \times \mathbf{F}$. $\mathbf{F} =$ $x(x^2 + y^2 + z^2)^{3/2}\mathbf{i} - y(x^2 + y^2 + z^2)^{3/2}\mathbf{j} +$ $x^2 \sin \pi y\, e^{\tan^{-1} z}\mathbf{k}$, so $(\nabla \times \mathbf{F})\cdot\mathbf{k} = Q_x - P_y =$ $-6xy(x^2 + y^2 + z^2)^{1/2}$. But this is an odd function in x and y, so $\int_{S'} (\nabla\times\mathbf{F})\cdot\mathbf{k}\, dS = 0$, by symmetry.

Appendices

A Real Numbers

1 $a + 3 < b + 3$

3 $5a < 5b$

5 $1/a > 1/b$

7 $2x + 7 < 4x + 9$, so $2x < 4x + 2$, $-2x < 2$, and $x > -1$; note that in the final step, division by a negative number reversed the sense of the inequality. The inequality is satisfied for all x in the interval $(-1, \infty)$.

9 $-3x + 2 > 5x + 18$, so $-3x > 5x + 16$, $-8x > 16$, and $x < -2$. The inequality is satisfied for all x in $(-\infty, -2)$.

11 $(x - 1)(x - 3)$ is positive when both factors have the same sign. As in Example 2, we indicate where the signs change in a table:

	1	3	
$x - 1$	−	+	+
$x - 3$	−	−	+
$(x - 1)(x - 3)$	+	−	+

Thus $(x - 1)(x - 3) > 0$ for x in $(-\infty, 1)$ and $(3, \infty)$.

13 $(x + 2)(x + 3)$ is negative whenever precisely one of the factors is negative:

	−3	−2	
$x + 2$	−	−	+
$x + 3$	−	+	+
$(x + 2)(x + 3)$	+	−	+

Thus $(x + 2)(x + 3) < 0$ for x in $(-3, -2)$.

15 $x(x - 1)(x + 1)$ is positive when all three factors are positive or when precisely one of the factors is positive (in which case the negative signs of the other two terms cancel). The sign changes are recorded in the following table:

	−1	0	1	
x	−	−	+	+
$x - 1$	−	−	−	+
$x + 1$	−	+	+	+
$x(x - 1)(x + 1)$	−	+	−	+

$x(x - 1)(x + 1) > 0$ for x in $(-1, 0)$ or $(1, \infty)$.

17 $x(x + 3)(x + 5)$ is positive whenever all three factors are positive or when precisely one of the factors is positive.

	−5	−3	0	
x	−	−	−	+
$x + 3$	−	−	+	+
$x + 5$	−	+	+	+
$x(x + 3)(x + 5)$	−	+	−	+

$x(x + 3)(x + 5) > 0$ for x in $(-5, -3)$ or $(0, \infty)$.

19 $(3x - 1)(2x - 1)$ is positive when both terms are negative or both are positive. $(3x - 1)(2x - 1) >$

	1/3	1/2	
$3x - 1$	−	+	+
$2x - 1$	−	−	+
$(3x - 1)(2x - 1)$	+	−	+

0 for x in $\left(-\infty, \frac{1}{3}\right)$ or $\left(\frac{1}{2}, \infty\right)$.

21 We have $|x - 3| < 2$, so $-2 < x - 3 < 2$, $-2 + 3 < x < 2 + 3$, or $1 < x < 5$.

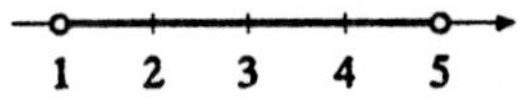

23 $|3(x - 1)| = 3|x - 1| < 6$, so $|x - 1| < 2$; hence $-2 < x - 1 < 2$, or $-1 < x < 3$.

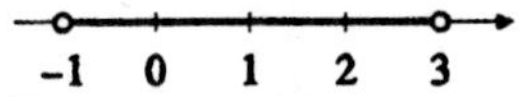

25 Note that $x^2 > 0$ for x except $x = 0$. Therefore $x^2(x - 3)$ is positive whenever $x - 3$ is positive; that is, for $x > 3$.

27 $x < x^2$ whenever $x - x^2 < 0$. But $x - x^2 = x(1 - x)$, and we are looking at only positive values of x, so $x(1 - x)$ is negative whenever $1 - x$ is negative; that is, for $x > 1$.

29 Let $a = -2$ and $b = 1$; then $a < b$, but $a^2 > b^2$.

31 Let $a = -1$, $b = 1$, $c = -2$, and $d = 1$; then $a < b$ and $c < d$, but $ac = 2 > 1 = bd$.

33 (a) $100a = 100(2.3474747\cdots) = 234.74747\cdots$

(b) $100a - a = 234.74747\cdots - 2.34747\cdots$
$= 232.40000\cdots = 232.4$

(c) By (b), $99a = 232.4$, so $a = \dfrac{232.4}{99}$.

(d) By (c), $a = \dfrac{232.4}{99} = \dfrac{232.4}{99}\cdot\dfrac{10}{10} = \dfrac{2324}{990}$
$= \dfrac{1162}{495}$.

B Graphs and Lines

B.1 Coordinate Systems and Graphs

1

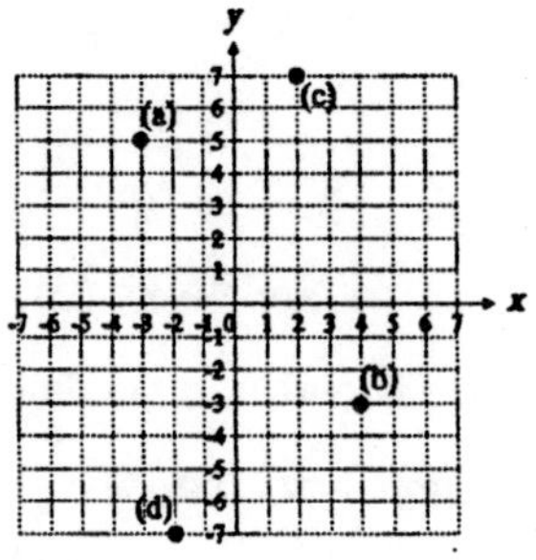

(a) $(-3, 5)$ is in the second quadrant.

(b) $(4, -3)$ is in the fourth quadrant.

(c) $(2, 7)$ is in the first quadrant.

(d) $(-2, -7)$ is in the third quadrant.

3 (a) $d^2 = (9 - 1)^2 + (-2 - 4)^2 = 8^2 + (-6)^2$
$= 100$, $d = \sqrt{100} = 10$

(b) $d^2 = (15 - 3)^2 + (1 - 6)^2 = 12^2 + (-5)^2$
$= 169$, $d = \sqrt{169} = 13$

(c) $d^2 = (0 - 0)^2 + (7 - 4)^2 = 0^2 + 3^2 = 9$,
$d = \sqrt{9} = 3$

5 $x^2 + y^2 = 49$ is the equation of the circle of radius $\sqrt{49} = 7$ with center at the origin.

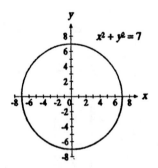

7

x	0	1	2	3	-1	-2	-3
$y = 2x^2$	0	2	8	18	2	8	18

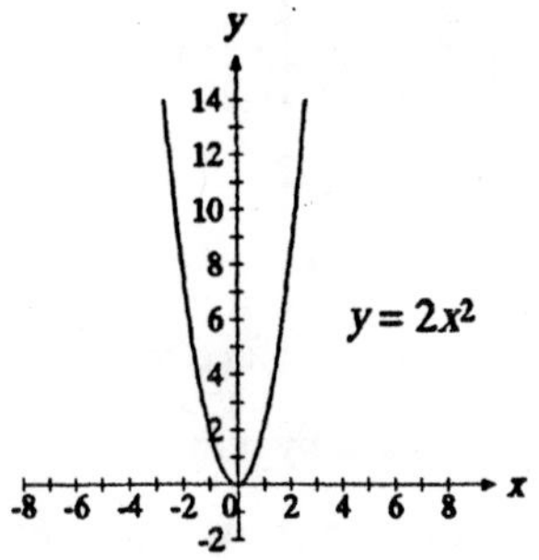

9

x	0	1	2	3	-1	-2	-3
$y = -x^2$	0	-1	-4	-9	-1	-4	-9

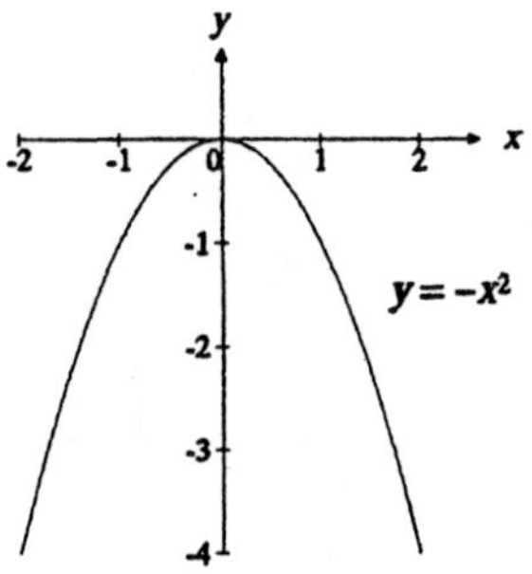

11

x	0	1	2	-1	-2
$y = x^3$	0	-1	-8	1	8

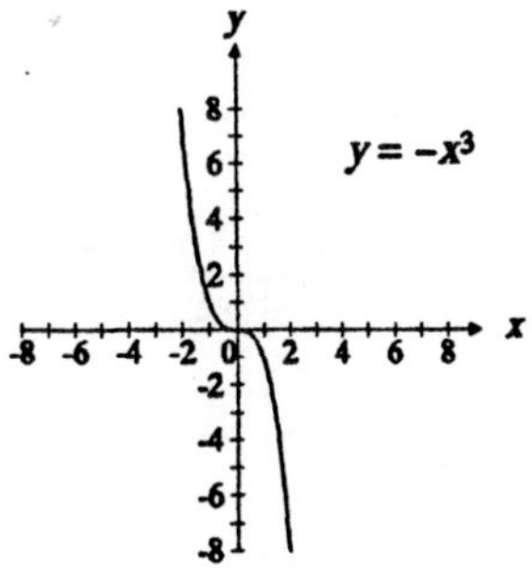

13

x	0	1/2	1	$-1/2$	-1
$y = x^4$	0	1/16	1	1/16	1

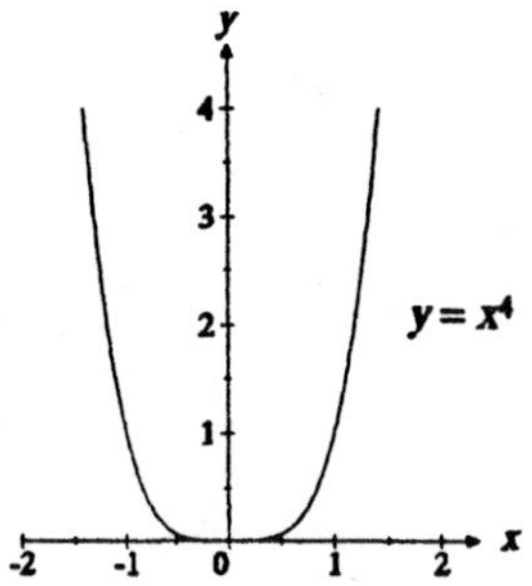

15 $y = x - 1$ is of the form $y = mx + b$, so it is the equation of the line. (See the review material in Sec. B.2.)

x	0	1	-1
$y = x - 1$	-1	0	-2

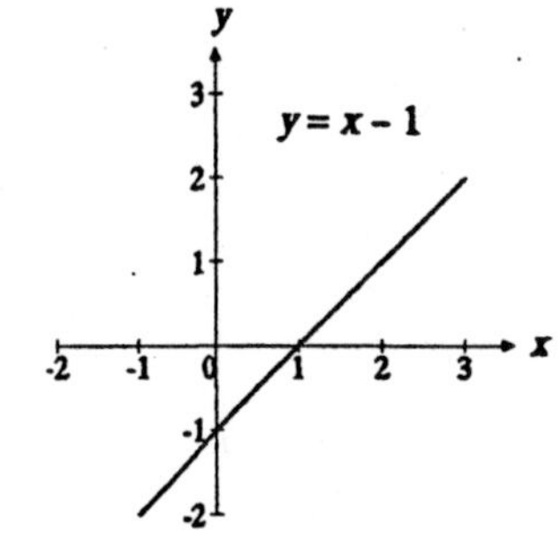

17 When $y = 0$, $0 = 2x + 6$, $-6 = 2x$, $x = -3$, so the x intercept is -3. When $x = 0$, $y = 2 \cdot 0 + 6 = 6$, so the y intercept is 6.

19 Since the product of x and y is 6, neither x nor y can be 0. Thus the hyperbola $xy = 6$ has no intercepts.

21 When $y = 0$, $0 = 2x^2 + 5x - 3$, $x = \dfrac{-5 \pm \sqrt{25 + 24}}{4} = \dfrac{-5 \pm \sqrt{49}}{4} = \dfrac{-5 \pm 7}{4}$. Thus

the x intercepts are $\dfrac{-5 - 7}{4} = -3$ and $\dfrac{-5 + 7}{4}$

$= \frac{1}{2}$. When $x = 0$, $y = 2 \cdot 0^2 + 5 \cdot 0 - 3 = -3$,

so the y intercept is -3.

23

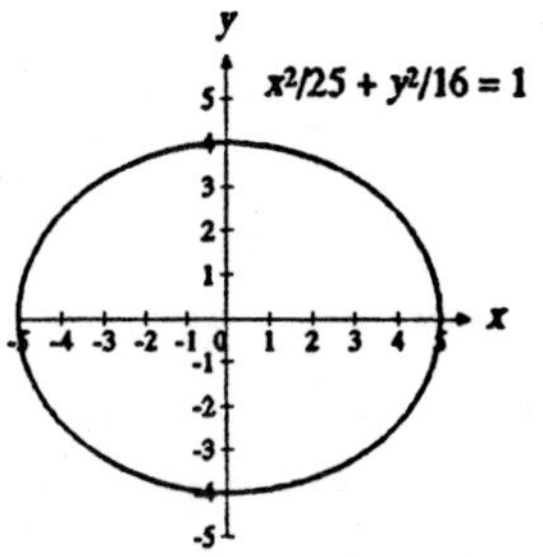

$\frac{x^2}{25} + \frac{y^2}{16} = 1$ is of the form $\frac{x^2}{a^2} + \frac{y^2}{b^2} = 1$, so it

is the equation of an ellipse. In this case $a^2 = 25$ and $b^2 = 16$, so $a = 5$ and $b = 4$. The x intercepts are therefore 5 and -5, while the y intercepts are 4 and -4.

25

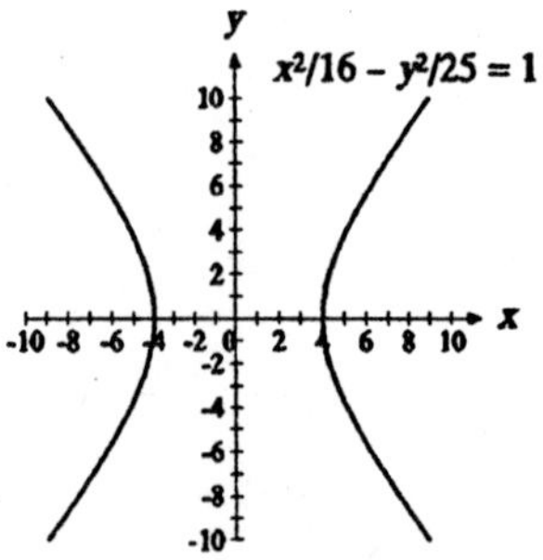

$\frac{x^2}{16} - \frac{y^2}{25} = 1$ is of the form $\frac{x^2}{a^2} - \frac{y^2}{b^2} = 1$, so it

is the equation of a hyperbola. For this case $a^2 = 16$ and $b^2 = 25$, so $a = 4$ and $b = 5$.

27

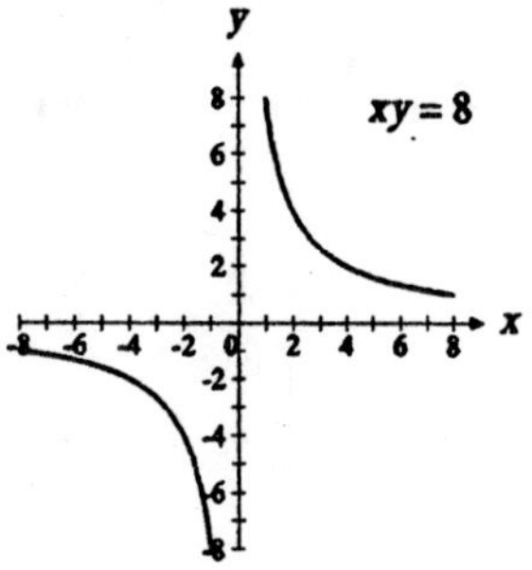

Since the product of x and y is 8, neither x nor y can be 0. Thus the hyperbola $xy = 8$ has no intercepts. For $x \neq 0$, we can solve for y in terms of x: $y = 8/x$.

x	1	2	3	4	8
$y = 8/x$	8	4	8/3	2	1

29

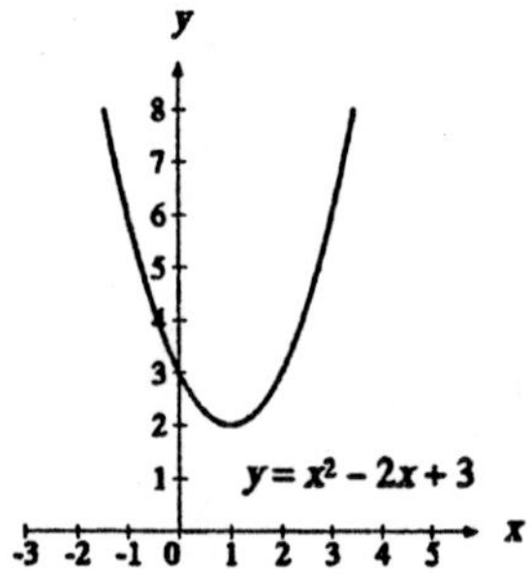

The graph is a parabola whose only intercept is a y intercept of (0, 3).

31

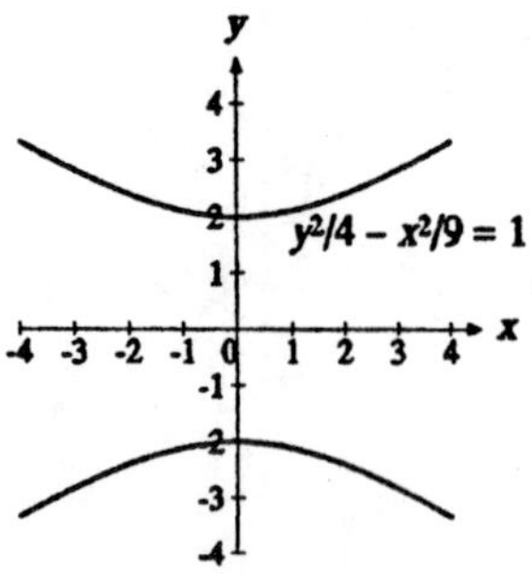

The graph is a hyperbola with intercepts on the y axis instead of the x axis.

33

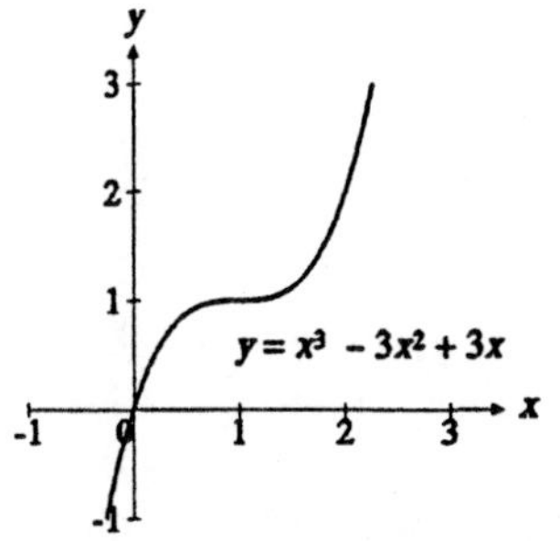

Note that this cubic passes through the origin and is symmetric about the point (1, 1).

35

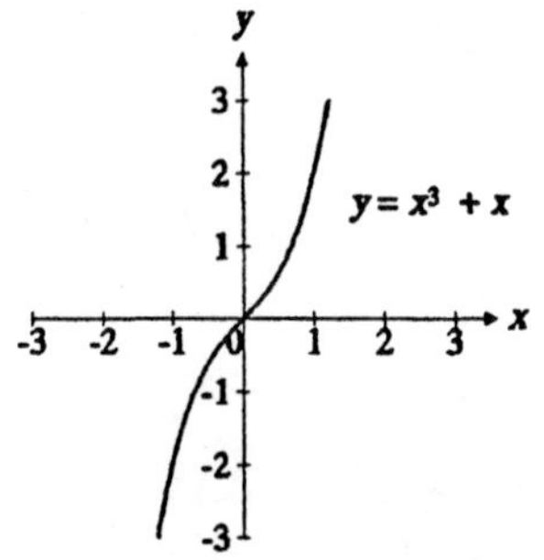

This cubic passes through the origin.

37

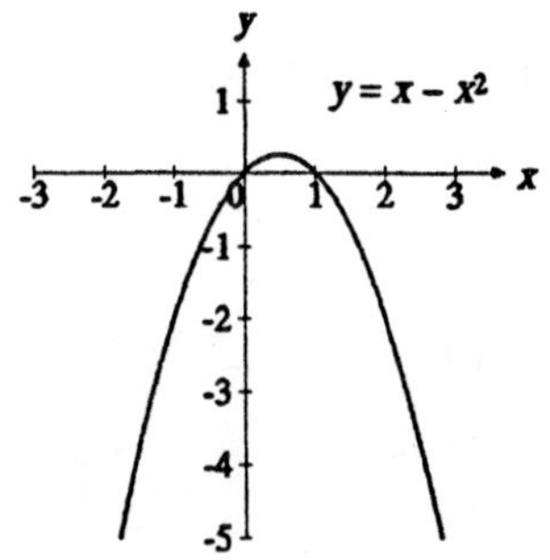

When $y = 0$, $0 = x - x^2 = x(1 - x)$, so 0 and 1 are x intercepts. When $x = 0$, $y = 0$, so 0 is the y intercept.

x	0	1/2	1	2	3	−1	−2
$y = x - x^2$	0	1/4	0	−2	−6	−2	−6

39

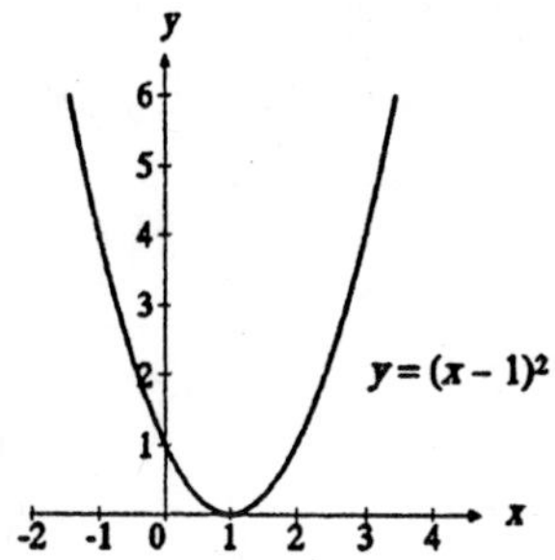

When $y = 0$, $0 = (x - 1)^2$, so 1 is the x intercept. When $x = 0$, $y = 1$, so 1 is the y intercept.

x	0	1	2	3	4	−1	−2
$y = (x - 1)^2$	1	0	1	4	9	4	9

41

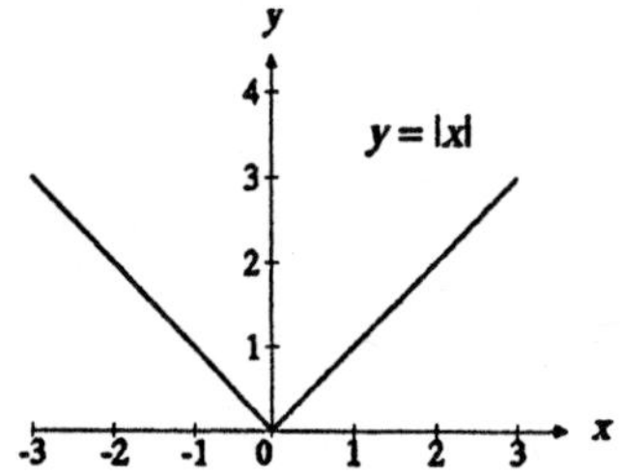

x	0	1	2	−1	−2
$y = \|x\|$	0	1	2	1	2

43

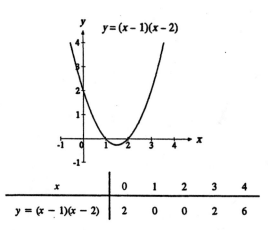

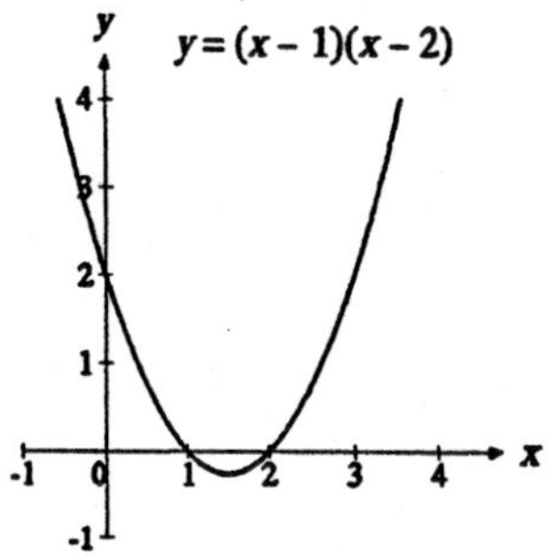

x	0	1	2	3	4
$y = (x - 1)(x - 2)$	2	0	0	2	6

When $y = 0$, $x = 1$ or $x = 2$, so 1 and 2 are the x intercepts. When $x = 0$, $y = (0 - 1)(0 - 2) = 2$, so 2 is the y intercept.

45 The circle whose center is (2, 1) and whose radius is 7 is the set of all points whose distance from (2, 1) is 7. For any point (x, y), the distance from (2, 1) is $\sqrt{(x - 2)^2 + (y - 1)^2}$. Thus, the equation for the circle is $\sqrt{(x - 2)^2 + (y - 1)^2} = 7$, which can be written as $(x - 2)^2 + (y - 1)^2 = 49$.

47 (a) Since $\frac{x^2}{a^2} - \frac{y^2}{b^2} = 1$, it follows that $y^2 = \frac{b^2}{a^2}(x^2 - a^2)$. Now a and b are assumed to be positive and we want to solve for the case where $y > 0$. Therefore $y = \frac{b}{a}\sqrt{x^2 - a^2}$.

(b) Note that $\frac{bx}{a} - \frac{b}{a}\sqrt{x^2 - a^2} = \frac{b\left(x - \sqrt{x^2 - a^2}\right)}{a}$. Following the hint, we have $\frac{b\left(x - \sqrt{x^2 - a^2}\right)}{a} \cdot \frac{x + \sqrt{x^2 - a^2}}{x + \sqrt{x^2 - a^2}} = \frac{b\left(x^2 - (x^2 - a^2)\right)}{a\left(x + \sqrt{x^2 + a^2}\right)} = \frac{ba^2}{a\left(x + \sqrt{x^2 - a^2}\right)} = \frac{ba}{x + \sqrt{x^2 - a^2}}$. For large values of x, the denominator $x + \sqrt{x^2 - a^2}$ is also a large number, making the quotient small. The bigger x is, the closer to 0 the quotient is. Since the quotient is the distance between the hyperbola and the line $y = bx/a$, this shows that the hyperbola approaches ever more closely to the line.

B.2 Lines and Their Slopes

1

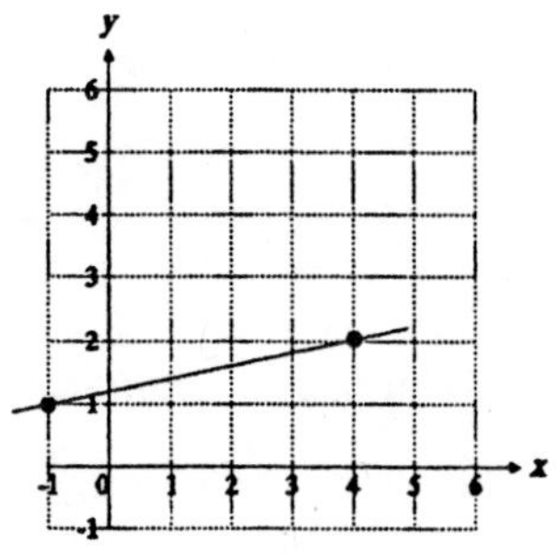

$$m = \frac{2 - 1}{4 - (-1)} = \frac{1}{5}$$

3

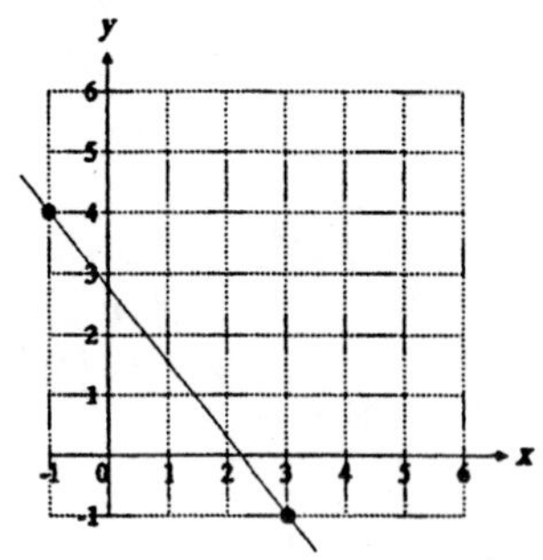

$$m = \frac{-1 - 4}{3 - (-1)} = -\frac{5}{4}$$

5

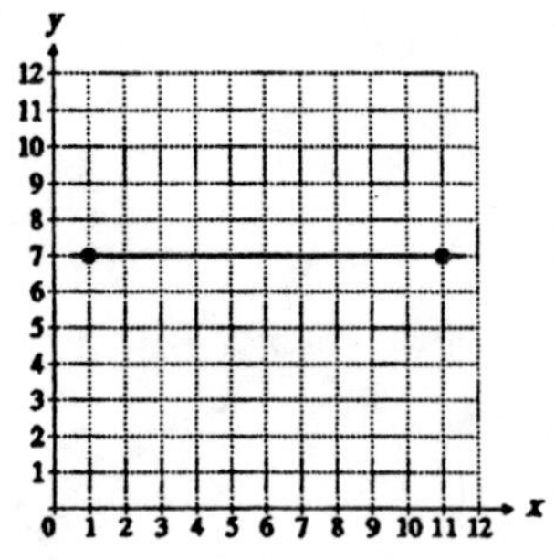

$$m = \frac{7 - 7}{11 - 1} = 0$$

7 (a) and (b) have negative slope since the y coordinate decreases as the x coordinate increases; (c) and (d) have positive slope since the y coordinate increases as the x coordinate increases; (e) has 0 slope, since it is parallel to the x axis.

9 (a) Any line parallel to L has the same slope as L, namely, 4.

(b) The slope of any line perpendicular to L is the negative reciprocal of the slope of L, namely, $-1/4$.

11 The slope of the first line is $\frac{7-4}{3-2} = 3$. The slope of the second line is $\frac{1-7}{-1-1} = 3$. Their slopes are the same, so the lines are parallel.

13 The slope of the first line is $\frac{6-0}{4-0} = \frac{6}{4} = \frac{3}{2}$.

The slope of the second line is $\frac{2-(-1)}{3-5} = -\frac{3}{2}$.

The product of the two slopes is $\frac{3}{2}\left(-\frac{3}{2}\right) = -\frac{9}{4}$ $\neq 1$, so the lines are not perpendicular.

15 (a) $b = 2$, $m = 3$, so the equation is $y = 3x + 2$.

(b) $b = -2$, $m = 2/3$, so the equation is $y = \frac{2}{3}x - 2$.

(c) $b = 0$, $m = -3$, so the equation is $y = -3x + 0$; that is, $y = -3x$.

17

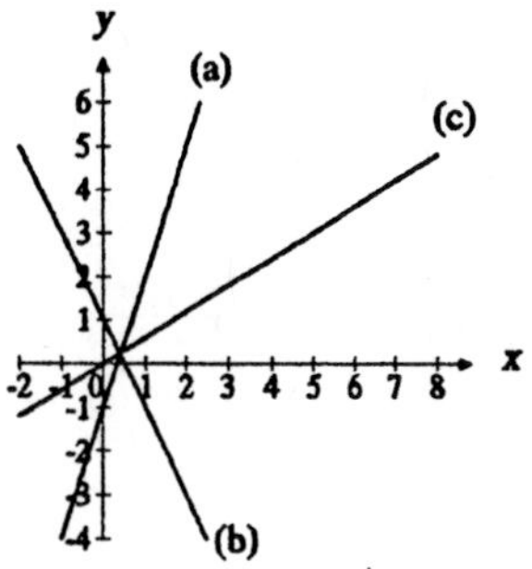

(a) The slope of $y = 3x - 1$ is 3 and the y intercept is -1.

(b) The slope of $y = -2x + 1$ is -2 and the y intercept is 1.

(c) The slope of $y = 3x/5$ is $3/5$ and the y intercept is 0.

19 (a) $y - 2 = 3(x - 1) = 3x - 3$, $y = 3x - 1$

(b) $y - (-1) = -2(x - 3) = -2x + 6$, $y + 1 = -2x + 6$, $y = -2x + 5$

21 (a) Slope $= \frac{3-2}{5-1} = \frac{1}{4}$, $y - 2 = \frac{1}{4}(x - 1) = \frac{1}{4}x - \frac{1}{4}$, $y = \frac{1}{4}x + \frac{7}{4}$

(b) Slope $= \frac{1-2}{3-(-1)} = -\frac{1}{4}$, $y - 2 = -\frac{1}{4}(x - (-1)) = -\frac{1}{4}x - \frac{1}{4}$, $y = -\frac{1}{4}x + \frac{7}{4}$

(c) Slope $= \frac{3-5}{2-4} = 1$, $y - 5 = 1(x - 4) = x - 4$, $y = x + 1$

23

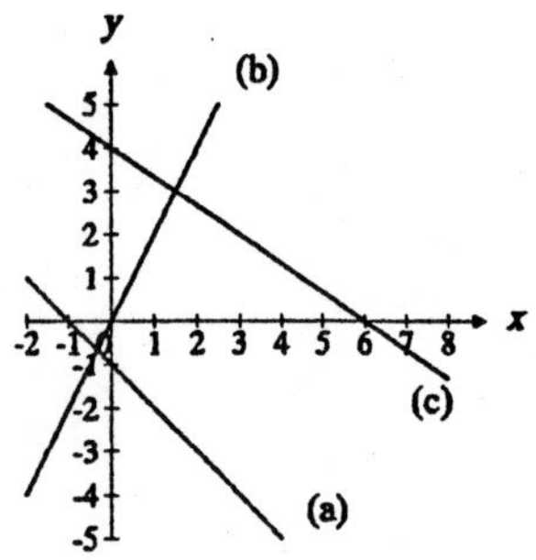

(a) $y = -x - 1$

(b) $y = 2x$

(c) $y = -\frac{2}{3}x + 4$

25 The line has slope $\frac{2-1}{7-4} = \frac{1}{3}$; the point-slope equation is $y - 1 = \frac{1}{3}(x - 4) = \frac{1}{3}x - \frac{4}{3}$, so $y = \frac{1}{3}x - \frac{1}{3}$. This line passes through a point (x, y) only if the above equation is satisfied.

(a) $\frac{1}{3}x - \frac{1}{3} = \frac{1}{3}\cdot 10 - \frac{1}{3} = \frac{9}{3} = 3 = y$, so the line passes through (10, 3).

(b) $\frac{1}{3}x - \frac{1}{3} = \frac{1}{3}\cdot 14 - \frac{1}{3} = \frac{13}{3} \neq 5$; the line does not pass through (14, 5).

27 The line through (4, 1) and (2, 5) has slope $\frac{5-1}{2-4} = -2$, so the slope of any line perpendicular to it is $-\frac{1}{-2} = \frac{1}{2}$. The point-slope formula yields $y - 3 = \frac{1}{2}(x - 1) = \frac{1}{2}x - \frac{1}{2}$, so $y = \frac{1}{2}x + \frac{5}{2}$.

29 Any line parallel to $y = 3x + 2$ has the same slope, namely 3. Using the slope-intercept form, the equation is $y = 3x + 5$.

30 When $x = 0$, we have $0/a + y/b = 1$, so $y/b = 1$, $y = b$. Thus, b is the y intercept. When $y = 0$, we have $x/a + 0/b = 1$, so $x = a$. Thus, a is the x intercept.

31 $\frac{x}{a} + \frac{y}{b} = 1$, $\frac{y}{b} = -\frac{x}{a} + 1$, $y = b\left(-\frac{x}{a} + 1\right) = -\frac{b}{a}x + b$, so the slope is $-\frac{b}{a}$.

33 (a) The equation of the line is $y = 4x - 7$. When $y = 0$, $0 = 4x - 7$, $4x = 7$, $x = 7/4$, so the x intercept is 7/4.

(b) The equation of the line is $(y - 3) = -\frac{2}{3}(x - 1) = -\frac{2}{3}x + \frac{2}{3}$, so $y = -\frac{2}{3}x + \frac{11}{3}$. When $y = 0$, $0 = -\frac{2}{3}x + \frac{11}{3}$, $\frac{2}{3}x = \frac{11}{3}$, $x = \frac{11}{2}$, so the x intercept is $\frac{11}{2}$.

(c) When $y = 0$, $\frac{x}{3} - \frac{0}{4} = 1$, $\frac{x}{3} = 1$, $x = 3$, so the x intercept is 3.

(d) When $y = 0$, $-2x + 3\cdot 0 = 12$, $-2x = 12$, $x = -6$, so the x intercept is -6.

35 (a) The slope of L is $\frac{y_2 - y_1}{x_2 - x_1}$.

(b) The slope of L' is $\frac{(y_2 + d) - (y_1 + d)}{x_2 - x_1} = \frac{y_2 - y_1}{x_2 - x_1}$. L and L' have the same slope.

37 (a)

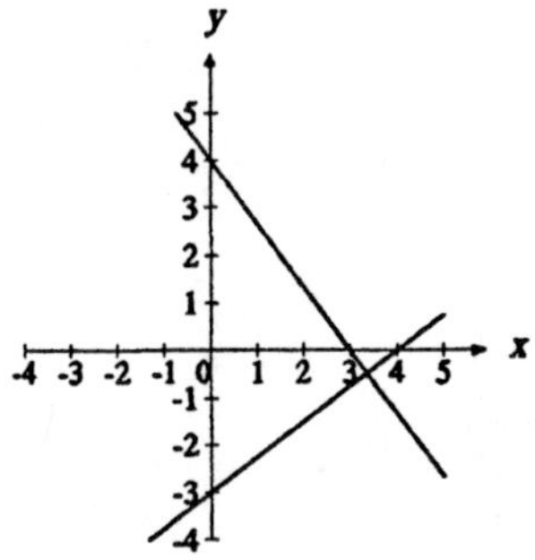

(b) By Exercise 31, the slopes are $-4/3$ and $3/4$, respectively.

(c) $\left(-\frac{4}{3}\right)\left(\frac{3}{4}\right) = -1$, so the lines are perpendicular.

39 (a) $|m|$ is very large for a line that is nearly vertical.

(b) $|m|$ is close to zero for a line that is nearly horizontal.

(c) m is negative.

(d) m is positive.

41 Let L_1 and L_2 be the lines through P and $(1, 1)$ and P and $(-3, 4)$, respectively. Find $P = (a, 0)$ so that L_1 is perpendicular to L_2. The slope of L_1 is $m_1 = \frac{1 - 0}{1 - a}$, and the slope of L_2 is $m_2 = \frac{4 - 0}{-3 - a}$.

We need $m_1 m_2 = -1$, so $\left(\frac{1}{1 - a}\right)\left(\frac{4}{-3 - a}\right) = -1$, or $4 = (1 - a)(3 + a) = 3 - 2a - a^2$, so $a^2 + 2a + 1 = (a + 1)^2 = 0$. Hence $a = -1$, so $P = (-1, 0)$ is the desired point.

C Topics in Algebra

1 $\frac{10}{\sqrt{5}} = \frac{10}{\sqrt{5}} \cdot \frac{\sqrt{5}}{\sqrt{5}} = \frac{10\sqrt{5}}{5} = 2\sqrt{5}$

3 $\frac{2 + 4\sqrt{2}}{\sqrt{2}} \cdot \frac{\sqrt{2}}{\sqrt{2}} = \frac{2\sqrt{2} + 4 \cdot 2}{2} = \sqrt{2} + 4$

5 $\frac{4}{3 - \sqrt{3}} = \frac{4}{3 - \sqrt{3}} \cdot \frac{3 + \sqrt{3}}{3 + \sqrt{3}} = \frac{12 + 4\sqrt{3}}{9 - 3}$

$= \frac{6 + 2\sqrt{3}}{3}$

7 $\frac{x}{1 - \sqrt{x}} = \frac{x}{1 - \sqrt{x}} \cdot \frac{1 + \sqrt{x}}{1 + \sqrt{x}} = \frac{x(1 + \sqrt{x})}{1 - x}$

9 $\frac{3 + \sqrt{2}}{5} = \frac{3 + \sqrt{2}}{5} \cdot \frac{3 - \sqrt{2}}{3 - \sqrt{2}} = \frac{9 - 2}{5(3 - \sqrt{2})}$

$= \frac{7}{5(3 - \sqrt{2})}$

11 $\frac{\sqrt{x} - \sqrt{5}}{x - 5} = \frac{\sqrt{x} - \sqrt{5}}{(\sqrt{x} + \sqrt{5})(\sqrt{x} - \sqrt{5})} = \frac{1}{\sqrt{x} + \sqrt{5}}$

13 (a) $x^2 + 8x + 13 = x^2 + 8x + 16 - 16 + 13$
$= (x + 4)^2 - 3$

(b) $x^2 - 8x + 23 = x^2 - 8x + 16 - 16 + 23$
$= (x - 4)^2 + 7$

(c) $x^2 - x + 2 = x^2 - x + \frac{1}{4} - \frac{1}{4} + 2$
$= \left(x - \frac{1}{2}\right)^2 + \frac{7}{4}$

15 (a) $x^2 + 3x - 2 = x^2 + 3x + \frac{9}{4} - \frac{9}{4} - 2$
$= \left(x + \frac{3}{2}\right)^2 - \frac{17}{4}$

(b) $x^2 + 3x + 7 = x^2 + 3x + \frac{9}{4} - \frac{9}{4} + 7$

$= \left(x + \frac{3}{2}\right)^2 + \frac{19}{4}$

(c) $x^2 + \frac{5}{2}x + 4 = x^2 + \frac{5}{2}x + \frac{25}{16} - \frac{25}{16} + 4$

$= \left(x + \frac{5}{4}\right)^2 + \frac{39}{16}$

17 (a) $2x^2 - 5x + 3 = 2\left(x^2 - \frac{5}{2}x\right) + 3$

$= 2\left(x^2 - \frac{5}{2}x + \frac{25}{16} - \frac{25}{16}\right) + 3 =$

$2\left(x - \frac{5}{4}\right)^2 - 2\cdot\frac{25}{16} + 3 = 2\left(x - \frac{5}{4}\right)^2 - \frac{1}{8}$

(b) $2x^2 + 6x + 7 = 2(x^2 + 3x) + 7$

$= 2\left(x^2 + 3x + \frac{9}{4} - \frac{9}{4}\right) + 7$

$= 2\left(x + \frac{3}{2}\right)^2 - 2\left(\frac{9}{4}\right) + 7 = 2\left(x + \frac{3}{2}\right)^2 + \frac{5}{2}$

(c) $3x^2 + 5x + 1 = 3\left(x^2 + \frac{5}{3}x\right) + 1$

$= 3\left(x^2 + \frac{5}{3}x + \frac{25}{36} - \frac{25}{36}\right) + 1 =$

$3\left(x + \frac{5}{6}\right)^2 - 3\left(\frac{25}{36}\right) + 1 = 3\left(x + \frac{5}{6}\right)^2 - \frac{13}{12}$

19 (a) $x^2 + x + 1 = 0$, so $x = \frac{-1 \pm \sqrt{1^2 - 4\cdot1\cdot1}}{2}$

$= \frac{-1 \pm \sqrt{-3}}{2}$, but $\sqrt{-3}$ is undefined, so the equation has no real roots.

(b) $x^2 + x - 1 = 0$, so $x =$

$\frac{-1 \pm \sqrt{1^2 - 4\cdot1\cdot(-1)}}{2} = \frac{-1 \pm \sqrt{5}}{2}$

(c) $x^2 + 2x + 1 = 0$, so $x = \frac{-2 \pm \sqrt{4 - 4\cdot1\cdot1}}{2}$

$= \frac{-2 \pm 0}{2} = -1.$

21 (a) $2x^2 + 5x + 6 = 0$, so $b^2 - 4ac =$ $5^2 - 4\cdot2\cdot6 = 25 - 48 = -23$, which is negative, so there exist no real roots.

(b) $2x^2 + 5x + 2 = 0$, so $b^2 - 4ac =$ $5^2 - 4\cdot2\cdot2 = 25 - 16 = 9$, which is positive, so there are two distinct real roots.

(c) $4x^2 - 12x + 9 = 0$, so $b^2 - 4ac =$ $(-12)^2 - 4\cdot4\cdot9 = 144 - 144 = 0$; hence there is precisely one real root.

23 (a) The coefficient of x^2 in $(1 + x)^5$ is

$\binom{5}{2} = \frac{5!}{2!3!} = \frac{5\cdot4}{1\cdot2} = 10.$

(b) The coefficient of x^2 in $(1 + x)^6$ is

$\binom{6}{2} = \frac{6!}{2!4!} = \frac{6\cdot5}{1\cdot2} = 15.$

(c) The coefficient of x^2 in $(1 + x)^{10}$ is

$\binom{10}{2} = \frac{10!}{2!8!} = \frac{10\cdot9}{1\cdot2} = 45.$

25 $(1 + x)^7 = \binom{7}{0} + \binom{7}{1}x + \binom{7}{2}x^2 + \binom{7}{3}x^3 +$

$\binom{7}{4}x^4 + \binom{7}{5}x^5 + \binom{7}{6}x^6 + \binom{7}{7}x^7 = 1 + 7x +$

$21x^2 + 35x^3 + 35x^4 + 21x^5 + 7x^6 + x^7$

27 The coefficient of a^8b^2 in $(a+b)^{10}$ is $\binom{10}{2}$

$$= \frac{10!}{2!8!} = \frac{10\cdot 9}{1\cdot 2} = 45.$$

29 (a) $1 + \frac{1}{3} + \frac{1}{3^2} + \frac{1}{3^3} + \cdots + \frac{1}{3^6}$

$$= \frac{1\cdot(1-(1/3)^7)}{1-1/3} = \frac{3}{2}\left(1-\left(\frac{1}{3}\right)^7\right)$$

$$= \frac{3}{2}\left(1-\frac{1}{2187}\right) = \frac{3}{2}\cdot\frac{2186}{2187} = \frac{1093}{729}$$

(b) $5 + 5\cdot 2 + 5\cdot 2^2 + \cdots + 5\cdot 2^7 = \frac{5\cdot(1-2^8)}{1-2}$

$$= -5(1-256) = 1275$$

(c) $6 - 6\left(\frac{1}{2}\right) + 6\left(\frac{1}{2}\right)^2 - 6\left(\frac{1}{2}\right)^3 + 6\left(\frac{1}{2}\right)^4$

$$= \frac{6\cdot\left(1-\left(-\frac{1}{2}\right)^5\right)}{1-\left(-\frac{1}{2}\right)} = 4\left(1-\left(-\frac{1}{32}\right)\right) = 4\cdot\frac{33}{32}$$

$$= \frac{33}{8}$$

31 By the technique of Example 11, if $a = b/c$ is a root of $x^3 + 2x^2 - x - 2 = 0$ and b and c have no common divisor larger than 1, then b is a divisor of 2 and c is a divisor of 1. Hence $a = -2, -1, 1,$ or 2. Upon substitution we find that $-2, -1,$ and 1 are roots, so the desired factors are $x - (-2)$, $x - (-1)$, and $x - 1$.

33 As in Exercise 31, we must have $a = -2, -1, 1,$ or 2. Upon substitution, only 1 proves to be a root, so the only linear factor of the desired form is $x - 1$.

35 Note that -1 is a root. (It's a good idea to look for simple roots like 0, 1, and -1 first.) Factoring out $x - (-1) = x + 1$, we find that $6x^3 - 7x^2 - 7x + 6 = (x+1)(6x^2 - 13x + 6)$. By the quadratic formula, the roots of $6x^2 - 13x + 6$ are

$$\frac{13 \pm \sqrt{13^2 - 4\cdot 6\cdot 6}}{2\cdot 6} = \frac{13 \pm 5}{12} = \frac{2}{3} \text{ and } \frac{3}{2}.$$

Hence the desired factors are $x - (-1)$, $x - 2/3$, and $x - 3/2$.

37 The roots are $\frac{-5 \pm \sqrt{5^2 - 4\cdot 1\cdot 2}}{2\cdot 1} = \frac{-5 \pm \sqrt{17}}{2}$

so $x^2 + 5x + 2$ is divisible by $x + \frac{5+\sqrt{17}}{2}$ and $x + \frac{5-\sqrt{17}}{2}$. In fact it is their product:

$$x^2 + 5x + 2 = \left(x + \frac{5+\sqrt{17}}{2}\right)\left(x + \frac{5-\sqrt{17}}{2}\right).$$

39 $\frac{3}{10} + \frac{3}{10^2} + \frac{3}{10^3} + \cdots + \frac{3}{10^8}$

$$= \frac{\frac{3}{10}\cdot(1-(1/10)^8)}{1-1/10} = \frac{1}{3}(1-10^{-8})$$

D Exponents

1 (a) $32^0 = 1$

(b) $32^1 = 32$

(c) $32^{1/5} = (2^5)^{1/5} = 2$

(d) $32^{-1/5} = \frac{1}{32^{1/5}} = \frac{1}{2}$

(e) $32^{3/5} = (2^5)^{3/5} = 2^3 = 8$

3

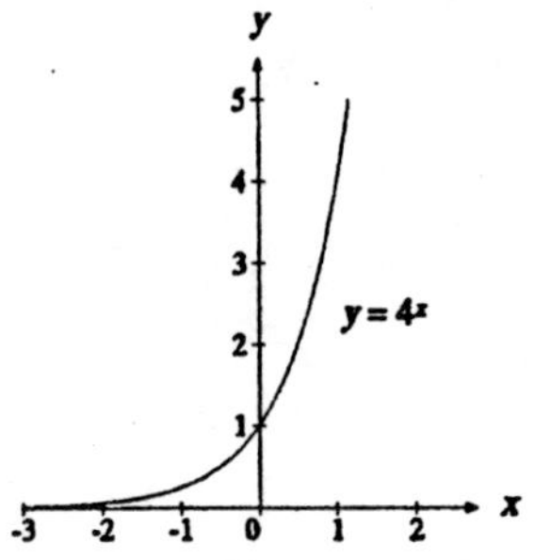

5

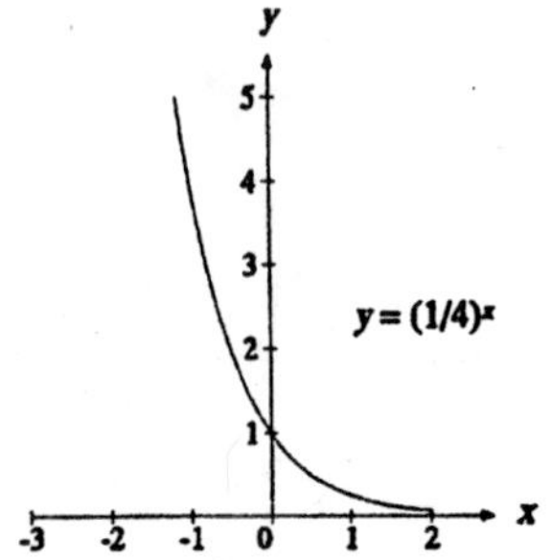

7 (a) $16 = 2^4$

(b) $1/8 = 1/2^3 = 2^{-3}$

(c) $\sqrt{2} = 2^{1/2}$

(d) $1 = 2^0$

(e) $0.25 = 1/4 = 1/2^2 = 2^{-2}$

9 (a) $b\sqrt{b} = b^1 b^{1/2} = b^{3/2}$

(b) $(\sqrt{b})^3 = (b^{1/2})^3 = b^{3/2}$

(c) $\dfrac{1}{\sqrt[3]{b}} = \dfrac{1}{b^{1/3}} = b^{-1/3}$

(d) $\sqrt[3]{b^2} = (b^2)^{1/3} = b^{2/3}$

(e) $\sqrt{\sqrt{b}} = (b^{1/2})^{1/2} = b^{1/4}$

11 (a) 2^x is defined for all x while its values consist of all positive numbers, so its domain is all x while its range is $y > 0$.

(b) $x^{1/3}$ is the cube root function; every number has a cube root and every number is a cube root, so the domain is all x and the range is all y.

(c) $x^{1/4}$ is the fourth-root function; only nonnegative numbers have fourth roots, and only nonnegative numbers can be obtained as fourth roots, so the domain is $x \geq 0$ and the range is $y \geq 0$.

(d) $x^{2/3}$ may be regarded as the square of the cube root function; since every number has a cube root and every number can be squared, the domain is all x. Only nonnegative numbers may be obtained as two-thirds powers, so the range is $y \geq 0$.

13 (a) $(-32)^{3/5} = ((-2)^5)^{3/5} = (-2)^3 = -8$; $(-1)^{3/5} = ((-1)^3)^{1/5} = (-1)^{1/5} = -1$; $0^{3/5} = 0$; $1^{3/5} = 1$; $32^{3/5} = (2^5)^{3/5} = 2^3 = 8$.

(b)

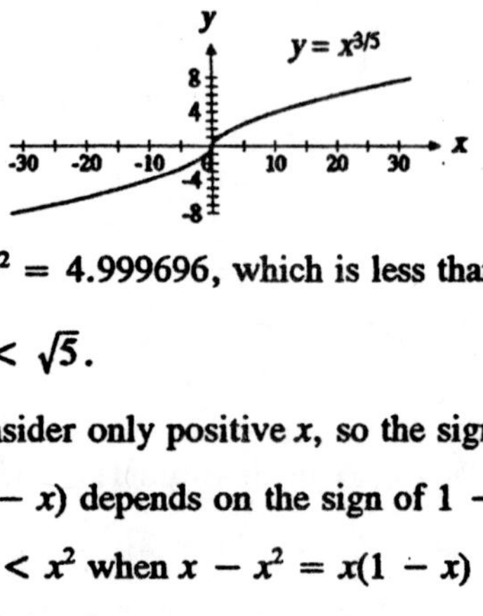

15 $(2.236)^2 = 4.999696$, which is less than 5, so $2.236 < \sqrt{5}$.

17 We consider only positive x, so the sign of $x - x^2 = x(1 - x)$ depends on the sign of $1 - x$.

(a) $x < x^2$ when $x - x^2 = x(1 - x) < 0$, which occurs for $x > 1$.

(b) $x = x^2$ when $x - x^2 = x(1 - x) = 0$, which occurs for $x = 1$.

(c) $x > x^2$ when $x - x^2 = x(1 - x) > 0$, which occurs for $0 < x < 1$.

19

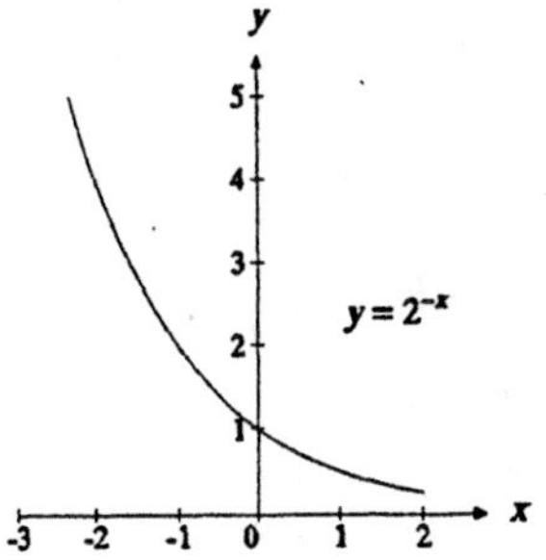

21 (a) $1000 = 10^3$

(b) $0.0001 = 10^{-4}$

(c) $1,000,000 = 10^6$

(d) $0.0000001 = 10^{-7}$

23 $(64^2)^{1/3} = (4096)^{1/3} = 16$, while $(64^{1/3})^2 = (4)^2 = 16$; most people would find the second calculation simpler.

25 (a) $\left(\frac{15}{11}\right)^2 = \frac{225}{121} < \frac{242}{121} = 2$, so $\frac{15}{11} < \sqrt{2}$.

$\left(\frac{16}{11}\right)^2 = \frac{256}{121} > \frac{242}{121} = 2$, so $\sqrt{2} < \frac{16}{11}$.

That is, $\frac{15}{11} < \sqrt{2} < \frac{16}{11}$.

(b) $(-1)^{15/11} = ((-1)^{15})^{1/11} = (-1)^{1/11} = -1$;
$(-1)^{16/11} = ((-1)^{16})^{1/11} = 1^{1/11} = 1$.

(c) As (b) demonstrates, $(-1)^{m/n} = -1$ when m and n are odd, while $(-1)^{m/n} = 1$ when m is even and n is odd. Therefore the "value" of $(-1)^{\sqrt{2}}$ jumps between two distinct possibilities for rational numbers near $\sqrt{2}$ and cannot be unambiguously defined.

E Mathematical Induction

1 Suppose that $S_k = 1 + 2 + 3 + \cdots + k = k(k+1)/2$. The formula works for $S_1 = 1$ and $S_2 = 1 + 2 = 2(3)/2 = 3$. Now $S_{k+1} = 1 + 2 + 3 + \cdots + k + (k+1) = \frac{k(k+1)}{2} + k + 1 =$

$\frac{k^2+k}{2} + \frac{2k+2}{2} = \frac{k^2+3k+2}{2} =$

$\frac{(k+1)(k+2)}{2}$, which shows that the formula works for $k + 1$. By mathematical induction, the formula is valid for all positive integers.

3 (a) By splitting the quadrilateral into two triangles by connecting any opposite pair of vertices, we see that the interior angles add up to 2π radians.

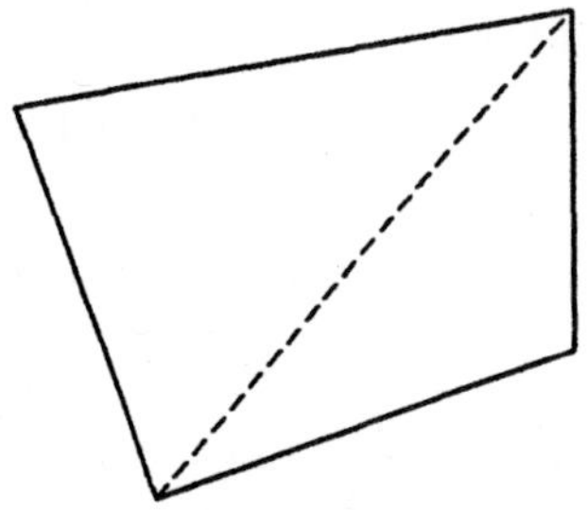

(b) An n-gon can be split into $n - 2$ triangles by connecting any chosen vertex to each of the other vertices, excepting its two immediate neighbors. The sum of the interior angles of the n-gon is equal to the sum of the angles of the $n - 2$ triangles, so it equals $(n - 2)\pi$.

(c) The formula from (b) works for $n = 3$. Assume it works for $n = k$. Now consider a $(k + 1)$-gon and pick one of its vertices.

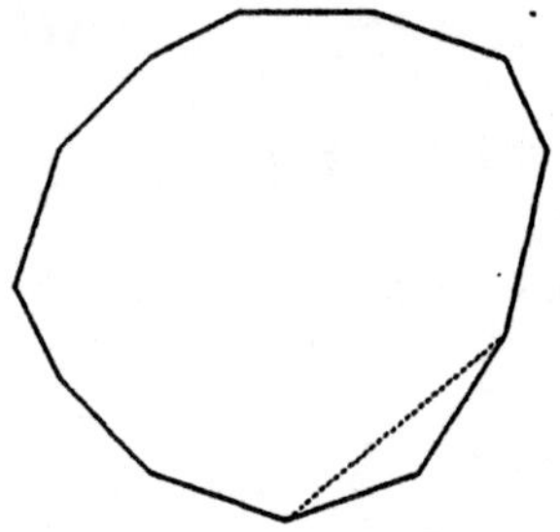

Connect the chosen vertex to a vertex immediately adjacent to either of the chosen vertex's two neighbors. (See the figure.) This splits the $(k + 1)$-gon into a k-gon and a triangle. The sum of the interior angles of the $(k + 1)$-gon equals the sum of the angles for the triangle, which we know to be π, and the angles for the k-gon, which by assumption is $(k - 2)\pi$. Hence the sum of the angles of the $(k + 1)$-gon is $\pi + (k - 2)\pi = (k - 1)\pi$, so that the formula works for $k + 1$. Hence it works for all $k \geq 3$.

5 Exploration exercises are in the *Instructor's Manual.*

7 Let P_k be the assertion that $D(1/x^k) = -k/x^{k+1}$.

Obviously P_0 is true. If P_k is true then $D\left(\frac{1}{x^{k+1}}\right) =$

$$D\left(\frac{1/x^k}{x}\right) = \frac{xD\left(\frac{1}{x^k}\right) - \frac{1}{x^k}D(x)}{x^2}$$

$$= \frac{x\left(-\frac{k}{x^{k+1}}\right) - \frac{1}{x^k}\cdot 1}{x^2} = \frac{1}{x^2}\left(-\frac{k}{x^k} - \frac{1}{x^k}\right) = -\frac{k+1}{x^{k+2}}$$

$= -\dfrac{k + 1}{x^{(k+1)+1}}$, so P_{k+1} is true. Hence P_k is true for all $k \geq 0$.

9 Let P_k be the statement that $D^k(e^{2x}) = 2^k e^{2x}$.

Clearly P_1 is true. If P_k is true, then $D^{k+1}(e^{2x}) = D(D^k(e^{2x})) = D(2^k e^{2x}) = 2^k D(e^{2x}) = 2^k \cdot 2e^{2x} = 2^{k+1}e^{2x}$, so P_{k+1} is true. Hence P_k is true for all positive k.

F The Converse of a Statement

1 (a) The converse is "If $a^2 = b^2$, then $a = b$." This is false: Let $a = 1$ and $b = -1$.

(b) The converse is "If $a^3 = b^3$, then $a = b$." This is true, since the function x^3 is increasing.

3 (a) The converse is "If $ab = ac$, then $b = c$." This is false: Let $a = 0$, $b = 1$, and $c = 2$.

(b) The converse is "If $a + b = a + c$, then $b = c$." This is true: If $a + b = a + c$, then $b = (a + b) - a = (a + c) - a = c$.

5 (a) The converse is "If ab is odd, then a and b are odd." This is true: If either a or b is even, then ab would be even.

(b) The converse is "If ab is even, then a and b are even." This is false: Let $a = 0$ and $b = 1$.

7 The converse is true. If the sides are a, b, and c and the angles all equal θ, then, by the law of sines, $\dfrac{a}{\sin\theta} = \dfrac{b}{\sin\theta} = \dfrac{c}{\sin\theta}$ so $a = b = c$.

9 The converse is true. Let the quadrilateral be $ABCD$. Since $\overline{AB} = \overline{CD}$, $\overline{BC} = \overline{AD}$, and $\overline{AC} = \overline{AC}$, triangles ABC and CDA are congruent. Hence angles CAD and ACB are equal, so AD is parallel

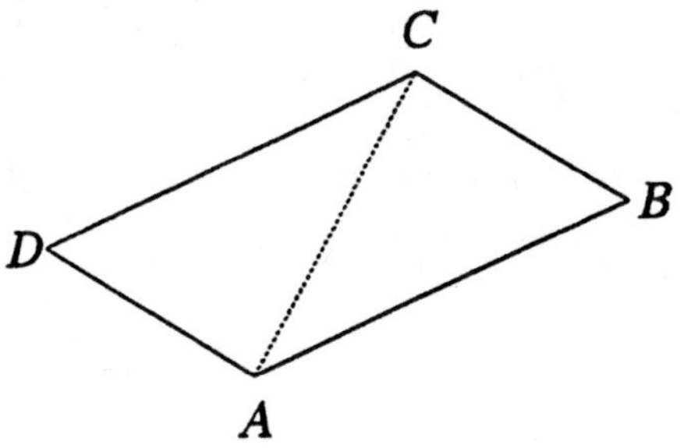

to BC. Similarly, AB is parallel to CD.

11 The converse is true. If $x - a$ divides $P(x)$, then $P(x) = (x - a)Q(x)$ for some polynomial Q. Hence $P(a) = (a - a)Q(a) = 0$.

13 The converse is false. Let $f(x) = x^2$ and $g(x) = e^x$. Then $(f \circ g)(x) = (e^x)^2 = e^{2x}$ is one-to-one, but f is not one-to-one.

15 The converse is true. If $f(x)$ is constant, then $f'(x) = 0$.

17 The converse is false. Let $f(x) = x + 1$. Then $f^{(3)}(x) = 0$ but f does not have degree 2. (It is true that if $f^{(3)}(x) = 0$ then f is a polynomial of degree *at most* 2.)

19 The converse is false. Let $f(x) = x^3$ and $a = 0$. Then $f'(a) = 0$, but f has neither a maximum nor a minimum at a.

21 The converse is false. Let $a_n = 1/n$. Then $\lim_{n\to\infty} a_n = 0$, but $\sum_{n=1}^{\infty} a_n = \sum_{n=1}^{\infty} \frac{1}{n}$ diverges.

23 The converse is true, by Theorem 2 in Sec. 16.6.

G Conic Sections

G.1 Conic Sections

1 $c = 2$ and $2a = 10$, so $a = 5$, $b^2 = a^2 - c^2 = 25 - 4 = 21$, and the equation is $\frac{x^2}{25} + \frac{y^2}{21} = 1$.

3

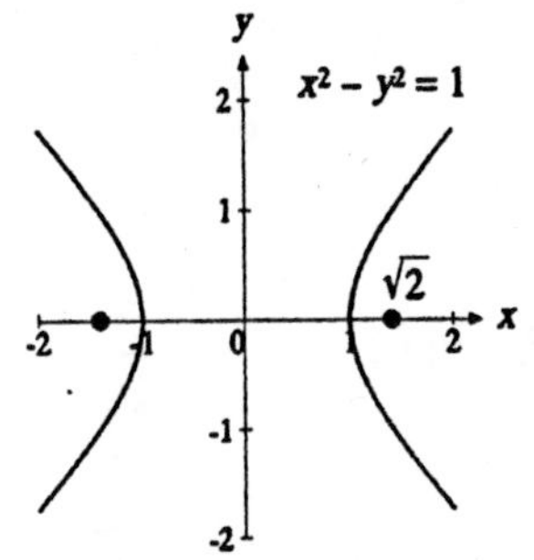

$x^2 - y^2 = 1$ is of the form $\frac{x^2}{a^2} - \frac{y^2}{b^2} = 1$ with $a = b = 1$, so $c = \sqrt{a^2 + b^2} = \sqrt{2}$.

5

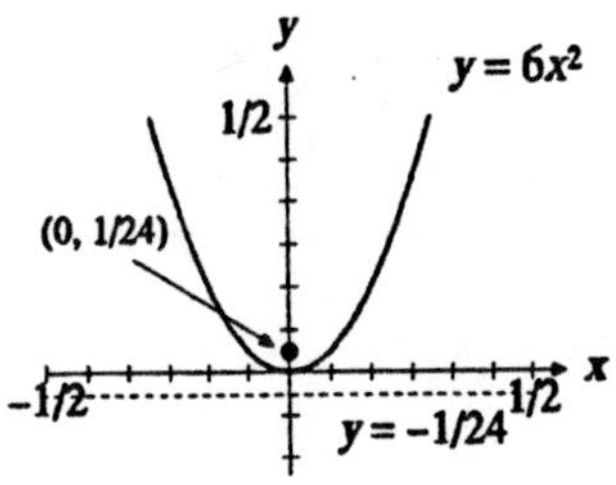

$y = 6x^2$, so $x^2 = \frac{1}{6}y = 2\left(\frac{1}{12}\right)y$; that is, $c = \frac{1}{12}$ in the equation $x^2 = 2cy$. Therefore the focus is at $\left(0, \frac{c}{2}\right) = \left(0, \frac{1}{24}\right)$ and the directrix is $y = -\frac{c}{2} = \frac{1}{24}$.

7 The equation is of the form $y^2 = 2cx$; the focus is at (3, 0), so $c/2 = 3$ or $c = 6$. Therefore $y^2 = 12x$.

9 Adding $\sqrt{(x + c)^2 + y^2}$ to both sides of (4) and squaring yields $(x - c)^2 + y^2 = (x + c)^2 + y^2 + 4a^2 \pm 4a\sqrt{(x + c)^2 + y^2}$. Simplifying leads to $xc + a^2 = \pm a\sqrt{(x + c)^2 + y^2}$; squaring now yields $x^2c^2 + 2a^2xc + a^4 = a^2x^2 + 2a^2xc + a^2c^2 + a^2y^2$. Cancelling $2a^2xc$ and subtracting $x^2c^2 + a^2c^2$ from both sides $a^4 - a^2c^2 = a^2x^2 - c^2x^2 + a^2y^2$; that is, $a^2(a^2 - c^2) = (a^2 - c^2)x^2 + a^2y^2$. Dividing by $a^2(a^2 - c^2)$ gives $1 = \dfrac{x^2}{a^2} + \dfrac{y^2}{a^2 - c^2}$, from which (5) follows.

11

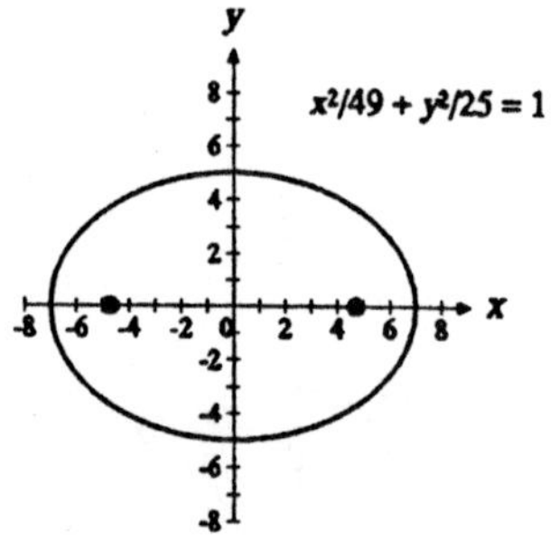

Since $49 > 25$, the foci are on the x axis at $x = \pm\sqrt{49 - 25} = \pm 2\sqrt{6}$.

13

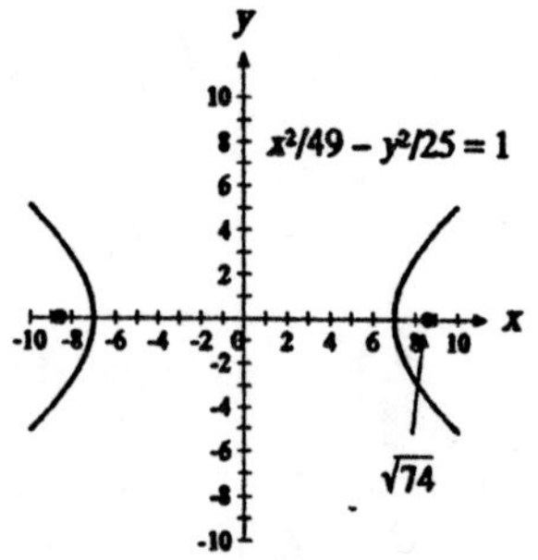

As in Example 3, the minus sign is with y^2, so the foci are on the x axis. Note that $a = 7$ and $b = 5$, so $c = \sqrt{7^2 + 5^2} = \sqrt{74}$. (It's helpful to sketch in the asymptotes, as in Example 6 of Sec. B.1, to get the shape of the hyperbola right.)

15

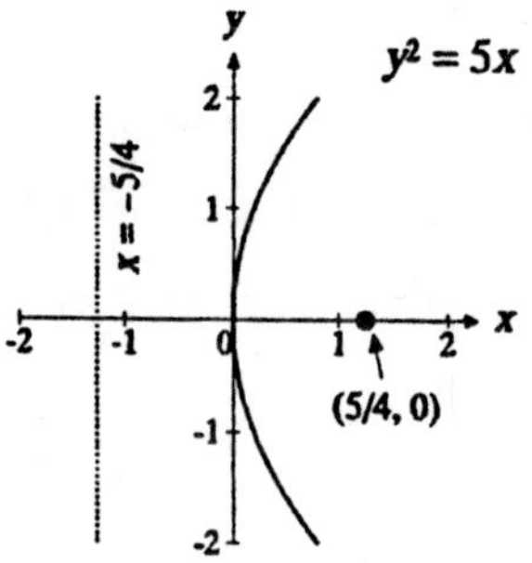

$y^2 = 2\cdot\frac{5}{2}x$, so $c = 5/2$. The focus is at (5/4, 0). The directrix is $x = -5/4$.

17

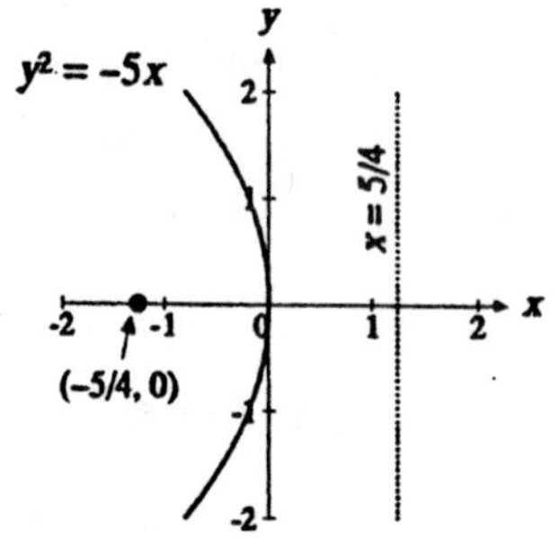

This is the parabola of Exercise 15 reflected in the y axis. The focus is (−5/4, 0). The directrix is $x = 5/4$.

19 (a) The distance from (x, y) to $(\sqrt{2}, \sqrt{2})$ is $\sqrt{(x - \sqrt{2})^2 + (y - \sqrt{2})^2}$ and from (x, y) to $(-\sqrt{2}, -\sqrt{2})$ it is $\sqrt{(x + \sqrt{2})^2 + (y + \sqrt{2})^2}$, so $2\sqrt{2} = \sqrt{(x - \sqrt{2})^2 + (y - \sqrt{2})^2} + \sqrt{(x + \sqrt{2})^2 + (y + \sqrt{2})^2}$. Thus

$2\sqrt{2} - \sqrt{(x - \sqrt{2})^2 + (y - \sqrt{2})^2} =$

$\sqrt{(x + \sqrt{2})^2 + (y + \sqrt{2})^2}$, so $8 - 4\sqrt{2}\sqrt{(x - \sqrt{2})^2 + (y - \sqrt{2})^2} + (x - \sqrt{2})^2 + (y - \sqrt{2})^2 = (x + \sqrt{2})^2 + (y + \sqrt{2})^2$.

Simplifying yields

$8 - 4\sqrt{2}\sqrt{(x - \sqrt{2})^2 + (y - \sqrt{2})^2} = 4\sqrt{2}(x + y)$, so $\sqrt{(x - \sqrt{2})^2 + (y - \sqrt{2})^2} = \sqrt{2} - x - y$. Squaring again gives $x^2 - 2\sqrt{2}x + 2 + y^2 - 2\sqrt{2}y + 2 = 2 + x^2 + y^2 - 2\sqrt{2}x - 2\sqrt{2}y + 2xy$, so $2 = 2xy$, or $xy = 1$, as claimed.

(b)

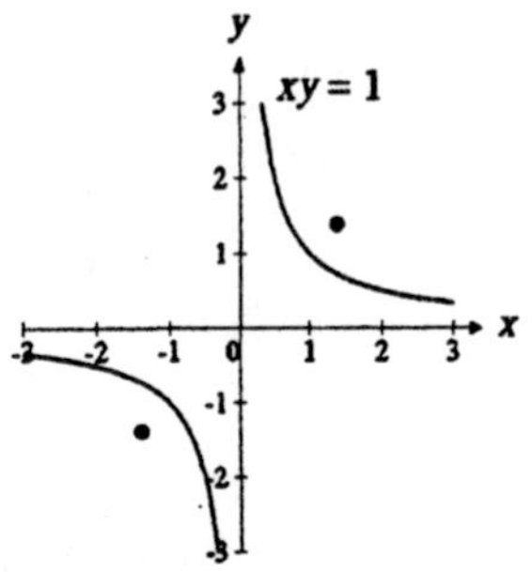

21 Refer to Figure 8. Suppose $2a > \overline{FF'}$ and that P is a point on the hyperbola. Then either $\overline{PF} - \overline{PF'} = 2a > \overline{FF'}$ or $\overline{PF'} - \overline{PF} = 2a > \overline{F'F}$. In the first case, $\overline{PF} > \overline{PF'} + \overline{F'F}$, violating the triangle inequality. In the second case, $\overline{PF'} > \overline{PF} + \overline{FF'}$, again violating the triangle inequality.

23 Assuming that sound travels at a known constant speed, the distances $d_A = \overline{DS} + \overline{SA}$, $d_B = \overline{DS} + \overline{SB}$, and $d_C = \overline{DS} + \overline{SC}$ can be computed from the known travel times, where S is the location of the submarine. Then S lies on three ellipses: one with foci D and A and "string length" d_A, another with foci D and B and "string length" d_B, and a third with foci D and C and "string length" d_C. In most cases, the three ellipses will uniquely determine S, which lies on their intersection.

25 As suggested, inscribe two spheres in the cylinder, S touching the plane at F and S' touching the plane

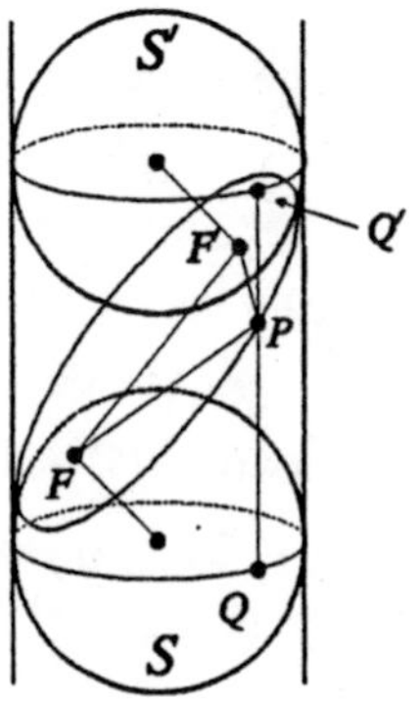

at F'. Let the distance between the centers of the spheres be $2a$. (This is also the distance between their equators as defined by their intersections with the cylinder.) Now let P be a point on the intersection of the plane and the cylinder. Observe that PF is tangent to S. Let Q be the point where the vertical line through P intersects the equator of S. Observe that PQ is another tangent to S that passes through the point P. Thus $\overline{PF} = \overline{PQ}$. Arguing similarly for the sphere S', we have $\overline{PF'} = \overline{PQ'}$. (See the figure.) Thus $\overline{PF} + \overline{PF'} = \overline{PQ} + \overline{PQ'} = \overline{QQ'} = 2a$ for all points P, which shows that P lies on an ellipse.

G.2 Translation of Axes and the Graph of $Ax^2 + Cy^2 + Dx + Ey + F = 0$

1

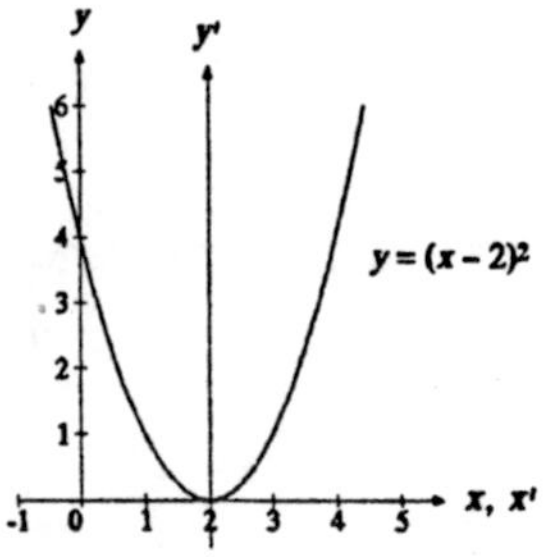

$y = (x - 2)^2$; let $x' = x - 2$ and $y' = y$. Then $y' = (x')^2$. The conic is a parabola. Since its equation is $(x')^2 = 2 \cdot \frac{1}{2} y'$, its focus and vertex relative to the $x'y'$ axes are $(0, 1/4)$ and $(0, 0)$, respectively. Relative to the xy axes, the focus is $(2, 1/4)$ and the vertex is $(2, 0)$. The directrix is $y = -1/4$. There are no asymptotes.

3

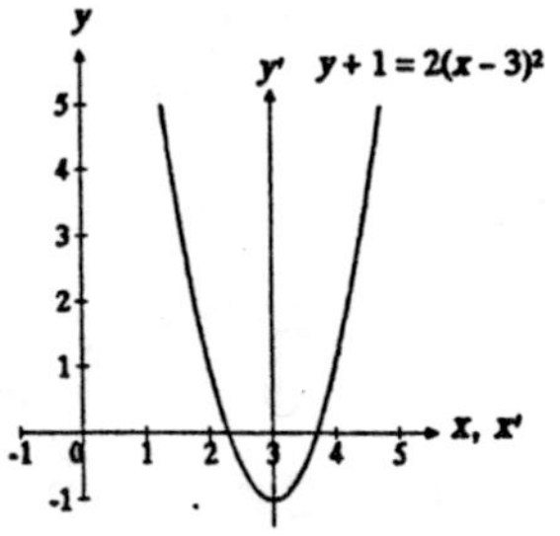

$y + 1 = 2(x - 3)^2$; let $x' = x - 3$ and $y' = y + 1$. Then we have $y' = 2(x')^2$. The conic is a parabola. Since $(x')^2 = 2 \cdot \frac{1}{4} y'$, the focus and vertex are at $(x', y') = (0, 1/8)$ and $(x', y') = (0, 0)$, respectively. In terms of x and y, the focus is $(3, -7/8)$ and the vertex is $(3, -1)$. The directrix is $y = -9/8$. There are no asymptotes.

5

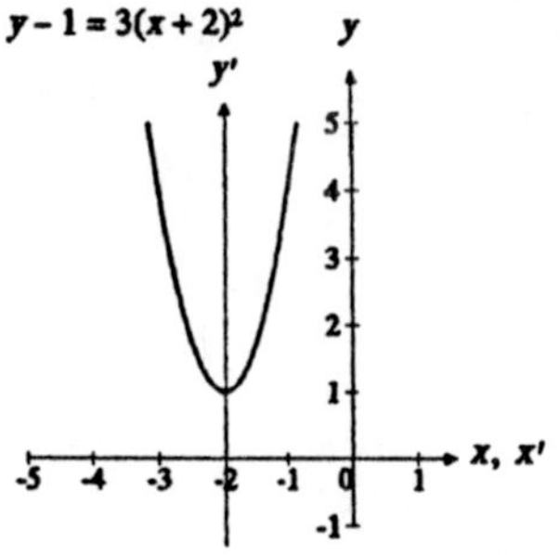

$y = 3x^2 + 12x + 13 = 3(x^2 + 4x + 4) + 1$, so $y - 1 = 3(x + 2)^2$; let $x' = x + 2$ and $y' = y - 1$. Then $y' = 3(x')^2$. The conic is a parabola. Since $(x')^2 = 2 \cdot \frac{1}{6} y'$, the focus and vertex are at $(x', y') = (0, 1/12)$ and $(x', y') = (0, 0)$. Hence the focus is $\left(-2, \frac{13}{12}\right)$, the vertex is $(-2, 1)$, the directrix is $y = \frac{11}{12}$, and there are no asymptotes.

7

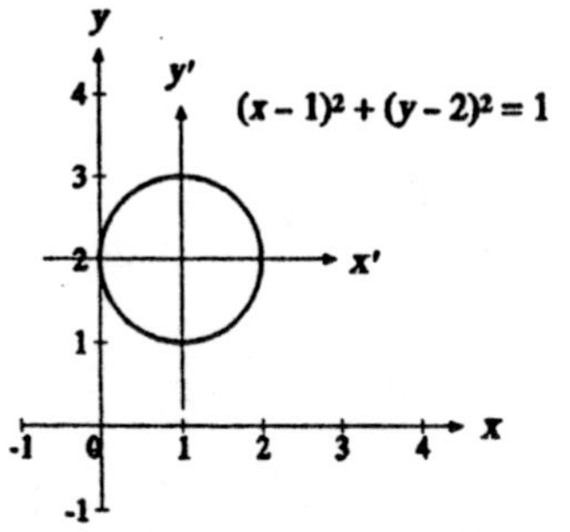

$0 = x^2 - 2x + y^2 - 4y + 4 = x^2 - 2x + 1 + y^2 - 4y + 4 - 5 + 4 = (x - 1)^2 + (y - 2)^2 - 1$; let $x' = x - 1$ and $y' = y - 2$. Then $(x')^2 + (y')^2 = 1$. The conic is a circle centered at $(1, 2)$.

9 $0 = x^2 - 4x - y^2 + 4y - 1 = x^2 - 4x + 4 - y^2 + 4y - 4 - 1 = (x - 2)^2 - (y - 2)^2 - 1$; let $x' = x - 2$ and $y' = y - 2$. Then $(x')^2 - (y')^2 = 1$. The conic is a hyperbola. Since $(x')^2 - (y')^2 = 1$, we have $c = \sqrt{1^2 + 1^2} = \sqrt{2}$; the foci are at

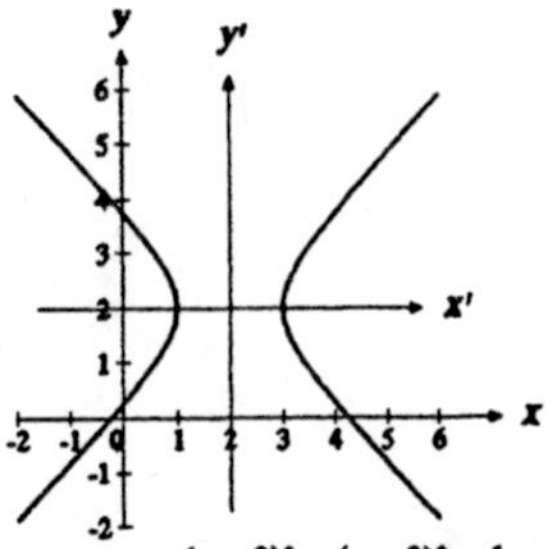

$(x - 2)^2 - (y - 2)^2 = 1$

$(x', y') = (\pm\sqrt{2}, 0)$. Setting $y' = 0$ gives the vertices: $(x', y') = (\pm 1, 0)$. The asymptotes are $y' = \pm x'$. (See page S-10 of the text for a discussion of asymptotes of hyperbolas.) In terms of x and y, the foci are $(2 - \sqrt{2}, 0)$ and $(2 + \sqrt{2}, 0)$, the vertices are (1, 0) and (3, 0), and the asymptotes are $y = x$ and $x + y = 4$.

11

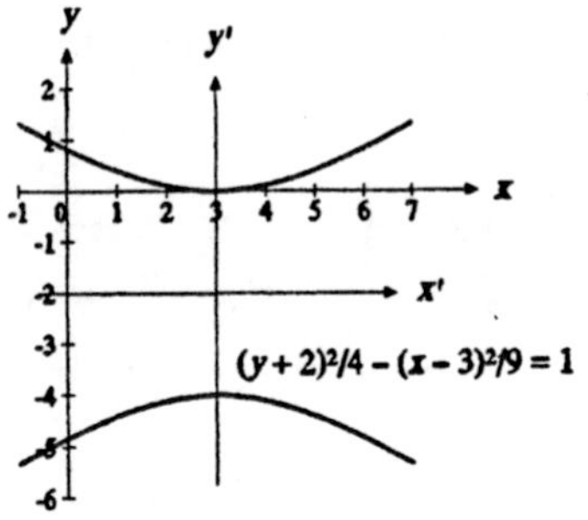

$0 = -4x^2 + 24x + 9y^2 + 36y - 36$
$= -4(x^2 - 6x) + 9(y^2 + 4y) - 36$
$= -4(x^2 - 6x + 9) + 9(y^2 + 4y + 4) - 36$
$= -4(x - 3)^2 + 9(y + 2)^2 - 36$; let $x' = x - 3$ and $y' = y + 2$. Then $-4(x')^2 + 9(y')^2 = 36$, or

$\frac{(y')^2}{4} - \frac{(x')^2}{9} = 1$. The conic is a hyperbola.

$\frac{(y')^2}{2^2} - \frac{(x')^2}{3^2} = 1$, so we have $c = \sqrt{2^2 + 3^2} = \sqrt{13}$; the foci are at $(x', y') = (0, \pm\sqrt{13})$. Setting $x' = 0$ yields the vertices: $(x', y') = (0, \pm 2)$. The asymptotes are $\frac{y'}{2} = \pm\frac{x'}{3}$; that is, $y' = \pm\frac{2}{3}x'$. In terms of x and y, the foci are $(3, -2 \pm \sqrt{13})$, the vertices are (3, 0) and (3, −4), and the asymptotes are $y + 2 = \pm\frac{2}{3}(x - 3)$; the equations simplify to $y = \frac{2}{3}x - 4$ and $y = -\frac{2}{3}x$.

13

$(x + 2)^2/4 + (y + 3)^2/25 = 1$

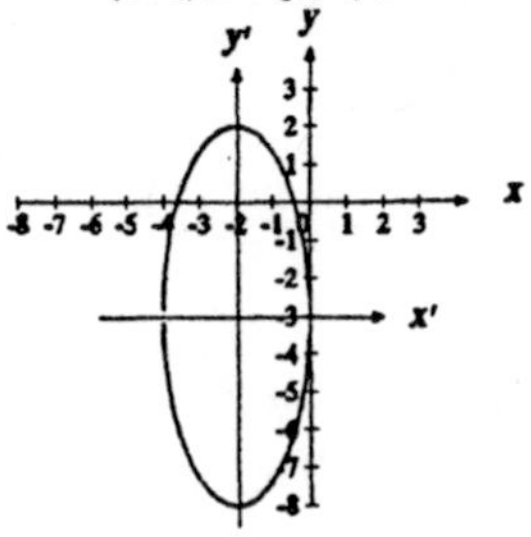

$0 = 25x^2 + 100x + 4y^2 + 24y + 36$
$= 25(x^2 + 4x) + 4(y^2 + 6y) + 36$
$= 25(x^2 + 4x + 4) + 4(y^2 + 6y + 9) - 100$
$= 25(x + 2)^2 + 4(y + 3)^2 - 100$; let $x' = x + 2$ and $y' = y + 3$. Then $25(x')^2 + 4(y')^2 = 100$, or

$\frac{(x')^2}{4} + \frac{(y')^2}{25} = 1$. The conic is an ellipse. Since

$\frac{(x')^2}{2^2} + \frac{(y')^2}{5^2} = 1$, $c = \sqrt{5^2 - 2^2} = \sqrt{21}$; the foci are at $(x', y') = (0, \pm\sqrt{21})$. Setting $x' = 0$ and $y' = 0$ (separately) gives the vertices: $(x', y') = (\pm 2, 0)$ or $(0, \pm 5)$. In terms of x and y, the foci are $(-2, -3 \pm \sqrt{21})$, the vertices are (−4, −3), (0, −3), (−2, −8), and (−2, 2).

15 The ellipse is centered at (1, 2), so let $x' = x - 1$ and $y' = y - 2$. In the $x'y'$ system, the vertices are (0, −2), (3, 0), (0, 2), and (−3, 0), so the

equation is $\frac{(x')^2}{9} + \frac{(y')^2}{4} = 1$.

17 The asymptotes meet when $\frac{2}{3}x + \frac{1}{3} = -\frac{2}{3}x + \frac{5}{3}$; that is, for $x = 1$. Thus the hyperbola's center of symmetry is (1, 1), so let $x' = x - 1$, $y' = y - 1$. In the $x'y'$ system, the asymptotes are $y' = \pm\frac{2}{3}x'$ and one focus is (2, 0). If the equation is $\frac{(x')^2}{a^2} - \frac{(y')^2}{b^2} = 1$, then $a^2 + b^2 = 2^2$ and $\frac{b}{a} = \frac{2}{3}$. Hence $4 = a^2 + \left(\frac{2}{3}a\right)^2 = \frac{13}{9}a^2$, $a = \frac{6}{\sqrt{13}}$ and $b = \frac{4}{\sqrt{13}}$. The equation is $\frac{(x')^2}{36/13} - \frac{(y')^2}{16/13} = 1$; that is, $\frac{13}{36}(x')^2 - \frac{13}{16}(y')^2 = 1$.

19 The center of the ellipse is the same as the center of the rectangle, namely, (4, 0). So let $x' = x - 4$, $y' = y$. In the $x'y'$ system, the vertices are $(\pm 3, 0)$ and $(0, \pm 2)$, so the equation is $\frac{(x')^2}{9} + \frac{(y')^2}{4} = 1$.

21 The vertex of the parabola is halfway between the focus and the directrix, at (4, 3). So let $x' = x - 4$, $y' = y - 3$. In the $x'y'$ system, the focus is (3, 0) and the directrix is $x' = -3$, so $c = 6$ and the equation is $(y')^2 = 12x'$.

23 (a) The equation becomes $\frac{x^2}{(\sqrt{-F/A})^2} + \frac{y^2}{(\sqrt{-F/C})^2} = 1$, which is an ellipse.

(b) If A, C, and F are positive, then $Ax^2 + Cy^2 + F \geq A\cdot 0 + C\cdot 0 + F = F$, which is never 0. The graph must therefore be empty.

(c) The equation becomes $\frac{x^2}{(-F/A)} - \frac{y^2}{(F/C)} = 1$. If $-F/A > 0$, then $F/C > 0$. The equation is then equivalent to $\frac{x^2}{(\sqrt{-F/A})^2} - \frac{y^2}{(\sqrt{F/C})^2} = 1$, which is a hyperbola. The other case is similar.

(d) When $A = C$ and their sign is opposite that of F.

25 We have $Cy^2 + Dx + Ey + F = 0$, so $x = -\frac{1}{D}(Cy^2 + Ey + F)$. We can complete the square of the right-hand side and shift axes to obtain $x' = 2c(y')^2$, so this is a parabola.

G.3 Rotation of Axes and the Graph of $Ax^2 + Bxy + Cy^2 + Dx + Ey + F = 0$

1

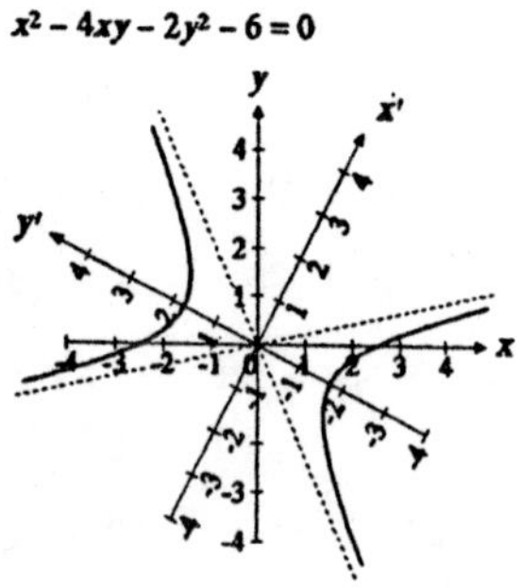

We have $A = 1$, $B = -4$, and $C = -2$, so $\tan 2\theta = \frac{B}{A - C} = -\frac{4}{3}$, $\cos 2\theta = -\frac{3}{5}$, $\sin\theta = \sqrt{\frac{1}{2}\left(1 - \left(-\frac{3}{5}\right)\right)} = \frac{2}{\sqrt{5}}$, and $\cos\theta =$

$\sqrt{\frac{1}{2}\left(1 + \left(-\frac{3}{5}\right)\right)} = \frac{1}{\sqrt{5}}$. Thus $x = \frac{x' - 2y'}{\sqrt{5}}$ and $y = \frac{2x' + y'}{\sqrt{5}}$. We can plug in and simplify to obtain $2(y')^2 - 3(x')^2 = 6$. (Observe that the rotation angle was approximately 63.4°. A rotation of −26.4° would also have eliminated the xy term, but x' and y' would switch roles.) In the $x'y'$ system, the asymptotes are given by $y' = \pm\sqrt{\frac{3}{2}}x'$.

3

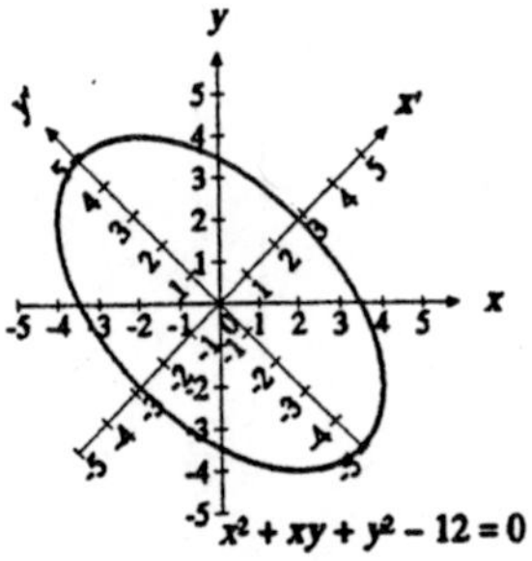

We have $A = B = C = 1$, so $\tan 2\theta = \frac{1}{1 - 1}$, which is undefined. Now $\tan 2\theta$ is undefined for $2\theta = \pi/2$; that is, for $\theta = \pi/4$. Then $x = \frac{x' - y'}{\sqrt{2}}$ and $y = \frac{x' + y'}{\sqrt{2}}$, so $\frac{(x')^2}{8} + \frac{(y')^2}{24} = 1$, an ellipse.

5

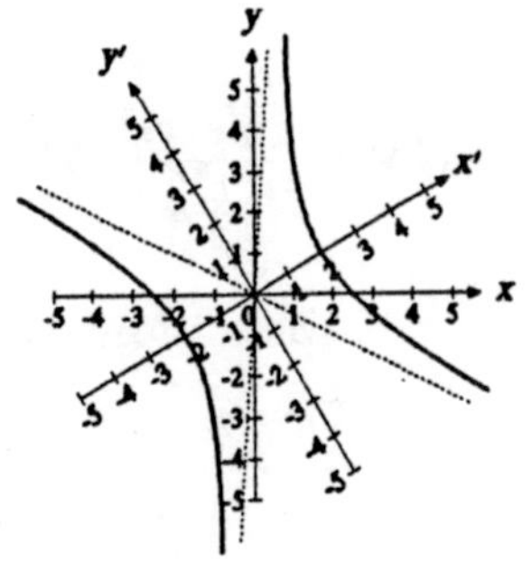

$A = 23$, $B = 26\sqrt{3}$, and $C = -3$, so $\tan 2\theta = \frac{26\sqrt{3}}{26} = \sqrt{3}$; hence $2\theta = \pi/3$ and $\theta = \pi/6$. Thus $x = \frac{\sqrt{3}x' - y'}{2}$, $y = \frac{x' + \sqrt{3}y'}{2}$, and $\frac{(x')^2}{4} - \frac{(y')^2}{9} = 1$. In the $x'y'$ system, the asymptotes are given by $\frac{(x')^2}{4} - \frac{(y')^2}{9} = 0$; that is, by $y' = \pm\frac{3}{2}x'$. Since $x' = \frac{\sqrt{3}x + y}{2}$ and $y' = \frac{-x + \sqrt{3}y}{2}$, these are equivalent to $\frac{-x + \sqrt{3}y}{2} = \pm\frac{3}{2}\cdot\frac{\sqrt{3}x + y}{2}$, or $y = \frac{13\sqrt{3} \pm 24}{3}x$.

7

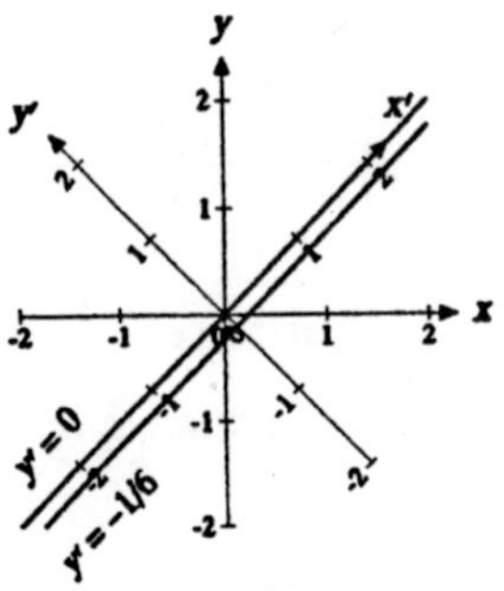

$A = 6$, $B = -12$, and $C = 6$, so $\tan 2\theta =$

$\frac{-12}{6-6}$ is undefined. As in Exercise 3, $\theta = \frac{\pi}{4}$, $x = \frac{x' - y'}{\sqrt{2}}$ and $y = \frac{x' + y'}{\sqrt{2}}$. We then obtain

$12(y')^2 + 2y' = 0$, which is equivalent to $2y'(6y' - 1) = 0$. Thus $y' = 0$ or $y' = -1/6$.

9

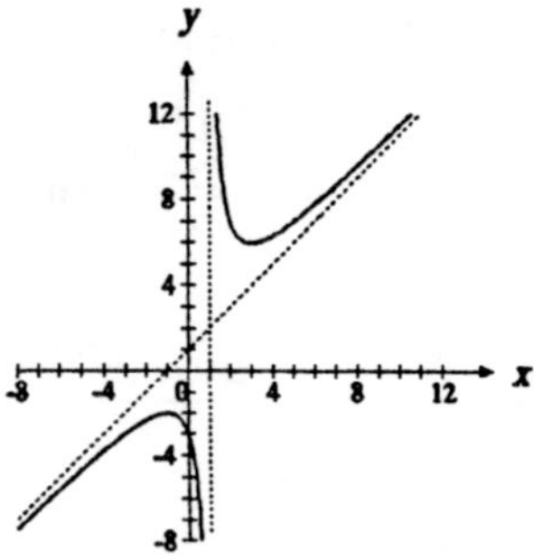

We have $y = x + 1 + \frac{4}{x-1}$, so the graph has a vertical asymptote $x = 1$ and a tilted asymptote $y = x + 1$. Plotting a few points produces the graph in the figure. To see that it is a conic section, rewrite the equation as $x^2 - xy + y + 3 = 0$. Since it is unbounded and disconnected, it must be a hyperbola.

11

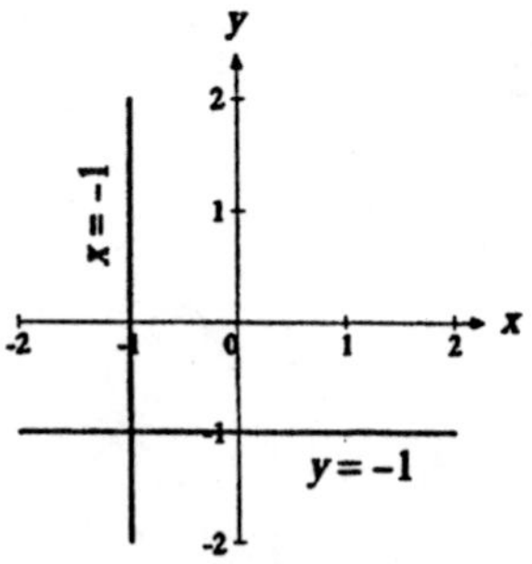

The equation factors as $(x + 1)(y + 1) = 0$, so the graph consists of two straight lines: $x = -1$ and $y = -1$.

13 $\mathcal{D} = B^2 - 4AC = 24^2 - 4(-1)(6) = 600 > 0$, so the conic is a hyperbola.

15 $\mathcal{D} = B^2 - 4AC = (-2)^2 - 4\cdot1\cdot1 = 0$, so the conic is a parabola.

17 (a)

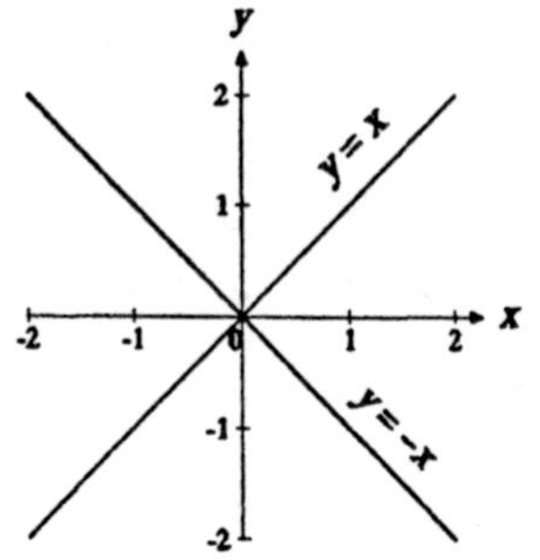

$x^2 - y^2 = (x + y)(x - y) = 0$

(b)

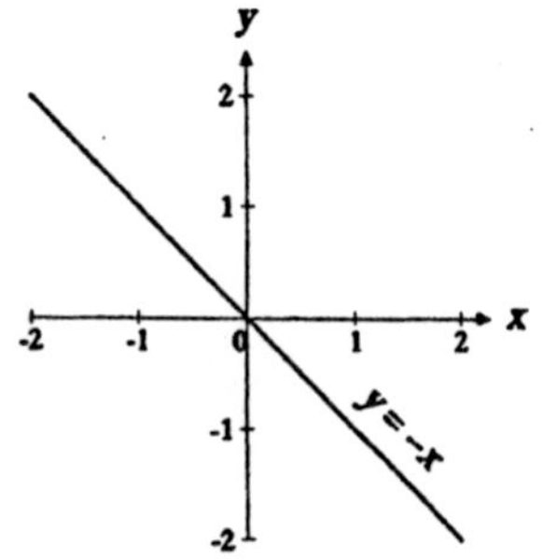

$x^2 + 2xy + y^2 = (x + y)^2 = 0$

(c)

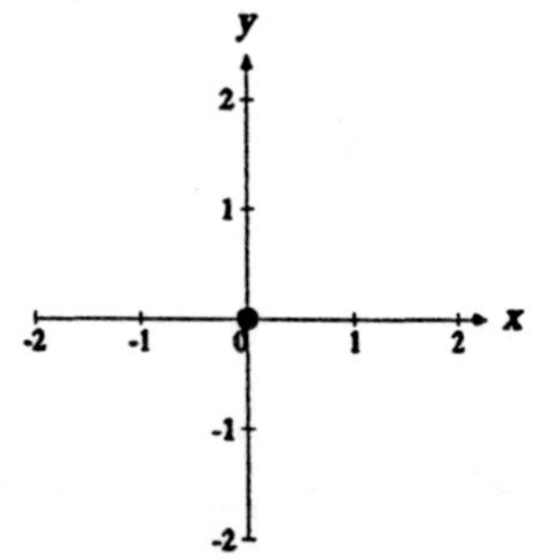

$3x^2 + 4y^2 = 0$ only at $(0, 0)$.

(d) $3x^2 + 2xy + 3y^2 + 1 = 0$ can be rotated by $\theta = \pi/4$ to obtain $4(x')^2 + 2(y')^2 = -1$, which has a completely empty graph, as shown below:

19 To save space, let $s = \sin\theta$ and $c = \cos\theta$. Then we have $(B')^2 = \left(2(C - A)sc + B(c^2 - s^2)\right)^2 =$

$4(C - A)^2s^2c^2 + 4(C - A)Bsc(c^2 - s^2) + B^2(c^2 - s^2)^2$ and $4A'C' = 4(Ac^2 + Bcs + Cs^2)(As^2 - Bcs + Cc^2) = 4A^2c^2s^2 - 4ABc^3s + 4ACc^4 + 4ABcs^3 - 4B^2c^2s^2 + 4BCc^3s + 4ACs^4 - 4BCcs^3 + 4C^2c^2s^2$. Subtracting and gathering like terms, we have $(B')^2 - 4A'C' = B^2((c^2 - s^2)^2 + 4c^2s^2) + AB(-4sc(c^2 - s^2) + 4c^3s - 4cs^3) + AC(-8s^2c^2 - 4c^4 - 4s^4) + BC(4sc(c^2 - s^2) - 4c^3s + 4cs^3) = B^2(c^2 + s^2) - 4AC(c^2 + s^2)^2 = B^2 - 4AC$.

21 Suppose that $Ax^2 + Bxy + Cy^2 + Dx + Ey + F = 0$ is the equation of a circle. By implicit differentiation, $2Ax + B\left(x\frac{dy}{dx} + y\right) + 2Cy\frac{dy}{dx} + D + E\frac{dy}{dx} = 0$, so $\frac{dy}{dx} = -\frac{2Ax + By + D}{Bx + 2Cy + E}$.

Since the two points on the circle at which $\frac{dy}{dx} = 0$ have the same x coordinate, we may call them (x, y_0) and (x, y_1). Then $0 = 2Ax + By_0 + D$ and $0 = 2Ax + By_1 + D$, so $B(y_0 - y_1) = 0$; now $y_0 \neq y_1$, so it follows that $B = 0$. Hence $Ax^2 + Cy^2 + Dx + Ey + F = 0$ and $\frac{dy}{dx} = -\frac{2Ax + D}{2Cy + E}$.

Also, the two points at which $dy/dx = 1$ lie on a line that passes through the center of the circle and thus is perpendicular to the tangents of slope 1; hence the line has an equation of the form $y = -x + k$, or $x + y = k$. Let the points where $dy/dx = 1$ be (x, y) and $(\bar{x}, \bar{y})$. Then $1 = -\frac{2Ax + D}{2Cy + E} = -\frac{2A\bar{x} + D}{2C\bar{y} + E}$ and $x + y = \bar{x} + \bar{y}$, so $2Ax + 2Cy + D + E = 0$, $2A\bar{x} + 2C\bar{y} + D + E = 0$, and $A(x - \bar{x}) + C(y - \bar{y}) = 0$. But $x - \bar{x} = \bar{y} - y = -(y - \bar{y})$, so we have $(A - C)(x - \bar{x}) = 0$; $x \neq \bar{x}$, so it follows that $A = C$.

23 The hyperbola has the equation $\frac{(x')^2}{a^2} - \frac{(y')^2}{b^2} = 1$, where $a = b = 1$. From Sec. G.1, $c = \sqrt{a^2 + b^2} = \sqrt{2}$, so the foci are at $(x', y') = (\pm\sqrt{2}, 0)$. By equation (13), $x = \frac{3}{5}(\pm\sqrt{2}) - \frac{4}{5}\cdot 0 = \pm\frac{3}{5}\sqrt{2}$ and $y = \frac{4}{5}(\pm\sqrt{2}) + \frac{3}{5}\cdot 0 = \pm\frac{4}{5}\sqrt{2}$, so the foci are at $(x, y) = \pm\left(\frac{3}{5}\sqrt{2}, \frac{4}{5}\sqrt{2}\right)$.

25 The ellipse has the equation $\frac{(x')^2}{8} + \frac{(y')^2}{24} = 1$, so $a = \sqrt{8}$, $b = \sqrt{24}$, and $c = \sqrt{b^2 - a^2} = 4$. The foci are at $(x', y') = (0, \pm 4)$. Hence $x = \frac{x' - y'}{\sqrt{2}} = \frac{0 - (\pm 4)}{\sqrt{2}} = \mp 2\sqrt{2}$ and $y = \frac{x' + y'}{\sqrt{2}} = \frac{0 + (\pm 4)}{\sqrt{2}} = \pm 2\sqrt{2}$. The foci are at $(x, y) = \pm(2\sqrt{2}, -2\sqrt{2})$.

G.4 Conic Sections in Polar Coordinates

1

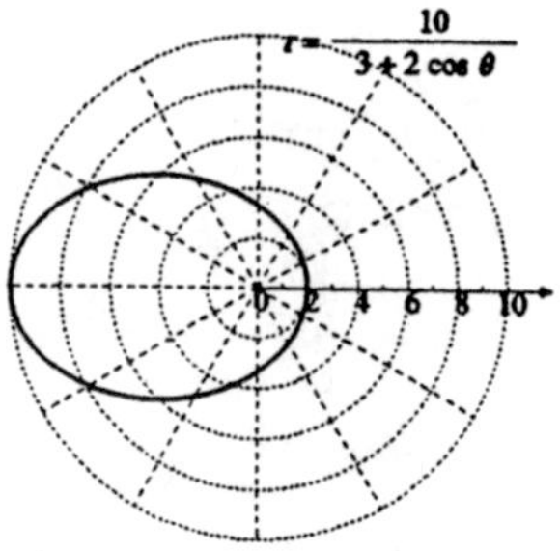

(a)

θ	0	$\pi/2$	π	$3\pi/2$
r	2	10/3	10	10/3

(b) We have $r = \dfrac{10}{3 + 2\cos\theta} = \dfrac{2/3 \cdot 5}{1 + 2/3\cos\theta}$, which is Eq. (2) with $e = 2/3$, $p = 5$, and $B = 0$.

3 (a) $r = \dfrac{5}{3 + 4\cos\theta} = \dfrac{5/3}{1 + 4/3\cos\theta}$, so $e = 4/3$.

(b) $r = \dfrac{5}{4 + 3\cos\theta} = \dfrac{5/4}{1 + 3/4\cos\theta}$, so $e = 3/4$.

(c) $r = \dfrac{5}{3 + 3\cos\theta} = \dfrac{5/3}{1 + \cos\theta}$, so $e = 1$.

(d) $r = \dfrac{5}{3 - 4\cos\theta} = \dfrac{5}{3 + 4\cos(\theta - \pi)} = \dfrac{5/3}{1 + 4/3\cos(\theta - \pi)}$, so $e = 4/3$.

5 (a) The distance from (x, y) to the focus and directrix are, respectively, $\sqrt{(x + 1)^2 + y^2}$ and $|x - 1|$. These are equal if and only if $(x + 1)^2 + y^2 = (x - 1)^2$; that is, $x^2 + 2x + 1 + y^2 = x^2 - 2x + 1$, or $y^2 = -4x$.

(b) The focus is at the pole and the directrix is $2 = x = r\cos\theta$, so (r, θ) is on the parabola if and only if $r = |x - 2| = |r\cos\theta - 2|$. Hence either $r = r\cos\theta - 2$ or $r = 2 - r\cos\theta$. In the first case, $r = \dfrac{-2}{1 - \cos\theta}$, which is negative. In the second case we have $r = \dfrac{2}{1 + \cos\theta}$. It is customary to take $r > 0$, so use the second equation.

(c) The pole is at the origin of the xy coordinate system, and the polar axis coincides with the positive x axis. We can therefore convert $y^2 = -4x$ into polar coordinates by the usual substitutions $x = r\cos\theta$ and $y = r\sin\theta$. We then have $r^2\sin^2\theta = -4\cos\theta$, so $r = \dfrac{-4\cos\theta}{\sin^2\theta} = -4\csc\theta\cot\theta$.

H Logarithms and Exponentials Defined through Calculus

H.1 The Natural Logarithm Defined as a Definite Integral

1 By the first fundamental theorem of calculus, the function $L(x) = \int_1^x \frac{1}{t}\,dt$ is differentiable. Therefore it is continuous.

3 (a) The width of each rectangle is 1/4. The heights of the rectangles are determined by

right endpoints. The total area enclosed by the rectangles is $\frac{1}{4}\left(\frac{1}{5/4} + \frac{1}{6/4} + \frac{1}{7/4} + \frac{1}{8/4} + \frac{1}{9/4} + \frac{1}{10/4} + \frac{1}{11/4} + \frac{1}{12/4}\right)$

$= \frac{1}{4}\left(\frac{4}{5} + \frac{4}{6} + \frac{4}{7} + \frac{4}{8} + \frac{4}{9} + \frac{4}{10} + \frac{4}{11} + \frac{4}{12}\right) = \frac{1}{5} + \frac{1}{6} + \frac{1}{7} + \frac{1}{8} + \frac{1}{9} + \frac{1}{10} + \frac{1}{11} + \frac{1}{12} = \frac{28271}{27720} \approx 1.0199.$

(b) $L(3)$ equals the area under the graph of $y = 1/x$ and above $[1, 3]$. Since this area contains the eight rectangles, $L(3) > \frac{28271}{27720} \approx 1.0199 > 1$. Thus $L(3) > 1$.

(c) $L'(x) = 1/x > 0$, so $L(x)$ is an increasing function; if $b > a$, then $L(b) > L(a)$ and vice versa. Since $L(3) > L(e) = 1$, $3 > e$.

5 $L(n)$ is the area under $y = 1/x$ above $[1, n]$. That area contains the rectangles whose total area is

$1 \cdot \frac{1}{2} + 1 \cdot \frac{1}{3} + \cdots + 1 \cdot \frac{1}{n} = \frac{1}{2} + \frac{1}{3} + \cdots + \frac{1}{n}.$

Thus $L(n) > \frac{1}{2} + \frac{1}{3} + \cdots + \frac{1}{n}.$

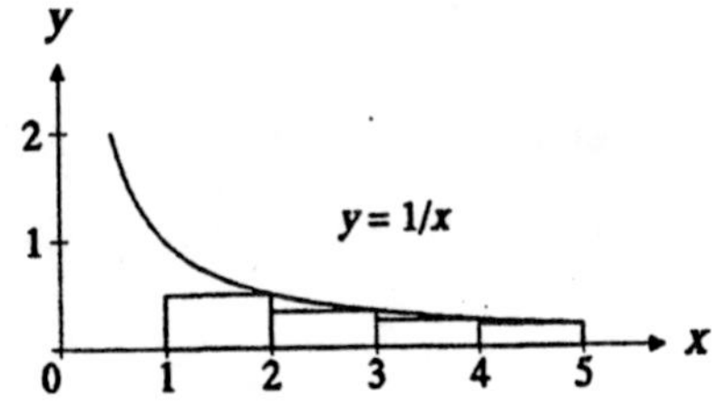

7 (a) Let $x_i = 1 + \frac{i}{n} = \frac{n+i}{n}$ for $i = 0, \cdots, n$.

This produces a partition of $[1, 2]$. Using right endpoints as sampling points, an underestimating sum $\sum_{i=1}^{n} f(x_i)(x_i - x_{i-1})$ is formed. Since $f(x_i) = \frac{1}{x_i} = \frac{n}{n+i}$ and $x_i - x_{i-1} = 1/n$, this sum becomes $\sum_{i=1}^{n} \frac{1}{n+i}$

$= \frac{1}{n+1} + \frac{1}{n+2} + \cdots + \frac{1}{2n}$. The area under the curve and above $[1, 2]$ is less than this sum; thus $L(2) > \frac{1}{n+1} + \frac{1}{n+2} + \cdots + \frac{1}{2n}$.

(b) Letting $n = 6$ in the above formula, $L(2) > \frac{1}{7} + \frac{1}{8} + \frac{1}{9} + \frac{1}{10} + \frac{1}{11} + \frac{1}{12} \approx 0.653.$

H.2 Exponential Functions Defined in Terms of Logarithms

1 $b^0 - 1 = 0$, so $\log_b 1 = 0$. $b^1 = b$, so $\log_b b = 1$. $b^{\log_b x + \log_b y} = b^{\log_b x} b^{\log_b y} = xy$, so $\log_b xy = \log_b x + \log_b y$. $b^{y \log_b x} = \left(b^{\log_b x}\right)^y = x^y$, so $\log_b x^y = y \log_b x$.

3 Let $y = \log_b x$, so $x = b^y$. Implicit differentiation and Theorem 7 yield $1 = b^y L(b) \frac{dy}{dx}$; thus $\frac{dy}{dx} = \frac{1}{b^y L(b)} = \frac{1}{xL(b)}$. Now $b^{\log_b e} = e$, so we apply Theorem 3 to $L\left(b^{\log_b e}\right) = L(e)$ to obtain

$(\log_b e)L(b) = 1$, hence $L(b) = \dfrac{1}{\log_b e}$. Thus

$$\frac{d}{dx}(\log_b x) = \frac{1}{xL(b)} = \frac{1}{x}\log_b e.$$

I The Taylor Series for $f(x, y)$

1 Let $f(x, y) = x^5y^7$. Then $f_x = 5x^4y^7$, $f_y = 7x^5y^6$, $f_{xx} = 20x^3y^7$, $f_{xy} = f_{yx} = 35x^4y^6$, and $f_{yy} = 42x^5y^5$. Therefore $f_{xxx} = 60x^2y^7$, $f_{xxy} = f_{xyx} = f_{yxx} = 140x^3y^6$, $f_{xyy} = f_{yxy} = f_{yyx} = 210x^4y^5$, and $f_{yyy} = 210x^5y^4$.

3 Let $f(x, y) = e^{2x+3y}$. Then $f_x = 2e^{2x+3y}$, $f_y = 3e^{2x+3y}$, $f_{xx} = 4e^{2x+3y}$, $f_{xy} = f_{yx} = 6e^{2x+3y}$, and $f_{yy} = 9e^{2x+3y}$. Therefore, $f_{xxx} = 8e^{2x+3y}$, $f_{xxy} = f_{xyx} = f_{yxx} = 12e^{2x+3y}$, $f_{xyy} = f_{yxy} = f_{yyx} = 18e^{2x+3y}$, and $f_{yyy} = 27e^{2x+3y}$.

5 The right-hand side of the equation is $2 + 4(x - 1) + (y - 2) + 2(x - 1)^2 + 2(x - 1)(y - 2) + (x - 1)^2(y - 2) = 2 + 4x - 4 + y - 2 + 2x^2 - 4x + 2 + 2xy - 4x - 2y + 4 + x^2y - 2xy + y - 2x^2 + 4x - 2$
$= (2 - 4 - 2 + 2 + 4 - 2) + (4 - 4 - 4 + 4)x + (1 - 2 + 1)y + 0\cdot x^2 + 0\cdot xy + x^2y$
$= 0 + 0x + 0y + 0x^2 + 0xy + x^2y = x^2y$.

7 (a) $f(x, y) = e^{x+y^2}$, $f(0, 0) = 1$; $f_x(x, y) = e^{x+y^2}$, $f_x(0, 0) = 1$; $f_y(x, y) = 2ye^{x+y^2}$, $f_y(0, 0) = 0$; $f_{xx} = e^{x+y^2}$, $f_{xx}(0, 0) = 1$; $f_{xy}(x, y) = 2ye^{x+y^2}$, $f_{xy}(0, 0) = 0$; $f_{yy} = 4y^2e^{x+y^2} + 2e^{x+y^2}$, $f_{yy}(0, 0) = 2$; the Taylor series is $f(x, y) = f(0, 0) + f_x(0, 0)x + f_y(0, 0)y + \frac{1}{2}f_{xx}(0, 0)x^2 + f_{xy}(0, 0)xy + \frac{1}{2}f_{yy}(0, 0)y^2$

$$= 1 + x + 0y + \frac{1}{2}x^2 + 0xy + \frac{2}{2}y^2 + \cdots$$

$$= 1 + x + \frac{1}{2}x^2 + y^2 + \cdots.$$

(b) We have $e^x = 1 + x + \frac{1}{2}x^2 + \cdots$ and $e^{y^2} = 1 + y^2 + \frac{1}{2}y^4 + \cdots$, so $e^{x+y^2} = (1 + x + \frac{1}{2}x^2 + \cdots)(1 + y^2 + \cdots) = 1 + x + \frac{1}{2}x^2 + y^2 + xy^2 + \frac{1}{2}x^2y^2 + \cdots$

$$= 1 + x + \frac{1}{2}x^2 + y^2 + \cdots.$$

9 $f(x, y) = f(0, 0) + f_x(0, 0)x + f_y(0, 0)y + \frac{1}{2}f_{xx}(0, 0)x^2 + f_{xy}(0, 0)xy + \frac{1}{2}f_{yy}(0, 0)y^2 + \cdots$

$$= 2 + 3x - 5y + 3x^2 + 7xy + \frac{1}{2}y^2 + \cdots.$$

11 $f(x, y) = (1 + x + y)^{1/2}$, $f_x = f_y = \frac{1}{2}(1 + x + y)^{-1/2}$, and $f_{xx} = f_{xy} = f_{yy} = -\frac{1}{4}(1 + x + y)^{-3/2}$.

Therefore, $f(x, y) = f(0, 0) + f_x(0, 0)x + f_y(0, 0)y + f_{xx}(0, 0)x^2/2 + f_{xy}(0, 0)xy + f_{yy}(0, 0)y^2/2 + \cdots$

$$= 1 + \frac{1}{2}x + \frac{1}{2}y - \frac{1}{4}\cdot\frac{x^2}{2} - \frac{1}{4}xy - \frac{1}{4}\cdot\frac{y^2}{2} + \cdots$$

$$= 1 + \frac{1}{2}x + \frac{1}{2}y - \frac{1}{8}x^2 - \frac{1}{4}xy - \frac{1}{8}y^2 + \cdots.$$

J Theory of Limits

1 We must show that, for $a \neq 0$, $\lim_{x\to a} \frac{1}{x} = \frac{1}{a}$. Let $\epsilon > 0$. We wish to find a number $\delta > 0$ such that, for $|x - a| < \delta$, $\left|\frac{1}{x} - \frac{1}{a}\right| < \epsilon$. Note that $\left|\frac{1}{x} - \frac{1}{a}\right| = \frac{|x-a|}{|x||a|}$. Now let δ be the lesser of $\frac{1}{2}|a|$ and $\frac{1}{2}\epsilon|a|^2$. For $|x - a| < \delta$, we have $|a| \le |x| + |x - a| \le |x| + \delta \le |x| + \frac{1}{2}|a|$, so $|x| \ge |a| - \frac{1}{2}|a| = \frac{1}{2}|a|$. It follows that $\left|\frac{1}{x} - \frac{1}{a}\right| = \frac{|x-a|}{|x||a|} \le \frac{|x-a|}{\frac{1}{2}|a||a|} = \frac{2}{|a|^2}|x - a| < \frac{2}{|a|^2}\delta \le \frac{2}{|a|^2}\frac{\epsilon|a|^2}{2} = \epsilon$, as required.

3 Let $f(x) = 1/x$ and $g(x) = x^2 + 1$. For any a, g is continuous by Theorem 7. Also, $g(a) = a^2 + 1 \ge 1$, so $g(a) \neq 0$. Hence, by Exercise 1, f is continuous at $g(a)$; by Exercise 2, $(f\circ g)(x) = \frac{1}{x^2+1}$ is continuous at a.

5 By Exercise 4, $1/g(x)$ is continuous at a. Applying Theorem 4 to f and $1/g$, we see that $f(x)/g(x)$ is continuous at a.

7 We must show that $\lim_{x\to a} f(x)g(x) = 0$. Let $\epsilon > 0$. Since $\lim_{x\to a} f(x) = A$, there is a δ_f such that, for $0 < |x - a| < \delta_f$, $|f(x) - A| < 1$. For such x, $|f(x)| \le |A| + |f(x) - A| < |A| + 1$. Since $\lim_{x\to a} g(x) = 0$, there is a δ_g such that, for $0 < |x - a| < \delta_g$, $|g(x) - 0| < \frac{\epsilon}{|A| + 1}$. Let δ be the lesser of δ_f and δ_g and suppose $0 < |x - a| < \delta$. Then $|f(x)g(x) - 0| = |f(x)||g(x)| < (|A| + 1)\cdot\frac{\epsilon}{|A|+1} = \epsilon$, so we are done.

9 Let $\epsilon = f(a)$. Then, by continuity, there is a $\delta > 0$ such that, for $|x - a| < \delta$, $|f(x) - f(a)| < f(a)$. For such x, $f(a) - f(x) \le |f(x) - f(a)| < f(a)$, so $f(x)$ is positive. Hence f is positive on the interval $(b, c) = (a - \delta, a + \delta)$.

11 f and g both possess limits at infinity, so $f(x) + g(x)$ is defined for sufficiently large x. Pick D_g so that $|g(x) - A| < 1$ for $x > D_g$. Then $|g(x)| < |A| + 1$ for $x > D_g$. Given E, choose D_f such that $f(x) > E + |A| + 1$ for $x > D_f$. Let D equal the greater of D_f and D_g; for $x > D$, $f(x) + g(x) > (E + |A| + 1) - (|A| + 1) = E$. Hence $\lim_{x\to\infty} (f(x) + g(x)) = \infty$.

13 f and g both possess limits at infinity, so $f(x)g(x)$ is defined for sufficiently large x. Choose D_f so that $x > D_f$ implies that $f(x) > 1$. Given E, pick D_g so that $g(x) > E$ for $x > D_g$. Now let D equal the greater of D_f and D_g. Then for $x > D$, $f(x)g(x) > 1\cdot E = E$, so $\lim_{x\to\infty} f(x)g(x) = \infty$.

15 f and g both possess limits at infinity, so $f(x)g(x)$ is defined for sufficiently large x. Given $\epsilon > 0$, choose D_f so that $x > D_f$ implies that $|f(x) - A| < \frac{\epsilon}{2(|B| + 1)}$. Choose D_g so that $x > D_g$ implies

that $|g(x) - B| < \dfrac{\epsilon}{2(|A| + 1)}$. Choose D_1 so that $x > D_1$ implies that $|f(x) - A| < 1$. (Hence, for $x > D_1$, $|f(x)| < |A| + 1$.) Let D be the greatest of D_f, D_g, and D_1. Then, for $x > D$, we have $|f(x)g(x) - AB| \le |f(x)||g(x) - B| + |B||f(x) - A| \le$

$$(|A| + 1)\frac{\epsilon}{2(|A| + 1)} + |B|\frac{\epsilon}{2(|B| + 1)} < \frac{\epsilon}{2} + \frac{\epsilon}{2}$$

$= \epsilon$. Hence $\lim_{x \to \infty} f(x)g(x) = AB$.

K The Interchange of Limits

K.2 The Derivative of $\int_a^b f(x, y)\, dx$ with Respect to y

1 We have $F(y) = \int_a^b x^3y^4\, dx = \dfrac{x^4y^4}{4}\Big|_{x=a}^{x=b}$

$= \dfrac{1}{4}(b^4 - a^4)y^4$, so $\dfrac{dF}{dy} = \dfrac{1}{4}(b^4 - a^4)(4y^3) =$

$(b^4 - a^4)y^3$. We also have $\dfrac{\partial f}{\partial y} = 4x^3y^3$, so

$\int_a^b \dfrac{\partial f}{\partial y}\, dx = \int_a^b 4x^3y^3\, dx = x^4y^3\Big|_{x=a}^{x=b}$

$= (b^4 - a^4)y^3 = \dfrac{dF}{dy}$.

3 At an extremum, we must have $\dfrac{dF}{dy} = 0$. By the theorem, $\dfrac{dF}{dy} = \int_0^{\pi/2} \dfrac{\partial}{\partial y}[(y - \cos x)^2]\, dx$

$= \int_0^{\pi/2} 2(y - \cos x)\, dx = (2xy - 2\sin x)\Big|_{x=0}^{x=\pi/2}$

$= \pi y - 2$. Hence $\dfrac{dF}{dy} = 0$ when $y = 2/\pi$. To see that this is a minimum, note that F is nonnegative because it is the integral of a square. It has no maximum because it is unbounded in y, so the extremum must be a minimum.

5 Let $H(u, v) = \int_0^u f(v, x)\, dx$. Then $G(u) = H(u, u)$, so $\dfrac{dG}{du} = \dfrac{\partial H}{\partial u}\cdot\dfrac{du}{du} + \dfrac{\partial H}{\partial v}\cdot\dfrac{du}{du} =$

$H_u(u, u) + H_v(u, u)$. But $H_u(u, v) = f(v, u)$ by the first fundamental theorem of calculus, and $H_v(u, v) = \int_0^u f_v(v, x)\, dx$ by the theorem of this section. It follows that $\dfrac{dG}{du} = f(u, u) + \int_0^u f_u(u, x)\, dx$.

7 (a) $\dfrac{dF}{dy} = \int_0^1 \dfrac{\partial}{\partial y}\left(\dfrac{x^y - 1}{\ln x}\right) dx = \int_0^1 x^y\, dx$

$= \dfrac{x^{y+1}}{y+1}\Big|_{x=0}^{x=1} = \dfrac{1}{1 + y}$

(b) F is an antiderivative of $\dfrac{1}{1 + y}$, so $F(y) = \int \dfrac{1}{1 + y}\, dy = \ln(1 + y) + C$ for some constant C.

(c) Note that $C = F(0) = \int_0^1 \dfrac{x^0 - 1}{\ln x}\, dx = \int_0^1 0\, dx = 0$, as claimed.

K.3 The Interchange of Limits

1 (a)

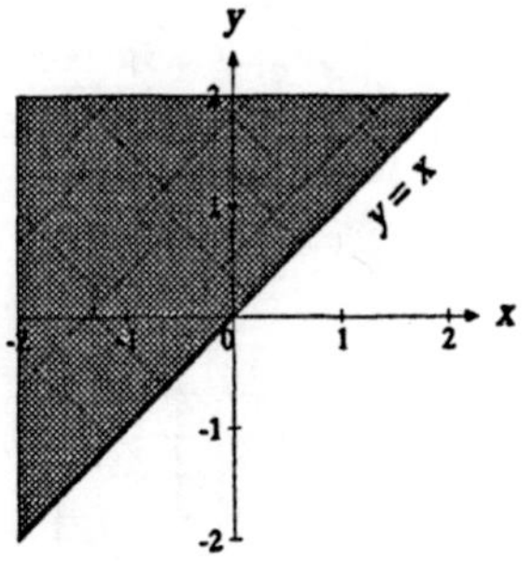

(b) $\lim_{y\to\infty} f(x, y) = 1$ because, for $y \geq x$, $f(x, y) = 1$; thus $\lim_{x\to\infty} \lim_{y\to\infty} f(x, y) = \lim_{x\to\infty} 1 = 1$.

(c) For $x > y$, $f(x, y) = 0$; hence $\lim_{y\to\infty} \lim_{x\to\infty} f(x, y) = \lim_{y\to\infty} 0 = 0$.

3 $\lim_{n\to\infty} f_n(x) = \lim_{n\to\infty} \dfrac{nx}{1 + n^2x^4} = 0$, because

$$\left|\frac{nx}{1 + n^2x^4}\right| = \frac{n|x|}{1 + n^2x^4} \leq \frac{n|x|}{n^2x^4} = \frac{1}{n|x|^3} \to 0$$

as $n \to \infty$ for $x \neq 0$ and $f_n(0) = 0$. Hence $\int_0^\infty \lim_{n\to\infty} f_n(x)\, dx = \int_0^\infty 0\, dx = 0$. For $n > 0$,

however, $\int_0^\infty f_n(x)\, dx = \int_0^\infty \dfrac{nx}{1 + n^2x^4}\, dx =$

$\lim_{b\to\infty} \dfrac{1}{2} \tan^{-1} nx^2 \Big|_0^b = \dfrac{1}{2}\left(\dfrac{\pi}{2} - 0\right) = \dfrac{\pi}{4}$, so

$\lim_{n\to\infty} \int_0^\infty f_n(x)\, dx = \lim_{n\to\infty} \dfrac{\pi}{4} = \dfrac{\pi}{4}$.

5 (a)

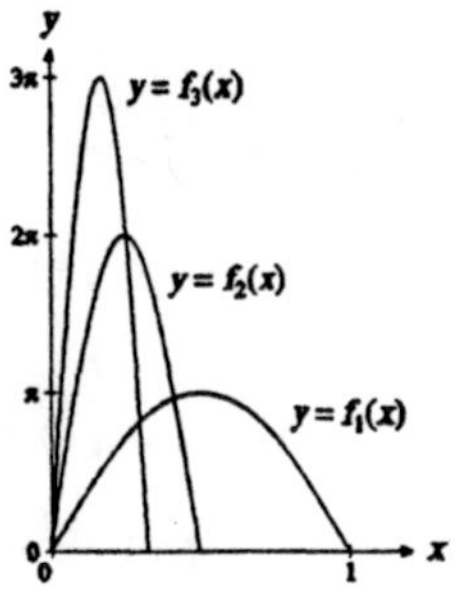

(b) $\int_0^1 f_n(x)\, dx = \int_0^{1/n} n\pi \sin n\pi x\, dx =$

$-\cos n\pi x \Big|_0^{1/n} = -(-1) + 1 = 2$, so

$\lim_{n\to\infty} \int_0^1 f_n(x)\, dx = \lim_{n\to\infty} 2 = 2$. Observe that for $0 \leq x \leq 1$, $\lim_{n\to\infty} f_n(x) = 0$ since for $n > 1/x$, $f_n(x) = 0$. Hence $\int_0^1 \lim_{n\to\infty} f_n(x)\, dx =$

$\int_0^1 0\, dx = 0$.

7 Note that $\dfrac{f_n(h) - f_n(0)}{h} = \dfrac{\frac{1}{n} \sin nh - 0}{h} =$

$\dfrac{\sin nh}{nh} \to 0$ as $n \to \infty$, although for $h \to 0$ it

approaches 1. Hence $\lim_{n\to\infty} \lim_{h\to 0} \dfrac{f_n(h) - f_n(0)}{h} = 1$

while $\lim_{h\to 0} \lim_{n\to\infty} \dfrac{f_n(h) - f_n(0)}{h} = 0$.

9 $\int_0^\infty \int_0^1 (2xy - x^2y^2)e^{-xy}\, dx\, dy$

$= \int_0^\infty \left(x^2ye^{-xy}\right)\Big|_{x=0}^{x=1} dy = \int_0^\infty ye^{-y}\, dy$

$= \lim_{b\to\infty} \left[-(y + 1)e^{-y}\Big|_{y=0}^{y=b}\right] = 1;$

$$\int_0^1 \int_0^\infty (2xy - x^2y^2)e^{-xy}\, dy\, dx$$

$$= \int_0^1 \lim_{b\to\infty} \left[(xy^2e^{-xy})\Big|_{y=0}^{y=b}\right] dx = \int_0^1 0\, dx = 0.$$

11 (a) f is continuous at a if and only if $\lim_{x\to a} f(x) = f(a)$. By the definition of f, this is the same as $\lim_{x\to a} \lim_{n\to\infty} f_n(x) = f(a) = \lim_{n\to\infty} f_n(a)$. But f_n is continuous at a, so we may replace $f_n(a)$ by $\lim_{x\to a} f_n(x)$, which leads to the desired result.

(b) $\lim_{n\to\infty} \lim_{x\to a} x^n = \lim_{n\to\infty} a^n$, which is 0 if $0 < a < 1$ and equals 1 if $a = 1$. Since $x^n \to 0$ as $n \to \infty$ for all x in $(0, 1)$, $\lim_{x\to a} \lim_{n\to\infty} x^n = \lim_{x\to a} 0 = 0$, so the limits differ at $a = 1$.

L The Jacobian

L.1 Magnification and the Jacobian

1 (a) We have $x = au + bv$ and $y = cu + dv$. Multiplying these equations by d and b, respectively, and subtracting yields $dx - by = (ad - bc)u$, so $u = \dfrac{dx - by}{ad - bc}$. Similarly, multiplying by c and a and subtracting yields $ay - cx = (ad - bc)v$, so $v = \dfrac{ay - cx}{ad - bc}$.

(b) If (u, v) is on the line and $(x, y) = f(u, v)$, then $0 = Au + Bv + C =$

$$A\cdot\frac{dx - by}{ad - bc} + B\cdot\frac{ay - cx}{ad - bc} + C = \frac{Ad - Bc}{ad - bc}x + \frac{Ba - Ab}{ad - bc}y + C.$$

So (x, y) is on the line $\alpha x + \beta y + C = 0$, where $\alpha = \dfrac{Ad - Bc}{ad - bc}$ and $\beta = \dfrac{Ba - Ab}{ad - bc}$.

3

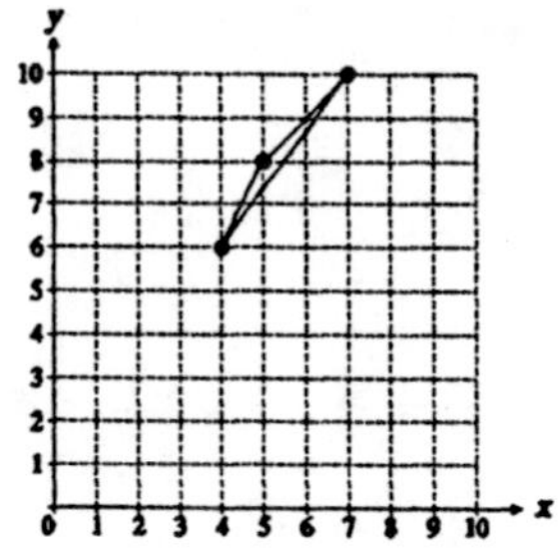

(a)

(u, v)	$(x, y) = (3u + v, 4u + 2v)$
(1, 1)	(4, 6)
(2, 1)	(7, 10)
(1, 2)	(5, 8)

The vertices of the image are (4, 6), (7, 10), and (5, 8).

(b) The triangle in the uv plane has base 1 and height 1, so its area is 1/2. The magnification of F is $|3\cdot 2 - 1\cdot 4| = 2$. Hence the image has area $1/2\cdot 2 = 1$.

5 (a)

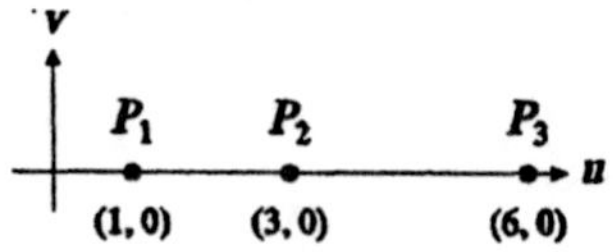

(b)

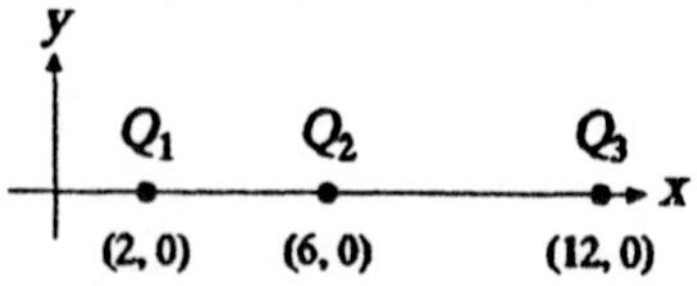

(c) The line $y = 0$.

7 (a)

(u, v)	$(x, y) = (2u, 3v)$
(0, 0)	(0, 0)
(2, 0)	(4, 0)
(1, 1)	(2, 3)

(b)

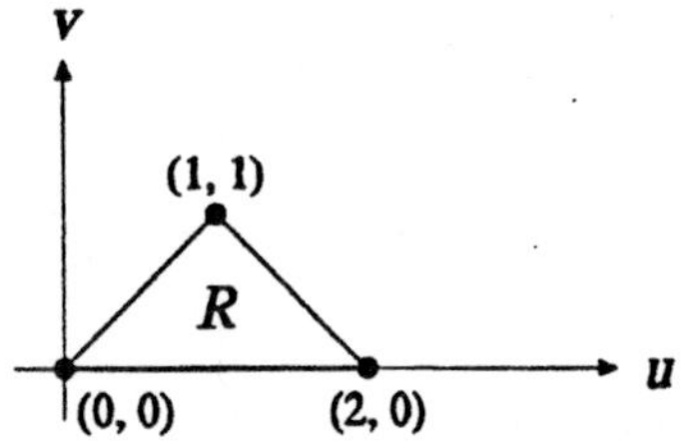

(c)

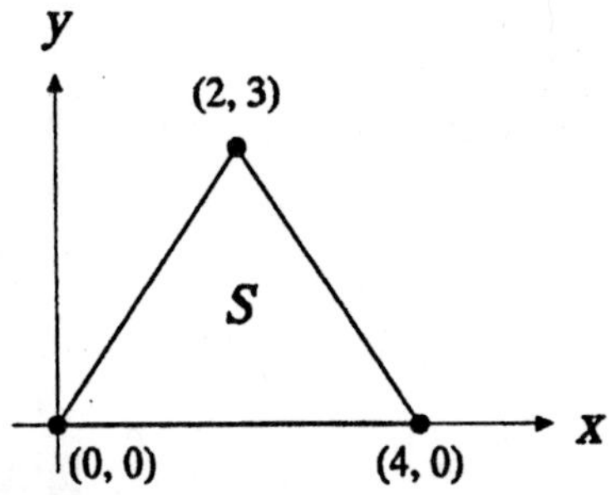

(d) See (b).

(e) See (c).

(f) The height of R is 1 and the base is 2, so the area is $\frac{1}{2} \cdot 2 \cdot 1 = 1$. S has base 4, height 3, and area equal to $\frac{1}{2} \cdot 4 \cdot 3 = 6$.

(g) The slope of the line joining (0, 0) and (1, 1) is 1. The slope of the line joining (1, 1) and (2, 0) is -1. Since $1 \cdot (-1) = -1$, the two lines are perpendicular, so R is a right triangle. The three sides of S are have slope 0, 3/2, and $-3/2$. No two slopes are negative reciprocals, so no two sides are perpendicular and S is not a right triangle.

9 (a)

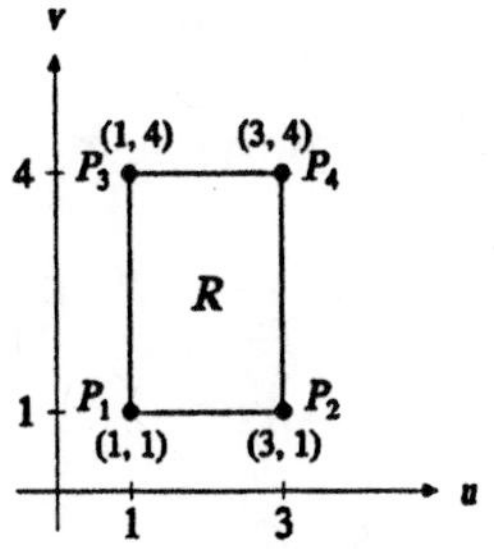

(b) $Q_1 = F(P_1) = F(1, 1) = (1 + 1, 1 - 1) = (2, 0)$

$Q_2 = F(P_2) = F(3, 1) = (3 + 1, 3 - 1) = (4, 2)$

$Q_3 = F(P_3) = F(1, 4) = (1 + 4, 1 - 4) = (5, -3)$

$Q_4 = F(P_4) = F(3, 4) = (3 + 4, 3 - 4) = (7, -1)$

[See the graph in (c).]

(c)

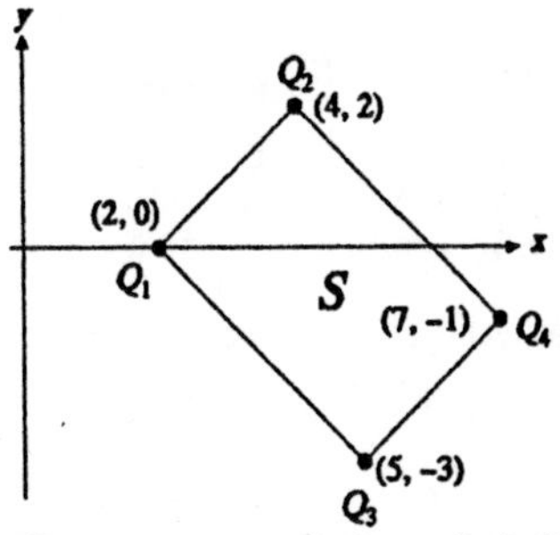

The line segments Q_1Q_2 and Q_3Q_4 have slope 1 while Q_1Q_3 and Q_2Q_4 have slope -1. Since adjacent sides of S are perpendicular, it follows that S is a rectangle.

11 (a)

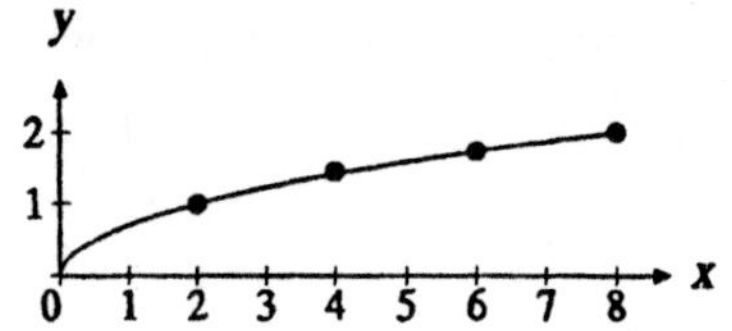

$F(1, 2) = (2, \sqrt{1}) = (2, 1)$; $F(2, 4) = (4, \sqrt{2})$; $F(3, 6) = (6, \sqrt{3})$; $F(4, 8) = (8, \sqrt{4}) = (8, 2)$

(b) $(x, y) = (v, \sqrt{u})$, so $x = v$ and $y = \sqrt{u}$; hence $v = x$ and $u = y^2$.

(c) $v = 2u$, so by (b), $x = 2y^2$; $x > 0$ because it is a positive multiple of a square, while $y > 0$ because $y = \sqrt{u}$.

(d) The image of the half–line $v = 2u$ $(u > 0)$ is just half the parabola $x = 2y^2$. (See the graph in (a).)

13 (a) $F(5, 0) = (5 \cos 0, 5 \sin 0) = (5, 0)$; $F(5, \pi/6) = (5 \cos \pi/6, 5 \sin \pi/6) = (5\sqrt{3}/2, 5/2)$; $F(5, \pi/2) = (5 \cos \pi/2, 5 \sin \pi/2) = (0, 5)$; $F(5, \pi) = (5 \cos \pi, 5 \sin \pi) = (-5, 0)$

(b) The image of $u = 5$ is a circle of radius 5 centered at the origin. Its equation is $x^2 + y^2 = 25$.

15 (a)

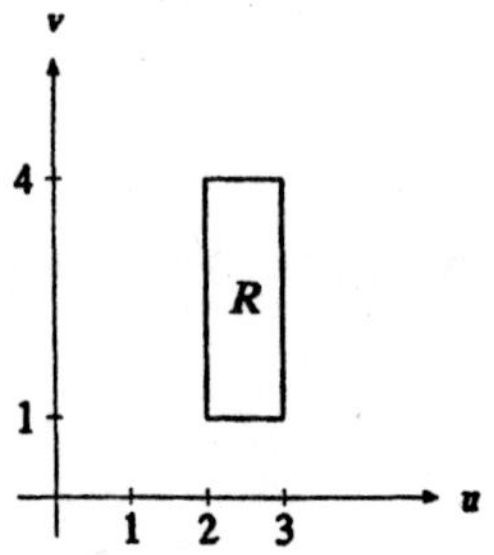

(b)

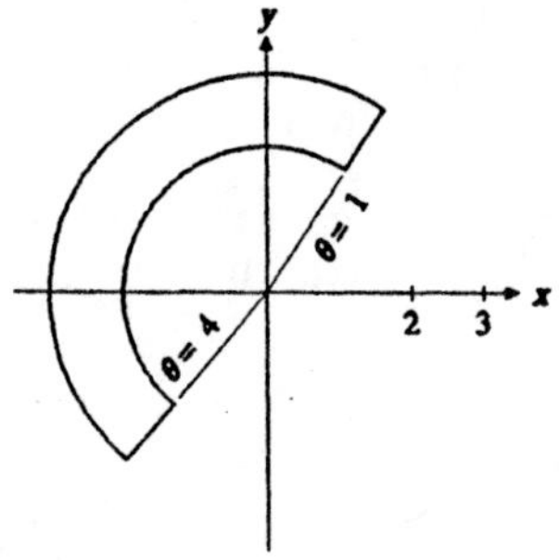

17 (a) Note that $u = x/y$ and $v = y$. If $u + v = 1$, then $x/y + y = 1$, so $x = y - y^2$. The image is that part of the parabola $x = y - y^2$ for which $0 < y < 1$. (For $y \le 0$, $v \le 0$; for $y \ge 1$, $u = x/y \le 0$.)

(b) Since $x = uv$, the image is the ray $x = 1$, $y > 0$.

19 (a) Area of $R = \Delta u \Delta v$

(b) Area of $S = \frac{1}{2}(u_0 + \Delta u)^2 \Delta v - \frac{1}{2}u_0^2 \Delta v$

$$= u_0 \Delta u \Delta v + \frac{1}{2}\Delta u^2 \Delta v$$

(c) $\lim_{\Delta u, \Delta v \to 0} \frac{\text{Area of } S}{\text{Area of } R}$

$$= \lim_{\Delta u, \Delta v \to 0} \frac{u_0 \Delta u \Delta v + \frac{1}{2}(\Delta u)^2 \Delta v}{\Delta u \Delta v}$$

$$= \lim_{\Delta u, \Delta v \to 0} \left(u_0 + \frac{1}{2}\Delta u\right) = u_0$$

L.2 The Jacobian and Change of Coordinates

1 $F(u, v) = (f(u, v), g(u, v)) = (uv, v^2)$, so $f_u = v$, $f_v = u$, $g_u = 0$, and $g_v = 2v$. By Theorem 2, the magnification of F at (u, v) is the absolute value of

the Jacobian: $\text{abs}\begin{vmatrix} f_u & f_v \\ g_u & g_v \end{vmatrix} = \begin{vmatrix} v & u \\ 0 & 2v \end{vmatrix} = 2v^2 - 0$

$= 2v^2$.

(a) The magnification at (1, 2) $= 2 \cdot 2^2 = 8$.

(b) The magnification at (3, 1) $= 2 \cdot 1^2 = 2$.

3 $f(u, v) = e^u \cos v$ and $g(u, v) = e^u \sin v$, so $f_u = e^u \cos v$, $f_v = -e^u \sin v$, $g_u = e^u \sin v$, and $g_v = e^u \cos v$: Magnification equals

$$\text{abs}\begin{vmatrix} e^u \cos v & -e^u \sin v \\ e^u \sin v & e^u \cos v \end{vmatrix}$$

$= |e^{2u} \cos^2 v + e^{2u} \sin^2 v| = e^{2u}$.

(a) The magnification at $(1, \pi/4)$ is $e^{2 \cdot 1} = e^2$.

(b) The magnification at $(2, \pi/6)$ is $e^{2 \cdot 2} = e^4$.

5 $x = f(u, v) = au + bv$ and $y = g(u, v) = cu + dv$, so $f_u = a$, $f_v = b$, $g_u = c$, and $g_v = d$. By definition, the Jacobian is given by $\begin{vmatrix} f_u & f_v \\ g_u & g_v \end{vmatrix} =$

$\begin{vmatrix} a & b \\ c & d \end{vmatrix} = ad - bc$.

7 (a) If $x = au$, $y = bv$, and $x^2/a^2 + y^2/b^2 = 1$, then $u = x/a$ and $v = y/b$, so $u^2 + v^2 = 1$. The mapping $f(u, v) = (au, bv)$ therefore maps the unit circle in the uv plane to an ellipse in the xy plane. (Furthermore, if $x^2/a^2 + y^2/b^2 \le 1$, then $u^2 + v^2 \le 1$, so the disk R is mapped to the interior S of the ellipse.)

(b) By Theorem 2, $\int_S y^2\, dA =$

$\int_R (bv)^2\, |J(P)|\, dA^*$. But, by Theorem 1, $J(P) = ab$, so $\int_S y^2\, dA = \int_R ab^3 v^2\, dA^*$.

(c) R is the unit circle, so $\int_R ab^3 v^2\, dA^* =$

$\int_0^{2\pi} \int_0^1 ab^3 (r \sin\theta)^2\, r\, dr\, d\theta =$

$ab^3 \int_0^{2\pi} \sin^2\theta \left. \frac{r^4}{4} \right|_0^1 d\theta =$

$\frac{ab^3}{4} \int_0^{2\pi} \sin^2\theta\, d\theta = \frac{ab^3}{4} \cdot \pi = \frac{\pi a b^3}{4}$. (In polar coordinates in the uv plane, $v = r \sin\theta$.)

9 The area of S is $\int_S 1\, dA$, which, by Theorem 2, equals $\int_R 1 \cdot |J|\, dA^* = \int_R |J|\, dA^*$, as desired.

11

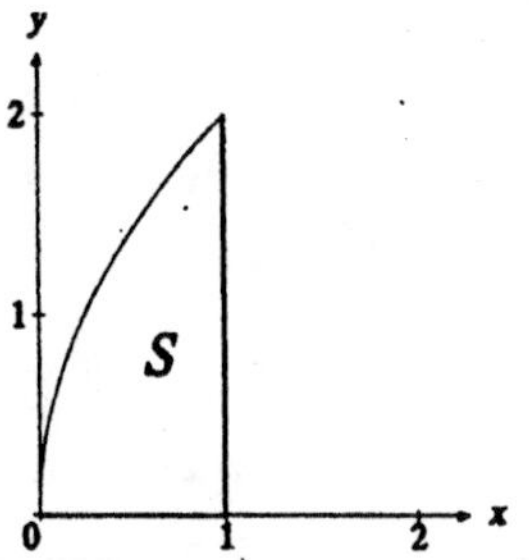

(a) For (u, v) in R, $u = \sqrt{x}$ and $v = y/2$. Hence the images of $u = v$, $v = 0$, and $u = 1$ are, respectively, $4x = y^2$, $y = 0$, and $x = 1$.

(b) $f(u, v) = u^2$ and $g(u, v) = 2v$, so $f_u = 2u$, $f_v = 0$, $g_u = 0$, and $g_v = 2$. Thus $|J| =$

$\text{abs}\begin{vmatrix} 2u & 0 \\ 0 & 2 \end{vmatrix} = |4u| = 4u$ and the area of S is

$\int_R |J|\, dA^* = \int_R 4u\, dA^* =$

$$\int_0^1 \int_0^u 4u\, dv\, du = \int_0^1 4u^2\, du = \frac{4u^3}{3}\Big|_0^1$$

$$= \frac{4}{3}.$$

(c) S is described by $0 \le y \le 2$, $y^2/4 \le x \le 1$.

Hence its area is $\int_S dA = \int_0^2 \int_{y^2/4}^1 dx\, dy =$

$$\int_0^2 \left(1 - \frac{y^2}{4}\right) dy = \left(y - \frac{y^3}{12}\right)\Big|_0^2 = \frac{4}{3}.$$

13 If $|J| = 1$, then Area of $S = \int_S dA =$

$\int_R |J|\, dA^* = \int_R dA^* =$ Area of R, so the image of the region has the same area as the region itself. The Jacobian of the mapping $f(u, v) = (4u + 3v, 3u + 2v)$ is $4 \cdot 2 - 3 \cdot 3 = 8 - 9 = -1$, so F is area-preserving.

15 (a) $x = u^{1/3}v^{2/3}$ and $y = u^{2/3}v^{1/3}$, so $x^2 = (u^{1/3}v^{2/3})^2 = u^{2/3}v^{4/3} = u^{2/3}v^{1/3}v = yv = vy$ and $y^2 = (u^{2/3}v^{1/3})^2 = u^{4/3}v^{2/3} = ux$.

(b) If $u = 1$, then $y^2 = x$. If $u = 2$, then $y^2 = 2x$. If $v = 3$, then $x^2 = 3y$. If $v = 4$, then $x^2 = 4y$. The region S is bounded by parabolas.

(c)

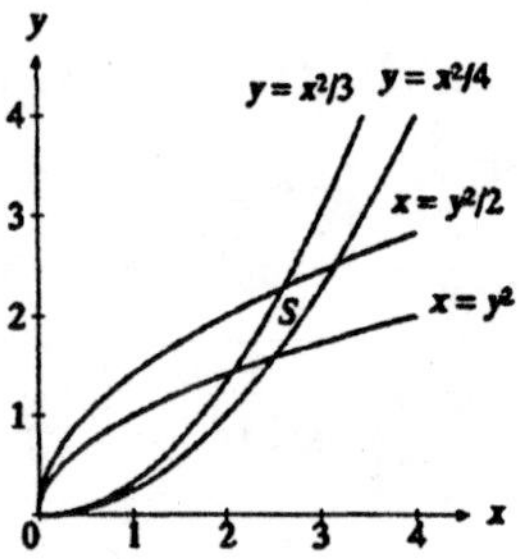

(d) $f(u, v) = u^{1/3}v^{2/3}$ and $g(u, v) = u^{2/3}v^{1/3}$, so $f_u = \frac{1}{3}u^{-2/3}v^{2/3}$, $f_v = \frac{2}{3}u^{1/3}v^{-1/3}$, $g_u = \frac{2}{3}u^{-1/3}v^{1/3}$,

and $g_v = \frac{1}{3}u^{2/3}v^{-2/3}$. Therefore, $J(P) =$

$$\begin{vmatrix} \frac{1}{3}u^{-2/3}v^{2/3} & \frac{2}{3}u^{1/3}v^{-1/3} \\ \frac{2}{3}u^{-1/3}v^{1/3} & \frac{1}{3}u^{2/3}v^{-2/3} \end{vmatrix} = \frac{1}{9} - \frac{4}{9} = -\frac{3}{9} =$$

$-\frac{1}{3}$. The area of S will be one-third of the area of R. R is a rectangle whose area is 1, so the area of S is 1/3.

17 S is described by $-2 \le x \le 2$, $-\frac{3}{2}\sqrt{4 - x^2} \le y \le \frac{3}{2}\sqrt{4 - x^2}$. Hence $\int_S x^2\, dA =$

$$\int_{-2}^2 \int_{-\frac{3}{2}\sqrt{4-x^2}}^{\frac{3}{2}\sqrt{4-x^2}} x^2\, dy\, dx = \int_{-2}^2 3x^2\sqrt{4 - x^2}\, dx =$$

$6 \int_0^2 x^2\sqrt{4 - x^2}\, dx$. Let $x = 2 \sin\theta$ for $0 \le \theta \le \pi/2$. Then $\sqrt{4 - x^2} = 2 \cos\theta$, $dx = 2 \cos\theta\, d\theta$, and the integral equals

$$6 \int_0^{\pi/2} (2 \sin\theta)^2\, 2 \cos\theta\, 2 \cos\theta\, d\theta$$

$$= 96 \int_0^{\pi/2} \sin^2\theta \cos^2\theta\, d\theta$$

$$= 24 \int_0^{\pi/2} (2 \sin\theta \cos\theta)^2\, d\theta =$$

$24 \int_0^{\pi/2} \sin^2 2\theta\, d\theta$. Let $u = 2\theta$; then $du = 2\, d\theta$, $d\theta = \frac{1}{2}\, du$, and the integral becomes

$$24 \int_0^{\pi} \sin^2 u\, \frac{1}{2}\, du = 12 \int_0^{\pi} \sin^2 u\, du = 6\pi.$$

19 $\int_2^5 \int_1^3 \frac{4u^2 - 4uv + v^2}{7^3}\, dv\, du$

$= \frac{1}{343} \int_2^5 \int_1^3 (4u^2 - 4uv + v^2)\, dv\, du$

$= \frac{1}{343} \int_2^5 \left(4u^2v - 2uv^2 + \frac{1}{3}v^3\right)\Big|_{v=1}^{v=3} du$

$= \frac{1}{343} \int_2^5 \left(8u^2 - 16u + \frac{26}{3}\right) du$

$= \frac{1}{343}\left(\frac{8}{3}u^3 - 8u^2 + \frac{26}{3}u\right)\Big|_2^5$

$= \frac{1}{343}\left(\frac{8}{3}\cdot 117 - 8\cdot 21 + \frac{26}{3}\cdot 3\right) = \frac{170}{343}.$

21 (a) The Jacobian is $\begin{vmatrix} \frac{\partial x}{\partial \rho} & \frac{\partial x}{\partial \theta} & \frac{\partial x}{\partial \phi} \\ \frac{\partial y}{\partial \rho} & \frac{\partial y}{\partial \theta} & \frac{\partial y}{\partial \phi} \\ \frac{\partial z}{\partial \rho} & \frac{\partial z}{\partial \theta} & \frac{\partial z}{\partial \phi} \end{vmatrix} =$

$$\begin{vmatrix} \sin\phi\cos\theta & -\rho\sin\phi\sin\theta & \rho\cos\phi\cos\theta \\ \sin\phi\sin\theta & \rho\sin\phi\cos\theta & \rho\cos\phi\sin\theta \\ \cos\phi & 0 & -\rho\sin\phi \end{vmatrix}$$

$= (\sin\phi\cos\theta)[(\rho\sin\phi\cos\theta)(-\rho\sin\phi) -$

$(\rho\cos\phi\sin\theta)\cdot 0] -$

$(-\rho\sin\phi\sin\theta)[(\sin\phi\sin\theta)(-\rho\sin\phi) -$

$(\rho\cos\phi\sin\theta)(\cos\phi)] +$

$(\rho\cos\phi\cos\theta)[(\sin\phi\sin\theta)\cdot 0 -$

$(\rho\sin\phi\cos\theta)(\cos\phi)] =$

$(\sin\phi\cos\theta)(-\rho^2\sin^2\phi\cos\theta) -$

$(-\rho\sin\phi\sin\theta)(-\rho\sin^2\phi\sin\theta -$

$\rho\cos^2\phi\sin\theta) +$

$(\rho\cos\phi\cos\theta)(-\rho\sin\phi\cos\phi\cos\theta)$

$= -\rho^2\sin^3\phi\cos^2\theta - \rho^2\sin^3\phi\sin^2\theta -$

$\rho^2\sin\phi\cos^2\phi\sin^2\theta -$

$\rho^2\sin\phi\cos^2\phi\cos^2\theta$

$= -\rho^2\sin\phi(\sin^2\phi\cos^2\theta + \sin^2\phi\sin^2\theta$

$+ \cos^2\phi\sin^2\theta + \cos^2\phi\cos^2\theta) =$

$-\rho^2\sin\phi(\sin^2\phi + \cos^2\phi)(\sin^2\theta + \cos^2\theta)$

$= -\rho^2\sin\phi.$

(b) Let R be a region in $\rho\phi\theta$ space; that is, in a space in which ρ, ϕ, and θ are rectangular coordinates. Let S be the image of R under the mapping F defined in (a). Let $f(x, y, z)$ be a real-valued function defined on S. Let dV denote the element of volume in S and let dV^* denote the element of volume in R. Note that $dV^* = d\rho\, d\phi\, d\theta$. Then $\int_S f(x, y, z)\, dV =$

$\int_R f(\rho\sin\phi\cos\theta, \rho\sin\phi\sin\theta, \rho\cos\phi)\, |J|\, dV^* =$

$\int_R f(\rho\sin\phi\cos\theta, \rho\sin\phi\sin\theta, \rho\cos\phi)\, \rho^2\sin\phi\, dV^*.$

The final integral is taken over a region in $\rho\theta\phi$ space, so it can be presented as a triple integral of the form

$$\int_{\theta_0}^{\theta_1} \int_{\phi_0(\theta)}^{\phi_1(\theta)} \int_{\rho_0(\theta,\phi)}^{\rho_1(\theta,\phi)} f(\rho\sin\phi\cos\theta, \rho\sin\phi\sin\theta, \rho\cos\phi)\, \rho^2\sin\phi\, d\rho\, d\phi\, d\theta,$$

where $\theta_0 \le \theta \le \theta_1$, $\phi_0(\theta) \le \phi \le \phi_1(\theta)$, $\rho_0(\theta, \phi) \le \rho \le \rho_1(\theta, \phi)$ describes the region of integration. (If the region is complicated, the description may have to be broken up into several parts.)

23 (a) The left–hand side equals $(ad - bc)(eh - fg) = adeh - adfg - bceh + bcfg$. The right–hand side equals $(ae + bg)(cf + dh) - (af + bh)(ce + dg) = (acef + adeh + bcfg + bdgh) - (acef + adfg + bceh + bdgh) = adeh + bcfg - adfg - bceh$. The two are equal.

(b) $$\frac{\partial(x, y)}{\partial(u, v)} \cdot \frac{\partial(u, v)}{\partial(x, y)} = \begin{vmatrix} \frac{\partial x}{\partial u} & \frac{\partial x}{\partial v} \\ \frac{\partial y}{\partial u} & \frac{\partial y}{\partial v} \end{vmatrix} \cdot \begin{vmatrix} \frac{\partial u}{\partial x} & \frac{\partial u}{\partial y} \\ \frac{\partial v}{\partial x} & \frac{\partial v}{\partial y} \end{vmatrix} =$$

$$\begin{vmatrix} \frac{\partial x}{\partial u}\cdot\frac{\partial u}{\partial x} + \frac{\partial x}{\partial v}\cdot\frac{\partial v}{\partial x} & \frac{\partial x}{\partial u}\cdot\frac{\partial u}{\partial y} + \frac{\partial x}{\partial v}\cdot\frac{\partial v}{\partial y} \\ \frac{\partial y}{\partial u}\cdot\frac{\partial u}{\partial x} + \frac{\partial y}{\partial v}\cdot\frac{\partial v}{\partial x} & \frac{\partial y}{\partial u}\cdot\frac{\partial u}{\partial y} + \frac{\partial y}{\partial v}\cdot\frac{\partial v}{\partial y} \end{vmatrix}.$$ Since x and y are functions of u and v, which are, in turn, functions of x and y, we may apply the chain rule to compute $\frac{\partial x}{\partial x}$, $\frac{\partial x}{\partial y}$, $\frac{\partial y}{\partial x}$, and $\frac{\partial y}{\partial y}$. We find that the last determinant above contains these chain rule expressions; hence it equals $\begin{vmatrix} \frac{\partial x}{\partial x} & \frac{\partial x}{\partial y} \\ \frac{\partial y}{\partial x} & \frac{\partial y}{\partial y} \end{vmatrix}$. But $\frac{\partial x}{\partial x} = \frac{\partial y}{\partial y} = 1$ and $\frac{\partial x}{\partial y} = \frac{\partial y}{\partial x} = 0$, so the determinant is $\begin{vmatrix} 1 & 0 \\ 0 & 1 \end{vmatrix} = 1$.

25 Let $u = y/x$ and $v = x + y$. As (x, y) varies over S, (u, v) varies over the rectangle R described by $1 \le u \le 2$, $3 \le v \le 4$. To solve for x and y, note that $y = ux$, so $v = x + ux = (1 + u)x$. Hence $x = \frac{v}{1 + u}$ and $y = \frac{uv}{1 + u}$. The Jacobian is

$$\frac{\partial(x, y)}{\partial(u, v)} = \begin{vmatrix} -\frac{v}{(1 + u)^2} & \frac{1}{1 + u} \\ \frac{v}{(1 + u)^2} & \frac{u}{1 + u} \end{vmatrix} = -\frac{v}{(1 + u)^2}.$$

The area of S is $\int_S dA = \int_R \left| -\frac{v}{(1 + u)^2} \right| dA^* =$

$$\int_1^2 \int_3^4 \frac{v}{(1 + u)^2}\, dv\, du = \int_1^2 \frac{1}{(1 + u)^2} \cdot \frac{v^2}{2}\Big|_3^4 du$$

$$= \frac{1}{2}(4^2 - 3^2)\left(-\frac{1}{1 + u}\right)\Big|_1^2 = \frac{7}{2}\left(-\frac{1}{3} + \frac{1}{2}\right) = \frac{7}{12}.$$

Hence the density is $\frac{M}{7/12} = \frac{12M}{7}$. The moment of inertia is $I_x = \int_S y^2 \cdot \frac{12M}{7}\, dA$

$$= \frac{12M}{7} \int_R \left(\frac{uv}{1 + u}\right)^2 \left| -\frac{v}{(1 + u)^2} \right| dA^*$$

$$= \frac{12M}{7} \int_1^2 \int_3^4 \frac{u^2 v^3}{(1 + u)^4}\, dv\, du$$

$$= \frac{12M}{7} \int_1^2 \frac{u^2}{(1 + u)^4} \cdot \frac{v^4}{3}\Big|_3^4 du$$

$$= \frac{12M}{7} \cdot \frac{1}{4}(4^4 - 3^4) \int_1^2 \frac{u^2}{(1 + u)^4}\, du$$

$$= \frac{12M}{7} \cdot \frac{175}{4} \int_2^3 \frac{(t - 1)^2}{t^4}\, dt,$$ where $t = 1 + u$, so we have $75M \int_2^3 (t^{-2} - 2t^{-3} + t^{-4})\, dt$

$$= 75M\left(-t^{-1} + t^{-2} - \frac{1}{3}t^{-3}\right)\Big|_2^3$$

$$= 75M\left[\left(-\frac{1}{3} + \frac{1}{9} - \frac{1}{81}\right) - \left(-\frac{1}{2} + \frac{1}{4} - \frac{1}{24}\right)\right]$$

$$= 75M\cdot\frac{37}{648} = \frac{925}{216}M.$$

27 (a)

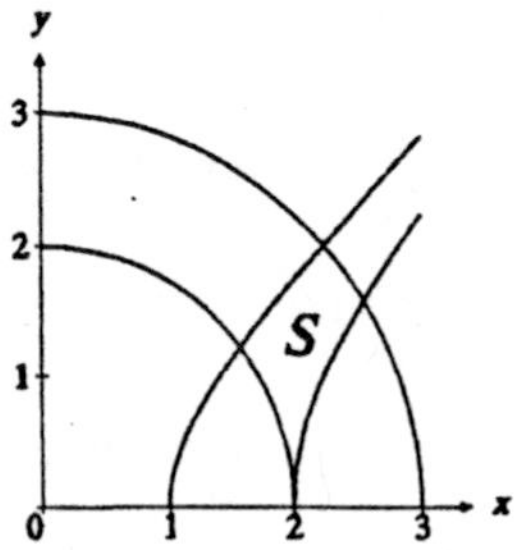

(b) Let $u = x^2 - y^2$, $v = x^2 + y^2$. Let R be the region in the uv plane given by $1 \le u \le 4$, $4 \le v \le 9$. Then the function is one–to–one on S and the image of S is R.

$$\frac{\partial(u, v)}{\partial(x, y)} = \begin{vmatrix} 2x & -2y \\ 2x & 2y \end{vmatrix} = 8xy$$

$= 8\sqrt{\dfrac{u+v}{2}}\sqrt{\dfrac{v-u}{2}}$. Therefore $\int_S x\, dA$

$$= \int_R \sqrt{\frac{u+v}{2}}\left(8\sqrt{\frac{u+v}{2}}\sqrt{\frac{v-u}{2}}\right)^{-1} dA^*$$

$$= \int_1^4\int_4^9 \frac{\sqrt{2}}{8}\cdot\frac{1}{\sqrt{v-u}}\, dv\, du$$

$$= \frac{\sqrt{2}}{8}\int_1^4 2\sqrt{v-u}\Big|_4^9\, du$$

$$= \frac{\sqrt{2}}{8}\int_1^4 \left(2\sqrt{9-u} - 2\sqrt{4-u}\right) du$$

$$= \frac{\sqrt{2}}{4}\left[-\frac{2}{3}(9-u)^{3/2} + \frac{2}{3}(4-u)^{3/2}\right]\Big|_1^4$$

$$= \frac{\sqrt{2}}{6}\left(-5^{3/2} + 0^{3/2} + 8^{3/2} - 3^{3/2}\right)$$

$$= \frac{\sqrt{2}}{6}\left(-5\sqrt{5} + 16\sqrt{2} - 3\sqrt{3}\right)$$

$$= -\frac{5\sqrt{10}}{6} + \frac{16}{3} - \frac{\sqrt{6}}{2}.$$

M Linear Differential Equations with Constant Coefficients

1 $y' + 2y = 0$; the equation is of the form $y' - ky = 0$, whose solution is $y = Ce^{kx}$. Therefore $y = Ce^{-2x}$.

3 $3y' + 12y = x$ is equivalent to $y' + 4y = x/3$. A particular solution to this equation should be of the form $y_p = Ax + B$, so $y_p' = A$ and thus $y_p' + 4y_p = A + 4Ax + 4B = x/3$. It follows that $4A = 1/3$, so $A = \dfrac{1}{12}$; $0 = A + 4B = \dfrac{1}{12} + 4B$, so $B = -\dfrac{1}{48}$. The general solution is $y = y_p + Ce^{-4x} = \dfrac{x}{12} - \dfrac{1}{48} + Ce^{-4x}$.

5 $y' - y = x^2$; the particular solution will be $y_p = Ax^2 + Bx + C$. Then $y_p' = 2Ax + B$, so $y_p' - y_p = 2Ax + B - Ax^2 - Bx - C = -Ax^2 + (2A - B)x + B - C = x^2$; thus $A = -1$, $2A - B = -2 - B = 0$, and $C = B = -2$. Then $y_p =$

$-x^2 - 2x - 2$; the general solution is $y = y_p + Ke^x = -x^2 - 2x - 2 + Ke^x$.

7 $y'' - 2y' - 3y = 0$, so the associated quadratic is $t^2 - 2t - 3 = 0$; this factors into $(t + 1)(t - 3) = 0$, so the roots are $r_1 = -1$ and $r_2 = 3$. By Theorem 2, $y = C_1e^{-x} + C_2e^{3x}$.

9 The associated quadratic is $2t^2 - t - 3 = (t + 1)(2t - 3)$, whose roots are 3/2 and -1; by Theorem 2, the general solution is $y = C_1e^{3x/2} + C_2e^{-x}$.

11 The associated quadratic is $4t^2 - 12t + 9 = (2t - 3)^2$, so $r_1 = r_2 = 3/2$; by Theorem 3, the general solution is $y = (C_1 + C_2x)e^{3x/2}$.

13 $y'' - 3y' + y = 0$, so the associated quadratic is $t^2 - 3t + 1 = 0$. By the quadratic formula, $r_1 = \dfrac{3 + \sqrt{5}}{2}$ and $r_2 = \dfrac{3 - \sqrt{5}}{2}$. By Theorem 2, $y =$

$C_1 \exp(r_1x) + C_2 \exp(r_2x)$

$$= C_1 \exp\left(\frac{3 + \sqrt{5}}{2}x\right) + C_2 \exp\left(\frac{3 - \sqrt{5}}{2}x\right).$$

15 $y'' - 6y' + 9y = 0$, so the associated quadratic is $t^2 - 6t + 9 = (t - 3)^2 = 0$, whose repeated root is $r = 3$. By Theorem 3, $y = (C_1 + C_2x)e^{3x}$.

17 $y'' - 2\sqrt{11}y' + 11y = 0$, so the associated quadratic is $t^2 - 2\sqrt{11}t + 11 = (t - \sqrt{11})^2 = 0$, and $r = \sqrt{11}$ is the repeated root. By Theorem 3, $y = (C_1 + C_2x)e^{\sqrt{11}x}$.

19 $y'' - 2y' - 3y = e^{2x}$; by Exercise 7, the solution of the homogeneous equation is $C_1e^{-x} + C_2e^{3x}$. A particular solution will be of the form $y_p = Ae^{2x}$, so $y_p' = 2Ae^{2x}$ and $y_p'' = 4Ae^{2x}$; then $y_p'' - 2y_p' - 3y_p = 4Ae^{2x} - 4Ae^{2x} - 3Ae^{2x} = -3Ae^{2x} = e^{2x}$, so $A = -1/3$. Therefore $y = -\frac{1}{3}e^{2x} + C_1e^{-x} + C_2e^{3x}$.

21 $y'' - 4y' + y = \cos 3x$, so the associated quadratic is $t^2 - 4t + 1 = 0$; by the quadratic formula, its roots are $r_1 = 2 + \sqrt{3}$ and $r_2 = 2 - \sqrt{3}$, so the solution to the homogeneous equation is

$C_1e^{(2+\sqrt{3})x} + C_2e^{(2-\sqrt{3})x} =$

$e^{2x}\left(C_1e^{\sqrt{3}x} + C_2e^{-\sqrt{3}x}\right)$. The particular solution will have the form $y_p = A \cos 3x + B \sin 3x$, so $y_p' = -3A \sin 3x + 3B \cos 3x$, $y_p'' = -9A \cos 3x - 9B \sin 3x = -9y_p$, and therefore $y_p'' - 4y_p' + y_p = -9y_p - 4y_p' + y_p = -8y_p - 4y_p' = (-8A - 12B) \cos 3x + (-8B + 12A) \sin 3x = \cos 3x$. Thus $-8A - 12B = 1$ and $-8B + 12A = 0$; these simplify to $3A = 2B$ and $1 = -8A - 12(3/2A) = -8A - 18A = 26A$, so $A = -\dfrac{1}{26}$ and $B = -\dfrac{3}{52}$. The general solution is $y =$

$-\dfrac{1}{26} \cos 3x - \dfrac{3}{52} \sin 3x +$

$e^{2x}(C_1e^{\sqrt{3}x} + C_2e^{-\sqrt{3}x})$.

23 (a) Let $g(x) = \int e^{ax}f(x)\,dx$; then $g'(x) = e^{ax}f(x)$. Let $y = e^{-ax}g(x)$; then $y' = e^{-ax}g'(x) + g(x)(-ae^{-ax}) = e^{-ax}e^{ax}f(x) - ae^{-ax}g(x) = f(x) - ay$; that is, $y' = f(x) - ay$, so $y' + ay = f(x)$, as claimed.

(b) $y' + y = \dfrac{1}{1 + e^x}$; thus $f(x) = \dfrac{1}{1 + e^x}$ and

$$g(x) = \int e^{ax}f(x)\,dx = \int \frac{e^x}{1 + e^x}\,dx =$$